PRINCIPLES OF

CELL BIOLOGY

George Plopper

Rensselaer Polytechnic Institute

JONES & BARTLETT
LEARNING

World Headquarters
Jones & Bartlett Learning
5 Wall Street
Burlington, MA 01803
978-443-5000
info@jblearning.com
www.jblearning.com

Jones & Bartlett Learning books and products are available through most bookstores and online booksellers. To contact Jones & Bartlett Learning directly, call 800-832-0034, fax 978-443-8000, or visit our website, www.jblearning.com.

Production Credits
Chief Executive Officer: Ty Field
President: James Homer
SVP, Editor-in-Chief: Michael Johnson
SVP, Chief Technology Officer: Dean Fossella
SVP, Chief Marketing Officer: Alison M. Pendergast
Publisher: Cathleen Sether
Senior Associate Editor: Megan R. Turner
Editorial Assistant: Rachel Isaacs
Production Editor: Amanda Clerkin
Senior Marketing Manager: Andrea DeFronzo
Associate Photo Researcher: Lauren Miller
V.P., Manufacturing and Inventory Control: Therese Connell
Cover Design: Kristin E. Parker
Composition: Circle Graphics
Cover Image: Courtesy of Dr. Chris Bjornsson and Madelyn May
Printing and Binding: Courier Kendallville
Cover Printing: Courier Kendallville

To order this product, use ISBN: 978-1-4496-3751-4

Library of Congress Cataloging-in-Publication Data
Plopper, George.
  Principles of cell biology / George Plopper.
      p. cm.
  Includes bibliographical references and index.
  ISBN 978-0-7637-5774-8 — ISBN 0-7637-5774-8
  1. Cytology. I. Title.
  QH581.2.P58 2012
  571.6—dc22

                                      2011009578

6048
Printed in the United States of America
16 15 14 13 12   10 9 8 7 6 5 4 3 2

This book is dedicated to Steve Carper, my education mentor and
the best teacher I have ever known; and to Stan Hillyard,
who showed me first-hand how to bring cells
to life in a classroom.

Brief Table of Contents

Table of Contents

Preface

As the rate of discovering facts in the sciences continues to escalate, both students and instructors must confront the age-old problem of deciding what material matters most, especially at the introductory level. In my own experience, even the most enthusiastic students have great difficulty distinguishing essential facts in cell biology from more technical details. Given the heavy emphasis on memorization in most K-12 programs, many introductory-level college students resort to storing as much information as possible in short-term memory, only to discover later that they've missed the underlying concepts that give these facts their significance. In the past 20 years of teaching, I have spent as much time helping students navigate a conceptual path through the dense web of facts in cell biology as I have explaining the meaning of those facts. The purpose of this book is to help students build a conceptual framework for cell biology that will persist long after their coursework is complete.

Our Approach to Learning Cell Biology

The field of introductory cell biology enjoys a wealth of well-written texts by outstanding authors. Why write yet another introductory text? This book is needed for two important reasons. First, it overtly focuses on some of the underlying principles that illustrate both how cells function as well as how we study them. While many textbooks reference "principles" in their fields, few specifically identify these principles or explore them in detail. In contrast, this book identifies 10 specific principles of cell biology, and devotes a separate chapter to illustrate each one of them. As a result:

- We intentionally shift away from the traditional focus on technical details and towards a more integrative view of cellular activity that can be tailored to suit students with a broad range of backgrounds.
- Instructors have great freedom to organize technical subjects as they see fit while permitting students to build their own conceptual view of how cells solve problems. In short, because every cellular activity discussed in the text is tied directly to an overlying principle, these activities can be arranged and taught in many different combinations, at varying depths of detail, without losing focus on the Big Picture.
- Students develop a framework for evaluating facts as they encounter them, and this invites them to critically evaluate information as they learn. The principles in this book are not intended to be treated as laws, and are thus always subject to criticism and review.
- Instructors can capitalize on this organizing style to seamlessly merge supplemental material with the text as the field changes, or to emphasize specific subjects in a topics course.
- Professionals in the field can use these principles as starting points for identifying additional principles in cell biology and in other related fields, for comparing these principles in other fields of biology, and for developing a more integrated curriculum across multiple scales of biological organization. For example, mapping

several courses to specific principles such as those identified in this text could assist in curriculum development and assessment at a department or program administrative level.

The second important distinguishing feature of this book is its informal, narrative writing style. This style is adopted to make even the most complex concepts accessible to students new to a scientific field, including stripping away some of the technical complexity that many introductory students find intimidating. Each chapter thus reflects my own lectures in introductory cell biology, in both style and content. Specifically, this includes:

- Liberal use of analogies that have proven effective over many years of teaching.
- Margin boxes throughout each chapter including studying tips, clarifications of apparent contradictions, explanation of naming schemes, FAQs, etc.
- Jargon is introduced gradually, after the concepts have been established, thereby de-emphasizing memorization of names.
- Ten principle-based chapters build on the foundation laid down in the first four chapters of the book, and include heavy emphasis on linking concepts across multiple chapters.
- Novel artwork is included, reflecting drawing exercises the author includes in his own lectures.

Audience

Principles of Cell Biology is written for introductory cell biology courses having an emphasis on eukaryotic cells, especially humans and other mammals. It is geared toward students in general biology, molecular biology, physiology, nursing, dental hygiene, and bioengineering. The book also provides a firm foundation for advanced programs in biological sciences, medicine, dentistry, and bioengineering.

Organization

The book consists of four chapters (1–4) that introduce the fundamental molecular building blocks of all cells: sugars, proteins, nucleic acids, and lipids. The remaining 10 chapters focus on explaining a single principle, supported by the topics typically discussed in cell biology courses. One important departure from most other books is that the topics of membrane transport and metabolism are covered in the same chapter (Chapter 10), though membrane transport is discussed in its own chapter subsection so it can be read independent of the metabolism sections if desired.

Through the use of cross-references, some chapters can be clustered into broader themes. Chapters 5 and 6 focus on the cytoskeleton and extracellular matrix, respectively, to explain how cells establish, maintain, and modify their shapes. Chapters 7–9 focus on DNA replication, transcription, translation, protein sorting, and the endomembrane system to illustrate the theme of information transfer from DNA to proteins. Chapters 11–13 use signal transduction as a unifying theme to illustrate the relationships between signaling pathways, control of gene expression, and cell growth/apoptosis. Finally, Chapter 14 evaluates the principles from earlier chapters at the tissue level, unifying the entire cell into a single functional unit of a multicellular organism.

Each chapter contains pedagogical features to assist instructors and facilitate student comprehension. These include:

- An introductory section, called the Big Picture, explains the learning objectives for the chapter, the relationship the chapter's subject matter shares with other chapters, and, in the last 10 chapters, introduces the Principle that the chapter illustrates.
- Every major section in each chapter includes a bullet list of key concepts.
- Each chapter contains Concept Check questions at the end of major sections to test comprehension of the section, with answers provided at the end of the chapter. The test bank includes exam questions that link to these concept check questions.
- End-of-chapter questions, provided on the companion website, ask students to integrate material across chapter sections and across different chapters.

Acknowledgments

Just as cell biology is a collaborative science, creating and publishing this textbook is the product of my extensive collaboration with the outstanding editorial and production team at Jones & Bartlett Learning, artists, my professional colleagues, and the external reviewers. I offer my deepest thanks to everyone who helped take this project from its very humble beginning to the finished product. This list includes, but is not limited to, the following individuals: Cathleen Sether, Megan Turner, Laura Almozara, Amanda Clerkin, and Lauren Miller. Special thanks are extended to Elizabeth Morales who developed the art program and to Fran Norflus and Melissa Marcucci for their contributions to the ancillary materials for this title. Special thanks to Dr. Jeffrey Pommerville, who let me use his "To the Student" section for this book.

Thanks is also extended to those who have reviewed various drafts of *Principles of Cell Biology:*

Eric Sikorski, University of Tampa
Stanley Hillyard, University of Nevada at Las Vegas
Charly Mallery, University of Miami
Helen Cronenberger, University of Texas at San Antonio
Scott Erdman, Syracuse University
Ken Lerea, New York Medical College
Lara Dowland, Mount Wachusett Community College
Jeff Hadwiger, Oklahoma State University
Janet Duerr, Ohio University
Rolf Prade, Oklahoma State University
Shelley Hepworth, Carleton University
Dr. Sarah Higbie, Saint Joseph College
Dr. Benjamin Johnson, Hardin-Simmons University
Brad Bryan, Worcester State College
Holger Lill, VU University, Amsterdam
Soohyoun Ahn, Arkansas State University
Carol Castaneda, Indiana University Northeast
Linda DeVeaux, Idaho State University
Ellen France, Georgia College & State University
Barbara Frank, Idaho State University
Barry Hoopengardner, Central Connecticut State University
Martin Kapper, Central Connecticut State University

Erik Larson, Illinois State University
Julia Lee, St. Joseph's University
Paolo Martini, Brandeis University
Jim Mulrooney, Central Connecticut State University
Hongmin Qin, Texas A&M University
Anne Grippo, Arkansas State University

Finally, I wish to thank my current and former students, who simultaneously ask the utmost of me and provide much of the inspiration and imaginative ideas to spur me on. I salute you and look forward to hearing your ideas for making this text as effective and enjoyable as possible.

George Plopper

Supplements to the Text

Jones & Bartlett Learning offers an array of ancillaries to assist instructors and students in teaching and mastering concepts in this text. Additional information and review copies of any of the following items are available through your Jones & Bartlett Learning sales representative or by going to www.jblearning.com.

For the Student

The website developed exclusively for *Principles of Cell Biology* by Fran Norflus of Clayton State University, offers a variety of resources to enhance understanding of cell biology. The site, located at http://biology.jbpub.com/book/principles, contains an interactive glossary, flashcards, crosswords, chapter outlines, weblinks, and video links.

For the Instructor

Compatible with Windows and Macintosh platforms, the Instructor's ToolKit provides instructors with the following traditional ancillaries:

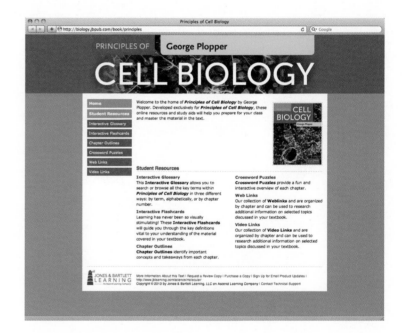

- The **PowerPoint Image Bank** provides the illustrations, photographs, and tables (to which Jones & Bartlett Learning holds the copyright or has permission to reproduce digitally) inserted into PowerPoint slides. You can quickly and easily copy individual images or tables into your existing lecture slides.
- The **PowerPoint Lecture Outline** presentation package, developed by Melissa Marcucci of Saint Joseph College, provides lecture notes and images for each chapter of *Principles of Cell Biology*. Instructors with the Microsoft PowerPoint software can customize the outlines, art, and order of presentation.

Finally, the Test Bank, also prepared by Fran Norflus of Clayton State University, is available as downloadable text files from our website.

About the Author

George Plopper is a Professor in the Department of Biology at Rensselaer Polytechnic Institute in Troy, New York. He received his Bachelor of Arts in General Biology from the University of California, San Diego. He completed his PhD in Cell & Developmental Biology at Harvard University, then completed his postdoctoral training in Cell Biology at The Scripps Research Institute in La Jolla, California. Dr. Plopper served as Assistant Professor at The University of Nevada, Las Vegas before moving to his present position at Rensselaer Polytechnic Institute. He has taught cell biology to undergraduate and graduate students since 1985, receiving four teaching awards. He was named a National Academies Education Fellow in the Life Sciences by the National Academy of Sciences in 2004.

To the Student—Study Smart

Your success in cell biology—or any college or university course—will depend on your ability to study effectively and efficiently. Therefore, this textbook was designed with you, the student, in mind. The text's organization will help you improve your learning and understanding and, ultimately, your grades. The learning design illustrated below reflects this organization. Study it carefully, and, if you adopt the flow of study shown, you should be a big step ahead in your preparation and understanding of cell biology—and for that matter any subject you are taking.

When I was an undergraduate student, I hardly ever read the "To the Student" section (if indeed one existed) in my textbooks because the section rarely contained any information of importance. This one does, so please read on.

In college, I was a mediocre student until my junior year. Why? Mainly because I did not know how to study properly, and, more importantly, I did not know how to read a textbook effectively. My textbooks were filled with highlighted sentences without any plan on how I would use this "emphasized" information. In fact, most textbooks *assume* you know how to read a textbook properly. I didn't and you might not, either.

Reading a textbook is difficult if you are not properly prepared. So you can take advantage of what I learned as a student and have learned from instructing thousands of students; I have worked hard to make this text user friendly with a reading style that is not threatening or complicated. Still, there is a substantial amount of information to learn and understand, so having the appropriate reading and comprehension skills is critical. Therefore, I encourage you to spend 20 minutes reading this section, as I am going to give you several tips and suggestions for acquiring those skills. Let me show you how to be an active reader.

Be a Prepared Reader

Before you jump into reading a section of a chapter in this text, prepare yourself by finding the place and time and having the tools for study.

Place. Where are you right now as you read these lines? Are you in a quiet library or at home? If at home, are there any distractions, such as loud music, a blaring television, or screaming kids? Is the lighting adequate to read? Are you sitting at a desk or lounging on the living room sofa? Get where I am going? When you read for an educational purpose— that is, to learn and understand something—you need to maximize the environment for reading. Yes, it should be comfortable, but not to the point that you will doze off.

Time. All of us have different times during the day when we best perform a certain skill, be it exercising or reading. The last thing you want to do is read when you are tired or simply not "in tune" for the job that needs to be done. You cannot learn and understand the information if you fall asleep or lack a positive attitude. I have kept all of the chapters in this text about the same length so you can estimate the time necessary for each and plan your reading accordingly. If you have done your preliminary survey of the chapter or chapter section, you can determine about how much time you will need. If 40 minutes is needed to read—and comprehend (see below)—a section of a chapter, find the place and time that will give you 40 minutes of uninterrupted study. Brain research suggests that most people's brains cannot spend more than 45 minutes in concentrated, technical reading. Therefore, I have avoided lengthy presentations and instead have focused on smaller sections, each with its own heading. These should accommodate shorter reading periods.

Reading Tools. Lastly, as you read this, what study tools do you have at your side? Do you have a highlighter or pen for emphasizing or underlining important words or phrases? Notice, the text has wide margins, which allow you to make notes or to indicate something that needs further clarification. Do you have a pencil or pen handy to make these notes? Or, if you do not want to "deface" the text, make your notes in a notebook. Lastly, some students find having a ruler is useful to prevent their eyes from wandering on the page and to read each line without distraction.

Be an Explorer Before You Read

When you sit down to read a section of a chapter, do some preliminary exploring. Look at the section head and subheadings to get an idea of what is discussed. Preview any diagrams, photographs, tables, graphs, or other visuals used. They give you a better idea of what is going to occur. We have used a good deal of space in the text for these features, so use them to your advantage. They will help you learn the written information and comprehend its meaning. Do not try to understand all the visuals, but try to generate a mental "big picture" of what is to come. Familiarize yourself with any symbols or technical jargon that might be used in the visuals.

The end of each chapter contains a **Summary** for that chapter. It is a good idea to read the summary before delving into the chapter, even though it is at the end. That way you will have a framework for the chapter before filling in the nitty-gritty information.

Be a Detective as You Read

Reading a section of a textbook is not the same as reading a novel. With a textbook, you need to uncover the important information (the terms and concepts) in the forest of words on the page. So, the first thing to do is read the complete paragraph. When you have determined the main ideas, highlight or underline them. However, I have seen students highlighting the entire paragraph in yellow, including every *a, the,* and *and.* This is an example of highlighting before knowing what is important. So, I have helped you out somewhat. Important terms and concepts are in **bold face** followed by the definition (or the definition might be in the glossary). So only highlight or underline with a pen essential ideas and key phrases—not complete sentences, if possible.

What if a paragraph or section has no boldfaced words? How do you find what is important here? From an English course, you may know that often the most important information is mentioned first in the paragraph. If it is followed by one or more examples, then you can backtrack and know what was important in the paragraph. In addition, I have added section "speed bumps" (called **Concept and Reasoning Checks**) to let you test your learning and understanding before getting too far ahead in the material. These checks also are clues to what was important in the section you just read.

Be a Repetitious Student

Brain research has shown that each individual can only hold so much information in short-term memory. If you try to hold more, then something else needs to be removed—sort of like a full computer disk. So that you do not lose any of this important information, you need to transfer it to long-term memory—to the hard drive, if you will. In reading and studying, this means retaining the term or concept; so, write it out in your notebook *using your own words*. Memorizing a term does not mean you have learned the term or understood the concept. By writing it out in your own words, you are forced to think and actively interact with the information. This repetition reinforces your learning.

Be a Patient Student

In textbooks, you cannot read at the speed that you read your e-mail or a magazine story. There are unfamiliar details to be learned and understood—and this requires being a patient, slower reader. Identifying the important information from a textbook chapter requires you to *slow down* your reading speed. Speed-reading is of no value here. It may help to go back and re-read sections as your general understanding of the topic improves. I use many cross-references in this book, and suggest you take the time to look up the referenced material in other chapters.

Know the What, Why, and How

Have you ever read something only to say, "I have no idea what I read!"? As I've already mentioned, reading a cell biology text is not the same as reading *Sports Illustrated* or *People* magazine. In these entertainment magazines, you read passively for leisure or perhaps amusement. In *Principles of Cell Biology*, you must read actively for learning and understanding—that is, for *comprehension*. This can quickly lead to boredom unless you engage your brain as you read—that is, be an active reader. Do this by knowing the *what, why,* and *how* of your reading.

- *What* is the general topic or idea being discussed? This often is easy to determine because the section heading might tell you. If not, then it will appear in the first sentence or beginning part of the paragraph.
- *Why* is this information important? If I have done my job, the text section will tell you why it is important or the examples provided will drive the importance home. These surrounding clues further explain why the main idea was important.
- *How* do I "mine" the information presented? This was discussed under "Be a Detective as You Read."

Have a Debriefing Strategy

After reading the material, be ready to debrief. Verbally summarize what you have learned. This will start moving the short-term information into the long-term memory storage—that is, *retention*. Any notes you made concerning confusing material should be discussed as soon as possible with your instructor. A lot of cell biology is represented visually, so allow time to draw out diagrams. Again, repetition makes for easier learning and better retention.

In many professions, such as sports, music, or the theater, the name of the game is practice, practice, practice. The hints and suggestions I have given you form a skill that requires practice to perfect and use efficiently. Be patient, things will not happen overnight; perseverance and willingness though will pay off with practice. You might also check with your college or university academic (or learning) resource center. These folks will have more ways to help you to read a textbook better and to study well overall.

Send me a Note

In closing, I would like to invite you to write me and let me know what is good about this textbook so I can build on it and what may need improvement so I can revise it. I can be reached at the Department of Biology, Rensselaer Polytechnic Institute, 110 8th St, Troy, NY, 12180. Feel free to email me at: ploppg@rpi.edu.

I wish you great success in your cell biology course. Welcome!
—Dr. Plopper

Website: http://www.rpi.edu/~ploppg

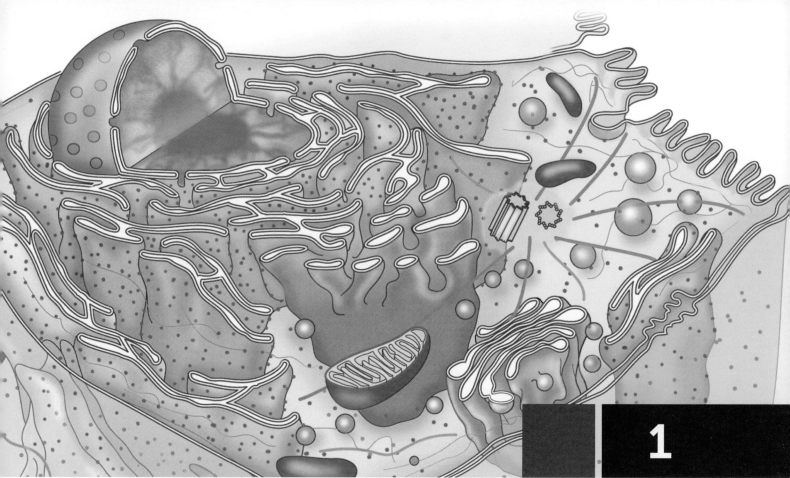

What Is a Cell?

■1.1■ The Big Picture

To begin implementing the study strategy outlined in the *To the Student* section of this book, let's start by exploring the overall organization of the text as a whole, and this chapter in particular. The first 4 chapters of the book address the question, "What are cells made of?" The remaining chapters of the book deal with 10 Principles of Cell Biology. While it is not essential to read the chapters in order, each builds on the material covered in earlier chapters.

This chapter is organized into five major sections, which serves two purposes. First, it provides an overall introduction to the fundamentals of cell biology, including the basic chemistry that helps define how cells are built and how they function. In particular, this introduction focuses on the structural organization of the simplest of the four cellular building blocks: sugars. Second, this chapter includes an overview of the common structures we will encounter in more detail in later chapters. The five major sections are as follows:

- The first section, *The Cell Is the Fundamental Unit of Life*, serves as an introduction to some of the essential concepts that form the foundation for supporting life. This includes the definition of life we will use for this book, as well as an examination of the three domains of living organisms. This section also introduces the concept that evolution by natural selection takes place at even the most basic levels of living organisms. We will return to this concept repeatedly throughout the book.

- The second section, *An Overview of the Molecular Building Blocks of Cellular Structures*, introduces some of the basic chemistry that governs how molecules in living organisms interact. This section also lists some of the most common chemical structures found in biomolecules. These form a critical part of the structure–function relationship, another essential concept in biology that we will refer to often.

  This section is also where we take our closest look at sugars. This information will be put to immediate use in Chapter 2, where we examine how they are used to synthesize nucleotides, another class of cellular building blocks. Pay particular attention to how sugars are assembled into polymers, as we will be referring to these polymers in several chapters.

- The third section, *Cells Contain Distinct Structures That Perform Specialized Functions*, focuses on the regions in eukaryotic cells called organelles. The descriptions of these structures are intended only to serve as an overview, since the remainder of the chapters will examine each of them in closer detail. It is important to keep in mind that these organelles are multitaskers, so most will appear in several chapters. Many students are tempted to map an organelle to one or two specific cellular functions (e.g., nucleus = DNA replication), when in fact every organelle can perform many functions at the same time. Thus, when we discuss DNA packing (Chapter 2), DNA replication (Chapter 7), nuclear pore transport (Chapter 8), and regulation of gene expression (Chapter 12), remember that these events are all occurring simultaneously in the nucleus, and they often influence one another. It is important to keep an eye on this big picture of a cell as we explore each of the Principles.

- The fourth section, *Cells in Multicellular Organisms Can Be Highly Specialized to Perform a Subset of the Functions Necessary for Life*, introduces another important concept in biology: that the organizing principles we will use to explain how individual cells function can be applied across many different scales of biological complexity. Here, we will take a quick look at the structure and function of tissues, the next highest order (after cells) of biological organization. We will need to keep this in mind as we make our way through the chapters, as it helps explain why cells expend so much effort to adhere to and communicate with their neighbors. In Chapter 14, we will look at this scaling issue in much greater detail.

■ The fifth section, *Model Organisms Are Often Studied to Understand More Complex Organisms*, explains where most of the information we will examine in later chapters comes from. While we will not be examining any of these organisms in detail in this book, it is important to recognize that at the cellular level, seemingly different organisms (such as bacteria, plants, flies, worms, and fish) share nearly as many similarities with us as does the familiar lab mouse, because they are all built on the same principles we will discuss in this book.

This chapter is one of the longest in this book, simply because there are many different topics to introduce. With the exception of sugars, we will review each of these topics in much greater detail in the chapters to come. While most of the remaining chapters will be somewhat shorter, they will contain a much higher density of critical details. Tackling these will provide students an opportunity to apply the study skills they will develop in this chapter. Welcome to the realm of cell biology! Let's get started.

1.2 ■ The Cell Is the Fundamental Unit of Life

KEY CONCEPTS

▸ Cells are self-replicating structures that are capable of responding to changes in their environment.
▸ Prokaryotes are the simplest forms of cells.
▸ Eukaryotes are complex cells capable of forming multicellular organisms.

For most cell biologists, the definition of the word *life* is fairly straightforward: it is a chemical system capable of Darwinian evolution. Objects that are alive must therefore have the ability to both *generate nearly exact copies of themselves* and *self-correct* (i.e., restore themselves to a defined state by repairing damage). In the hierarchy of biological complexity, individual molecules, such as proteins or DNA, are not considered to be alive, even if they exist inside a living organism, because they lack one or both of these abilities. Although many molecules (e.g., enzymes) have the ability to catalyze chemical reactions, no single chemical reaction can both replicate *and* repair these molecules.

From these observations, it is clear that life is a trait possessed only by groups of molecules that work together. Biologists believe that the earliest biological molecules may have been simple molecules capable of self-replication. Once these molecules were enclosed in a selective barrier, called a **membrane**, they were capable of forming teams that cooperated to maintain a fairly constant internal environment, even when conditions outside the membrane varied greatly. Molecular teams that developed the additional ability to repair or replace damaged team members were then able to generate nearly exact replicas of themselves, including the membrane. This membrane-enclosed team of molecules is now called a **cell**. The membrane is often referred to as the **plasma membrane** or **cell membrane**, and it encloses a compartment called the cytosol or **cytoplasm**. Because a cell is the simplest living structure, it is also referred to as the fundamental unit of life (**BOX 1-1**). All living beings are composed of cells. The simplest cells are called **prokaryotes**, to distinguish them from more complex cells known as eukaryotes. Examples of different types of cells are shown in **FIGURE 1-1**.

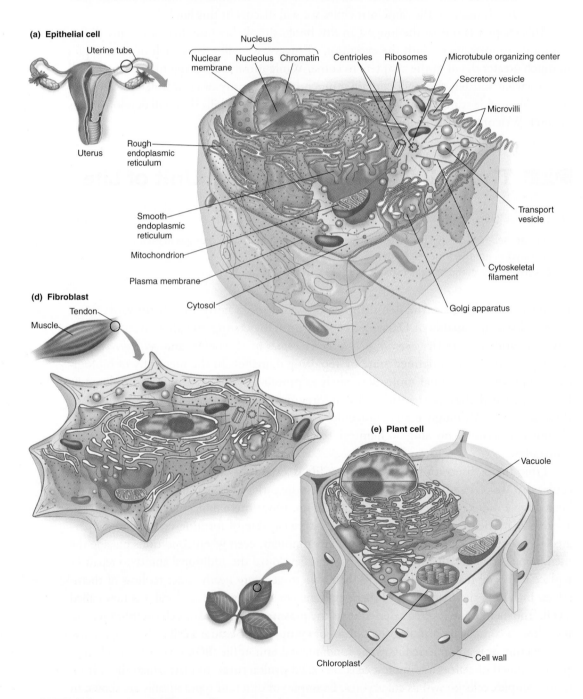

FIGURE 1-1 Examples of different cell types, including structures specific to each.

CHAPTER 1   What Is a Cell?

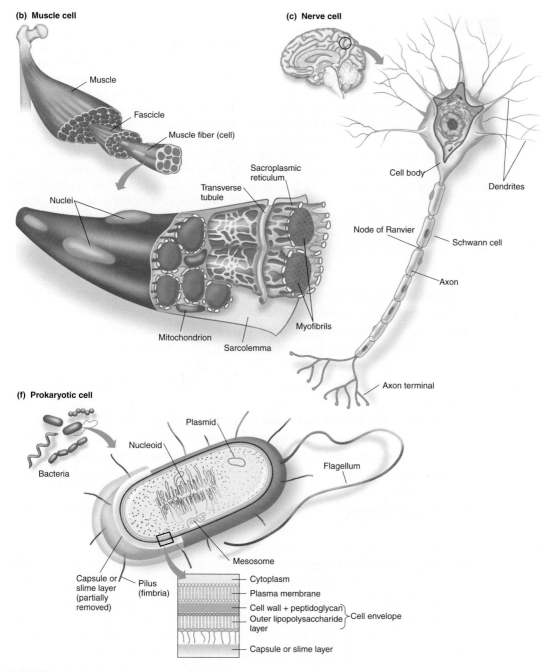

(b) Muscle cell

Muscle

Fascicle

Muscle fiber (cell)

Nuclei

Sacroplasmic reticulum

Transverse tubule

Mitochondrion

Sarcolemma

Myofibrils

(c) Nerve cell

Cell body

Dendrites

Node of Ranvier

Schwann cell

Axon

Axon terminal

(f) Prokaryotic cell

Plasmid

Nucleoid

Bacteria

Flagellum

Mesosome

Capsule or slime layer (partially removed)

Pilus (fimbria)

Cytoplasm

Plasma membrane

Cell wall + peptidoglycan

Outer lipopolysaccharide layer

Cell envelope

Capsule or slime layer

FIGURE 1-1 (*Continued*)

Cells Are Self-Replicating Structures That Are Capable of Responding to Changes in Their Environment

Division of labor is a common theme in cell biology. Since no single molecule is capable of both self-replication and self-maintenance, these tasks are tackled by cooperating groups of molecules that specialize in specific activities. All cells contain molecular teams responsible for accomplishing these essential tasks (**BOX 1-2**):

- **Maintenance of the internal environment.** Living organisms must capture and store energy, and this is accomplished by forming and maintaining chemical disequilibrium with the external environment. To remain alive, cells must continually

The "cell as a busy city" analogy. Because most students new to cell biology are unfamiliar with cells, they could try thinking of a cell as something similar to what is already familiar: a very busy city. Consider how to describe the concept of a city to someone who has never been in (or even seen) one. It's a daunting task. Hovering over any large, busy city in a helicopter and looking down at it, we might start by describing the general function(s) of the largest structures we can see (factories, office buildings, schools, roads, etc.) without delving into how they all work together to keep the city functioning. Likewise, we might describe the general concept of "people" in our initial description of a city, but we certainly would not want to start describing every single person in that city. The purpose of this chapter, then, is to (1) introduce the building blocks of the "big buildings" in cells (organelles, proteins, and other structures), (2) survey some of the most common "big buildings" (nucleus, plasma membrane, mitochondria, etc.), and (3) introduce the concept that not all "cities" are alike (i.e., they can be clustered into different regions—tissues—that work together). We'll return to this analogy in other chapters, and as we add more detail to the analogy, we can bring this city/cell to life, with the goal of arriving at a clear picture of how cells function at the molecular level.

adjust their internal activities to maintain a consistent environment that differs from conditions outside the cell membrane.

- **Sensing the external environment**. It is essential that cells be "aware" of changes in the external environment that may impact their own internal environment (e.g., changes in temperature, acidity, nutrient levels, osmotic pressure). Cells contain sensors for relevant environmental conditions such as these and ignore the rest.

- **Controlling the flow of molecules into and out of the cell**. Cells communicate with their external environment mainly through selective transport of molecules (e.g., cells import nutrients and export metabolic waste). Controlling this molecular traffic also helps cells maintain chemical disequilibrium and sense their surroundings.

- **Catalyzing chemical reactions**. In order to maintain a consistent internal environment, cells must be able to control the chemical reactions taking place within them. Molecules called enzymes play an important role in regulating these reactions.

- **Generating useful energy**. To catalyze most chemical reactions and do any form of work, cells must expend energy. Many molecules in cells are devoted to capturing energy from outside of the cell (e.g., sunlight and "food") and converting it into a small number of energy forms that cells can use directly. A well-known example of a useful energy form is adenosine triphosphate, otherwise known as ATP.

- **Storing genetic information**. Cells contain instructions for manufacturing most of the biological molecules necessary to stay alive. These instructions are stored in the form of a simple molecular polymer called deoxyribonucleic acid, or DNA.

- **Synthesis of biological molecules**. A considerable amount of the energy captured by cells is used to construct new biological molecules inside cells. These molecules may serve to replace damaged molecules, permit new functions in the cell, or generate sufficient copies of a molecule for the cell to replicate. To generate a nearly exact copy of itself, a cell must ensure that all information stored in its DNA is present in the newly created daughter cell. Cells possess molecular teams responsible for accurately replicating DNA and properly segregating it during cell division.

- **Regulating information flow**. Much like teams of people that communicate with one another to accomplish a complex task, molecular teams in a cell communicate with one another as well. Some molecules specialize in transferring information from one team to another.

Prokaryotes Are the Simplest Form of Cells

At the most fundamental level, all organisms are classified into one of three domains, as shown in **FIGURE 1-2**. Two of these domains, bacteria and archaea, are composed of prokaryotes, while the third (eukarya) contains eukaryotes. Three features are unique to prokaryotes:

- **Prokaryotes have only one membrane, the plasma membrane.** Because all of their internal chemical reactions take place in the cytosol, the degree of specialization that these cells can achieve is limited. Some prokaryotes have evolved elaborate modifications of the plasma membrane, such as stacks of membrane folds that provide them with some degree of compartmentalization. Likewise, the cytosol of prokaryotes appears heterogeneous when viewed with an electron microscope (**FIGURE 1-3**), suggesting that it may be partially organized.

- **All prokaryotic organisms are unicellular.** Prokaryotes do not assemble into multicellular organisms, although some cluster together to form enormous structures called **biofilms**.

- **Prokaryotes do not divide by mitosis.** Most genetic information in prokaryotic cells is contained in a single circular DNA molecule called the **chromosome**, while eukaryotic cells contain several chromosomes. Mitosis, which is an elaborate mechanism designed to ensure proper segregation of chromosomes during cell division, evolved in eukaryotic organisms only.

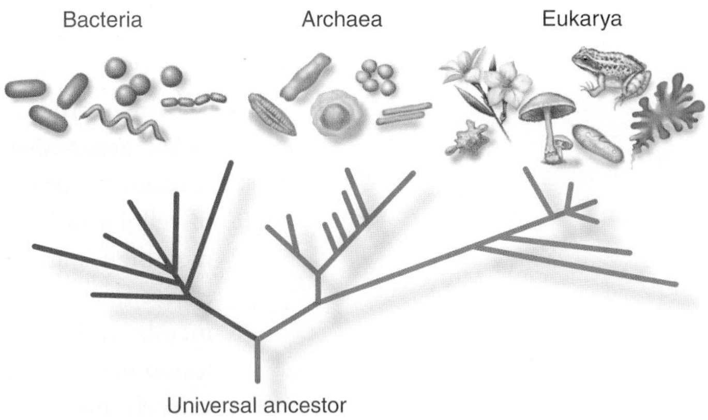

Despite their relatively simple structure, prokaryotes can occupy some of the harshest environments on earth. These include extreme heat and cold, tremendous atmospheric pressure, little or no atmospheric oxygen, and pH values ranging from 2 to 12. This is likely because modern prokaryotes are the direct descendents of the Earth's earliest cells, which are estimated to have first appeared about 3.5 billion years ago, when the Earth's atmosphere was much different from today. Due in part to their high degree of adaptability, prokaryotes are also by far the most abundant organisms on earth.

Prokaryotic Cells Are Protected by a Cell Wall

Most, if not all, prokaryotes contain an additional layer of protection outside the plasma membrane. The general term for this structure is **cell wall**, and the portion of it that is directly connected to the plasma membrane is often called the **capsule** (**FIGURE 1-4**). The cell wall is composed largely of sugar molecules linked together to form a thick mesh. Aside from providing protection against physical trauma, the cell wall also retains water to help ensure that the cell is properly hydrated.

Eukaryotes Are Complex Cells Capable of Forming Multicellular Organisms

Eukaryotic cells comprise the third, and most recent, domain of biological organisms shown in Figure 1-2. When viewed under a microscope, the most striking feature of

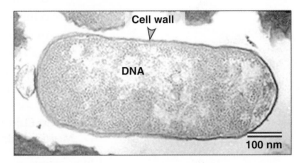

FIGURE 1-2 The three domains of organisms are all derived from a common ancestor, but have distinct structural properties. This is a phylogenetic tree showing the likely sequence of evolution leading from the common ancestor (bottom) to eukaryotes (top).

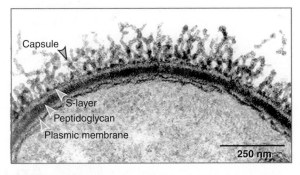

FIGURE 1-3 An electron micrograph of a bacterial cell, which is a prokaryote. Photo courtesy of Jonathan King, Massachusetts Institute of Technology.

FIGURE 1-4 The capsule in prokaryotes is connected to the plasma membrane. An electron micrograph of the capsule of a strain of bacteria called *Bacillus anthracis*. Reproduced from J. Bacteriol., 1998, vol. 180, pp. 52–58, DOI and reproduced with permission from American Society for Microbiology. Photo courtesy of Agnés Fouet, Pasteur Institute.

eukaryotic cells is that their cytosol is highly organized. Even the simplest microscope reveals the presence of a large, often oval-shaped structure called the **nucleus**. In fact, the presence of a nucleus is the defining feature of eukaryotic cells. Closer examination with more powerful microscopes allows us to see additional distinct structures in the cytosol (**FIGURE 1-5**). These structures are generally classified into two groups:

- Cytosolic structures that are surrounded by at least one distinct membrane are called **organelles**. The presence of these membranes allows the cell to create specialized compartments in the cytosol that are devoted to performing a subset of cellular tasks under optimized conditions. Like the plasma membrane, these membranes are selectively permeable, which helps to create a unique internal environment optimally suited to the molecules contained inside. Because it is membrane bound, the nucleus is classified as an organelle. Additional organelles found in eukaryotic cells include the **endoplasmic reticulum**, **Golgi apparatus**, **mitochondria**, **chloroplasts**, **endosomes**, **lysosomes**, and **peroxisomes**. Each of these organelles contains unique molecules and performs a separate set of functions.

- Large molecular complexes that are not enclosed in a separate membrane do not have a generic name (like *organelle*), but they do share one important trait: they represent specialized regions in the cytosol that are devoted to a subset of cellular tasks. For example, eukaryotic cells contain three different types of fibers that serve as a scaffold for the organization of the cytosol, thereby earning them the name **cytoskeleton**. Careful arrangement of the cytoskeleton is essential for proper cell function; without it, muscle cells would not contract and nerve cells would fall silent.

◼◼ Some Eukaryotic Cells Contain Structures That Allow Them to Form Multicellular Clusters

One obvious advantage to possessing such a high degree of compartmentalization is that eukaryotic cells can customize their cytosol to generate a tremendous number of different cell types. Approximately one billion years after cells first appeared, some eukaryotic cells acquired the ability to work together in groups, giving rise to multicellular organisms. The structural basis for multicellularity is revealed by a relatively small number of structures that hold adjacent cells together. Many of these structures are also linked to the cytoskeleton, thereby creating a supercytoskeleton that spans multiple cells and helps integrate them into a single functioning unit.

CONCEPT CHECK #1

Explain why all living organisms are enclosed by at least one membrane.

◼ 1.3 ◼ An Overview of the Molecular Building Blocks of Cellular Structures

KEY CONCEPTS

▸ The structure–function relationship is a powerful tool for understanding cellular organization.
▸ Water is the most common compound found in cells.
▸ The study of cellular chemistry begins with an examination of the carbon atom.
▸ Complex biomolecules are mostly composed of chemical building blocks called functional groups.
▸ Lipids, sugars, amino acids, and nucleic acids are the most common biomolecules in cells.

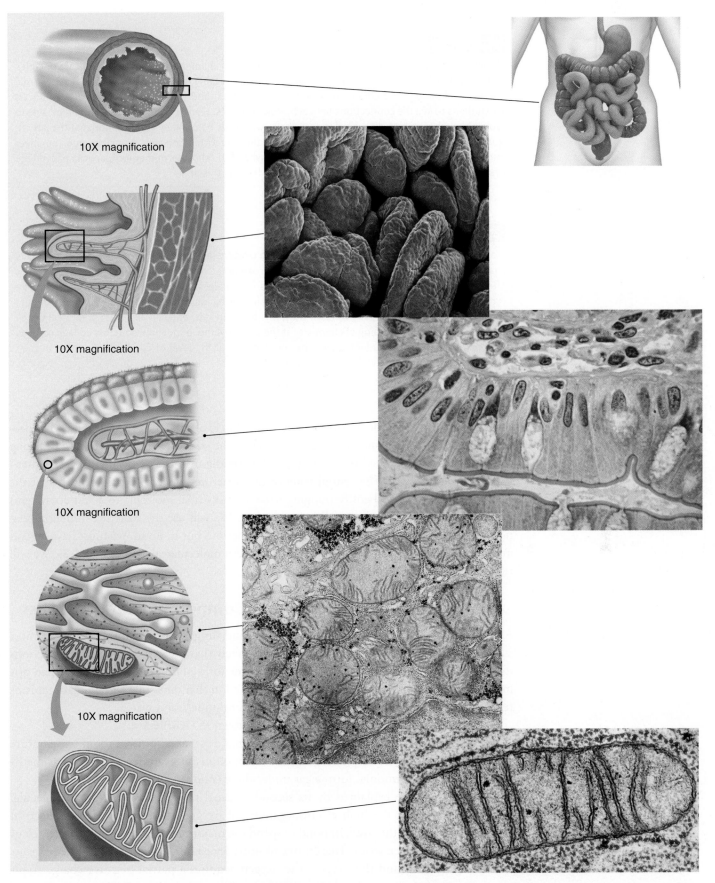

FIGURE 1-5 Subcellular structures can be visualized with light microscopes or electron microscopes. © Sebastian Kaulitzki/ShutterStock, Inc., © Dr. Thomas Deerinck/Visuals Unlimited, Inc., © Peter Arnold, Inc./Alamy, © CNRI/Photo Researchers, Inc., © Keith R. Porter/Photo Researchers, Inc.

10X magnification

10X magnification

10X magnification

10X magnification

Much of the remainder of this book is devoted to exploring the organization and molecular function of membranes, organelles, and cytosolic structures in eukaryotic cells. Along the way, we will be guided by a very powerful paradigm in biology called the **structure–function relationship**. The central tenet of the structure–function relationship is that the *function* of any biological entity (ranging from an individual molecule to a vast ecosystem) can be determined by understanding its *structure*. We will use the structure–function relationship to understand how molecules in cells function, but to do so, we must have a good understanding of the chemical principles that govern molecular structure (**BOX 1-3**). We will review the fundamentals here (**BOX 1-4**).

Water Is the Most Common Compound Found in Cells

All living cells, including the driest plant seeds and fungal spores, contain water. In most cells, water is the most abundant molecule. It is likely that the very first chemical reactions that lead to the formation of cells occurred in some form of liquid water (the so-called primordial soup). Consequently, every chemical reaction that takes place in living organisms reflects this ancestry. To understand life, we need to understand water.

Water is unusual in at least five aspects, as shown in **FIGURE 1-6**:

- First, it is the only molecule of its size that exists as a liquid at room temperature and normal atmospheric pressure (compounds of a similar mass, such as methane and ammonia, form a gas under these conditions).
- This can be explained by its second unusual trait: it is a very polar molecule. Specifically, the high electronegativity of oxygen causes the hydrogen electron involved in the covalent bond to spend the majority of its time circling the nucleus of the oxygen atom. This creates an imbalanced bond; because the electron orbits mostly around the oxygen, the oxygen acquires a partial negative charge (typically represented by the symbol δ–), while each hydrogen atom becomes partially positive (represented as δ+). The imbalance in charge is what holds water

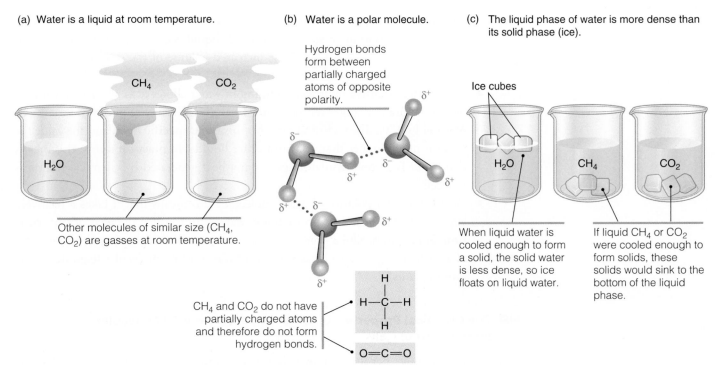

(a) Water is a liquid at room temperature.

H_2O

CH_4 CO_2

Other molecules of similar size (CH_4, CO_2) are gasses at room temperature.

(b) Water is a polar molecule.

Hydrogen bonds form between partially charged atoms of opposite polarity.

δ^- δ^+
δ^-
δ^+ δ^+
δ^+ δ^-
δ^+

CH_4 and CO_2 do not have partially charged atoms and therefore do not form hydrogen bonds.

$$\begin{array}{c} H \\ | \\ H-C-H \\ | \\ H \end{array}$$

$$O=C=O$$

(c) The liquid phase of water is more dense than its solid phase (ice).

Ice cubes

H_2O CH_4 CO_2

When liquid water is cooled enough to form a solid, the solid water is less dense, so ice floats on liquid water.

If liquid CH_4 or CO_2 were cooled enough to form solids, these solids would sink to the bottom of the liquid phase.

(d) Water has a high specific heat and is a good insulator.

H_2O CH_4 CO_2

It takes more heat energy to raise the temperature of water than is needed to raise the temperature of CH_4 or CO_2.

(e) Water has a high heat of evaporation.

CH_4 CO_2

H_2O

FIGURE 1-6 Five unusual traits of water.

molecules together so tightly: the $\delta-$ of the oxygen atom attracts the $\delta+$ of the hydrogen atoms in nearby water molecules, and vice versa. This phenomenon occurs in other molecules as well, and it is called **hydrogen bonding**.

■ A third unusual trait of water is that its liquid phase has a higher density than its solid phase in standard conditions (i.e., ice floats in water). Again, this can be explained by the ubiquitous hydrogen bonding that occurs in liquid water, as these bonds pack water molecules more closely together than the regular, repeating arrangement found in crystals such as ice. If liquid water forms its most common solid, known as ice (I_h) in cells, the increased volume of the solid water can tear membranes apart and rupture cells.

- Fourth, the extensive hydrogen bonding in water allows it to absorb a great deal of heat before it changes temperature: 1 calorie of heat is required to heat 1 mL of water by 1°C. The technical term for this is the **specific heat**, and the value for water is much higher than for most other liquids. In practical terms, this means that water is a good insulator against any heat generated by chemical reactions in a cell.

- Finally, the high polarity of water molecules also means that it takes a relatively large amount of heat to vaporize liquid water; this is called the **heat of vaporization**. Some multicellular organisms take advantage of this property by using water as a coolant (e.g., sweat) or as a means of transporting molecules (e.g., transpiration in plants).

■■■ The Chemical Properties of Water Impact Nearly All Molecular Interactions in Cells

Why does this matter? Because it allows us to better understand why molecules behave as they do in cells. Since water is polar, every other molecule that interacts with it can be classified as compatible (hydrophilic) or incompatible (hydrophobic). Hydrophilic molecules include ions (H^+, Na^+, Cl^-, etc.) as well as other polar molecules (e.g., sugars, ammonia); the charge imbalance in these compounds attracts the polar atoms in water. Hydrophobic molecules are nonpolar (and contain no charged atoms); therefore, they do not attract water. This can be easily demonstrated by placing a drop of oil in a cup of water: the oil does not disperse, but instead remains clustered together in an attempt to avoid contact with the water molecules.

At first glance, one might expect that every molecule in a cell *must* be hydrophilic if life began in water. However, this is not true, and cells can in fact benefit from having hydrophobic molecules. There are three major advantages that hydrophobic molecules offer cells, as shown in **FIGURE 1-7**:

- First, hydrophobic molecules spontaneously cluster together in water. This principle is the underlying reason why phospholipids, which form the bulk of biological membranes, spontaneously assemble into a bilayer when submerged in water: the fatty acid tails in phospholipids are hydrophobic and thus cluster together. In short, cells have to expend little or no energy to organize phospholipids into a membrane.

- A second advantage is that this spontaneous assembly process permits membranes to automatically reseal if they are punctured (remember that self maintenance is a critical feature of all cells).

- Finally, a membrane composed of hydrophobic molecules repels most hydrophilic molecules, thereby creating an effective barrier to hydrophilic molecules between the two sides of that membrane. This is the main reason why cells can create their own internal environment.

The advantages of water's hydrogen bonding to other hydrophilic molecules are quite clear: polar (and ionic) molecules dissolve readily in water, so cells can concentrate a large number of them on one side of a membrane. As we will see in later chapters, the concentration and movement of ions across membranes is a common activity in all cells, and is responsible for conducting the electrical signals in our nervous system.

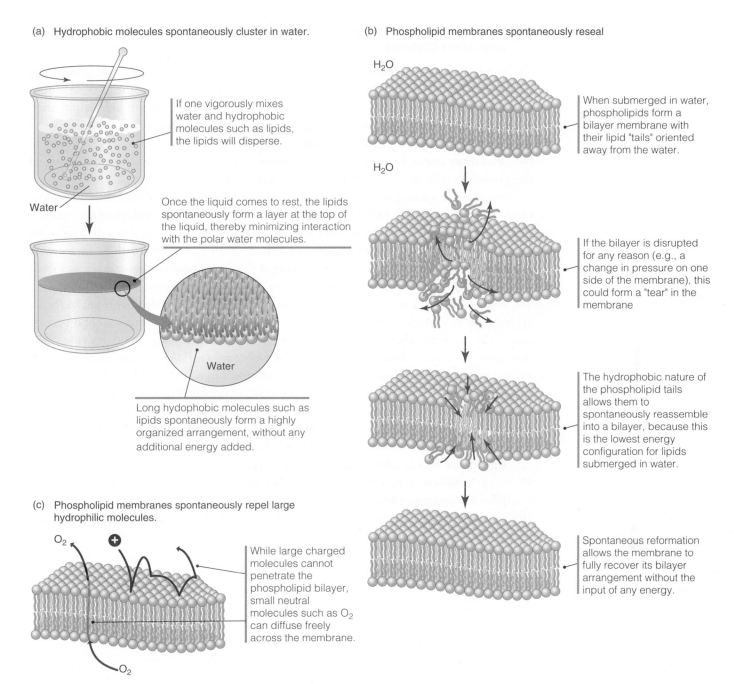

(a) Hydrophobic molecules spontaneously cluster in water.

If one vigorously mixes water and hydrophobic molecules such as lipids, the lipids will disperse.

Water

Once the liquid comes to rest, the lipids spontaneously form a layer at the top of the liquid, thereby minimizing interaction with the polar water molecules.

Water

Long hydophobic molecules such as lipids spontaneously form a highly organized arrangement, without any additional energy added.

(b) Phospholipid membranes spontaneously reseal

H_2O

When submerged in water, phospholipids form a bilayer membrane with their lipid "tails" oriented away from the water.

H_2O

If the bilayer is disrupted for any reason (e.g., a change in pressure on one side of the membrane), this could form a "tear" in the membrane

The hydrophobic nature of the phospholipid tails allows them to spontaneously reassemble into a bilayer, because this is the lowest energy configuration for lipids submerged in water.

Spontaneous reformation allows the membrane to fully recover its bilayer arrangement without the input of any energy.

(c) Phospholipid membranes spontaneously repel large hydrophilic molecules.

O_2

While large charged molecules cannot penetrate the phospholipid bilayer, small neutral molecules such as O_2 can diffuse freely across the membrane.

O_2

FIGURE 1-7 Three advantages of using hydrophobic molecules in cells.

The Study of Cellular Chemistry Begins with an Examination of the Carbon Atom

Carbon is an especially important element in cells, for three reasons:

1. Aside from water, carbon-containing compounds are the most abundant molecules in cells.
2. These compounds exist in a dizzying array of variations: of all known molecules, those containing carbon vastly outnumber all of the rest, combined.
3. Carbon atoms can attach to one another more readily than the atoms of any other element, giving rise to molecules of tremendous size (some contain several thousand carbons) and structural complexity.

No wonder, then, that organisms are often referred to as carbon-based life forms.

Carbon Forms Characteristic Bonds with Hydrogen, Oxygen, Nitrogen, and Other Carbons

The number of carbon-containing compounds is so vast that they are classified into groups (or families) according to their structure. We will have a closer look at some of these groups later in this chapter, but first we need to recognize an important concept: that despite their tremendous complexity, carbon-containing compounds are typically constructed from a small number of basic chemical shapes. In cells, carbon atoms are typically covalently bound to only four other atoms: hydrogen, oxygen, nitrogen, and other carbons. Most carbon-based compounds in cells are built with these simple structures.

For those interested in the details, let's examine these carbon building blocks more carefully. (As discussed in **BOX 1-4**, readers are encouraged to check with their instructor if this material is new; some courses may not require this level of detail.) Since each contains at least one carbon, we have to understand how covalent bonds with carbon are formed. The carbon atom contains six electrons, arranged in three orbital configurations, as shown in **FIGURE 1-8**. Two electrons are present in the 1s orbital (the innermost shell), filling it. The other four are located in the second (valence, or outermost) shell: two are in the 2s orbital, and in a single (unbound) carbon atom, the other two are unpaired and occupy two of the three 2p orbitals, as shown in Figure 1-8A. In chemistry, the **octet rule** states that *nonmetallic atoms tend to gain, lose, or share electrons until they are surrounded by eight valence electrons.* Because carbon has only four electrons in its valence shell, it "needs" four additional electrons. It fills the remaining positions in the two 2p orbitals by forming four covalent bonds with other atoms. This results in a rearrangement of the valence shell, as shown in Figure 1-8B: one electron in the 2s shell is "borrowed" by the 2p orbitals, resulting in the formation four orbitals called sp3 hybrids.

These four bonds adopt characteristic shapes for each configuration, as seen in **FIGURE 1-9**. When carbon binds to four other atoms, these four bonds are arranged in a tetrahedral pattern, with bond angles of 109.5°. When carbon binds to three other atoms, one of these atoms forms a double bond, and this forces the other two atoms

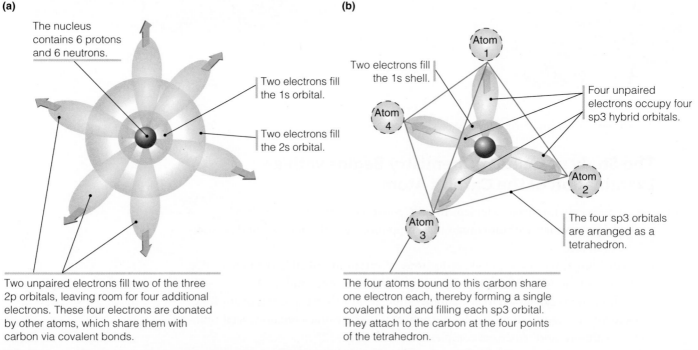

(a)

The nucleus contains 6 protons and 6 neutrons.

Two electrons fill the 1s orbital.

Two electrons fill the 2s orbital.

Two unpaired electrons fill two of the three 2p orbitals, leaving room for four additional electrons. These four electrons are donated by other atoms, which share them with carbon via covalent bonds.

(b)

Atom 1

Two electrons fill the 1s shell.

Atom 4

Four unpaired electrons occupy four sp3 hybrid orbitals.

Atom 2

Atom 3

The four sp3 orbitals are arranged as a tetrahedron.

The four atoms bound to this carbon share one electron each, thereby forming a single covalent bond and filling each sp3 orbital. They attach to the carbon at the four points of the tetrahedron.

FIGURE 1-8 (a) Model of a single carbon atom. **(b)** Model of a carbon atom bound to four other atoms.

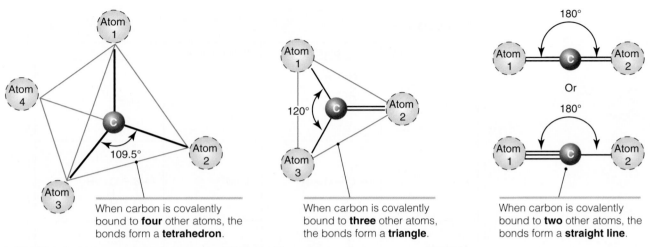

When carbon is covalently bound to **four** other atoms, the bonds form a **tetrahedron**.

When carbon is covalently bound to **three** other atoms, the bonds form a **triangle**.

When carbon is covalently bound to **two** other atoms, the bonds form a **straight line**.

FIGURE 1-9 The orientation of covalent bonds formed by carbon.

into a triangular, planar configuration. When carbon binds to two other atoms, the three atoms adopt a linear arrangement. Carbon can be connected to two atoms by a pair of double bonds (e.g., carbon dioxide) or by a triple bond and a single bond (e.g., cyanide). Carbon does not form four covalent bonds with a single atom.

Complex Biomolecules Are Mostly Composed of Chemical Building Blocks Called Functional Groups

Because carbon always binds to at least two other atoms, these atoms and their associated binding partners can be easily combined to create a wide variety of structures. The field of organic chemistry, which is devoted to the study of chemical compounds in organisms, subdivides them according to their chemical structure. The different classes are called **functional groups**, and **TABLE 1-1** lists some of the more common functional groups we will encounter throughout this book. Note that while not all functional groups contain carbon, those that do are by far the most abundant in cells.

Lipids Are Carbon-Rich Polymers That Are Insoluble in Water

When carbon forms covalent bonds with oxygen or nitrogen, these bonds are classified as polar (oxygen and nitrogen are more electronegative than carbon). By contrast, carbon–carbon bonds are nonpolar, and covalent bonds between carbon and hydrogen have so little polarity that most molecules consisting only of carbon and hydrogen do not attract water. These compounds, often called hydrocarbons, are hydrophobic and do not dissolve in water.

Lipids are a class of hydrocarbons commonly found in cells; **FIGURE 1-10** shows the generalized structures of some common cellular lipids (see also **BOX 1-5**). Because they are insoluble in the cytosol, most lipids in cells are attached to hydrophilic functional groups that confer some degree of water solubility. These modified lipids are sometimes referred to as being **amphipathic** (derived from Greek, meaning "having both properties").

Common modified lipids include:

- **Phospholipids**, which are by far the most common form of modified lipids in cells. They constitute most of the mass of cellular membranes, as we shall see in Chapter 4.
- A second modified lipid common in cells is **cholesterol**. Cholesterol is an essential component of cellular membranes. Because it is largely hydrophobic, it is most commonly found in the middle (hydrophobic) zone of membranes, where

TABLE 1-1

Functional Group	Structure	Functional Group	Structure
Acetyl	$-\overset{\overset{O}{\|\|}}{C}-\overset{\overset{H}{\|}}{\underset{H}{C}}-H$	Ether	$-O-$
Alcohol	$-\overset{\|}{\underset{\|}{C}}-O-H$	Ethyl	$-\overset{\overset{H}{\|}}{\underset{H}{C}}-\overset{\overset{H}{\|}}{\underset{H}{C}}-H$
Aldehyde	$-\overset{\overset{O}{\|\|}}{C}-H$	Ester	$-\overset{\overset{O}{\|\|}}{C}-O$
Alkenyl	$\rangle C=C\langle$	Imino	$\rangle C=N\rangle$
Alkyl	$-\overset{\overset{H}{\|}}{\underset{H}{C}}-$	Ketone	$-\overset{\overset{O}{\|\|}}{C}-$
Amide	$-\overset{\overset{O}{\|\|}}{C}-\overset{\|}{N}-$	Methyl	$-\overset{\overset{H}{\|}}{\underset{H}{C}}-H$
Amino	$-\overset{\overset{H}{\|}}{N}-H$ $-\overset{\overset{H}{\|}}{\underset{H}{N^+}}-H$	Phenyl	(phenyl ring structure)
Carbonyl	$-\overset{\overset{O}{\|\|}}{C}-$	Phosphate	$O-\overset{\overset{O}{\|\|}}{\underset{O^-}{P}}-O^-$
Carboxylic acid (aka Carboxyl)	$-\overset{\overset{O}{\|\|}}{C}-O-H$	Sulfide/Sulfydryl	$-S-H$
Carboxylate	$-\overset{\overset{O}{\|\|}}{C}-O^-$	Disulfide	$-S-S-$

Common functional groups found in biological molecules. In this abbreviated version of chemical structure, the bond angles for most atoms are ignored, and the atoms are usually arranged at right angles.

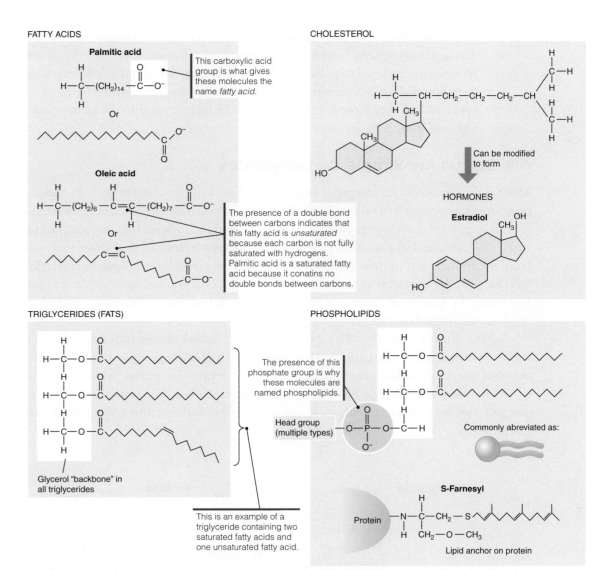

FATTY ACIDS

Palmitic acid

H—C—(CH₂)₁₄—C—O⁻

This carboxylic acid group is what gives these molecules the name *fatty acid*.

Or

Oleic acid

H—C—(CH₂)₆—C=C—(CH₂)₇—C—O⁻

Or

The presence of a double bond between carbons indicates that this fatty acid is *unsaturated* because each carbon is not fully saturated with hydrogens. Palmitic acid is a saturated fatty acid because it conatins no double bonds between carbons.

CHOLESTEROL

Can be modified to form

HORMONES

Estradiol

TRIGLYCERIDES (FATS)

Glycerol "backbone" in all triglycerides

This is an example of a triglyceride containing two saturated fatty acids and one unsaturated fatty acid.

PHOSPHOLIPIDS

The presence of this phosphate group is why these molecules are named phospholipids.

Head group (multiple types)

Commonly abreviated as:

S-Farnesyl

Protein —N—C—CH₂—S

Lipid anchor on protein

FIGURE 1-10 Common types of lipids in cells. Common abbreviations of organic structures are shown.

it interacts with phospholipids to change the mechanical properties of the membrane. In some animals, derivatives of cholesterol are used as hormones that circulate in the body to permit communication between even very distant cells.

■ A third form of modified lipid is a **triglyceride**, commonly known as **fat**. Triglycerides serve as an important form of energy storage in animals. Because they are largely insoluble in water, they form distinct droplets that are, in some cells,

---

BOX 1-5 TIP

Understanding molecular structure diagrams. Most of the diagrams that biologists and chemists use to depict molecular structures are shorthand simplifications of the real three-dimensional structures. For example, a carbon atom's four bonds are often drawn in an up–down–left–right orientation, even though we know that the bonds are never at 90° angles to one another. In other cases, the bonds aren't even drawn: a methyl group is often abbreviated as –CH₃. In still other instances, even the letters are missing; in most drawings of large organic molecules such as lipids, simple lines are used to represent bonds between carbons, and the carbon–hydrogen bonds aren't included at all. Thus, a zigzag line can represent a series of alkyl (–CH₂–) groups. There are mixtures of many different versions of chemical structure shorthand in cell biology, but don't let it be confusing. An alkyl group is always an alkyl group, no matter how it is drawn.

easy to identify with a microscope. Often, triglycerides are transported through the circulatory system, bound to proteins to form a **lipoprotein**. Some lipids are permanently attached to proteins, where they play an important role in targeting these proteins for cell membranes; these structures are sometimes referred to as lipid tails, and they serve as a form of anchor by interacting with the hydrophobic region of the membrane to keep the protein in place.

Sugars Are Simple Carbohydrates

Many molecules in cells are composed entirely of carbon, hydrogen, and oxygen, and often, these atoms are present in a ratio of $C_nH_{2n}O_n$. Chemists who first characterized these compounds speculated that they might be arranged such that the carbon atoms are attached to water molecules, and so they referred to them as **carbohydrates** (we now know that this is not the arrangement of these compounds, but the name stuck). Carbohydrates typically form rings or linear strands of linked carbon atoms; the oxygen and hydrogen atoms are present in the functional groups (with names such as hydroxyls, carbonyls, aldehydes, alkanes, etc.) formed by the carbons.

Sugars, illustrated in **FIGURE 1-11**, are common carbohydrates found in cells. All sugars contain at least three carbons. (For those interested in the technical terms organic chemists use to describe them, one carbon forms a *carbonyl* group, which exists as either an *aldehyde* or a *ketone*, and the remainder of the carbons are attached to *hydroxyl* groups.) The most important sugars in cellular metabolism contain between three and seven carbons.

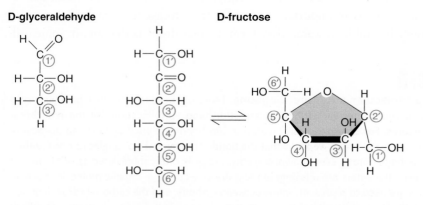

FIGURE 1-11 Common monosaccharides in cells. The carbons are numbered by convention with "primes" as shown.

CHAPTER 1   What Is a Cell?

The Common Sugars Glucose and Ribose Serve Several Different Functions in Cells

For many students, mentioning sugar immediately brings to mind its important role as a source of metabolic energy. But sugars are far more than food. For example, **ribose** is a five-carbon sugar found in all cells. Ribose and its derivative, known as **deoxyribose**, form the backbone of the nucleic acids RNA and DNA, respectively. We will discuss nucleic acids in more detail. **Glucose** is a six-carbon sugar that serves as the building block for complex molecules such as starch, cellulose (a major component of the cell wall in plants), and chitin (found in the exoskeleton of arthropods).

Many Sugars Exist as Disaccharides in Cells

Glucose and ribose are examples of the simplest form of a sugar, called a **monosaccharide**. In cells, sugars are commonly found in linked pairs called **disaccharides**. For example, the common table sugar, sucrose, consists of glucose and another six-carbon sugar, fructose. Lactose and maltose are other common disaccharides. Two important properties of disaccharides help determine their function in cells. The first of these is simply the types of monosaccharides they contain.

The second is how they are connected, as illustrated in **FIGURE 1-12**. All monosaccharides are linked by a **glycosidic bond** (a form of ether bond specific to sugars) between the carbon atoms on each sugar, but these bonds are not all the same. All disaccharides are formed between sugars in a ring form. The two sides of the ring are never identical, because each carbon in the ring has two different atoms (–H, –OH, or –CHO)

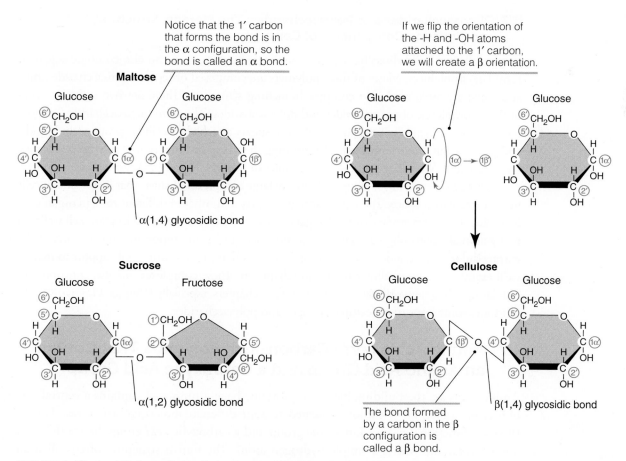

FIGURE 1-12 α and β glycosidic bonds in common disaccharides.

projecting outward. This means that the bond formed between two carbons oriented in the same direction is different from the bond that joins two carbons facing in opposite directions, even if the same carbons are involved. Organic chemists have developed a shorthand way of defining these bonds. First, each carbon in a monosaccharide is assigned a number, followed by an apostrophe to designate the term "prime" (1': one prime, 2': two prime, 3': three prime, etc.). Next, the bond between carbons in two monosaccharides is identified by simply listing the involved carbons, separated by a comma (e.g., 1', 4'). Finally, these bonds are classified as either α or β according to the specific orientation of the 1' carbon. Notice, for example, the α and β orientations of glucose in Figure 1-12, and the resulting disaccharides formed by α1,4 and β1,4 bonds between the two glucose molecules.

A seemingly minor point such as the orientation of a single atom has profound effects for cells. All of the chemical reactions necessary to make and break bonds between sugars are catalyzed by enzymes, which are proteins that contain a binding site where these chemical reactions take place. Each enzyme has its own specific binding site; this means that an enzyme that catalyzes the formation of an α1,4 glycosidic bond cannot also catalyze the formation of an α1,6 bond or a β1,4 bond because these bonds are shaped differently. Likewise, the enzymes that degrade a specific bond cannot degrade the others. Practically speaking, this difference determines which disaccharides cells can make and degrade, based on the type of enzymes they contain. This fact also explains why humans can digest many α1,4 bonds (e.g., in sucrose and maltose) but not most β1,4 bonds (e.g., in chitin and cellulobiose): human cells contain few enzymes capable of binding and breaking β1,4 glycosidic bonds. The enzyme lactase, which breaks the β1,4 bond in the disaccharide lactose, is expressed only in infancy in most mammals.

■■ Oligosaccharides and Polysaccharides: The Storage, Structural, and Signaling Components of Cells

Just as cells can build disaccharides with monosaccharides, they can also combine sugars to form larger polymers. Many of these polymers are composed of repeating disaccharide units, and some are assembled into complex branching structures. There are two different types of these complexes: oligosaccharides and polysaccharides. The term oligosaccharide ("oligo-" means *many*) is typically used to describe the sugars attached to cellular proteins and lipids, and they play an important role in determining the shape and function of the molecules they are attached to. We will see numerous examples of this in Chapter 9. Polysaccharides ("poly-" means *even more than oligo*) are tremendously large complexes of sugar that lie outside of cells in the extracellular space. These polysaccharides play a number of different roles in organisms, including long-term storage of food sugars and as a reinforcing material in plant cell walls. In many animals, both oligosaccharides and polysaccharides are important components of the **extracellular matrix**, a dense network of molecules that provide structural support to tissues; we'll examine the extracellular matrix in Chapter 6. These complexes can also contribute to cellular signaling networks, as we will see in later chapters, especially Chapter 11. **FIGURE 1-13** illustrates some important features of oligo- and polysaccharides.

■■ Amino Acids Form Carbon-Rich Polymers That Contain an Amino Acid Group and a Carboxylic Acid Group

Amino acids are the building blocks of proteins. Each amino acid contains a central carbon atom, called an α-carbon, attached to four different molecular structures. Two of these are functional groups (an *amino* group and a carboxylic *acid* group, hence the term *amino acid*), and one is a simple hydrogen atom. The fourth structure, often called an **amino acid side chain** (or **"R" group**), differs in each different amino acid. The proteins in cells are constructed from 20 different amino acids; these 20 amino acids are classified

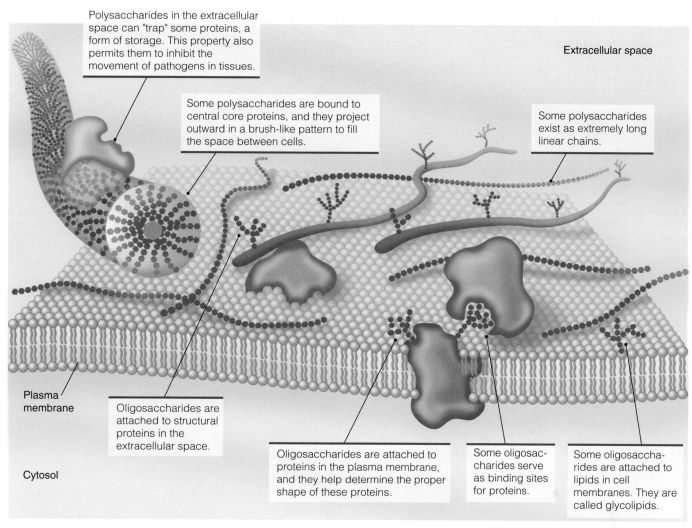

Polysaccharides in the extracellular space can "trap" some proteins, a form of storage. This property also permits them to inhibit the movement of pathogens in tissues.

Some polysaccharides are bound to central core proteins, and they project outward in a brush-like pattern to fill the space between cells.

Extracellular space

Some polysaccharides exist as extremely long linear chains.

Plasma membrane

Oligosaccharides are attached to structural proteins in the extracellular space.

Cytosol

Oligosaccharides are attached to proteins in the plasma membrane, and they help determine the proper shape of these proteins.

Some oligosaccharides serve as binding sites for proteins.

Some oligosaccharides are attached to lipids in cell membranes. They are called glycolipids.

FIGURE 1-13 Oligosaccharides and polysaccharides play many important roles in cells. Most oligosaccharides and polysaccharides are located on the cell surface and in the extracellular space.

into three groups, according to the chemical nature of the side chains (nonpolar, polar, or ionic), as shown in **FIGURE 1-14**.

Proteins are composed of long, linear sequences of amino acids. These sequences are created by forming covalent bonds (here called **peptide bonds**) between the carboxylic acid group of one amino acid and the amino group of another. Depending upon the number of amino acids linked together, the names of the polymers differ; two and three amino acids linked together are called dipeptide and tripeptide, respectively, while the term **polypeptide** typically refers to polymers of 10 or more amino acids (some polypeptides contain over 1,500 amino acids). When polypeptides fold into a stable configuration that is useful to cells, they are called **proteins**. A more in-depth discussion of the naming schemes for polypeptides, proteins, and protein subunits is provided in Chapter 3.

■■ Chemical Modifications of Amino Acids Help Control Protein Function

The structure–function relationship predicts that any condition that alters the shape of a protein also impacts its function. Changes in a protein's environment (temperature, pH, ionic strength, etc.) have global effects on protein folding and function that are difficult to control. In contrast, cells use targeted chemical modifications of individual amino

(a) General amino acid structure

The carboxylic acid group can also exist as a negatively charged carboxylate (-COO⁻) group.

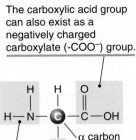

α carbon

The amino group can also exist as a positively charged -NH₃⁺ group.

Proline is an unusual amino acid because the end of its side chain loops back to form a covalent bond with the nitrogen in the amino group. Note that this causes the nitrogen to become positively charged.

(b) Hydrophobic (nonpolar) side chains

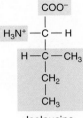

Gylcine
(Gly or G)

Alanine
(Ala or A)

Valine
(Val or V)

Leucine
(Leu or L)

Isoleucine
(Ile or I)

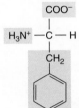

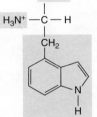

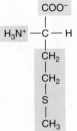

Proline
(Pro or P)

Phenylalanine
(Phe or F)

Tryptophan
(Trp or W)

Methionine
(Met or M)

(c) Polar hydrophilic side chains

Serine
(Ser or S)

Threonine
(Thr or T)

Tyrosine
(Tyr or T)

Asparagine
(Asn or N)

Glutamine
(Gln or Q)

Cysteine
(Cys or C)

(d) Charged hydrophilic side chains

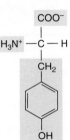

Lysine
(Lys or L)

Arginine
(Arg or A)

Histidine
(His or H)

Aspartic acid
(Asp or D)

Glutamic acid
(Glu or E)

FIGURE 1-14 The 20 most common amino acids are classified into 3 classes based on the structure of their side chains. Examples of chemically modified side chains are shown in part E.

(e) Modified side chains

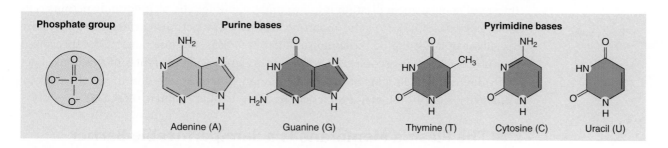

Phosphoserine

Phosphothreonine

Phosphotyrosine

Acetyllysine

Methyllysine

FIGURE 1-14 (*Continued*)

acids to slightly alter protein shape; these modifications are carried out by other proteins, and so are comparatively easy to control. Examples of these modifications include the addition of phosphate, methyl, and acetyl groups to the side chains of amino acids, as shown in Figure 1-14. These modifications are easily reversible, which allows cells to fine-tune the shape of individual proteins with great precision. Other modifications are permanent, and essential for protein function. Two good examples of permanent modifications are the creation of disulfide bonds between the sulfhydryl groups in cysteine amino acid side chains and the addition of oligosaccharides to some asparagine and serine amino acid side chains.

Nucleotides Are Complex Structures Containing a Sugar, a Phosphate Group, and a Base

Nucleotides are the building blocks for DNA and RNA, the genetic material of cells. The nucleotides that compose RNA contain a five-carbon sugar called ribose, and those found in DNA contain a modified ribose called deoxyribose, as shown in **FIGURE 1-15**. The sugar

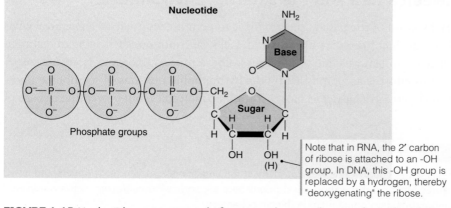

FIGURE 1-15 Nucleotides are composed of a sugar, a base, and one to three phosphate groups.

portion of nucleotides is always in a ring conformation. One, two, or three phosphates are attached to the 5′ carbon, and a nitrogen-containing base is attached to the 1′ carbon. Polymers of nucleotides are assembled by joining the 3′ carbon of one nucleotide to the single phosphate attached to the 5′ carbon of another. We will examine the structure of DNA and RNA more closely in Chapters 2 and 7.

CONCEPT CHECK #2

Imagine for a moment that life is discovered on another planet. While it is neither carbon nor water based, in many ways it resembles terrestrial life forms, including the use of equivalents of lipids, sugars, amino acids, and nucleotides. What critical chemical properties would you expect these alien molecules to have?

1.4 Cells Contain Distinct Structures That Perform Specialized Functions

---

KEY CONCEPTS

▸ To accomplish the tremendous number of tasks necessary to remain alive, cells specialize and practice division of labor.
▸ Membranes separate cellular contents into distinct regions that perform specific functions.
▸ All cells contain thousands of different proteins, and each protein performs a small number of cellular tasks.
▸ Some proteins cluster together to form complex structures on the plasma membrane or in the cytosol.
▸ Eukaryotic cells contain membrane-bound organelles in the cytosol that are highly specialized.

Cells expend a tremendous amount of energy to synthesize the polymers we have been discussing. To ensure that these molecules function optimally, cells also devote a great deal of effort toward assembling them into functional teams. Both prokaryotes and eukaryotes organize cytosolic proteins into clusters; eukaryotes however, go a step further by enclosing some of these teams in membranes, thereby creating compartments called organelles. This concept of compartmentalization is key to understanding cells, as it not only permits cells to delegate related tasks to subcompartments, but it also permits entire cells to specialize in a subset of tasks required in a multicellular organism. Because most of the specialized structures occur in eukaryotic cells, we will focus largely on eukaryotes in this book. Let's briefly discuss the highlights of the most common structures found in these cells as a way of surveying some of the major topics covered in the other chapters (**BOX 1-6**).

▪▪ The Plasma Membrane Is a Semipermeable Barrier between a Cell and the External Environment

The primary function of the plasma membrane is to permit cells to create a chemical environment that is distinct from the outside world. Cells are continuously striving to maintain this internal environment at some optimal set point, a process known as **homeostasis**. A large part of this effort is accomplished by the hydrophobic portions of membrane phospholipids, which serve as a barrier to most hydrophilic molecules. But preventing passage of

---

BOX 1-6 TIP ▪

Because this is a quick survey, it leaves out many of the details necessary to really understand how these structures function; if the short descriptions here are somewhat confusing, rest assured we will spend more time developing them in the later chapters.

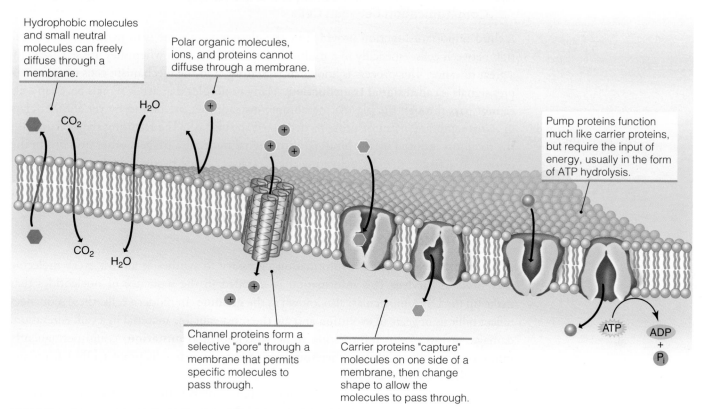

Hydrophobic molecules and small neutral molecules can freely diffuse through a membrane.

Polar organic molecules, ions, and proteins cannot diffuse through a membrane.

Pump proteins function much like carrier proteins, but require the input of energy, usually in the form of ATP hydrolysis.

CO_2

H_2O

CO_2

H_2O

ATP

ADP + P_i

Channel proteins form a selective "pore" through a membrane that permits specific molecules to pass through.

Carrier proteins "capture" molecules on one side of a membrane, then change shape to allow the molecules to pass through.

FIGURE 1-16 Common methods for controlling transport of molecules across/through a cell membrane.

molecules through the plasma membrane is not enough to ensure homeostasis, because cells must also have a way to allow molecules to enter and exit the cytosol in a controlled fashion.

Proteins in the plasma membrane control nearly all molecular traffic between a cell and its external environment (**FIGURE 1-16**). Many of these proteins actually span the membrane, and are therefore called **transmembrane proteins**. Classes of proteins called **channels**, **carriers**, and **pumps** act as selective passageways that permit transit of ions and small molecules such as sugars and amino acids. Most large molecules entering a cell are captured by transmembrane **receptor proteins** at the cell surface, and are then clustered into a patch on the plasma membrane and internalized by a process known as **endocytosis** (see Figure 9-1, in Chapter 9). Similarly, molecules synthesized by cells can be expelled by a related process called **exocytosis**. These subjects are discussed in Chapters 9 and 10.

▉▉ Protein Complexes in and near the Plasma Membrane Control a Cell's Attachment to the External Environment and to Other Cells

A second important function of the plasma membrane is to provide a means for attaching to and interacting with molecules in the external environment (often referred to as the **extracellular space**). This task is handled entirely by proteins in or near the plasma membrane that form stable complexes with proteins capable of generating and/or resisting mechanical force. The bulk of the proteins that perform this task are present in the cytosol, and they bind to the cytoplasmic portion of a specific set of transmembrane receptor proteins. In multicellular organisms, similar complexes form around different types of transmembrane proteins that act as receptors for neighboring cells. Both types of protein complexes can associate with the cytoskeleton, which is discussed in greater detail in Chapter 5.

■■ The Plasma Membrane Contains a Wide Variety of Proteins That Allow Communication between Cells

A third important function provided by the plasma membrane is to permit communication between cells, especially in a multicellular organism, even when they are separated by a great distance. This process is known as **signaling**, and the mechanism cells use to interpret signals is called **signal transduction**. Many molecules that act as signals between cells cannot pass through the plasma membrane; instead, they are bound by yet another class of transmembrane receptor proteins dedicated to this task. The resulting shape change in the receptors induced by the binding of the signal molecules triggers other proteins in the cytosol to activate, thereby delivering the messages carried by the signal molecules. Cell signaling is discussed in greater detail in Chapter 11.

■■ The Nucleus Is the Central Storehouse of Genetic Information

All cells carry with them the information necessary to synthesize almost every molecule they need to survive. This information is encoded in the sequence of nucleotides that make up the DNA molecules, also known as the **genome**. In modern cells, these sequences reflect billions of years of evolution and are the most valuable material in a cell. Alteration, damage, or loss of this information (otherwise known as **mutation**) could permanently injure or even kill a cell. It is therefore very important that cells protect DNA from any potentially dangerous agents.

As we will discuss in Chapters 6 and 7, in eukaryotic cells, the strategy for protecting DNA has many elements:

■ First, and perhaps the most obvious, is that most of a cell's DNA is enclosed in an organelle committed entirely to its use and maintenance, the nucleus. Like the mitochondrion and chloroplast, each of which contains its own small genome, the nucleus is an organelle surrounded by two membranes, as shown in **FIGURE 1-17**. The double membrane surrounding these organelles is an excellent barrier to most insults, and many cell biologists believe that these shared traits reflect a common ancestry: each arose from a cell that was engulfed, but not destroyed, and then developed a symbiotic relationship with its host; this scenario is known as the **endosymbiotic theory**.

■ A second means of protecting DNA is to limit access through the use of highly selective portals that penetrate the nuclear double membrane. These structures, known as **nuclear pore complexes**, are some of the most sophisticated devices in cells. They permit entry into and exit from the interior of the nucleus (sometimes called the **nucleoplasm**) through an intricate system of proteins devoted solely to this task.

■ A third strategy for protecting DNA is to carefully package it in a form that restricts access to only those portions of the DNA that are currently needed. The strands of DNA are tightly wrapped around a protein scaffold to form a complex called **chromatin**. Chromatin can be so tightly twisted that it reduces the length of a chromosome by approximately 20,000-fold! (For an understanding of exactly what this means, consider the effort that would be required to twist a rope 20 km [over 12 miles] long into a structure only 1 meter long.) By carefully unwinding only those portions necessary at any given time, cells help secure the structural integrity of chromosomes.

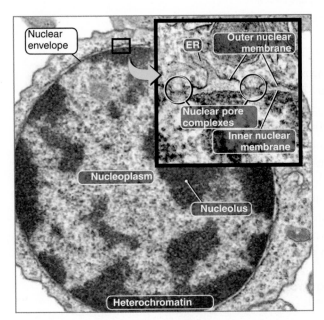

FIGURE 1-17 The nucleus contains several features that distinguish it from other cellular structures. Photo courtesy of Terry Allen, University of Manchester.

- Fourth, cells make RNA copies of DNA sequences (a process known as **transcription**), then use them to assemble the machinery necessary for cell function. This strategy is effective because any mistakes that arise from misreading the RNA, damage to the RNA, and so on, take place outside of the nucleus. With the exception of ribosomes (which are formed by the **nucleolus**, a specialized sub-region in the nucleus), all macromolecules other than DNA and RNA are assembled outside the nucleus.

Chloroplasts Build Food Molecules from CO_2 and H_2O Using Light Energy

The function of chloroplasts (illustrated in **FIGURE 1-18** and discussed in Chapter 10) is to harness light energy from the sun to create the carbon polymers necessary for building all of the other macromolecules in cells. Only **autotrophic cells** (literally, *cells that nourish themselves*), such as those found in the leaves of plants, contain chloroplasts. The biochemical

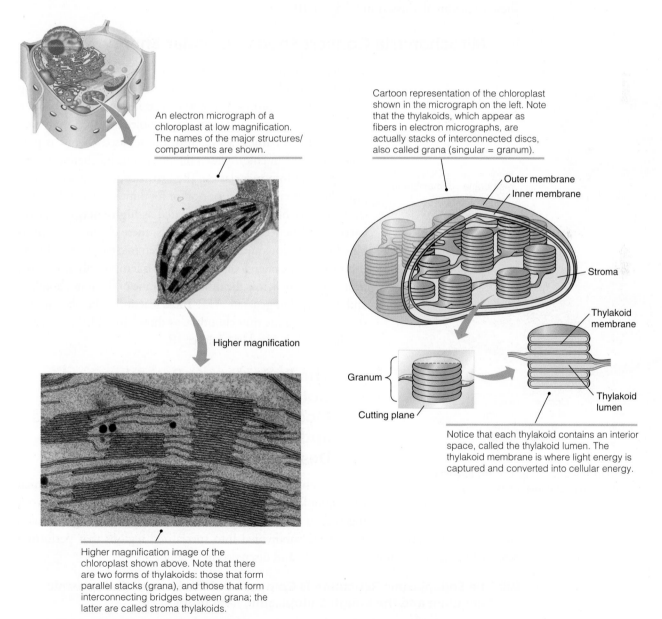

An electron micrograph of a chloroplast at low magnification. The names of the major structures/compartments are shown.

Higher magnification

Higher magnification image of the chloroplast shown above. Note that there are two forms of thylakoids: those that form parallel stacks (grana), and those that form interconnecting bridges between grana; the latter are called stroma thylakoids.

Cartoon representation of the chloroplast shown in the micrograph on the left. Note that the thylakoids, which appear as fibers in electron micrographs, are actually stacks of interconnected discs, also called grana (singular = granum).

Outer membrane
Inner membrane
Stroma
Thylakoid membrane
Thylakoid lumen

Granum
Cutting plane

Notice that each thylakoid contains an interior space, called the thylakoid lumen. The thylakoid membrane is where light energy is captured and converted into cellular energy.

FIGURE 1-18 The chloroplast is specialized to convert CO_2, water, and light energy into complex carbon compounds. © Dr. Jeremy Burgess/Photo Researchers, Inc., © Dr. Kenneth R. Miller/Photo Researchers, Inc.

steps required are numerous and organized into two classes. The first, sometimes called the light reactions, is concerned with capturing photons of light and using this energy to strip electrons from water molecules and add them to carbon dioxide (CO_2) molecules, which results in the release of molecular oxygen (O_2) and the accumulation of two types of three-carbon sugars: glyceraldehyde 3-phosphate and dihydroxyacetone phosphate. The second (dark reactions) is concerned with converting these sugars into starch and sucrose. In turn, these provide the necessary molecules for building all of the other biomolecules necessary for cell function.

Chloroplasts are also an excellent example of a common theme in cell biology: that cells capitalize on the selective permeability of membranes to build **molecular gradients** across them as a form of intermediate **energy storage**. Many of these are ion gradients, built by ion pump proteins. The potential energy stored in these gradients can later be used by membrane proteins to drive energy-dependent chemical reactions, such as the synthesis of ATP from ADP and inorganic phosphate that occurs in chloroplasts and mitochondria. These topics are discussed in Chapter 10.

Mitochondria Convert Food to Cellular Energy

Whereas chloroplasts are committed to the creation of many complex biomolecules, mitochondria are committed to breaking them apart. This process is not as easy as it might seem, because mitochondria must dismantle these molecules in a way that allows them to harness as much of the released energy as possible. And, like in chloroplasts, the numerous biochemical steps are organized into two classes. The first is concerned with stripping the high-energy electrons off of food molecules and storing them in a set of molecules designated as **high-energy electron carriers**. In the second set of reactions, these electrons are combined with protons and molecular oxygen to create water; during this process, the energy of the electrons is temporarily stored as a proton gradient across the inner membrane of the mitochondria, and then used to generate ATP. These reactions take place in distinct regions of the mitochondria, as shown in **FIGURE 1-19**, and are discussed in detail in Chapter 10.

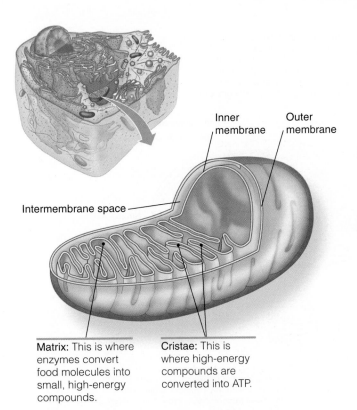

Inner membrane   Outer membrane

Intermembrane space

Matrix: This is where enzymes convert food molecules into small, high-energy compounds.

Cristae: This is where high-energy compounds are converted into ATP.

FIGURE 1-19 The mitochondrion is specialized to convert complex biomolecules into useful metabolic energy.

The Endoplasmic Reticulum and Golgi Apparatus Collaborate to Modify, Sort, and Transport Proteins and Phospholipids to Their Final Destinations

An excellent example of teamwork in cells is the way in which the endoplasmic reticulum (ER) and Golgi apparatus work together to handle much of the membrane trafficking that takes place inside cells. These structures are subdivided into specialized regions that perform a subset of tasks, as shown in **FIGURE 1-20** and discussed in Chapter 9.

The Endoplasmic Reticulum Is Composed of the Smooth Endoplasmic Reticulum and the Rough Endoplasmic Reticulum

The membrane defining the endoplasmic reticulum is attached to the outer membrane of the nucleus and forms two distinct structures that perform different functions. One

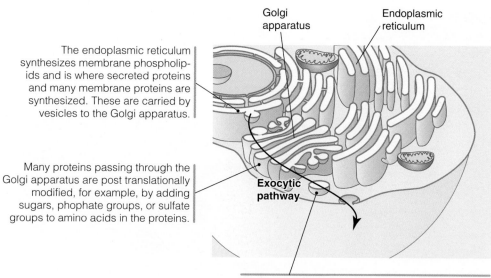

The endoplasmic reticulum synthesizes membrane phospholipids and is where secreted proteins and many membrane proteins are synthesized. These are carried by vesicles to the Golgi apparatus.

Golgi apparatus

Endoplasmic reticulum

Many proteins passing through the Golgi apparatus are post translationally modified, for example, by adding sugars, phophate groups, or sulfate groups to amino acids in the proteins.

Exocytic pathway

The molecules leave the Golgi apparatus in vesicles destined for a variety of intracellular locations. The term *exocytosis* refers to the targeting of these vesicles to the plasma membrane, which causes the contents of the vesicles to be secreted into the extracellular space.

FIGURE 1-20 The endoplasmic reticulum and Golgi apparatus collaborate to control protein and phospholipid traffic in cells.

portion, called the **smooth endoplasmic reticulum** (or **SER**), is specialized to perform four important functions:

- It is the site of phospholipid biosynthesis. Phospholipids are synthesized on the cytoplasmic face of the SER membrane, and then distributed to the rest of the cell membranes.
- The SER serves as a storage site for calcium ions. These ions are released in brief bursts to control the activities of calcium-dependent proteins, and then quickly pumped back into the SER to shut these proteins off.
- In some cells, such as liver and muscle cells, proteins associated with the SER membrane play an important role in converting stored energy (in the form of glycogen, a polymer of glucose held together by $\alpha 1,4$ bonds) into individual glucose molecules for immediate use by cells.
- The SER also contains enzymes that inactivate biochemical toxins; these are found in great abundance in liver cells, which remove these toxins from the bloodstream.

The **rough endoplasmic reticulum** (or **RER**) is responsible for controlling the synthesis, modification, and assembly of two important classes of cellular proteins. When viewed with an electron microscope (**FIGURE 1-21**), the RER can be easily distinguished from the SER by the presence of numerous ribosomes, which dot the cytoplasmic surface of the RER, giving it a comparatively "rough" appearance. Ribosomes attached to the surface of the RER are actively synthesizing polypeptides that pass from the cytosol through a complex of membrane proteins called the translocon and into the interior, or **lumen**, of the RER. Many of these polypeptides do not pass entirely into the lumen, because they contain one or more regions of hydrophobic amino acids that remain in the RER membrane. Polypeptides like these that have one or more transmembrane sequences form transmembrane proteins, the first class of proteins. The second class of polypeptides synthesized by ribosomes on the RER consists of those that pass entirely through the translocon without attaching to the RER membrane. A generic term for proteins synthesized in this fashion is

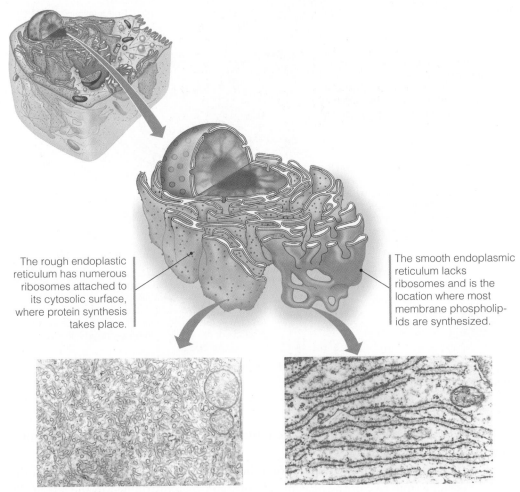

The rough endoplasmic reticulum has numerous ribosomes attached to its cytosolic surface, where protein synthesis takes place.

The smooth endoplasmic reticulum lacks ribosomes and is the location where most membrane phospholipids are synthesized.

FIGURE 1-21 The endoplasmic reticulum is organized into two linked subcompartments. Both images by © Dr. Donald Fawcett/Visuals Unlimited, Inc.

soluble proteins; a subset of these are ultimately released into the extracellular space, and are therefore called **secreted** (or **secretory**) **proteins**.

Most proteins synthesized by ribosomes on the RER undergo a series of chemical modifications (often called **posttranslational modifications**) catalyzed by enzymes in the RER. The most striking of these modifications is the addition of branched oligosaccharides to some asparagine and serine side chains, a process known as glycosylation. As discussed previously, these oligosaccharides play an important role in controlling the structure and function of the proteins they are attached to. A second type of modification is the formation of a covalent bond between the sulfur atoms in two cysteine amino acid side chains, called a **disulfide bond**. Because it is covalent, it is very difficult to break and essentially locks the two cysteines together, which helps stabilize protein structure. Some proteins receive a third form of modification, the addition of a phospholipid to the carboxyl terminus of the polypeptide, which anchors the protein to the luminal face of the endoplasmic reticulum membrane.

◼◼ Traffic between the RER and Golgi Apparatus Is Carried by Membrane-Bounded Vesicles

With the exception of those proteins destined to remain in the RER to assist with the translocation and modification of newly synthesized polypeptides (often called ER-resident proteins), all other polypeptides synthesized by RER-associated ribosomes are carried to the Golgi apparatus. The mechanism cells use to accomplish this is quite complex and only

CHAPTER 1   What Is a Cell?

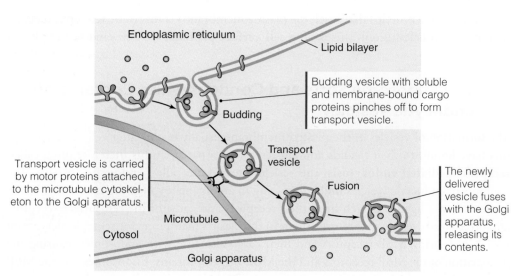

FIGURE 1-22 Vesicles transport proteins and other molecules between different compartments in the cell. In this example, a vesicle carries molecules from the endoplasmic reticulum to the Golgi apparatus.

partially understood. With the assistance of proteins designated specifically for the task, small patches of the RER membrane are pinched off to form a small, membrane-bound compartment called a **vesicle** (which contains both transmembrane proteins and soluble proteins), as seen in **FIGURE 1-22**. This vesicle is then transported to the Golgi apparatus by **motor proteins** associated with the cytoskeleton, and a different set of proteins assist in the fusion of the vesicle with the Golgi membrane. Similar mechanisms are used to shuttle vesicles from the Golgi back to the RER and between different compartments in the Golgi. This transportation of vesicles between organelles is called **membrane trafficking**.

■■ The Golgi Apparatus Modifies and Sorts Proteins Destined for the Plasma Membrane and Lysosomes

The Golgi apparatus is the only organelle with a defined cellular orientation (i.e., a "front" and a "back"). It consists of a single membrane-bounded compartment that is highly folded to generate separate compartments, known as **cisternae** (Figure 1-19). Vesicles arriving from the RER are targeted to fuse with one compartment, called the **cis Golgi network**, or CGN. **CGN-resident proteins** further modify the newly arrived proteins by adding and subtracting sugars to/from their oligosaccharides. Posttranslational modifications continue as the proteins are carried by more vesicles to the **medial Golgi stacks** and then to the **trans-Golgi network**, or TGN.

One of the most important functions of the TGN is to sort membrane and soluble proteins into vesicles specifically targeted to other organelles or to the plasma membrane. While many of the mechanisms responsible for this sorting are not well known, one very clear case concerns the targeting of proteins to the lysosome. In mammalian cells, one set of modifications to the protein-linked oligosaccharides results in the formation of a sugar called **mannose 6-phosphate** (or M6P). A protein in the TGN called the **mannose 6-phosphate receptor** binds to any protein bearing an M6P tag and directs it into a vesicle destined for the lysosome.

Other vesicles budding from the TGN are targeted to the plasma membrane. The process of synthesizing proteins in the RER, shuttling them through the different compartments of the Golgi apparatus, and sending them to the plasma membrane is called **exocytosis**. Soluble proteins that are exocytosed are released into the extracellular space, while membrane proteins that reach the plasma membrane via this route remain associated with it after vesicle fusion. This is how **cell surface proteins** are delivered to the plasma membrane.

The Endosome Sorts and Condenses the Contents of Endocytic Vesicles

The formation of vesicles at the plasma membrane usually occurs after cell surface receptors have bound to their ligands and clustered into a patch, and is therefore often called **receptor-mediated endocytosis**; the vesicles created are called **endocytic vesicles**. The contents of endocytic vesicles undergo sorting in an organelle called the endosome. In the endosome, the ligands detach from their receptors and some of the cell surface receptors are sorted into a vesicle that returns them to the plasma membrane for reuse. Vesicles from the TGN bearing M6P receptors and their cargo also fuse with endosomes, causing the dissociation of the M6P receptors and the M6P-tagged proteins. Endosomes sort the M6P receptors into a vesicle that returns them to the TGN so they can be reused.

The mechanisms responsible for this sorting by endosomes are not well understood, but it is clear that acidification of the endosome lumen plays an important role. Some of the M6P-tagged proteins arising from the TGN are **proton pumps**, and once they arrive in the endosome, they pump protons into the lumen. The resulting drop in pH alters the shape of the proteins in the endosome, allowing them to change their functions: cell surface receptors and M6P receptors release their cargo and cluster into separate vesicles, leaving behind only M6P-tagged proteins and any material captured by the cell surface receptors.

The Lysosome Digests Proteins, Lipids, and Nucleic Acids

The end product of endosomal sorting is called a lysosome. Due to the activity of the proton pumps they contain, lysosomes are characteristically very acidic. Most of the other M6P-tagged proteins that originate in the RER are **hydrolytic enzymes**. These are proteins that break large molecules (virus particles, proteins, lipids, nucleic acids, oligosaccharides, etc.), or sometimes even entire cells, into their simpler building blocks (amino acids, fatty acids, sugars) by simply adding back the water that was removed during the dehydration reactions used to construct them. Glycosidic bonds, phosphoester bonds, and peptide bonds are common targets of these enzymes. The resulting building blocks are then transported into the cytosol for reuse. Lysosomes are especially prevalent in many cells that are active in the immune system, such as macrophages and lymphocytes, and literally engulf invaders such as bacteria and viruses.

The Peroxisome Contributes to a Number of Metabolic Activities in Cells

As its name suggests, the peroxisome is an organelle that oxidizes molecules and generates hydrogen peroxide as a byproduct. Hydrogen peroxide (H_2O_2) is a potentially dangerous reactant in cells, which may explain why it is generated in its own organelle. Peroxisomes contain an enzyme called catalase that efficiently degrades hydrogen peroxide into water and molecular oxygen. Many of the oxidation reactions taking place in peroxisomes convert long-chain fatty acids into simpler compounds that can be more easily metabolized by cells. Peroxisomes are also capable of oxidizing toxic chemicals, thereby inactivating them. At least some membrane proteins in peroxisomes are synthesized in the RER and delivered by vesicles, while proteins destined for the interior of the peroxisome (called the **peroxisomal matrix**) are synthesized by free ribosomes in the cytosol and pass through the peroxisomal membrane after their synthesis is complete.

The Plasma Membrane, Endoplasmic Reticulum, Golgi Apparatus, Endosomes, Lysosomes, and Peroxisomes Form a Protein-Trafficking Network Called the Endomembrane System

Despite the fact that each different organelle performs its own specific set of cellular activities, most of the organelles found in eukaryotic cells are linked to one another via the vesicles that shuttle between them. This network, which consists of the plasma membrane, endoplasmic reticulum, Golgi apparatus, endosomes, lysosomes, and peroxisomes, is called the **endomembrane system**. (This system is the primary subject of Chapter 9.) The only organelles that do not participate in this system are the nucleus, mitochondrion, and chloroplast. Recall that the endosymbiotic theory suggests these three organelles arose from the engulfment of one cell by another, and that prokaryotes sometimes exhibit elaborate extensions of the plasma membrane. Taken together, these observations suggest that all organelles are derived from specialized regions of the plasma membrane of an ancestral prokaryotic cell.

The Cytoskeleton and Motor Proteins Determine the Shape and Motion of Prokaryotic and Eukaryotic Cells

Recall from our discussion of membrane structure that the hydrophobic property of phospholipids allows them to spontaneously assemble in an aqueous environment. In the absence of any other force, these membranes will form a sphere to minimize the surface-to-volume ratio. But it is important that these membranes actively participate in the selective transport of molecules across them, and this favors an elaboration of the membrane surface. Cells resolve this apparent paradox by employing proteins that generate force and stabilize the resulting deformations in membranes. These proteins are the subject of Chapter 5.

The Eukaryotic Cytoskeleton Is Composed of Three Types of Filamentous Proteins

The structural components of this system are called cytoskeletal proteins, or simply the cytoskeleton, as illustrated in **FIGURE 1-23**. Prokaryotic cells contain a cytoskeleton that permits them to adopt a myriad of shapes, and forms the **cilia** and **flagella** they use to move through their environment. In eukaryotic cells, these relatively primitive cytoskeletal proteins evolved into a larger number of proteins that are grouped into three classes (though all three form filamentous polymers). **Actin filaments** are used in many ways, including cell crawling and the final division of a cell body after mitosis (called **cytokinesis**), and they also permit cells to flatten and spread to achieve a tremendous variety of shapes. **Microtubules** serve as tracks for the transport of cytosolic components, including the vesicles that link the endomembrane system. Microtubules also form the **mitotic spindle** that ensures the proper segregation of replicated chromosomes during mitosis. Proteins in the third class of filaments are called **intermediate filaments**; these filaments serve as a strong reinforcing material to hold a cell in place once it has adopted its preferred shape. Members of all three classes of cytoskeletal proteins are present in each eukaryotic cell.

The proteins that generate force in this system are called motor proteins. Motor proteins bind to actin filaments or microtubules, but

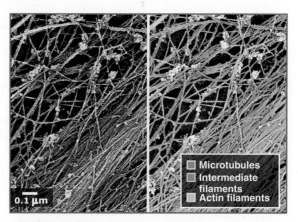

FIGURE 1-23 Cytoskeletal proteins form a dense meshwork in the cytosol. See a small region of a fibroblast cell viewed by electron microscopy (left panel). Numerous filaments are visible. In the right panel, the three types of cytoskeletal polymers present in eukaryotic cells have been colored so that they can be easily distinguished from one another. Reprinted from J. Struct. Biol., vol. 115, M. T. Svitkina, A. B. Verkhovsky, and G. G. Borisy, Improved Procedures for Electron Microsopic Visualization..., pp. 290–303, Copyright (1995) with permission from Elsevier [http://www.sciencedirect.com/science/journal/10478477]. Photos courtesy of Tatyana Svitkina, University of Pennsylvania.

not intermediate filaments. Actin-associated motor proteins generate the tension necessary to deform the plasma membrane, and are therefore responsible for most of the motility of eukaryotic cells. The interaction of the muscle protein **myosin** with the actin cytoskeleton, which causes the contraction of skeletal muscle cells, is a good example. Microtubule-associated motor proteins generate a different kind of force, one that allows them to drag membrane-bound structures in the cytosol along the tracks defined by microtubules. In mitosis, these proteins are also responsible for separating the two halves of the mitotic spindle during anaphase. **Dynein** and **kinesin** are common microtubule-associated motor proteins.

■ 1.5 ■ Cells in Multicellular Organisms Can Be Highly Specialized to Perform a Subset of the Functions Necessary for Life

KEY CONCEPTS

- ▸ Groups of cells that collaborate to perform a complex function are called tissues.
- ▸ Tissues in animals are classified into four types: epithelial, connective, muscle, and nervous.
- ▸ Tissues in plants are classified into three groups: dermal, vascular, and ground.
- ▸ Most diseases impact tissue structure and function.

Just as the evolution of organelles was a major step that allowed eukaryotic cells to diversify and specialize, the advent of multicellularity introduced division of labor at the level of cell clusters. Multicellularity permitted organisms to occupy environmental niches that had previously been unavailable, and introduced a new level of natural selection. Molded by this natural selection, effective teams of cells became ever more interdependent, giving rise to larger and more sophisticated organisms.

■ Specialized Subsets of Cells Are Called Tissues

The technical term for these specialized teams of cells is **tissue**. Combinations of tissues that work closely together are called **organs**. All of the cells in animals are classified into four tissue types: epithelial tissue, **connective tissue**, **muscle tissue**, and **nervous tissue**. Each tissue type can be customized for different organs: the epithelial tissue in the lung is quite different from the epithelial tissue in the intestines, for example. The field of **physiology** is devoted to the study of tissue and organ function, and is beyond the scope of this text. For now, we will focus only on the general properties of the constituent cells that make these distinct tissue types, and a more in-depth discussion of this topic can be found in Chapters 13 and 14.

■■ Cell Differentiation Gives Rise to a Wide Variety of Cell Types

Because each tissue type is distinct, the cells composing each tissue are likewise quite different. This distinction arises from the types of proteins they express and how they are used. The process of modifying protein expression en route to the formation of a mature tissue is called **development**. In some animals (e.g., humans), development continues for many years. Understanding the mechanisms that control development (also known as **differentiation**) is one of cell biology's principal objectives.

■■ Histology Is the Study of Tissues

The field of cell biology was born when scientists developed microscopes powerful enough to see cells, and microscopes are still important tools today. A great deal of study was devoted to inventing better ways to study cells in tissues, and one area grew so rapidly that biologists coined a new term for this discipline: **histology**. The word histology literally means the *study of tissues*, but it actually refers to a specific set of microscope-based techniques. Histologists treat cell and tissue samples with special contrast agents, called **stains**, to better see structural organization in the samples with light and electron microscopes. While the heyday of histology has largely passed, the lessons learned form the structural foundation for modern cell biology.

Let's now have a brief look at each of the four tissue types in animals, using Figures 1-24, 1-25, 1-26, and 1-27 as visual aids.

■■ Epithelial Tissues Form Selectively Permeable Barriers between Distinct Body Compartments

The most striking feature of epithelial tissues is that they are arranged as sheets, as shown in **FIGURE 1-24**. On one side of the sheet, called the **basal surface**, the cells are in contact with a special form of fibrous material called the **basement membrane**. On the opposite side of the sheet, called the **apical surface**, the cells are in contact with an entirely different environment, which varies from organ to organ. In the lung, for example, the apical surface of the epithelium is in contact with air, while the apical surface of the epithelium in intestines interacts with digested food from the stomach. Many epithelial tissues are composed of multiple layers of epithelial cells, each lying directly on top of the layer below it. The function of epithelial tissues in organs is analogous to the function of the plasma membrane in a single cell: to act as a selectively permeable barrier that separates and protects the tissues on either side of the epithelial cell layer. And, just as individual cells use proteins to regulate traffic across the plasma membrane, epithelial sheets use protein complexes to regulate the trafficking of molecules between cells and across the cell layer. A general term used to describe these protein complexes is the **junctional complex**.

■■ Connective Tissues Provide Structural Stability to Organs

For organs to be effective, cells in tissues must be organized in a specific three-dimensional arrangement. Connective tissues, such as those shown in **FIGURE 1-25**, supply the strong material necessary to maintain this arrangement, much like the cytoskeleton maintains the shape and organization of a single cell. This structural material is concentrated in the space between cells, and is called the extracellular matrix. The extracellular matrix is a complex mixture of organic and

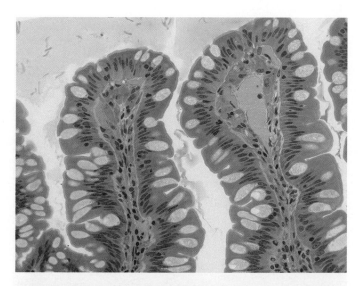
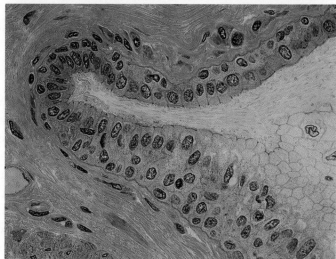
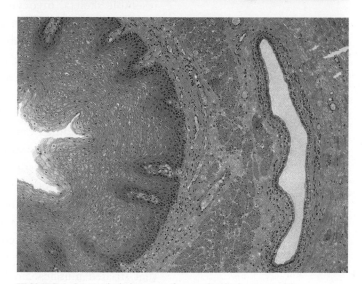

FIGURE 1-24 Epithelial tissues form selectively permeable barriers between distinct body compartments. In these photos, the epithelial tissues lie between the bright spaces (top of top panel, right of middle panel, left of bottom panel) and the dark pink spaces. All three images by © Donna Beer Stolz, PhD, Center for Biologic Imaging, University of Pittsburgh Medical School.

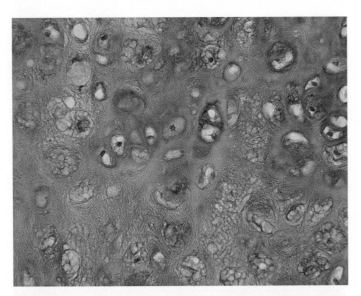

FIGURE 1-25 Connective tissues provide structural stability to organs. © Donna Beer Stolz, Ph.D., Center for Biologic Imaging, University of Pittsburgh Medical School.

inorganic molecules that completely fills the space between cells in connective tissues and also forms the basement membrane under sheets of epithelial cells. One characteristic common to all connective tissues is that they consist largely of extracellular matrix molecules, populated with a relatively low density of cells. Familiar examples of connective tissues include bone, tendon, and cartilage.

◼◼ Muscle Tissue Provides the Force Necessary to Move the Body and Pump Blood

There are three different types of muscle tissue in animals, each composed of a different type of muscle cell, as shown in FIGURE 1-26. **Skeletal muscles** are composed of muscle cells called myocytes, which have a characteristic organization of their cytoskeleton. The bulk of the cytosol in these cells is occupied by overlapping fibers of actin and myosin, all oriented in one direction; these regions of overlap are called **sarcomeres**. Typically, hundreds of sarcomeres are lined up to form a **myofibril** that extends the entire length of the muscle cell. When viewed with a microscope, the sarcomeres appear as alternating stripes (or striations) of light and dark material, so these cells are also called **striated muscle**. Upon stimulation by a nerve cell, the actin and myosin proteins slide past one another to shorten the length of the sarcomere. When hundreds of sarcomeres contract in this fashion, the muscle cell shortens considerably; when thousands of sarcomeres contract together, a skeletal muscle can generate tremendous force. Note that because all of the actin and myosin filaments are aligned in the same direction, skeletal muscles generate force only in that direction. Thus, to move a bone in many different directions, several skeletal muscles must be employed, each oriented in a different direction.

Cardiac muscle cells, also called **cardiomyocytes**, are found in the heart and resemble skeletal myocytes in many ways, including a striated appearance. However, one important difference is that these cells can branch, so the sarcomeres can be oriented in more than one direction in a single cell. A second important difference is that these cells can stimulate their own contraction, without the need of a nerve. Third, cardiomyocytes are connected by one of the special protein complexes found in epithelial cells, called the **gap junction**, which permits them to pass small molecules directly from the cytosol of one cell into the cytosol of a neighboring cell. These junctions permit rapid spreading of any contraction signal, so the entire heart works together during each beat to pump blood in one direction.

The third type of muscle cell, called a **smooth muscle cell**, is quite different from the other two. Smooth muscle cells contract using actin and myosin, but do not organize these proteins into sarcomeres. Instead, these proteins are oriented in every direction, so these cells can contract in any direction necessary. Their contractions can also last much longer than those in striated muscle cells. This versatility comes at a cost, however. Smooth muscle cells typically take longer to contract, and do not generate the magnitude of force that striated muscles can. Smooth muscle tissue is commonly found beneath the basal surface of epithelia. Many epithelial tissues line tubes (airways in the lung, blood vessels, intestines), and the smooth muscle layer that surrounds them helps to control the diameter of these tubes. In intestines, the periodic contraction of the smooth muscle, called **peristalsis**, is essential to move food through the digestive tract.

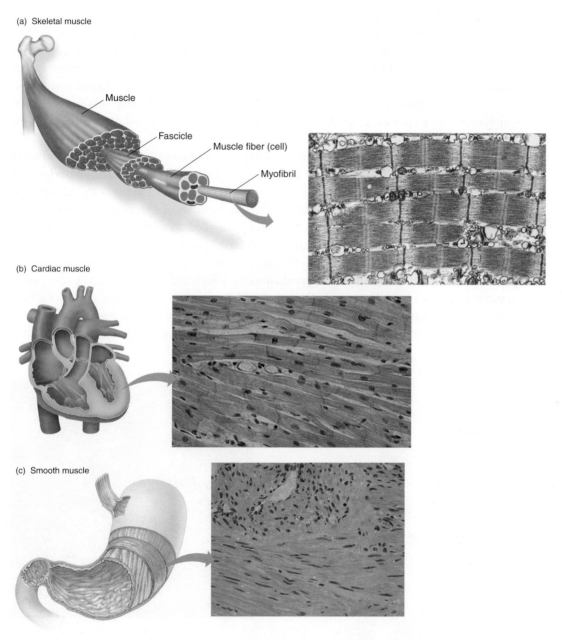

(a) Skeletal muscle

Muscle

Fascicle

Muscle fiber (cell)

Myofibril

(b) Cardiac muscle

(c) Smooth muscle

FIGURE 1-26 Muscle tissue provides the force necessary to move the body and pump blood.
© Donna Beer Stolz, PhD, Center for Biologic Imaging, University of Pittsburgh Medical School.,
© Biology Pics/Photo Researchers, Inc., © Donna Beer Stolz, PhD, Center for Biologic Imaging,
University of Pittsburgh Medical School.

■■ Nervous Tissue Provides for Rapid Communication between Body Parts in Animals

The function of nervous tissue is to transmit signals from one part of the body to another. This is not the only means of communication between cells, but it is certainly the fastest. These signals are transmitted along the length of a single nerve cell much like an electrical signal passes through a wire. The imbalance of ions across the plasma membrane (called a **resting potential**) is used as a tool to transmit signals. At one end of a nerve cell (or **neuron**), a stimulus causes proteins to permit a brief exchange of ions (called a **depolarization**) across the plasma membrane, and this triggers nearby proteins to do so as well. This propagation of ion movement, called an **action potential**, travels to the opposite end of the nerve cell at rapid speed (10–100 m/s) and is then converted into a chemical signal that diffuses to a nearby nerve cell and initiates a new action potential.

Anatomists classify nerve tissues into two types, as shown in **FIGURE 1-27**. The **central nervous system**, or CNS, is composed of the brain and spinal cord, while the **peripheral nervous system**, or PNS, consists of the rest of the nervous tissues in the body. In humans, this includes nerves in the arms and legs, as well as bundles of nerves that lie outside but near the spinal cord, called ganglia. Nerves can be classified according to the direction their action potential travels with respect to the central nervous system: efferent nerves carry action potentials toward the CNS, while afferent nerves (also called **motor nerves**) carry action potentials away from the CNS and toward effectors, which include muscles. The term *nerve* actually refers to a bundle of two cell types. Not all cells in nervous tissue conduct action potentials. Support cells, called **glial cells**, actually outnumber the conducting cells. Their function is to insulate, protect, and nourish the nerve conducting cells.

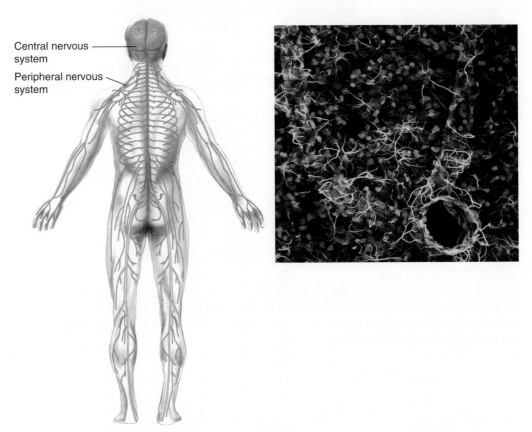

Central nervous system

Peripheral nervous system

FIGURE 1-27 Nervous tissue is organized into two types and is composed of many different cell types working together to process and transmit nerve impulses. Courtesy of Dr. Chris Bjornsson.

■■ Plants Contain Three Classes of Tissues

Plant cells are organized into three tissues that are considerably different from those in animals, as illustrated in FIGURE 1-28. **Dermal tissue** forms the outer layers of plants, and is specialized for each plant organ (leaves, stems, and roots). It consists of **epidermal cells** (usually one layer thick) that provide the outermost layer of protection, and **guard cells** that form stomata on the underside of leaves to permit gas exchange. In roots, dermal tissue regulates uptake of ions and water. **Vascular tissue**, composed of phloem and xylem, provides a vascular system to move water and nutrients throughout the plant. Phloem and xylem form channels that conduct water, organic solutes, and minerals. They are typically organized into parallel channels called vascular bundles. **Ground tissue** contains three different cell types that together supply new cells for growth, and provide rope-like fibers for strength.

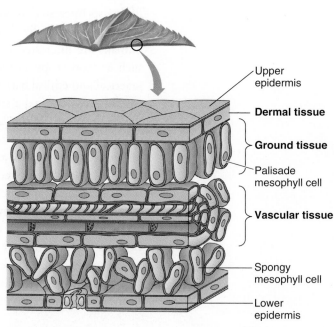

FIGURE 1-28 Plants contain three classes of tissues.

Upper epidermis
Dermal tissue
Ground tissue
Palisade mesophyll cell
Vascular tissue
Spongy mesophyll cell
Lower epidermis

■■ Most Diseases Are Manifested at the Level of Tissues

Understanding tissue structure and function remains a significant challenge in biology. Cancer and heart disease, the two leading causes of death in the United States, are not dangerous when they only affect individual cells. These diseases typically are not diagnosed until they begin to affect the teams of cells in tissues, as shown in FIGURE 1-29. For example, cancer cells multiply and invade surrounding tissues, disrupting their organization and function. Often, by the time cancer is diagnosed, the tumor cells have invaded a tissue and damaged its cells so badly they no longer function properly. Similarly, heart disease causes the death of heart muscle cells; loss of a few cells is not typically noticed, but when enough cells die such that the heart no longer contracts properly, this damage is largely irreversible. Treating both of these diseases remains a significant challenge because we do not yet fully understand how cells cooperate to form a tissue.

CONCEPT CHECK #3

Biologists commonly refer to different scales, or levels, of biological organization. One common organizing scheme, based on increasing complexity, is: molecules ⇨ organelles ⇨ cells ⇨ tissues ⇨ organs, and so on. Aside from their obvious differences in size, what other important differences are there between cells and tissues?

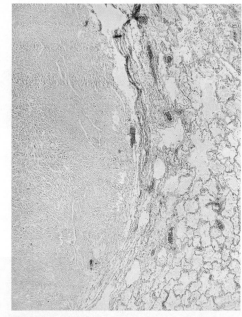

FIGURE 1-29 Most diseases are manifested at the level of tissues. Leiomyosarcoma cancer metastatic to lung is shown. © Garry DeLong/ Photo Researchers, Inc.

1.6 Model Organisms Are Often Studied to Understand More Complex Organisms

KEY CONCEPTS

▸ A small number of organisms, called model organisms, are often studied as representatives of all living organisms.

▸ Because model organisms are composed of cells, the mechanisms controlling their behavior are very similar to those in other organisms.

Biologists interested in understanding the mechanisms controlling cellular and organismal behavior theoretically have a multitude of organisms from which to choose. In many cases, humans are interested in studying organisms that most immediately impact their lives, such as food crops, domesticated animals, and even other humans. But there are often practical and ethical barriers to studying these organisms. To overcome this problem, biologists have identified a small set of organisms, known as **model organisms**, to represent the diversity of life (**FIGURE 1-30**). The most commonly used model organisms include rabbits, rats, and mice to represent mammals; the zebrafish *Danio rerio*, the fruit fly *Drosophila melanoaster*, and the nematode worm *Caenorhabditis elegans* to represent less complex animals; the weed *Arabidopsis thaliana* to represent plants; and the yeast *Saccharomyces cerevisiae* and the bacterium *Escherichia coli* to represent single-celled eukaryotes and prokaryotes, respectively.

The justification for focusing on this set of organisms is relatively simple (**BOX 1-7**). Because the fundamental unit in each one of these (and all other) organisms is a cell, many of the mechanisms we seek to understand in more complex organisms can be found in model organisms. Mice and rats can develop cancers that closely resemble those in humans, so they are commonly used to test potential anticancer therapies. Fruit flies do not get

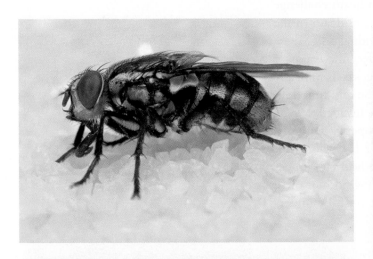
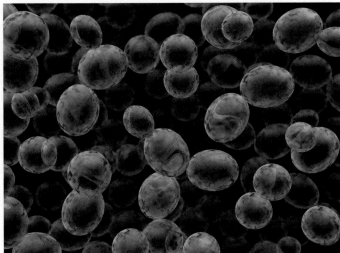

FIGURE 1-30 Model organisms used in biology. Clockwise from top left: fruitflies, yeast, mice, and arabidopsis represent primitive animals, simple eukaryotes, complex animals, and plants, respectively. © iStockphoto/Thinkstock, © Knorre/ShutterStock, © Hemere/Thinkstock, © iStockphoto/Thinkstock.

CHAPTER 1 What Is a Cell?

cancer, but many of the mechanisms governing the growth and differentiation of their cells are closely related to those found in more complex animals. Model organisms also have relatively short life cycles, which permits researchers to study the mechanisms governing all stages of life with relative ease; this also helps researchers understand the genetic basis of organismal variation, because large numbers of offspring can be maintained over several generations. Finally, the cost of maintaining model organisms in a laboratory setting is typically much lower than that for more complex organisms.

1.7 Chapter Summary

Cells, which are the fundamental units of life, are complex yet highly organized structures that possess two traits common to all living beings: the ability to self-replicate and the ability to self-correct. The remaining chapters in this book focus on the cellular mechanisms responsible for these two abilities, but before we explore the structure and function of cells in greater detail, it is important to understand that cells share common features. All cells contain water and are composed primarily of carbon-based molecules that can be categorized into four structural classes: lipids, sugars, amino acids, and nucleic acids. These simple building blocks form a tremendous number of different molecules, each of which performs a very small number of the tasks necessary for cells to remain alive. Over the course of evolution, cells acquired a common strategy to simplify these tasks: they clustered these molecules into teams that cooperate to perform a subset of cellular activities. In many cases, the structure–function relationship helps explain how these teams work. For example, eukaryotic cells acquired the ability to enclose many of these teams in membrane-bound compartments called organelles, and many of the structural features found in an organelle play an important role in that organelle's function. Other functionally related teams of molecules are not in organelles, but play critical roles in cell behavior, such as attachment to extracellular molecules and neighboring cells, or replicating and separating chromosomes during cell growth. These ideas form the foundation for understanding how and why cells look and act the way they do.

CONCEPT CHECK ANSWERS #1

Membranes are essential to all living things because they create the boundary that permits cells to establish chemical disequilibrium, and thereby store energy. Without this energy, life is not possible.

If an alien life form contains equivalents of cellular building blocks, then they would likely have the following properties:

▸ Alien lipids would spontaneously aggregate in the prevailing solvent in which this organism lives, and this aggregate would form a semipermeable barrier surrounding the contents of an alien equivalent of a cell. They need not be amphipathic—they simply need to create some form of selective barrier.

▸ Alien sugars would be linked to one another to generate structurally complex polymers that could provide structural support to the living organism and generate structural diversity in the other molecules to which they are attached.

▸ Alien amino acids would link together in linear polymers that are chemically very stable and structurally flexible, and the polymer would have the ability to bind to other biologically important molecules with great specificity.

▸ Alien nucleotides would serve to store information in a linear sequence. The nucleotides need not be as structurally complex as terrestrial nucleotides are; they simply need to form stable polymers that can store information.

We can address this question by first briefly defining the two words:

▸ Cells are the smallest unit of life, and are defined by the presence of at least one biological membrane that separates them from the external environment. In the hierarchy of biological complexity, the cell level refers to single cells only, so this means a cell must be able to perform all the necessary functions to remain alive.

▸ Tissues are clusters of interconnected cells that cooperate to perform a function that individual cells cannot. It is assumed that the cells composing the tissue are capable of remaining alive.

Thus, an important difference between cells and tissues is the *functions* they perform. If a function can be performed by an individual cell (e.g., growth, generating ATP, building proteins), by our definition it is not a tissue property. Likewise, functions performed only by teams of cells (e.g., transmitting an electrical signal from one part of the body to another, contracting to shrink the diameter of a tube, forming a barrier to protect against physical trauma that would kill any single cell) belong in the tissue-level properties.

A second important difference is that cells are assumed to be self-sustaining, while tissues are not. Once a cluster of cells forms a specialized team, it requires help from other cells to remain alive. This is why a cluster of single-celled organisms (e.g., a group of bacteria in a drop of water) is not considered a tissue. Tissues are far more than a group of similar cells.

? CHAPTER STUDY QUESTIONS

Q. Using the bulleted list of "essential tasks" performed by cells as a guide, what additional tasks do cells have to perform? What additional tasks might be especially important for unicellular organisms, or for prokaryotes, or for tissue cells (e.g., skin, heart muscle, brain)?

A. Some additional tasks cells might perform include (1) determining where they are with respect to other cells, (2) moving from one location to another, and (3) modifying their external environment. Unicellular and/or prokaryotic organisms might need to (1) ingest and digest food, (2) evade predators, or (3) fight off attacks by viruses. Tissue cells must be able to (1) communicate with other cells in the same tissue and in the rest of the body, (2) grow and differentiate when needed, and (3) die at the appropriate time to keep the tissue healthy. Typically, muscle and nerve cells in the adult do not die intentionally, but skin cells do on a regular basis.

Q. Most prokaryotes are generally considered to be less complex than most eukaryotic cells, yet they are the most abundant cells on earth. How would one explain why simple prokaryotes are so abundant?

A. These organisms are representative of some of the oldest cells on earth, and thus have been subject to evolution by natural selection for the longest time. Because their life cycle is typically shorter than most eukaryotes, they are capable of generating larger numbers of offspring with greater phenotypic variation than eukaryotes, and thus are able to adapt to environmental changes more rapidly than more complex cells.

Q. The human body is composed of at least 100 different cell types. If one were to compare a skin (epithelial) cell with a heart (cardiac) muscle cell, what structural and functional similarities and differences would one expect to find?

A. Functional similarities would include the essential tasks, and structural similarities include the presence of the organelles. Because epithelial tissue and muscle tissue perform distinct functions in the body, the differences between the two cell types would reflect these roles. Skin epithelial cells separate the body from the outside environment and bind tightly to one another to form a strong protective barrier against harmful agents in the environment (chemicals, viruses, bacteria, etc.). Cardiac (heart) muscle cells work together to pump blood through the body, and therefore have a very different structural organization: their cytoskeleton is organized to maximize the contraction of each cell in a coordinated fashion to pump blood through the heart and into the circulatory system. As such, these cells also form strong connections with one another, but are not organized to form a barrier; instead, they deliver concentrated force.

Q. Explain the structure–function relationship. Then, develop an example of the structure–function relationship in everyday (non-biological) life. How applicable is this concept to the nonbiological structures that people interact with during a typical day?

A. A simple way of describing the structure–function relationship is that a device is built to accomplish a task, so if one understands the structure of the device, it is possible to guess the task it performs. Almost any simple device in our daily lives follows this line of reasoning. For example, kitchen utensils have different shapes because they perform different tasks—one would not use a spoon to cut food into pieces, or a knife to eat soup. This concept pervades our everyday lives, and is equally as powerful in cell biology. A good habit to form when learning cell biology is to ask, "What can the structure of this item (protein, organelle, cell) tell me about the tasks it performs?"

Q. Apply the structure–function relationship to the nucleus: how do the functions of the nucleus influence its structure, and vice-versa?

A. One of the most important functions of the nucleus is to safely store DNA. This is reflected by the fact that it is membrane bound (as all organelles are), and that the membrane tightly regulates molecular traffic in and out through nuclear pore complexes. The nucleus is also the starting point for transfer of genetic information in eukaryotes; hence it is the site where DNA is transcribed to RNA. It shares its outer membrane with the endoplasmic reticulum, where many mRNAs are translated into proteins. Thus, the DNA→RNA→protein progression of genetic information is physically linked together.

An important structural property of the nucleus is that it is enclosed by a double membrane. Rather than being simply redundant, the two membranes function quite differently. The inner nuclear membrane contains proteins that connect to chromosomes and help control their spatial organization within the nucleoplasm. The outer nuclear membrane extends beyond the inner membrane, into the cytosol to include the endoplasmic reticulum, where many of the mRNA products of gene transcription are translated. This links transcription and translation together.

Q. What properties of the carbon atom permit it to form a wider variety of molecules than any other atom?

A. According to the octet rule in organic chemistry, carbon must add four electrons to fill its valence shell. This encourages carbons to form *four* covalent bonds, which is a larger number of bonds than is found in the other abundant atoms in cells (hydrogen, nitrogen, oxygen). This allows carbon to build a variety of complex molecules. In addition, carbon has low electronegativity, which permits it to form large chains held together by nonpolar covalent bonds; nitrogen and oxygen have comparatively high electronegativity, and thus form polar covalent bonds that do not easily polymerize.

Q. Carbon is one of the most abundant atoms in sugars and lipids, but sugars and lipids have very different chemical properties. How would one explain this?

A. Sugar is very soluble in water, but lipids are not. One explanation for this is that most carbon atoms in sugars are covalently bound to oxygen, creating a polar bond that attracts water. Most carbon atoms in lipids are covalently bound to hydrogen, creating nonpolar bonds that are hydrophobic.

Q. Cows and people are mammals, but cows can use most grasses (which contain the β1,4 linked polysaccharide cellulose) as a food source, while humans cannot. What special abilities must cows have to enable them to eat grass? (Hint: cows and humans have different bacteria growing in their gastrointestinal tract.)

A. Many mammals do not posses the enzymes necessary for breaking β1,4 glycosidic bonds in cellulose, and therefore cannot break cellulose into the glucose monomers cells use for generating metabolic energy. Cows overcome this limitation by having bacteria in their stomachs that are capable of digesting these bonds; the bacteria that live in humans do not have this ability.

Q. All organelles are bound by membranes. What makes each type of organelle distinct from the rest?

A. Each organelle possesses its own complement of distinct proteins. Proteins facilitate most of the chemical reactions taking place in organelles, and this helps each organelle perform a distinct set of tasks in cells.

Q. The cytoskeleton is a good example of proteins working as a team. What are some other teams of proteins?

A. Other teams include the proteins that replicate and repair DNA in the nucleus, the proteins that capture sunlight and convert its energy into organic compounds in chloroplasts, and the proteins that oxidize these organic compounds in mitochondria.

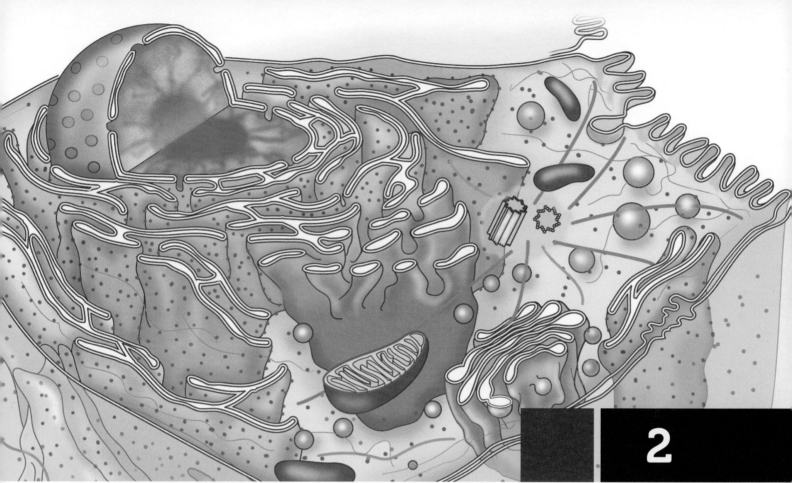

Nucleic Acids

2.1 ■ The Big Picture

Having discussed the structure of saccharides (sugars) in Chapter 1, we will use that information in this chapter to help understand the second building block of cells, called a **nucleotide**. The nucleotides we will focus on contain derivatives of the monosaccharide ribose, so it is important to understand the structure of ribose as we work through this chapter. Polymers of nucleotides in cells are called **nucleic acids.**

One of the most familiar nucleic acids to many students is **deoxyribonucleic acid (DNA)**, because it serves as the genetic material in all cells. Most of this chapter is devoted to the structure and function of DNA molecules. Before we start discussing the details, here is a brief preview the four major ideas of this chapter:

- First, DNA is a storehouse of "cellular information." This concept is somewhat abstract; pay attention to how a relatively simple linear code can produce an enormous number of different products.
- Second, the structure of a DNA molecule is relatively simple. It can often be learned largely by memorization.
- Third, DNA organization in a cell is relatively complex. To assist in working through this subject, I have divided the organization into five levels, from poorly organized to very highly organized.
- Fourth, cells have developed strategies for protecting the structure and organization of DNA against damage. This is the logical consequence of the first and third ideas: if something is valuable, it is worth protecting. This idea is so important that we will see it reappear in other chapters throughout the book. Use this chapter as an introduction to how this concept applies to cells.

2.2 ■ All of the Information Necessary for Cells to Respond to Their External Environment Is Stored as DNA

KEY CONCEPTS

- ▸ Cells store the information necessary to build ribonucleic acid (RNA) and proteins as DNA, a simple linear polymer found in all living organisms.
- ▸ When a cell divides, each of the resulting daughter cells contains an almost exact copy of the parental cell DNA.
- ▸ DNA must transmit its information to other molecules to be useful.
- ▸ The functional unit of DNA information is called a gene.
- ▸ The DNA information in genes is converted into matching, single strands of RNAs by a process called transcription.
- ▸ The sequence information in messenger RNAs (mRNA) is converted into a sequence of amino acids by a process called translation.
- ▸ Mutations in DNA are passed on to RNAs and proteins, resulting in variations that are acted upon by natural selection.

As we stated in Chapter 1, to be alive cells must have the ability to self-replicate and self-correct. This statement implies that cells are able to sense their surroundings so they know when it is safe to replicate, and that they are aware of their own functional state and know when they are damaged. To "know" these things, cells must contain information (**BOXES 2-1** and **2-2**).

What is cellular information? The concept of *information* is fairly abstract—while we are all quite familiar with its use in our daily lives, we can't point to it. Most often when

Here is a studying tip for every chapter in this book: before reading the chapter in detail, flip through the pages and note how many figures are devoted to each of the major subdivisions. Remember that it takes considerably more effort to generate a figure than it does to write a corresponding volume of text, so if an idea warrants a figure, it is likely worth paying attention to. Having a look at the figures before reading reveals a visual roadmap of some of the most important concepts in the chapter. By noting how many figures are used for each subdivision, and what they contain, students will have a good idea of what to expect as they work their way through a chapter.

BOX 2-2

The Library of Congress analogy, part 1. The information stored in DNA is analogous to the letters, numbers, and other symbols used in books. Units of these books, such as chapters, are the genes in DNA. For an information storage system to be useful, someone has to actually open and read it; in a library, this is done by people, and in cells, this is done by proteins. A nucleus is similar to the Library of Congress in that entry is restricted and the removal of information is prohibited. So, proteins in the nucleus "photocopy" the genes in the DNA "books" by making RNA copies; this is called transcription because the language of a DNA "book" is transcribed to a new form. **Box Figure 2-1** shows how these concepts apply to both cells and our hypothetical library. While the source materials (books or genes) must remain within the structure (library or nucleus), several copies can be made rather quickly and distributed to numerous locations outside the library building. Note that in a cell's "library," many of the books are actually "instruction manuals," not novels or textbooks. Thus, photocopies of these manuals are only useful if someone reads the instructions and builds something useful from them; the reading of mRNA photocopies to build polypeptides and proteins is called translation for this reason: words are converted into physical objects (proteins) that do the actual work. The reading of mRNA is done by ribosomes, which can be thought of as protein factories (several proteins working together as a team). Note that this strategy protects the books quite well, while also making them useful to the population.

FIGURE BF2-1 The Library of Congress analogy for information storage and transmission in cells. © iStockphoto/Thinkstock

Where the analogy fails somewhat is that many cells divide, and therefore copy their DNA; this is similar to copying the entire contents of the Library of Congress to build two new libraries. However,

(continued on next page)

this idea also demonstrates how mutations can affect cells. Imagine that someone (or even entire teams of people) actually tried to reproduce every letter of every book in the library—chances are good they would make several mistakes, and the new books would have different words in them. If these mistakes are not corrected, all subsequent photocopies of these books would have different information, and many of the new words would be unrecognizable; this is why many mutations are harmful to cells. On very rare occasion, the mistakes yield changes that are actually useful, such that the RNA (and possibly proteins) resulting from these new books help the cell better adapt to their environment. These cells have a slightly higher probability of surviving long enough to divide and pass on this new information to the next generation of cells. This is how evolution by natural selection works.

we refer to information, we are in fact referring to the way it is represented in physical form: as written text, recorded sound, or images, for example. Cells don't use any of these forms to store information, but they nonetheless capture and store a tremendous amount of information about their surroundings and their own physical state. This information is stored in the form of molecules. Just as we use different physical forms to store different kinds of data, cells use different types of molecules to store the different types of information they need to stay alive, as illustrated in **FIGURE 2-1**.

One of the most important molecules used to store cellular information is deoxyribonucleic acid, or DNA. Cells use DNA to store information for constructing other complex biological molecules, such as proteins and ribonucleic acid (RNA), which in turn are responsible for constructing nearly every molecule found in cells. (How cells use DNA to construct these other molecules is discussed in Chapter 8). DNA stores the most fundamental information necessary for life; every cell, and therefore every living organism, uses DNA for this purpose.

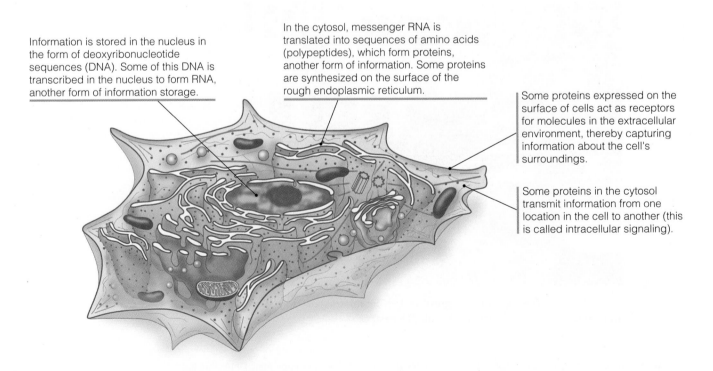

Information is stored in the nucleus in the form of deoxyribonucleotide sequences (DNA). Some of this DNA is transcribed in the nucleus to form RNA, another form of information storage.

In the cytosol, messenger RNA is translated into sequences of amino acids (polypeptides), which form proteins, another form of information. Some proteins are synthesized on the surface of the rough endoplasmic reticulum.

Some proteins expressed on the surface of cells act as receptors for molecules in the extracellular environment, thereby capturing information about the cell's surroundings.

Some proteins in the cytosol transmit information from one location in the cell to another (this is called intracellular signaling).

FIGURE 2-1 Some forms of information storage in cells.

A Cell's DNA Is Inherited

DNA stores information in the form of a linear sequence of repeating units, somewhat analogous to how a linear sequence of letters or characters in an alphabet can be used to store information in the form of words, sentences, or paragraphs. DNA uses a very simple language, composed of only four molecules, commonly known as A, C, T, and G. As we will see below, each of these letters is a molecule known as a **deoxyribonucleotide**. These deoxyribonucleotides are attached end-to-end to form very large structures in cells.

When cells replicate, they divide into two parts and each of the daughter cells inherits its own complete copy of the parental cell's DNA. This requires the parental cell to replicate its DNA prior to cell division. The mechanisms of DNA replication are discussed in Chapter 7. In multicellular organisms such as humans, most cells (called **somatic cells**) only pass their DNA on to the cells that replace them during that organism's lifetime; a specialized set of cells, often called **germ cells** (e.g., eggs and sperm), is usually responsible for passing DNA from one organism to its offspring.

Mutations in DNA Are Passed from Generation to Generation

When cells replicate their DNA, they frequently make mistakes, as seen in **FIGURE 2-2**. Some of these mistakes result in changes to the deoxyribonucleotide sequence. This is understandable, considering how many DNA replication events take place during an organism's lifetime. For example, take a look at the human body. Each human cell contains approximately 12 billion (12×10^9) deoxyribonucleotides in its DNA. The number of cells in an average human body at any given time is estimated to be 3×10^{13}, which replicate to form approximately 10^{16} cells over the course of a person's lifetime. Generating 10^{16} cells from a single fertilized egg therefore requires $(12 \times 10^9) \times 10^{16} = 12 \times 10^{25}$ nucleotides to be replicated *in the correct order*. Nothing in nature can achieve the level of accuracy necessary to replicate all of these nucleotides perfectly. In fact, DNA replication errors are actually quite common: in humans, the rate of inserting the wrong deoxyribonucleotide in a DNA sequence (often called a **point mutation**) is approximately 1 per every 2×10^{10} deoxyribonucleotides, or about 6 errors each time a cell replicates.

Other changes in the original DNA template sequence can also occur: extra deoxyribonucleotides can be inserted, some can be left out—even large pieces of DNA can be accidentally deleted, added, and/or moved to another location in the DNA sequence. The net effect of these changes is that every cell, and every organism, is at least slightly different from its ancestors, siblings, and other relatives. This heterogeneity in DNA sequences contributes a great deal to the variation found in populations of organisms that undergo evolution by natural selection.

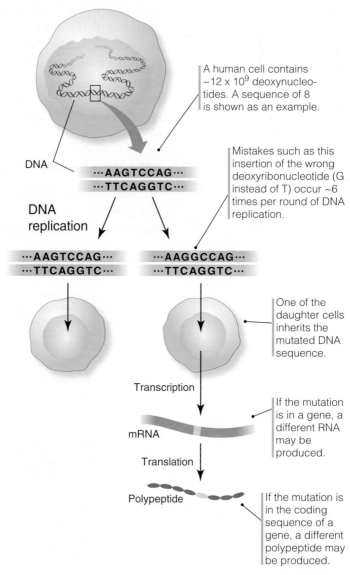

A human cell contains ~12 x 10⁹ deoxynucleotides. A sequence of 8 is shown as an example.

DNA

···AAGTCCAG···
···TTCAGGTC···

DNA replication

Mistakes such as this insertion of the wrong deoxyribonucleotide (G instead of T) occur ~6 times per round of DNA replication.

···AAGTCCAG···
···TTCAGGTC···

···AAGGCCAG···
···TTCAGGTC···

One of the daughter cells inherits the mutated DNA sequence.

Transcription

mRNA

If the mutation is in a gene, a different RNA may be produced.

Translation

Polypeptide

If the mutation is in the coding sequence of a gene, a different polypeptide may be produced.

FIGURE 2-2 Mistakes in DNA replication may cause mutations.

DNA Must Be Read to Be Useful

Like other forms of coded information, DNA is not useful in isolation; for example, information in a book is only meaningful if it is read and put to use. The same principle applies to DNA in a cell: only the portions of DNA that are "read" are meaningful. Cells don't literally read a DNA sequence, of course. Instead, they use proteins that bind to specific deoxyribonucleotide sequences in the DNA, and this binding changes the behavior of the proteins. The information is then converted into a useful form (an RNA molecule); this is illustrated in **FIGURE 2-3**. Some of these proteins are responsible for *transcribing* deoxyribonucleotide sequences in DNA into RNA sequences; some of these RNA sequences are then *translated* into amino acid sequences in proteins, a topic we will cover in much greater detail in Chapter 8. We will see numerous other examples of how DNA information is used throughout the book.

For many years, scientists thought that a large percentage of the DNA in most eukaryotic cells was useless, because they could find no evidence that proteins would bind to these regions. More recently, they discovered that this so-called junk DNA contains characteristic patterns of repeating DNA sequences, and that many of these sequences either bind to proteins directly or control the shape of neighboring DNA sequences that bind proteins. Much of this DNA contains deoxyribonucleotides that are chemically altered (e.g., by methylation). These chemical modifications can have a profound impact on which portions of DNA are read by a cell, and are so important that a new field of biology, called **epigenetics**, has emerged for studying how these modified sequences impact the phenotypes and behaviors of organisms. We will see more specific examples of epigenetic modifications later in this chapter (see Figure 2-23) and in other chapters.

DNA Information Is Packaged into Units Called Genes

The most familiar form of DNA information is a segment of deoxyribonucleotides known as a **gene** (**FIGURE 2-4**). However, despite its common usage in many fields of biology, there is no universally accepted definition for this term. In this book, we will define a gene

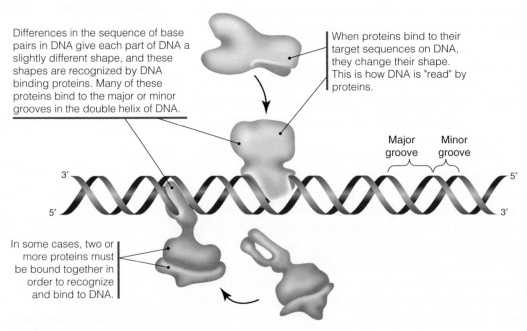

Differences in the sequence of base pairs in DNA give each part of DNA a slightly different shape, and these shapes are recognized by DNA binding proteins. Many of these proteins bind to the major or minor grooves in the double helix of DNA.

When proteins bind to their target sequences on DNA, they change their shape. This is how DNA is "read" by proteins.

Major groove   Minor groove

3′   5′

5′   3′

In some cases, two or more proteins must be bound together in order to recognize and bind to DNA.

FIGURE 2-3 DNA information is "read" by proteins.

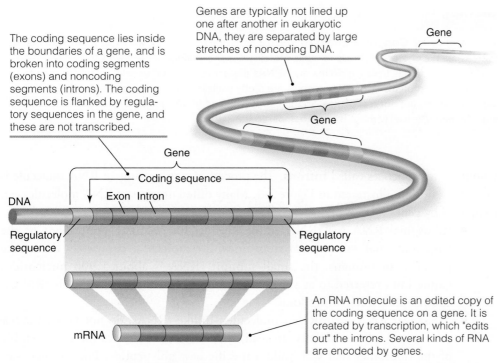

The coding sequence lies inside the boundaries of a gene, and is broken into coding segments (exons) and noncoding segments (introns). The coding sequence is flanked by regulatory sequences in the gene, and these are not transcribed.

Genes are typically not lined up one after another in eukaryotic DNA, they are separated by large stretches of noncoding DNA.

Gene

Gene

Gene

Coding sequence

Exon Intron

DNA

Regulatory sequence

Regulatory sequence

mRNA

An RNA molecule is an edited copy of the coding sequence on a gene. It is created by transcription, which "edits out" the introns. Several kinds of RNA are encoded by genes.

FIGURE 2-4 The smallest function unit of DNA is a gene. A sample of eukaryotic DNA is illustrated.

as the linear sequence of deoxyribonucleotides necessary for converting a portion of that sequence into a complementary sequence of ribonucleotides inside a cell. Less formally, we can say that a gene is a portion of DNA that can be converted into RNA, plus some additional sequences that are absolutely necessary for this conversion to take place. A gene is *always* a single linear sequence of deoxyribonucleotides on a single piece of DNA; a gene cannot be fragmented into different portions of DNA scattered throughout different DNA molecules. The average length of a human gene is 10,000–15,000 nucleotides (often abbreviated as **base pairs**, or bp), though there is considerable variation in size.

Genes are the best-known units of biological inheritance. When Gregor Mendel first discovered the principles of genetic inheritance in the mid-19th century, he was studying how individual genes of pea plants in his garden were passed from generation to generation. A few decades later, Heinrich Wilhelm Gottfried Waldeyer-Hartz discovered linear strands in the nucleus that changed color when he added stain to the cells, and he called these structures chromosomes (derived from Greek, meaning *colored bodies*). Another 20 years passed before Thomas Hunt Morgan determined that genes are arranged on chromosomes. A single chromosome may contain thousands of genes lined up one after another, each with its own distinct packet of information (see Figure 2-4). The modern definition of a chromosome is a genetic element containing genes essential to cell function. **Genetics**, the field of study devoted to uncovering the mechanisms governing the inheritance and expression of genes, has contributed greatly to our understanding of cell behavior. We will discuss the mechanisms controlling gene expression in greater detail in Chapter 12.

■■ Genes Are Transcribed into RNAs

All genes share one important trait: some portion of their deoxyribonucleotide sequence can be converted by a cell into a complementary sequence of ribonucleotides (also known as RNA). The portion of a gene that is replicated as RNA is called the **coding sequence**. In eukaryotes, the coding sequence is often broken up into segments, called **exons**, separated by noncoding sequences called **introns**. The process of synthesizing an RNA molecule is called transcription, illustrated in Figure 2-4. Many different types of RNA molecules can be encoded by genes (**BOX 2-3**):

- Ribosomal RNA (or rRNA) molecules are an essential component of the large and small subunits of ribosomes, the molecular complexes that make proteins. In humans, the small subunit rRNA is about 1,900 nucleotides (sometimes referred to as simply bases) long, and the large subunit rRNA is about 5,000 nucleotides long.

- Messenger RNA (or mRNA) molecules, unlike rRNAs, do not play an *active* role in any cellular activity. Instead, they serve as templates for the assembly of the ribosomes that will build a specific new polypeptide. Put another way, mRNAs are short-lived, intermediate copies of DNA information that is translated into proteins. Human mRNAs average 2,500 nucleotiodes in length.

- Transfer RNA (or tRNA) molecules are bridge molecules that link amino acids to a specific three-nucleotide sequence on mRNA. They specifically deliver the correct amino acids to ribosomes, where they are added to the polypeptides being synthesized. Compared to the rRNAs and mRNAs they interact with, tRNAs are comparatively tiny (typically about 73–93 nucleotides long).

- A variety of less well-known RNA molecules, including small nuclear RNAs (snRNAs) (100–200 nucleotides long), **short interfering RNAs (siRNAs)** (20–25 nucleotides in length), and microRNAs (miRNAs) (19–30 nucleotides long) have a wide range of functions. They are called noncoding RNAs because they are transcribed, but not converted into proteins, as is discussed below. Some play a role in controlling gene transcription and chromosome condensation during mitosis, while others help determine the location of mRNA in the cytosol. Identification and characterization of these RNAs is currently a very exciting field in cell biology.

Messenger RNAs Are Translated into Proteins

The combined product of rRNAs, tRNAs, and mRNAs working together is a newly synthesized polypeptide, as shown in **FIGURE 2-5**. During this process, three-nucleotide **codons** in the coding sequence of mRNA are matched with **anticodons** in tRNA by ribosomes (including rRNA) to determine the sequence of amino acids in the resulting polypeptide. Despite the fact that 64 different codons are possible (4 different nucleotides can fill each position, therefore 4^3 or $4 \times 4 \times 4 = 64$), each does not code for a unique amino acid. Instead, 3 codons (UAG, UAA, UGA) are designated as "stop" codons, which halt translation, and in the remaining 61 codons, redundancies ensure that only 20 different tRNAs (and amino acids) are specified by an mRNA. The average size of a human gene coding sequence is about 500–600 codons, yielding a polypeptide of 500–600 amino acids in length. Therefore, an average-sized human gene can theoretically encode at least 20^{500} different polypeptide sequences (20 different amino acids per each of

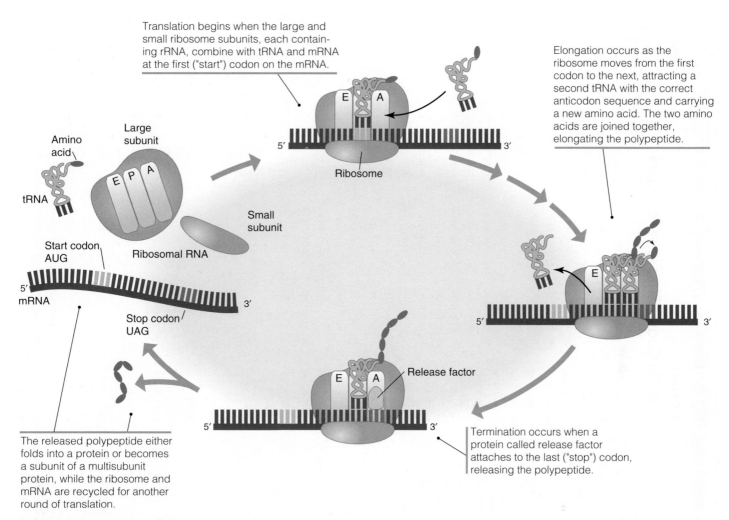

Translation begins when the large and small ribosome subunits, each containing rRNA, combine with tRNA and mRNA at the first ("start") codon on the mRNA.

Elongation occurs as the ribosome moves from the first codon to the next, attracting a second tRNA with the correct anticodon sequence and carrying a new amino acid. The two amino acids are joined together, elongating the polypeptide.

Amino acid

tRNA

Large subunit

Ribosomal RNA

Start codon AUG

5′ mRNA

3′

Small subunit

Ribosome

Stop codon UAG

Release factor

The released polypeptide either folds into a protein or becomes a subunit of a multisubunit protein, while the ribosome and mRNA are recycled for another round of translation.

Termination occurs when a protein called release factor attaches to the last ("stop") codon, releasing the polypeptide.

FIGURE 2-5 An overview of translation in eukaryotes.

the 500 codons, see **TABLE 2-1**). In reality, the actual number of polypetides produced by any single organism is far less than this, for a very good reason: most of these polypeptides would not be useful to cells. Polypeptides (either as individuals or as assembled groups) form proteins, some of the most common molecular structures in cells, and proteins have some strict requirements for how their polypeptides must behave. I call these requirements the three traits of proteins, and we'll discuss them in detail in Chapter 3.

The information stored in DNA can undergo two transformations to become useful to cells. The first, transcription, simply changes the DNA informational sequence into a complementary RNA informational sequence. This has a profound benefit for cells: relative to the typically very large DNA molecules, RNA molecules are quite small and can be transported to different regions in a cell relatively easily. Also, multiple RNA molecules can be generated from the same DNA sequence, so a cell can configure its RNA profile by simply changing the number of RNA molecules it copies from each of its genes. This is one easy way for cells to become specialized.

The second transformation of DNA information occurs when mRNA information is used to build polypeptides. The process of converting the ribonucleotide sequence of mRNA into an amino acid sequence in a polypeptide is called **translation**. We will discuss the molecular events responsible for translation in much greater detail in Chapter 8. Note that mRNA is the only known type of RNA that is translated. This, too, offers great benefits to cells. DNA and RNA molecules are each composed of only 4 different subunits

TABLE 2-1 A comparison of the information storage capability of the genetic code, ASCII computer code, and musical notation.

Genetic code	ASCII code	Musical notation*

Genetic code

```
GATCCTGCC   TGGGCTTTGC  CTCTGCAGCC  CCCCGCCCCA  CAGGTTCACA  CCTCGGGTCT
TCTCCACCGC  TGCCACACGC  CAGAGCCTGT  CAGAGCTTGG  GGGAACCTTG  GAGGTGGGAC
TCCTGCACCT  CAGCCATCAT  CAGACCCATG  GGGCCACCCA  GGGAACCTTG  GCAGGGACCA
TTACCAGTGA  CCTGCCGAGG  CCCCGGACTC  CTGTGCCCAG  CTGTGCCCCC
CGGACAGTGT  CGGTTCATGT  GGGAACTAGG  GGACGATGTG  GTTCTTCCGA  TCTGATGATG
AAGGCCCTGG  GCCACTTGGC  ACGGGCGGGC  GCTCCCAGGA  AGTCCTGCGG  GAGCCCCCTC
TGCCCAACTC  CCAGAAAGGC  CGAGCCTCTG  CAGCGGGAGG  CTCTTCCTGG  CCAGCCCTCC
GGGCAGCAGC  GCAGGGCACA  GGGACAGCCC  CCCTCCACAG  CCGACCAGC   CCAGGGGTCC
CCACTATCTG  CCAGGAGGTT  GCTTCTTCCA  GCAGGCTTTT  CCCGACCAGC  CCAGGGGTCC
AGGGTCTGGG  GCTCCCAGCT  GCTGTGAGTG  CTGCACATTC  TCTTGAGGAC  AGCCCCCTCC
CTCCCCACC   CACTTCTGGT  GCCCACTGTG  GCCACAGCAA  GCACTGGGGC  CTGCACTCAG
GGACCTCGGG  GCCTCCTGGG  GAGCTGCTGA  CCCTAGGCAG  AGAGATTGCA  CATCCCTAAG
AGTCTACAGA  CACCCCAGTG  TTTGCCCGTG  CTGGGTGCAA  GAGTGACTGG  TTTGCCAGTG
TTTGCCAGTA  TTTGCTCGCC  ACTTGTCCCT  CTGGGTGCAA  GAGTGACTGG
GTTTGGGCGG  GAAGTTGCAG  GTCCCTCCAG  GACAGTTGGC  GATGACGTG   GCGGTGCTCA
CACCCCCCAA  TCCTGGCTCC  CTGCAGGACG  CGGGGCCCCC  CGAGATCCTG  GCGGTGCTCA
GCACGACGGG  CACCTCCGTG  TTCACCAGTC  CAATGGGCAC  GGAGCGTGGC  TTTATTTGCA
TGTCTGGATT  CCTAACGACT  TCAGCCTCTG  CACCTCCTGG  GTTTTCCCTG  CTGCAAATTG
CCATTTGGCG  TCGTCCCCAA  TTTCCGGCCA  AGGCCCGCCC  GTCGTGCTGC  TGTGTAATTT
GATGTGTGGA  GTTCTAGATA  CCAAGTGTCT  GTCGGTTTTA  GACATCGCAA  ACGTCCTTCC
CAGTGTGGCC  CGTCCATTCG  CTTCTGTGCA  GCAAAATCTT  TAATTATTTG  ATGGCATCAA
AATGTGTGTC  CAGTTTTACC  TTCTAGTTTA  CATTTGTTTG  AGAAATCTTT
CTCCCACCTG  TGGCTGATAG  TGACGTCTTC  TTTACTATGT  TACATTCAGA
CCCATCATCT  TCAGGAAGAC  GCTTGTGTGC  TGAGGCCCCC  ACACCCCGCC
TCAGGACCAC  TGTCCATGGT  TCCACCCCTG  ACCCCGGACT  CCGCTCCCCA  GACCTCCTAA
```

ASCII Code - Character to Binary

0 0011 0000	I 0100 1001	b 0110 0010	v 0111 0110
1 0011 0001	J 0100 1010	c 0110 0011	w 0111 0111
2 0011 0010	K 0100 1011	d 0110 0100	x 0111 1000
3 0011 0011	L 0100 1100	e 0110 0101	y 0111 1001
4 0011 0100	M 0100 1101	f 0110 0110	z 0111 1010
5 0011 0101	N 0100 1110	g 0110 0111	
6 0011 0110	O 0100 1111	h 0110 1000	: 0011 1010
7 0011 0111	P 0101 0000	i 0110 1001	; 0011 1011
8 0011 1000	Q 0101 0001	j 0110 1010	? 0011 1111
9 0011 1001	R 0101 0010	k 0110 1011	. 0010 1110
	S 0101 0011	l 0110 1100	, 0010 1111
A 0100 0001	T 0101 0100	m 0110 1101	! 0010 0001
B 0100 0010	U 0101 0101	n 0110 1110	' 0010 0111
C 0100 0011	V 0101 0110	o 0110 1111	" 0010 0010
D 0100 0100	W 0101 0111	p 0111 0000	(0010 1000
E 0100 0101	X 0101 1000	q 0111 0001	) 0010 1001
F 0100 0110	Y 0101 1001	r 0111 0010	space 0010 0000
G 0100 0111	Z 0101 1010	s 0111 0011	
H 0100 1000	a 0110 0001	t 0111 0100	
		u 0111 0101	

Comparison

	Genetic code	ASCII code	Musical notation*
Name of basic message	polypeptide	sentence	song
Name of smallest functional unit	codon	byte	measure
Length of functional unit	3 nucleotides	8 bits	4 notes
Number of different values in each position of functional unit	4 (A, C, T, G)	2 (0, 1)	36 (12 notes × 3 octaves)
Number of different functional units possible	Theory: $4^3 = 64$ Actual: 20 (due to redundancy)	$2^8 = 256$	$36^4 = 1{,}679{,}616$
Estimated number of units in an average "message"	500 codons in average polypeptide	140 bytes in average sentence	80 measures per average four-minute song
Number of different average-size messages possible	Theory: 64^{500} polypeptides Actual: 20^{500}	256^{140} sentences	$1{,}679{,}616^{80}$ songs

*Traditional musical notation uses sequences of 12 different notes (C, C#, D, etc.) to designate different pitches in an octave, and most modern music uses a range of about three octaves, or 36 notes (we will ignore music-specific rules such as tempo, key, and scales, which influence the choice of notes in a given musical piece). The standard unit of musical notation is a *measure*, and for the sake of simplicity we assume each measure contains only four notes.

Images used with permission from © Suzanne Long/ShutterStock, Inc., © cristi18o884j/ShutterStock, Inc., and © AXL/ShutterStock, Inc.

(nucleotides), so that the number of variations available is relatively small. By comparison, proteins are composed of 20 different subunits (amino acids), which permit them to form much greater numbers of different molecules. Translation allows the information in DNA to be expressed in a myriad of different proteins.

■■■ Mutations in DNA Give Rise to Variation in Proteins, Which Are Acted Upon by Natural Selection

Mutations such as those discussed previously have the potential to alter the structure and function of RNAs and/or proteins. If a mutation occurs in the coding sequence of a gene, this change in the DNA informational sequence is reflected by a corresponding change in the RNA informational sequence that arises from it through transcription. In some cases, point mutations have little or no effect on the structure and function of the RNA, but in other cases, the effects of mutations can be profound, resulting in the formation of a dramatically different informational RNA or even no RNA at all.

Because all three types of RNAs play a role in synthesizing proteins, mutations in their sequences have the potential to alter the sequence of amino acids created from an mRNA. This is especially true when a sequence in the coding region of an mRNA is altered, because this region of mRNA determines the order of tRNAs it binds to, and thus of amino acids in a new polypeptide. If the coding region of an mRNA is changed, the order of tRNAs that bind to it will be changed accordingly, and the resulting polypeptide will have a different sequence of amino acids, as shown in **FIGURE 2-6**.

A classic example is the point mutation in the hemoglobin gene that causes **sickle-cell disease**. In this case, *a single deoxyribonucleotide change* in the DNA coding sequence of the hemoglobin gene (the 17th nucleotide is changed from A to T) causes a change in the mRNA and a *single amino acid change* in the hemoglobin protein (the 6th amino acid is changed from glutamic acid to valine). This tiny change causes red blood cells to adopt a characteristic "sickle" shape when the concentration of oxygen in the blood drops, since

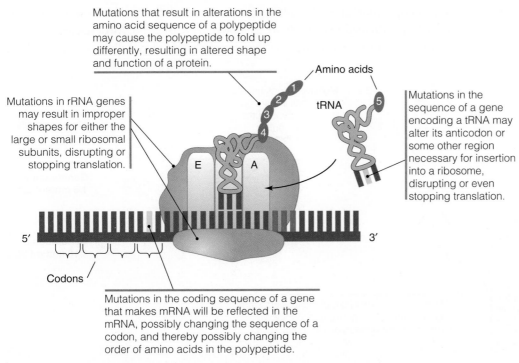

Mutations that result in alterations in the amino acid sequence of a polypeptide may cause the polypeptide to fold up differently, resulting in altered shape and function of a protein.

Mutations in rRNA genes may result in improper shapes for either the large or small ribosomal subunits, disrupting or stopping translation.

Amino acids

tRNA

Mutations in the sequence of a gene encoding a tRNA may alter its anticodon or some other region necessary for insertion into a ribosome, disrupting or even stopping translation.

Codons

Mutations in the coding sequence of a gene that makes mRNA will be reflected in the mRNA, possibly changing the sequence of a codon, and thereby possibly changing the order of amino acids in the polypeptide.

FIGURE 2-6 Mutations can alter amino acid sequence and protein function.

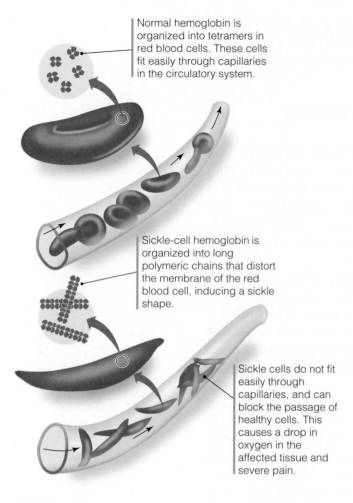

Normal hemoglobin is organized into tetramers in red blood cells. These cells fit easily through capillaries in the circulatory system.

Sickle-cell hemoglobin is organized into long polymeric chains that distort the membrane of the red blood cell, inducing a sickle shape.

Sickle cells do not fit easily through capillaries, and can block the passage of healthy cells. This causes a drop in oxygen in the affected tissue and severe pain.

FIGURE 2-7 A single point mutation causes sickle-cell disease.

the hemoglobin protein in the red blood cells changes from its normal tetrameric configuration to long polymer strands that distort the membrane of red blood cells. Resulting sickle-shaped cells can get stuck in capillaries and/or hemolyzed (ripped open), thereby interfering with proper circulation and causing a great deal of pain (**FIGURE 2-7**).

Mutations in tRNA or rRNA mitochondrial genes can also have significant effects in cells (recall from Chapter 1 that mitochondria and chloroplasts both have their own DNA, and therefore perform transcription and translation of genes). Several mutations have been found in the gene encoding the tRNA that carries the amino acid leucine to ribosomes in mitochondria, causing a wide range of problems (such as diabetes, degeneration of muscle fibers and nerves, and stroke-like episodes). Likewise, a single point mutation—the substitution of G for A at position 1,555—in the coding sequence of the gene for the rRNA found in the small ribosomal subunit (often called the 12S subunit) is associated with a form of deafness. The molecular mechanisms linking these mutations to their associated physiological problems are still not clear.

There are as many different possible combinations of mutations in an organism as there are nucleotides in its genotype. Some can be fatal, such as those that cause healthy cells to become cancer cells, but most mutations do not cause serious problems for living cells and organisms. The reason for this is fairly simple: typically, alterations in a DNA sequence either (1) have little to no effect on the structure and function of the RNAs and/or proteins produced by a cell, or (2) have such a drastic impact on the cell that it (and possibly the entire multicellular organism in which it lives) quickly dies. Because of this, most cells in a multicellular organism, or in a population of single-celled organisms, are *subtly* different from the previous generation of cells that divided to form them, as shown in **FIGURE 2-8**. Most scientists believe that a slow, steady rate of mutation persists for several rounds of cell division until enough

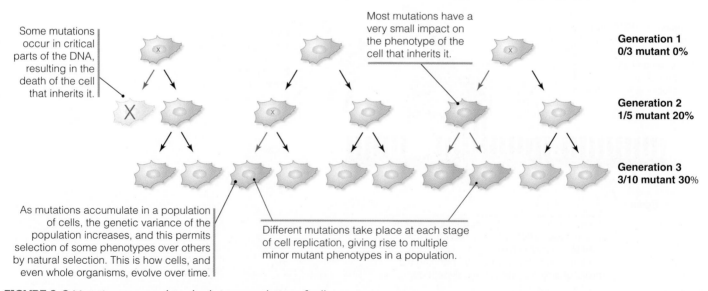

Some mutations occur in critical parts of the DNA, resulting in the death of the cell that inherits it.

Most mutations have a very small impact on the phenotype of the cell that inherits it.

Generation 1
0/3 mutant 0%

Generation 2
1/5 mutant 20%

Generation 3
3/10 mutant 30%

As mutations accumulate in a population of cells, the genetic variance of the population increases, and this permits selection of some phenotypes over others by natural selection. This is how cells, and even whole organisms, evolve over time.

Different mutations take place at each stage of cell replication, giving rise to multiple minor mutant phenotypes in a population.

FIGURE 2-8 Mutations accumulate slowly in a population of cells.

Is cancer inherited or not? The answer is *both*. Most cancers occur as mutations that accumulate in a somatic cell, such as a skin or lung cell; these cancers are not inherited, because skin and lung cells are not passed from parent to offspring. But, if a germ cell (e.g., egg or sperm cell) has a mutation that increases the likelihood of developing cancer, this can be passed onto offspring like any other genetic trait.

mutations accumulate to generate a noticeably different cell type, and possibly even a new type of organism, if these mutations are passed on from one generation to the next.

In the short term, this can have dire consequences for an organism: most cancers arise from cells that have acquired multiple mutations from their ancestral cells over the course of an individual's lifetime (**BOX 2-4**). However, the same principles can have a positive impact over the long term. Each generation of cells and organisms is subtly different from its ancestors, and it is this variation that permits populations of organisms to adapt to changing environmental conditions. Evolution by natural selection, the term coined by Charles Darwin to describe the impact of the environment on the survival of individuals, only functions if a population of organisms contains some inheritable variation. The persistent mutation rate resulting from errors in DNA replication can therefore be viewed as an important tool to help ensure the survival of a species. In effect, every member of a population of organisms, including humans, can be viewed as an experiment in natural selection. We all speak the same genetic language, but we are all mutants, in one way or another.

CONCEPT CHECK #1

Table 2-1 shows a comparison between the genetic code and two other codes commonly used in our everyday lives. Personal computers typically store numbers and symbols as magnetized particles on a hard drive, organized into bits and bytes, while sheet music stores musical performance instruction as symbols representing notes. What are the similarities and differences between these forms of information storage? In what ways is DNA a better or worse system than the others for storing information?

2.3 | DNA Is Carefully Packaged inside Cells

KEY CONCEPTS

▸ The fundamental structural unit of DNA is a deoxyribonucleotide; combinations of the four possible deoxyribonucleotides are arranged sequentially to form a linear strand of DNA.

▸ The simplest form of stable DNA in a cell is a called a DNA double helix, formed by two strands of DNA oriented antiparallel to one another and held together by hydrogen bonds between the atoms in the base portion of the deoxyribonucleotides.

▸ Three different forms of double helix DNA have been observed, suggesting that the shape of the double helix may vary in different regions of a DNA molecule.

Because DNA is heritable (i.e., passed on from a cell's ancestors), it reflects the tremendous amount of information that has been gathered over the course of billions of years of evolution by natural selection. Even the simplest cells have hundreds of thousands of nucleotides in their DNA. One of the smallest known genomes, that of the microbe *Nanoarchaeum equitans*, contains nearly 500,000 nucleotides. Remember that to be useful as a template, these nucleotides need to be accessible on demand. This presents a special challenge for cells: condensing DNA into a manageable size, while still permitting access to each nucleotide.

The solution to this challenge is complex. To better understand it, we will address the problem one level at a time. In this section of the chapter, we will focus on the structure of the DNA double helix, the most fundamental unit of DNA structural organization.

DNA Is a Linear Polymer of Deoxyribonucleotides

Let's have a closer look at DNA and how it is built (**BOX 2-5**). We will start by examining the structure of a deoxyribonucleotide, the simplest building block in DNA, then move up in complexity until we reach a complete double-stranded DNA molecule (**BOX 2-6**). Refer to **TABLE 2-2** to keep track of the names of the different structures as we increase in complexity.

A deoxyribonucleotide is a fairly simple structure. One good way to become familiar with this structure is to practice drawing it (use a pencil because some erasing will be required). There are three steps to this, and they are outlined in **FIGURE 2-9**. Let's start by drawing ribose, a five-carbon sugar in a ring configuration (**BOX 2-7**). Be sure to number each of the carbons in this sugar in a clockwise fashion as shown.

- Remember that DNA is a **deoxy**ribonucleotide, so in step 1 we will take one oxygen away by replacing the hydroxyl group (–OH) on the 2′ carbon with hydrogen, yielding deoxyribose. Erase the hydroxyl group attached to the 2′ carbon, and replace it with a single hydrogen atom.

TABLE 2-2 The bases, nucleosides, and nucleotides of RNA and DNA.

	RNA		DNA	
Bases	Nucleoside	Nucleotide	Deoxynucleoside	Deoxynucleotide
Purines				
Adenine (A)	Adenosine	Adenosine mono-phosphate (AMP)	Deoxyadenosine	Deoxyadenosine mono-phosphate (dAMP)
Guanine (G)	Guanosine	Guanosine mono-phosphate (GMP)	Deoxyguanosine	Deoxyguanosine
Pyrimidines				
Cytosine (C)	Cytidine	Cytidine mono-phosphate (CMP)	Deoxycytidine	Deoxycytidine mono-phosphate (dCMP)
Uracil (U)	Uridine	Uridine mono-phosphate (UMP)	—	—
Thymine (T)	—	—	Deoxythymidine	Deoxythymidine mono-phosphate (dTMP)

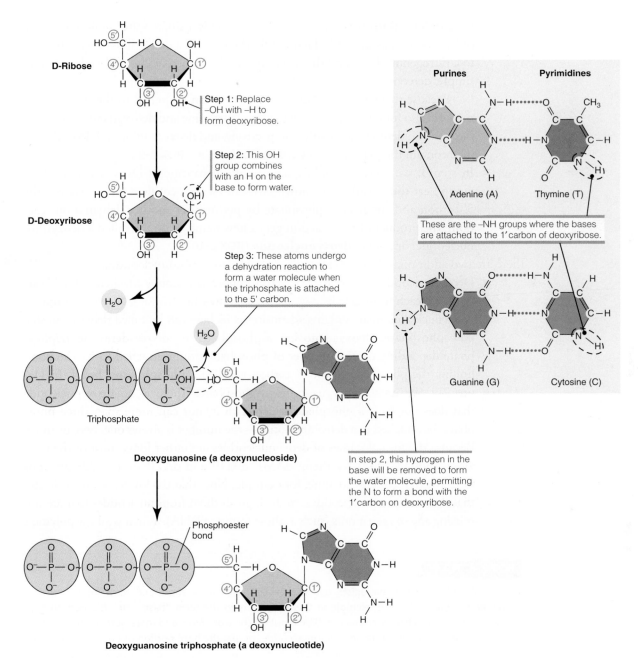

FIGURE 2-9 A step-wise method for drawing a deoxyribonucleotide.

- In step 2, we will attach a **base** to the deoxyribose. Draw a base near the 1′ carbon. We have four choices: two **purines** (adenine or guanine) or two **pyrimidines** (thymine or cytosine). Notice that all four of these bases contain a nitrogen atom bonded to a hydrogen atom that is pointing downward in the diagram; this is where we will attach the base to the ribose. We can join our base to the rest of our structure by creating a covalent bond between the 1′ carbon on the ribose molecule and the nitrogen atom on the base. This reaction is called a **dehydration**

BOX 2-7 TIP

Study Figure 2-9 carefully; there is a lot of information here. The basic structure and nomenclature of sugars and nucleotides are reviewed in Chapter 1.

(or **condensation**) reaction because it also yields a single water molecule as one of its products, as shown in Figure 2-9. The result is a deoxyribose sugar attached to a nitrogenous base, and this structure is called a deoxyribonucleoside, or more simply, **deoxynucleoside** (BOX 2-8).

Because there are four different bases that can be joined to the deoxyribose, there are also four deoxynucleosides. Deoxyadenosine and deoxyguanosine contain the purines adenine and guanine, respectively, and deoxycytidine and deoxythymidine contain the pyrimidines cytosine and thymine (BOX 2-9).

- In step 3, we attach a triphosphate to the deoxyribose. Draw a triphosphate group near the 5′ carbon as shown. We can create a covalent bond between the 5′ carbon and the nearest phosphate by performing another dehydration reaction. The product of this reaction gets a new name: it is now called a deoxyribonucleotide, or simply **deoxynucleotide** (BOX 2-10).

There are two additional things to remember about deoxy(ribo)nucleotides:

- Deoxynucleotides have one, two, or three phosphate groups attached to the 5′ carbon, and each form gets its own name, as shown in FIGURE 2-10. For example, a deoxynucleotide that contains adenosine as its base can be called deoxyadenosine *mono*phosphate, deoxyadenosine *di*phosphate, or deoxyadenosine *tri*phosphate depending on the number of phosphate groups attached to it. The bond linking the phosphate to the 5′ carbon is called a **phosphoester bond**. In Figure 2-9, we drew deoxyguanosine triphosphate. There is no deoxynucleotide that does not have a phosphate attached to it, nor can more than three phosphates be attached to a deoxynucleotide—the number is always one, two, or three. Because the formal names of deoxynucleotides are rather long, most of the time we use abbreviations for them: dAMP, dADP, and dATP are the abbreviations for those that contain adenine, for example. Note that the lowercase "d" indicates that these are *deoxy*nucleotides, to distinguish them from nucleotides that are not missing any oxygen atoms (such as those found in RNA). When we draw polymers

---

BOX 2-8 TIP

Many cell and molecular biologists use the word "base" as a shorthand term for the entire deoxyribonucleotide. A good example in this chapter is in the term "base pair." It's easy to get confused when this is first encountered. Watch out for the word *base*, and make sure to know which structure is being referred to (a deoxyribonucleotide, or just the base portion of one) when we use that term.

BOX 2-9 TIP

Notice the difference in spelling between the base and the corresponding deoxynucleoside: although it would be convenient to simply put the term "deoxy" in front of the base to yield the corresponding name of the nucleoside, the rules of chemistry don't allow it. One way to remember the difference in spelling is that for the purines, one inserts the letters "os" before the "-ine" when referring to the nucleoside (or deoxynucleoside), and for the pyrimidines one inserts the letters "id" before the "-ine." It helps that the word *pyrimidine* also follows this rule. Remember to think about this only *after* fully understanding the structures these words refer to.

BOX 2-10 TIP

Notice that the only difference in spelling between deoxynucleotide and deoxynucleoside is a single letter: s or t. Because "s" comes before "t" in the alphabet, it is easy to remember that the smaller structure contains the "s" and the bigger structure contains the "t".

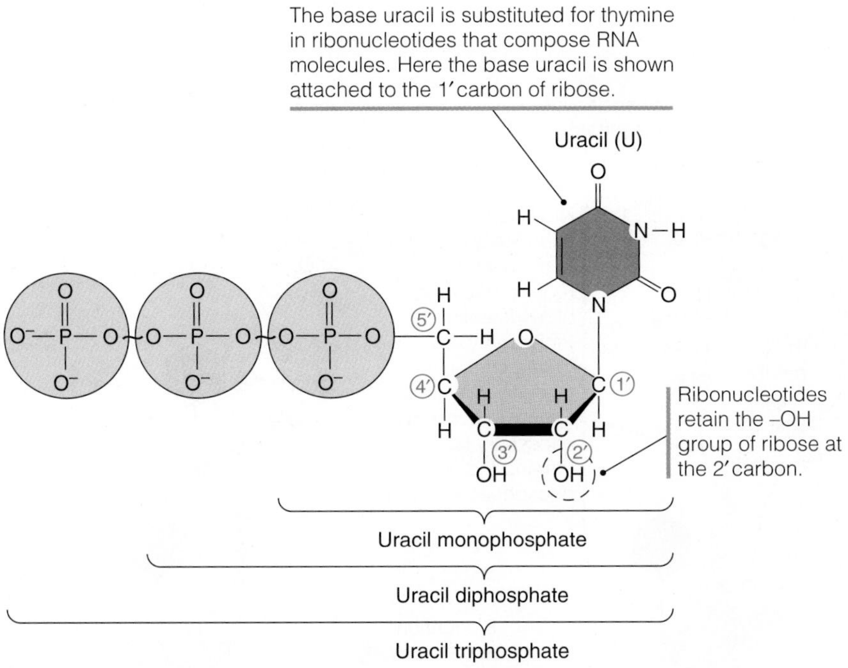

The base uracil is substituted for thymine in ribonucleotides that compose RNA molecules. Here the base uracil is shown attached to the 1' carbon of ribose.

Uracil (U)

Ribonucleotides retain the –OH group of ribose at the 2' carbon.

Uracil monophosphate

Uracil diphosphate

Uracil triphosphate

FIGURE 2-10 Distinctinctive features of ribonucleotides. The mono-, di-, and triphosphate forms of uracil triphosphate are indicated.

of deoxynucleotides, we simplify the names even more by resorting to the familiar single letter abbreviations (A, G, C, T).

- Ribonucleic acids (commonly known as RNAs) are composed of subunits that very closely resemble the deoxynucleotides in DNA (see Figure 2-10). The two most important differences between the deoxynucleotides found in DNA and the corresponding nucleotides found in RNAs are (1) the pyrimidine uracil is used in place of thymine, and (2) the nucleotides in RNAs contain ribose rather than deoxyribose (hence the absence of the term "deoxy" when naming these structures). Similar to their deoxyribonucleotide counterparts, AMP, ADP, and ATP are the abbreviations for adenine ribonucleotides containing one, two, or three phosphates. We will have a much closer look at RNAs in Chapter 13. In addition to serving as building blocks for RNA, ATP and GTP serve as sources for metabolic energy and help control the function of proteins, as we will see in later chapters.

■■ A Single Strand of DNA Is Held Together by Phosphodiester Bonds

Deoxynucleotides can be joined together in a linear fashion, via ester bonds, by performing dehydration reactions between the phosphate group of one and the hydroxyl group on the 3' carbon of another. (We will discuss the exact mechanism of this reaction in much greater detail in Chapter 8) The result is a string of deoxynucleotides linked together by an alternating sequence of phosphates and deoxyribose sugars (hence the familiar name sugar-phosphate backbone) held together by **phosphodiester bonds**, with the nitrogenous bases extending out to the side, as shown in **FIGURE 2-11**. Note also that this string of deoxynucleotides has different structures at the two ends: no matter how long the string is, one end always has an unbound 5' carbon (no additional nucleotides attached to its 5' carbon) and the other end has an unbound 3' carbon. These are called the 5' and 3' ends of DNA (**BOX 2-11**).

Just as the simple building block (monomer) of DNA has a name, so too does the polymer: our strand is now called a deoxyribonucleic acid. Note that this is a generic name, and that the sequence of deoxynucleotides in the strand makes no difference—any linear sequence of

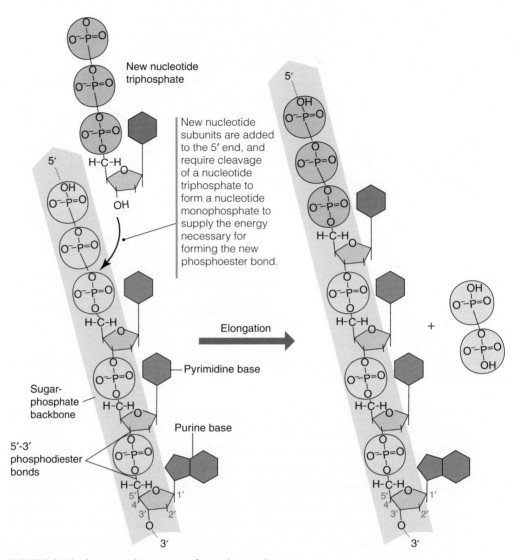

New nucleotide triphosphate

New nucleotide subunits are added to the 5′ end, and require cleavage of a nucleotide triphosphate to form a nucleotide monophosphate to supply the energy necessary for forming the new phosphoester bond.

Elongation

Sugar-phosphate backbone

5′-3′ phosphodiester bonds

Pyrimidine base

Purine base

FIGURE 2-11 The general structure of a nucleic acid.

the four deoxynucleotides, arranged in this 5′ to 3′ fashion, is called DNA. When one wants to discuss both DNA and RNAs as a group, we use the more general term **nucleic acids**.

DNA Forms a Double-Stranded Helix

Single strands of DNA, such as the one in Figure 2-11, are not stable in cells. Cells stabilize DNA by organizing it. To keep track of how, we'll identify five different levels and number them consecutively. The simplest form of stable DNA in cells, which we'll call Level 1, is a double-stranded DNA molecule where the two strands run antiparallel to one another (one strand of 5′ to 3′ is alongside one that runs 3′ to 5′) and are held together by hydrogen bonds between oxygen and nitrogen atoms in the **complimentary bases** to

---

BOX 2-11

The holiday lights analogy. The linear string of nucleotides resembles a string of holiday lights: the sugar-phosphate backbone is represented by the wires and the bases are represented by the lightbulbs. In this analogy, there would be four colors of bulb to represent each of the four bases. One end would represent the 5′ end, and the other end would represent the 3′ end.

---

form base pairs. The absence of the –OH group on the 2′ carbon of deoxyribose allows the two DNA strands to twist around one another to form a helix. **FIGURE 2-12** shows three common representations of double-stranded DNA to highlight different structural features of the molecule. Figure 2-12A is a simple line drawing illustrating how complementary base pairing holds the two strands of DNA together. Figure 2-12B uses a three-dimensional drawing to demonstrate that the two strands in the double helix are stabilized by the hydrogen bonds between complementary bases.

Practice drawing the ribbon-ladder form of DNA as shown in Figure 2-12A. When doing so, make sure that the double helix contains two different grooves. The wider of

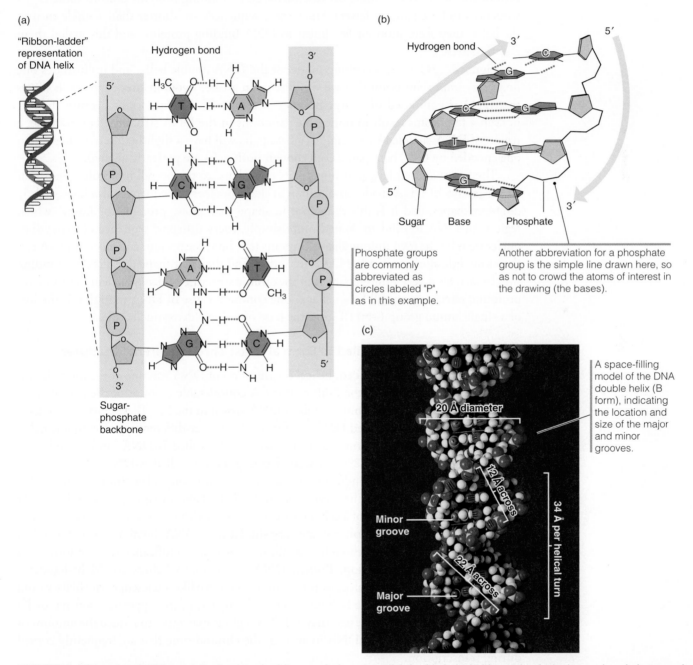

FIGURE 2-12 Level 1 of DNA organization is a double-stranded, antiparallel double helix held together by hydrogen bonds between base pairs. (a) A simple line drawing of the double-stranded double helix. A "ribbon-ladder" representation of the double helix is shown for reference. (b) A 3-D drawing showing the spatial arrangement of the nucleotides in a DNA double helix. (c) A space-filling model of the DNA double helix (B form), indicating the location and size of the major and minor grooves. © Photodisc.

these grooves is called the **major groove**, while the narrower is called the **minor groove**. These grooves are important because they form attachment sites for DNA binding proteins. DNA binding proteins often contain finger-like structures that fit into these grooves, and these structures allow DNA binding proteins to slide back and forth in the grooves as they search for the specific sequence of deoxynucleotides they are targeting, as shown in Figure 2-3. Note that the twisting of the DNA strands results in a periodicity of approximately 3.4 nm (or 34 Å); this means that there are approximately 10.5 base pairs per turn of the helix. This matters because many DNA binding proteins bind to short DNA sequences (6 base pairs or fewer). Since these sequences are shorter than a single turn of the helix, they seem more-or-less linear to DNA binding proteins, and this makes them easy to detect.

Figure 2-12C is a space-filling model of the DNA double helix, and it illustrates one final, very important point that we will discuss in greater detail in later chapters. It shows that DNA is composed of a large number of atoms. This is important because changes in even a few atoms result in noticeable variations in the DNA structure. For example, a region of DNA encoded by several A–T base pairs will have a slightly different shape than one encoded by G–C base pairs, even though both pairs are perfectly aligned.

The fact that each base pair imparts its own shape to the overall molecule means that two segments of DNA made up of different deoxynucleotide sequences also have slightly different shapes, and it is this difference in shape that allows proteins to "know" which regions of DNA to bind to. Stated more simply, every different sequence of deoxyribonucleotides has its own unique shape. Proteins that bind to specific sequences of DNA can therefore slide along a strand of DNA until they find the exact shape that fits their binding site. In fact, even very minor changes in the atomic structure of deoxynucleotides can have profound effects on overall DNA shape: one common cause of DNA mutation is the loss of a single amino group ($-NH_2$) from the base of a single deoxynucleotide.

■■■ DNA Can Be Supercoiled to Form at Least Three Different Structures

Discovering the double-stranded, helical organization of DNA was one of the most significant advances in biology of the 20th century. A considerable amount of the data used to deduce this structure came from crystals of DNA grown in the lab. These crystals also demonstrated that double-stranded DNA can form at least three different types of double helix. DNA adopts the configuration shown in **FIGURE 2-13**, called B-DNA, under conditions of high relative humidity (92%). A second configuration, called A-DNA, appears when DNA is crystallized under conditions of lower relative humidity. (The third type of double helix is called Z-DNA.) All three types of double helix have a major groove and a minor groove. Because the interior of a cell is entirely saturated with water, it is likely that most of the DNA in a living cell adopts a shape very similar to B-DNA. Small regions of Z-DNA have been detected in cells, though it is still unclear what significance this conformation plays in chromosome function. Both A-DNA and B-DNA helices are "right-handed," meaning that if one holds a piece of it in front of their eye like a telescope, the helix would appear to turn in a clockwise fashion. Z-DNA is twisted in the opposite ("left-handed") direction, so it has been suggested that Z-DNA regions may serve to reduce the amount of effort required to unwind B-DNA in areas of the chromosome that are frequently copied during transcription.

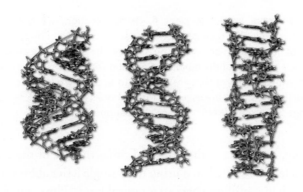

FIGURE 2-13 *Three different forms of double helical DNA (L to R: A form, B form, Z form).* Note that all three forms have major and minor grooves. Courtesy of Richard Wheeler.

CONCEPT CHECK #2

This section discusses the physical form of DNA as a repository of information. A common expression is that humans are currently living in an age of information overload. Are these two ideas at all related? Can cells suffer from information overload? What would be the consequences if they did? Use the vocabulary in this section to answer these questions.

2.4 DNA Packaging Is Hierarchical

KEY CONCEPTS

- The length of DNA in most cells is so great that complex packaging strategies are necessary to make the DNA fit inside the cells.
- Cells construct a protein/RNA scaffold that protects and supports DNA; in prokaryotes the resulting DNA-scaffold structure is called nucleoid, and in eukaryotes it is called chromatin. These structures are folded and twisted to further condense the DNA inside cells.
- The first level of packaging is a beads-on-a-string configuration, wherein the DNA double helix is wrapped around a "spool" made up of histone proteins; this shortens the length of a DNA molecule by 7-fold. Chemical modification of the histone proteins is an important mechanism for controlling the expression of genes in eukaryotes.
- The second level of packaging is the formation of 30- to 40-nm-thick fibers, made by twisting the beads-on-a-string structure into a coil. This shortens the length of a DNA strand approximately 42-fold.
- The third level of packaging is called looped domains, formed by periodic attachment of 30- to 40-nm fibers to a protein/RNA complex that shortens DNA approximately 750-fold.
- Compaction beyond the third level silences gene expression in DNA; this hypercondensed DNA is only found in eukaryotes and is called heterochromatin.

Now that we have a better understanding of what a DNA double helix looks like, it's time to confront a new problem: storing lots of information, accumulated over billions of years of evolution, in the form of linear sequences of deoxyribonucleotides means that these sequences are very, very long. For example, the roughly 12 billion deoxyribonucleotides in an average human cell form about 6 billion base pairs that, if laid end-to-end, would be over 2 meters long, hundreds of thousands of times the size of the cell containing it. If each of the deoxynucleotides were listed using their single letter abbreviations, these letters would occupy more than a million pages of a typical book. How can all that information be packed into a single cell without it getting all jumbled up (**BOX 2-12**)? Here is where we describe Levels 2 through 5 of DNA organization.

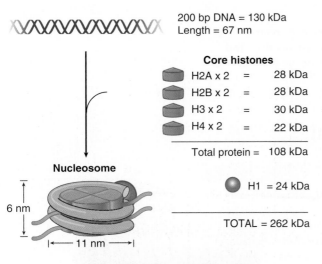

200 bp DNA = 130 kDa
Length = 67 nm

Core histones

H2A x 2	=	28 kDa
H2B x 2	=	28 kDa
H3 x 2	=	30 kDa
H4 x 2	=	22 kDa

Total protein = 108 kDa

H1 = 24 kDa

TOTAL = 262 kDa

Nucleosome

6 nm

11 nm

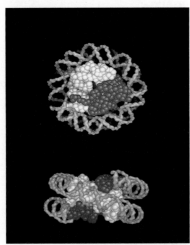

FIGURE 2-14 Two representations of the nucleosome. Top panel shows octameric arrangement of histones in core particle, with DNA wrapped around it. Histone H1 pins the DNA to the core particle. Lower panel shows computer model of DNA wrapped around core particle. Photos courtesy of E. N. Moudrianakis, John Hopkins University.

DNA Is Bound to a Protein/RNA Scaffold

The first part of the strategy for packaging DNA in cells is to support it with an elaborate infrastructure made of proteins and RNA. The proteins and RNA don't store any information, but without them the DNA would be hopelessly tangled and altogether useless. The complex formed by these proteins, RNA, and their associated DNA is called chromatin in the nuclei of eukaryotes, and nucleoid in mitochondria, chloroplasts, prokaryotes, and Archea. Proteins account for at least 50% of the mass of chromatin and are likely as abundant in nucleoid.

Double-Stranded DNA Is Wrapped around Histone Proteins to Form a Small Particle

The best-known set of structural proteins belongs to the **histone** family. These proteins are found in all organisms and are thought to be some of the earliest proteins to appear during evolution. When associated with DNA, they form spools, similar to those used to store thread, string, wire, etc. This is Level 2 of DNA organization. In eukaryotic cells, these spools are composed of two copies each of four different histones (named H2A, H2B, H3, and H4). The histones contain many positively charged amino acids, which attract them to the negatively charged backbone of DNA in a way that is largely sequence independent. The double-stranded DNA molecule is wrapped around a nucleosome approximately 1.7 times, to form a **core particle** as shown in **FIGURE 2-14**. An additional histone, either H1 or H5, "pins" the DNA to the core particle, resulting in a structure called a **nucleosome** (see Figure 2-14). In eukaryotes, nucleosomes containing around 167 base pairs of DNA are separated by short stretches (20–50 base pairs) of DNA called **linker DNA**, resulting in a structure that looks like a string of beads when viewed with an electron microscope (**FIGURE 2-15**). Similar structures are found in prokaryotic cells, where the DNA is wrapped around a different set of histone protein spools. Wrapping DNA in this fashion causes the DNA double-helical strand to become shorter and thicker: the

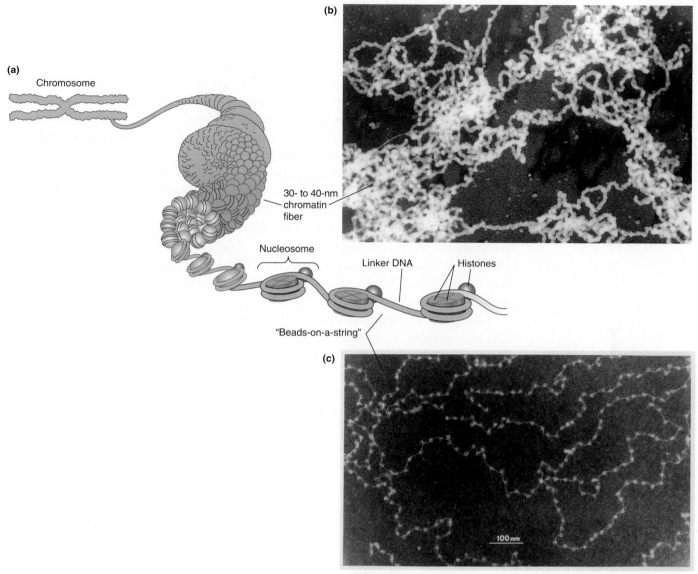

FIGURE 2-15 The 30- to 40-nm fiber is made by coiling the "beads-on-a-string" arrangement of nucleosomes. Electron micrographs show evidence for both forms in isolated chromatin. Courtesy of Ada L. Olins and Donald E. Olins, Bowdoin College. © Dr. Donald Fawcett, H. Ris/ Visuals Unlimited, Inc.

length is reduced approximately 7-fold, and the width increases from 2 nm to about 11 nm. These "beads-on-a-string structures" also contain proteins in addition to histones in both prokaryotes and eukaryotes.

▨▪ Nucleosome Structure Can Be Modified by Cells

Does wrapping DNA around a spool have any negative consequences? Remember that the goal is to compact DNA without compromising a cell's ability to access the genetic information. Because so much of the DNA comes into contact with the spool, most of it is inaccessible to other DNA-binding proteins, thereby negating the beneficial effects of these spools. Cells use three different mechanisms to address this problem:

- First, in eukaryotes, DNA can be partially unwrapped from the nucleosome by members of the **SWI/SNF** family of proteins. These proteins use ATP energy to move the core particle a short distance along the DNA, thereby freeing up any base pair sequences that may have been buried in the core particle. At least two other families of proteins participate in this type of **chromatin remodeling** as well; this is illustrated in **FIGURE 2-16**.

■ **Histone remodeling** is a second, very important way to modify the shape of the nucleosome, and it is based on chemically modifying the histone proteins in the nucleosome. A variety of proteins are capable of attaching relatively small molecules (methyl groups, acetyl groups, or phosphate groups) to the tails of histones, which causes them to change their shape, as shown in **FIGURE 2-17**. In some cases, the modifications make it easier to access the core particle DNA, and in other cases the opposite is true.

■ Third, methyl groups may also be added directly to the bases adenine (in prokaryotes) or cytosine and guanine (in eukaryotes) in the DNA, a process called **DNA methylation**. This is shown in **FIGURE 2-18**.

Together, these three mechanisms modify the structure and three-dimensional shape of chromatin without altering the sequence of deoxynucleotides. This is an important topic in the field of epigenetics, because it is believed that patterns of these modifications help control the expression of genes. We will discuss histone and DNA methylation in much greater detail in Chapter 12.

The beads-on-a-string structure we see in the microscope only appears when we break cells open and spread out their contents. In intact cells, nucleosomes are clustered together in a highly ordered fashion to form a series of similar configurations, all of which are called the **30-nm fiber** (see Figure 2-15). (In prokaryotes, similar **40-nm fibers** form from clusters of the "beads" in the nucleoid.) These fibers represent Level 3 of DNA organization. The reason that multiple configurations have the same name is that scientists are still not entirely sure how many of the configurations are present in living cells (the electron microscope, which is a common tool for observing chromatin, cannot be used to view living cells. These fibers will form spontaneously in a test tube if the salt concentration of the buffer is kept low enough, demonstrating that no additional proteins or metabolic energy

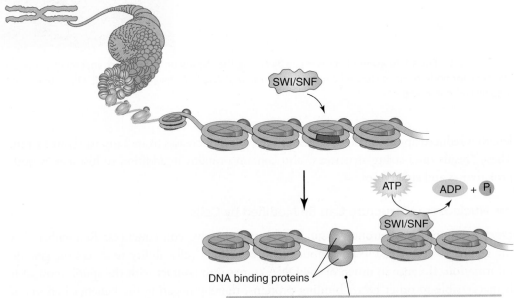

The SWI/SNF proteins slide the core particle, thereby exposing the previously wrapped DNA (purple) to DNA-binding proteins. Note that this requires metabolic energy: ATP is cleaved to ADP and inorganic phosphate (P$_i$).

FIGURE 2-16 SWI/SNF proteins participate in chromatin remodeling by partially unwrapping DNA around nucleosomes, allowing them to slide the core particle a short distance.

are required. The 30-nm fiber is held together by electrostatic interactions between different histones. For example, a negatively charged region on histone H4 binds to a positive region in a histone H2A/H2B complex in another nucleosome, drawing the two nucleosomes together and further compacting the chromosome. This results in additional shortening and thickening of the chromosome, and increases the packing density to about 42 times that of double-stranded DNA alone.

At Level 4 of DNA organization, the 30-nm fibers are attached to a protein-RNA **scaffold** (also called a **matrix**) that keeps them organized (**FIGURE 2-19**). Specifically, the fibers are attached to the scaffold at intervals of 10–30 μm, thereby forming so-called **loop domains** of approximately 60 kilobases in length. Exactly how these loops are attached to the scaffold is not known, but the attachment occurs at DNA sequences called MARs or SARs (for matrix or scaffold attachment regions, respectively). These sequences typically contain a large number of A–T base pairs, but are otherwise not very similar. A number of proteins have been isolated from the chromosome scaffold, and the structure is sensitive to enzymes that digest an unknown form of RNA, but it is not yet clear how these molecules assemble to form the mature structure. Loop domains are approximately 750-fold more compact than B-DNA.

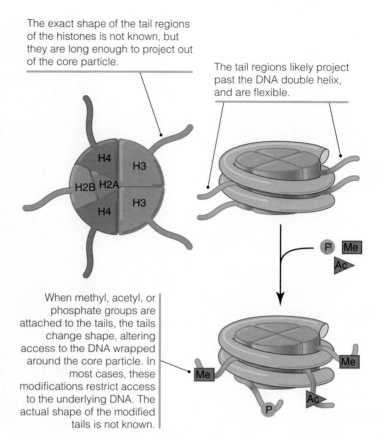

The exact shape of the tail regions of the histones is not known, but they are long enough to project out of the core particle.

The tail regions likely project past the DNA double helix, and are flexible.

When methyl, acetyl, or phosphate groups are attached to the tails, the tails change shape, altering access to the DNA wrapped around the core particle. In most cases, these modifications restrict access to the underlying DNA. The actual shape of the modified tails is not known.

FIGURE 2-17 Chemical modification of histone tails changes the shape of nucleosomes.

Chromatin Is Packaged into Highly Condensed Chromosomes

Eukaryotic cells adopt an additional means for organizing DNA that most prokaryotes do not use: they cut their DNA up into several chromosomes. Human beings are considered to be diploid, meaning they contain two copies of each chromosome that range in length from approximately 47 million base pairs to nearly 250 million base pairs. These are organized as 2 copies each of 22 autosomes and 1 pair of sex chromosomes. We can tell by simply looking in a microscope that chromosomes are very large bundles of DNA with distinctive shapes that change over the course of a cell's life. During mitosis, these chromosomes condense to form their familiar X-shaped structures, and they decondense once mitosis is complete, as shown in **FIGURE 2-20**. This condensation/decondensation is analogous to packing one's belongings into a small, compact space (e.g., a suitcase for a trip), then unpacking once the journey is complete.

These observations reveal two important things. First, DNA organization is dynamic: chromosomes can be tightly bundled or loosely bundled as necessary during a cell's lifetime. Second, this implies that there must be some machinery responsible for controlling this bundling. Consider how important this machinery is. During mitosis, a eukaryotic chromosome can be condensed into a structure that is about 15,000 to 20,000 times shorter than its unwound length. Changes in the length of a chromosome result from a complex series of folding and twisting events designed to prevent the individual strands from becoming tangled.

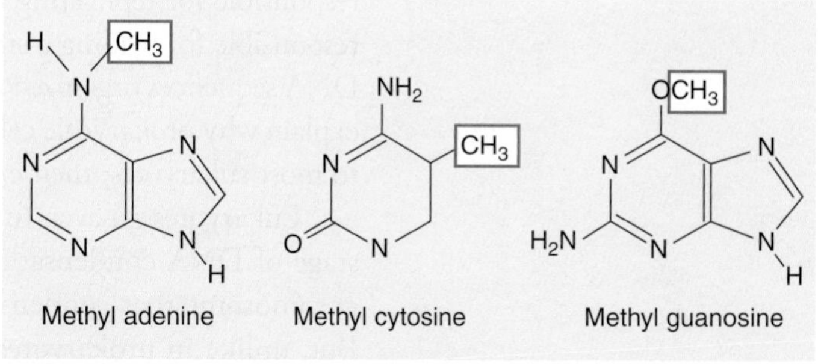

Methyl adenine   Methyl cytosine   Methyl guanosine

FIGURE 2-18 Methylated bases found in DNA. The methyl groups are indicated in red boxes.

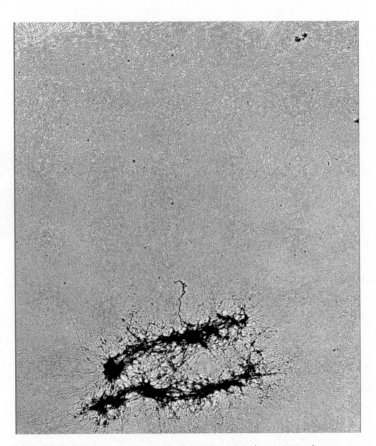

FIGURE 2-19 Level 4 of DNA organization can be seen in this electron micrograph of loop domains projecting outward from the protein/RNA scaffold (dark material at bottom). Photo courtesy of Ulrich K Laemmli, University of Geneva, Switzerland. Photo © Biophoto Associates/Photo Researchers, Inc. Reproduced from Cell, vol. 12, Paulson, J. R., and Laemmli, U. K., The structure of histone..., pp. 817–828. Copyright 1977, with permission from Elsevier. Photo courtesy of Ulrich K. Laemmli, University of Geneva, Switzerland.

Heterochromatin Is a Form of Tightly Packed DNA in Eukaryotic Cells

Thus far, we have been describing only the first few levels of DNA folding: wrapping around a spool, twisting of the spooled DNA into 30- to 40-nm fibers, and further folding to yield loop domains. (The prokaryotic chromosome is thought to consist entirely of these structures.) Collectively, Levels 1–4 of DNA organization are called **euchromatin** in eukaryotes, because they all share one important property: they can be easily accessed by proteins responsible for replicating the chromosomes in preparation for cell division, or by proteins responsible for reading a strand of DNA to make RNA (i.e., transcription). In other words, DNA sequences organized as a form of euchromatin are *easy to use* (**FIGURE 2-21**). This helps explain why prokaryotic cells can alter their gene expression patterns fairly rapidly, compared to most eukaryotes: their entire DNA is easily accessible.

Eukaryotes go even further in compacting their chromosomes. Advancing to the next stage of DNA condensation, Level 5, is a very big step for these cells. Any portion of a chromosome that condenses past the point of loop domains becomes essentially inactive. But, unlike in prokaryotes, a considerable amount of DNA in eukaryotic chromosomes is actually rather useless to a cell. This DNA may contain genes or fractions of genes that were important for our distant ancestors, but are no longer useful. Or it may contain

(a)

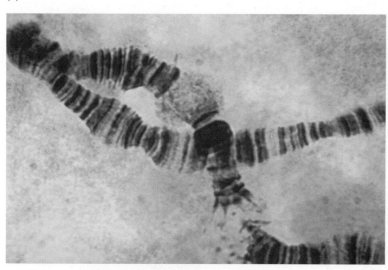

(b)

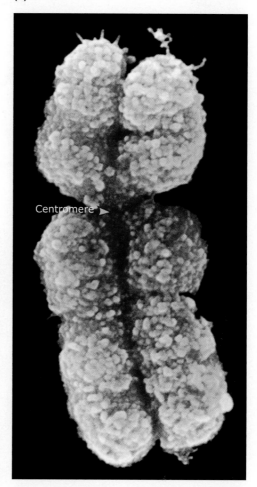

Centromere

FIGURE 2-20 (a) Chromosomes at different levels of DNA compaction. These chromosomes have banded regions containing tightly wound chromatin, and loose "puffs" where genes are easier to access. © Dr. Robert Calentine/Visuals Unlimited, Inc. (b) This chromosome is fully condensed, as it would look during mitosis. © Biophoto Associates/Photo Researchers, Inc.

genes that are only useful during early embryonic development or in specialized cells (e.g., humans form gill-like structures only during early development, usually only nerve and muscle cells make neurotransmitter receptors while only bone cells make bone proteins). Once a cell commits to a specific developmental fate (e.g., nerve, muscle, or bone), it no longer needs access to at least some of the genes required for other fates (e.g., skin or liver). A nerve cell can afford to "pack up" the portions of its DNA containing instructions necessary for functioning as a liver cell, because it has no expectation of ever using them.

■■ Twisting DNA into Heterochromatin Requires Metabolic Energy

The additional condensation of DNA is accomplished by twisting loop domains into shorter and thicker filaments. Several different structures are possible, depending on the extent of twisting used. The degree of this compaction has been estimated to be between 250- and 10,000-fold. We will break this range into two parts: those areas of condensed DNA found in cells that are not actively dividing (also called interphase) are Level 5A; more compact chromosomes are required for cells to undergo the mitotic or meiotic phase of cell division, and this extra degree of compaction is Level 5B. Regardless of their size, all of these Level 5 structures are called **heterochromatin**, both to differentiate them from euchromatin and because they appear as blobs of varying darkness in an electron microscope

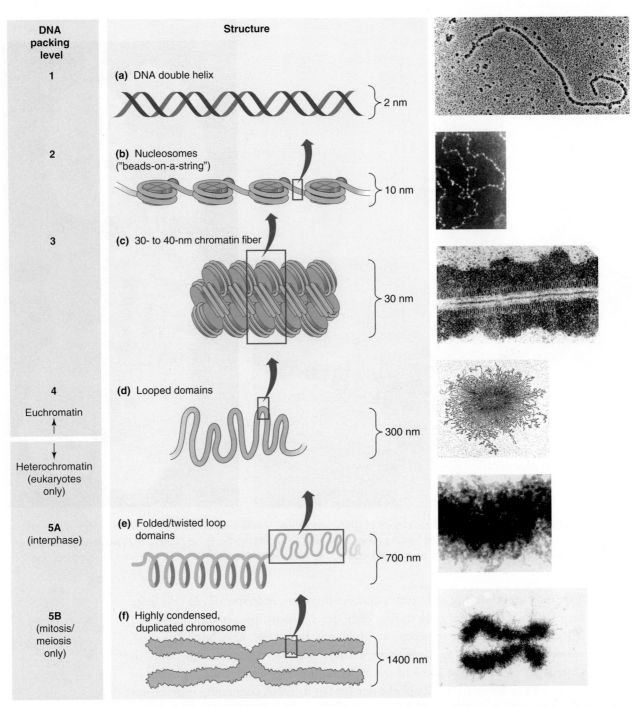

DNA packing level	Structure	
1	**(a)** DNA double helix	2 nm
2	**(b)** Nucleosomes ("beads-on-a-string")	10 nm
3	**(c)** 30- to 40-nm chromatin fiber	30 nm
4 ↑ Euchromatin ↑ ↓ Heterochromatin (eukaryotes only)	**(d)** Looped domains	300 nm
5A (interphase)	**(e)** Folded/twisted loop domains	700 nm
5B (mitosis/ meiosis only)	**(f)** Highly condensed, duplicated chromosome	1400 nm

FIGURE 2-21 The five different levels of DNA organization in eukaryotic cells. Note that two types of level 5 (heterochromatin) are shown: 5A is found in nondividing (interphase) cells, and 5B is found only in cells undergoing cell division (mitosis or meiosis). © Science Source/Photo Researchers, Inc.© Donald Fawcett/Visuals Unlimited, Inc. Reproduced from D. von Wettstein. 1971. Proc. Natl. Acad. Sci. USA. 68: 851–855. Courtesy of D. von Wettstein, Washington State University. Courtesy of Bruno Zimm and Ruth Kavenoff. Used with permission of Georgianna Zimm, University of California, San Diego. Courtesy of the Cell Image Library.

(**FIGURE 2-22**). Unlike the condensation of nucleosomes to form 30-nm fibers, this condensation requires additional proteins.

Two good examples are those belonging to the **SMC** (s̲tructural m̲aintenance of c̲hromosomes) family of proteins. One group, called **condensins**, is responsible for general chromosome structure, along with chromosome condensation at mitosis. A second group, called **cohesins**, is less well understood, but plays an important role in controlling

CHAPTER 2   Nucleic Acids

chromosome replication and segregation. Some of these require energy to function, so formation of heterochromatin is metabolically costly. In both groups, ATP hydrolysis is thought to supply the energy necessary for them to assist in condensation, though it is still not clear exactly what role it plays. Condensins and cohesins are discussed in more detail in Chapter 7.

Recently, scientists discovered another ATP-dependent group of proteins that plays a crucial role in chromosome condensation prior to mitosis. These proteins are responsible for reading highly repetitive sequences of DNA to synthesize special forms of double-stranded RNA called short interfering RNA (siRNA). This came as something of a surprise, because until then most researchers believed siRNAs only controlled gene transcription. While the exact mechanism has yet to be determined, the siRNA-forming proteins recruit and bind to still another group of proteins that modify histones (**FIGURE 2-23**). This too is an essential step in heterochromatin formation.

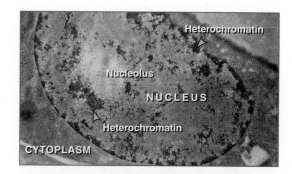

FIGURE 2-22 Heterochromatin appears as dark patches in the nucleus during interphase. This is Level 5A of DNA organization. Photo courtesy of Edmund Puvion, Centre National de la Recherche Scientifique.

■■■ DNA Is "Silenced" in Heterochromatin

Though the exact mechanism of heterochromatin formation is not known, the best-known method is "silencing" DNA in a nucleosome. So far, at least two different mechanisms for silencing DNA have been identified.

- In the first (illustrated in Figure 2-23), an enzyme called histone deacetylase removes an acetyl group (found in many actively transcribing genes) from histone H3. Following this, a protein called histone methyltransferase attaches a methyl group to the ninth amino acid (a lysine, abbreviated K) on histone H3 (this is sometimes abbreviated as H3K9me). Finally, a third protein (the names

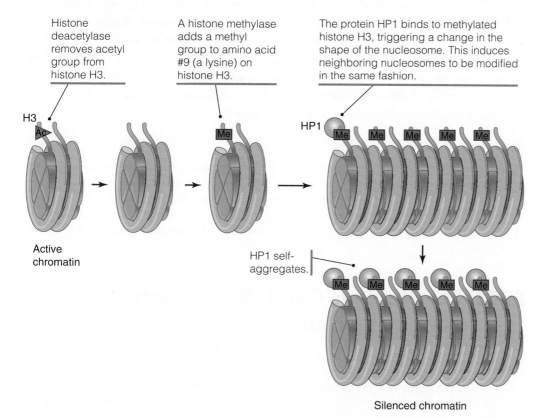

Histone deacetylase removes acetyl group from histone H3.

A histone methylase adds a methyl group to amino acid #9 (a lysine) on histone H3.

The protein HP1 binds to methylated histone H3, triggering a change in the shape of the nucleosome. This induces neighboring nucleosomes to be modified in the same fashion.

Active chromatin

HP1 self-aggregates.

Silenced chromatin

FIGURE 2-23 Histone modification can silence DNA to form heterochromatin.

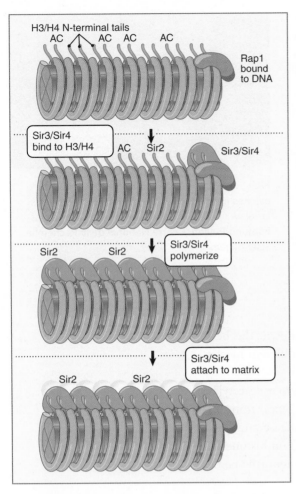

H3/H4 N-terminal tails
AC   AC AC   AC

Rap1 bound to DNA

Sir3/Sir4 bind to H3/H4   AC   Sir2   Sir3/Sir4

Sir2   Sir2   Sir3/Sir4 polymerize

Sir2   Sir2   Sir3/Sir4 attach to matrix

FIGURE 2-24 Rap1, Sir3, and Sir4 can silence DNA to form heterochromatin.

vary in different species; in mammals it is called heterochromatin protein 1, or HP1) attaches to the newly methylated histone H3. When this histone is deactylated and methylated, its shape changes. This triggers a conformational change in the entire core particle, such that the DNA attached to the core particle can no longer be transcribed; it is now silenced. As additional nucleosomes next to the newly silenced section undergo the same modifications, the silencing can extend to larger stretches of DNA.

■ The second mechanism, as seen in **FIGURE 2-24**, begins when a protein called Rap1 binds to a portion of DNA. The resulting change in Rap1's shape allows two additional proteins named Sir3 and Sir4 to attach to it, and they in turn bind to histones H3 and H4. This creates a change in the shape of the core particle and facilitates the binding of additional Sir3/Sir4 complexes to adjacent nucleosomes. Note that in both strategies, changing the structure of the nucleosome core particle by altering the configuration of histones is a key step.

■■ Some Regions of Eukaryotic Chromosomes Are Always Silenced

In eukaryotes, portions of the chromosomes are never active, and are therefore called constitutive heterochromatin (in biology, *constitutive* means "constantly produced"). A minimal amount of transcription, such as for the siRNAs mentioned previously, takes place in these regions, but none of the resulting RNA transcripts are ever translated. Instead, these regions play important roles in maintaining the structure and organization of a chromosome during mitosis. For example, the centromere region of the chromosome is essential for proper attachment of the chromosome to the microtubule spindle during mitosis, and the telomere regions protect the ends of the chromosome from damage. Both of these regions bind to a number of different proteins, but only unwind completely during DNA replication. They are discussed in greater detail in Chapter 7.

CONCEPT CHECK #3

To test understanding of the main ideas in this section of the chapter, consider the following question: People have a natural tendency to accumulate items, and therefore are always devising ways to keep their belongings organized. Think of how information is organized in two very different-seeming contexts: in a library and on a personal computer. How are these storage strategies similar and different from the strategies we've just discussed for organizing DNA?

2.5 The Nucleus Carefully Protects a Eukaryotic Cell's DNA

KEY CONCEPTS

▶ Because DNA stores all of the information necessary to survive, cells protect it from damage. This is especially evident in eukaryotic cells, which typically have much larger DNA molecules than prokaryotes.

▶ In eukaryotes, the primary function of the nucleus is to sequester DNA in its own chemically specialized compartment. Passage of large molecules (e.g., proteins) into and out of the nucleus is carefully regulated by a structure called the nuclear pore complex.

▶ A secondary function of the nucleus is to physically shield the DNA from damage. Nuclear lamins are cytoskeletal proteins that form a tough, fibrous network attached to the inner surface of the nuclear membrane. This network protects DNA from mechanical force.

Yet another strategy eukaryotic cells use to organize their DNA is to place it in its own organelle, the nucleus. When viewed from the perspective of information storage and transfer, the nucleus has a very tough job: it must simultaneously allow access to a large number of proteins responsible for reading, replicating, repairing, and remodeling the DNA (e.g., histones), while also protecting the DNA from a host of environmental dangers. How might a cell do this? The answer is sufficiently intricate that we will discuss it over the course of several chapters, but we'll start with an introduction to the complexity of cellular structure and function.

The Nuclear Pore Complex Restricts Access to the Interior of the Nucleus

For now, let's consider two major structures in nuclei that protect DNA. One of the most striking features of the nucleus is the nuclear pore complex, or NPC (**FIGURE 2-25**), which penetrates both membranes surrounding the nucleus. Each nucleus contains hundreds

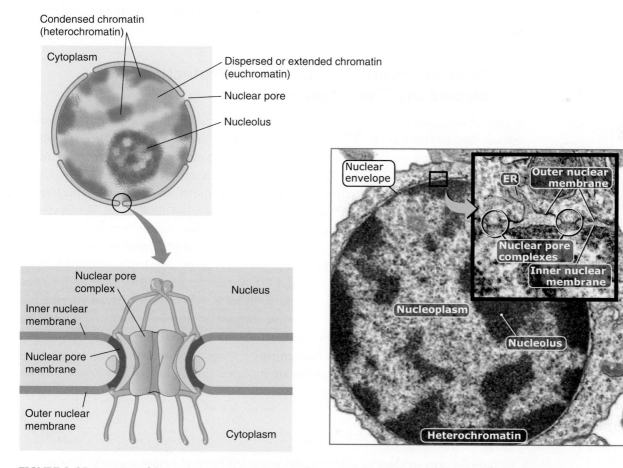

FIGURE 2-25 Overview of the nucleus. Nuclear pore complexes are channels that penetrate the inner and outer nuclear membrane and regulate the traffic into and out of the nucleus. The inset in the micrograph at right is a magnification of the nuclear membrane indicated by the box. Photo courtesy of Terry Allen, University of Manchester.

to thousands of NPCs, the primary function of which is to control the molecular traffic (of objects >30 kilodaltons in size) between the nucleus and the cytoplasm. Several different types of molecules pass through these pores, including large structures such as RNA molecules (synthesized in the nucleus and exported) and proteins (recycled between the cytosol and nucleoplasm), as well as the large and small ribosomal subunits (synthesized in nucleoli, regions in the nucleoplasm specialized for this purpose; see the electron micrograph in Figure 2-25), and even viruses. The key to passing through the NPC is a specific signal that is attached to these molecules; without the signal, the molecules are blocked by the complex. Transport of these molecules requires the expenditure of metabolic energy; the mechanisms governing how these molecules are marked and transported through NPCs are quite complicated, and are discussed in greater detail in Chapter 8. Molecules smaller than 30 kilodaltons, such as nucleotides, sugars, ions, and water, can diffuse freely through the NPC. This strategy protects DNA because the kinds of molecules most likely to damage it (e.g., proteins that degrade nucleic acids) are restricted from entering the nucleus unless they are needed.

Nuclear Lamin Proteins Form a Protective Cage around the Chromosomes

A second structure protecting chromosomes is the network formed by proteins called nuclear lamins (**BOX 2-13**). They are members of the intermediate filament family of proteins (discussed in detail in Chapter 5) and are characterized by their tremendous resistance to mechanical force. They are found in most multicellular eukaryotes (apart from plants) and in some unicellular eukaryotes (including yeast). Lamin proteins bind to both NPCs inside the nucleus and to chromatin, as shown in **FIGURE 2-26**. When viewed with an electron microscope, nuclear lamins give the inner nuclear membrane a rough, fibrous appearance. Nuclear lamin complexes are composed of three different lamin proteins, called lamins A, B, and C. Mounting evidence suggests that lamin proteins interact directly with both chromatin and even the DNA itself (at the MARs/SARs regions discussed previously). One of the chromatin proteins is HP1 (also discussed previously), which binds lamin B. Collectively, the lamin-chromatin complex is part of a network of proteins called the nuclear matrix, which coats chromosmes and plays a role in regulating gene expression (e.g., lamins also help organize chromosome unwinding during telophase of mitosis). When nuclear lamins are mutated and no longer functional, this has dramatic consequences for most cells, especially muscle cells that are subject to great mechanical force.

CONCEPT CHECK #4

What are the costs and benefits of having DNA sequestered away in its own organelle? Would prokaryotes be better off if they had a nucleus? Why or why not?

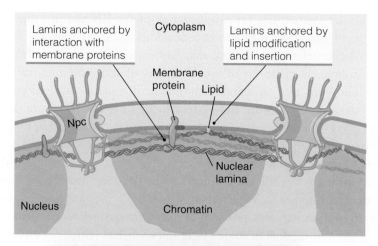

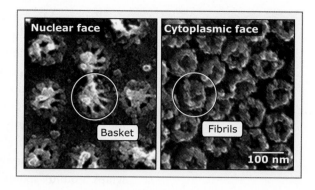

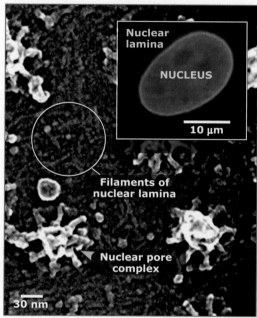

FIGURE 2-26 The nuclear lamina form a fibrous network attached to the inner nuclear membrane. These fibers lie between nuclear pore complexes, as shown. Reprinted from J. Struct. Biol., vol. 140, B. Fahrenkrog, et al., Domain-specific antibodies reveal multiple-site topology..., pp. 254-267, Copyright (2002) with permission from Elsevier [http://www.sciencedirect.com/science/journal/10478477]. Photos courtesy Ueli Aebi, University of Basel. Main photo courtesy of Terry Allen, University of Manchester.

2.6 ▮ Chapter Summary

To remain alive, cells must do two things: respond appropriately to external signals and internal programs, and maintain their internal environment. Nearly all of the molecules responsible for these activities are either RNA or proteins, and the instructions for creating these molecules are stored in a relatively simple polymer, DNA. Each time a cell divides, the daughter cells inherit a copy of the parental cell DNA, which is slightly modified in each successive generation of cells by mistakes (mutations) made during the replication process. The instructions in DNA are organized into units called genes; cells read these genes to produce several types of RNAs (by transcription) and proteins (by translation of messenger RNA). Mutated genes produce altered RNAs and proteins when they are transcribed and translated, and it is these differences in RNAs and proteins that yield the variation in cellular phenotype that is acted upon by natural selection in each generation of cells and organisms. DNA is, therefore, the heritable material acted upon by evolution.

Because DNA encodes the instructions for producing all of the RNAs and proteins a cell will require during its lifetime (and may also encode additional unused genes), it is typically an enormous molecule relative to the cell that harbors it. The complete DNA molecule, called a chromosome, is made up of combinations of four subunits, called deoxyribonucleotides, which form two antiparallel strands held together by hydrogen bonds. One of these two strands contains the coding sequence of a gene. To ensure that the genes can be easily accessed while also compacting them enough to fit into a cell, DNA is supported by an elaborate protein/RNA scaffold, called nucleoid in prokaryotes and chromatin in eukaryotes. Histone proteins are at the heart of these scaffolds, and DNA wraps around "spools" of histones. Chemical modifications of histones and DNA bases play an important role in controlling which sections of DNA molecules are read by the transcription machinery. In some eukaryotes, portions of the chromosomes are condensed and modified so that they are not capable of undergoing transcription. These regions are called heterchromatin, to distinguish them from the transcriptionally accessible regions called euchromatin. Prokaryotes do not form heterochromatin.

Eukaryotic cells store DNA in the nucleus and restrict access to it with nuclear pore complexes, highly selective structures that control the traffic of proteins, nucleic acids, and other large molecules into and out of the nucleus. Some eukaryotes also contain proteins in their nucleus, called lamins, as an added form of support and protection for the DNA. Lamin proteins are especially strong proteins that resist mechanical forces imposed on the nucleus, and recent evidence suggests they help control gene expression by binding chromatin and specific DNA sequences.

CONCEPT CHECK ANSWERS #1

All three codes have the potential to generate an unimaginable number of different products, far more than could ever possibly be used. But the genetic code has something the others don't: error correction, also known as functional redundancy. In cells, 64 different codons code for only 20 amino acids; this means that, on average, every amino acid is encoded by 3 different codons. This reduces the efficiency of the code tremendously, but given the vast number of possibilities available, it has no real impact on the workings of a cell. The redundancy is there to combat the effects of mutation: if 1 of the 3 nucleotides in a codon changes, there is a decent chance the amino acid it encodes will remain the same and the mutations will have no impact on the encoded proteins. ASCII has no such redundancy, and is completely vulnerable to even the slightest change: switching a single bit in a word processor file from 0 to 1 can render a word unrecognizable. Musical notation generates even more possibilities per unit than the other two codes, because (1) it expands the number of choices for each position approximately 10- to 20-fold, (2) it increases the length of the functional unit to 4 (i.e., 4 quarter-notes is a measure in common time), and (3) it has no functional redundancy. For cells to achieve this level of complexity, they would have to (1) build 10 times as many different deoxyribonucleotides, which would be terribly complex and energetically costly, and (2) increase the length of their codons to 4 instead of 3. This would increase the mass of the DNA by 25% and require still more energy consumption. Thus, the logic of the genetic code is like a compromise between the efficiency of ASCII and the complexity of musical notation, with the added advantage of having a built-in protection against small errors. A misspelled word or wrong musical note may be unpleasant, or even costly for the individual who makes the mistake, but an altered polypeptide can be fatal to a cell.

CONCEPT CHECK ANSWERS #2

The term *information overload* typically refers to two problems. The first is the multitude of *forms* of information we encounter: television, videos, movies, newspapers, radio, music, books, telephones, magazines, text messages, billboard advertising, and so on plus the wealth of information on the internet. Keep in mind that using a common *language* (e.g., English, Spanish, Mandarin) across all of these forms does not significantly diminish the sense of overload. The second is the enormous *volume* of information we encounter. In this chapter, we have described only three forms of cellular information (DNA, RNA, and proteins), with a heavy emphasis on DNA as the central repository of the information necessary to create the other two. But, from the perspective presented in this chapter, cells could potentially suffer from both problems. The first stems from the fact that DNA information is shape dependent; changing the shape of a DNA molecule alters the kinds of proteins that may bind to it, so every different shape of DNA is a different form of information, even though they are all presented in the language of DNA. A, B, and Z forms of the triple helix are good examples of macro-changes in DNA forms, and differences in the sequences of DNA, arising from mutations, are more subtle changes. Likewise, the major and minor grooves in the DNA double helix are distinct ways of representing the sequence of deoxyribonucleotides in a DNA double helix. Conversion of a DNA sequence to an RNA sequence by transcription is another example of a different form of information. Further conversion of an mRNA to a polypeptide is a dramatic change in the information form.

Many cells also have a *volume* of information problem. For example, we humans have leftover genes in our DNA that were useful to our evolutionary ancestors (e.g., for making gills, fins, and tails), but that are essentially useless now. When a cell needs to transcribe a gene, it has to sort through all of the useless material to find the important information in the DNA. Also, cells are quite good at making multiple RNA copies of the same gene, adding to the volume of information.

Put together, this spells out a problem for cells. Just like humans, if cells suffer from information overload, they get confused and become less effective in solving problems. For example, if protein X is damaged and needs to be replaced, the mRNA that encodes it must be found and translated by a ribosome, or a new copy of the mRNA will have to be made from the X gene. Consider that thousands of different genes are transcribed by a human cell over the course of its lifetime—the transcription machinery in cells is rarely idle, so getting a new copy of the X gene takes time, and it's quite possible that other proteins need to be replaced at the same time (especially if the cell was recently damaged). So, we can imagine a damaged cell trying to sort through thousands of genes to find those it needs to convert into proteins, while also handling the day-to-day tasks (like making ATP). The potential for chaos is fairly clear. Information overload is a very real threat to cells. We will address part of the solution later in this text.

CONCEPT CHECK ANSWERS #3

Let's go through each level of DNA packaging in turn for both examples. First, double-stranded DNA is wrapped around a spool of histone proteins and held in place with an additional histone to form chromatin/nucleoid. The similarities to storing thread, rope, hoses, wires, and other long filaments are obvious, except we typically don't wrap a single long filament around several spools. Why the difference? Because unlike a spool of thread, where the only part that is being used is the loose end on the outer edge of the spool, *every* part of a DNA strand has to be accessible without having to unwind the entire molecule (recall that when a cell divides, it has to access and copy every single part of the DNA).

In the library example, this is similar to clustering words into groups called books: the cover, binding, and glue in the book contribute no additional meaning, but they keep the pages of the book organized. Next, the twisting of the beads-on-string arrangement to form a 30- to 40-nm fiber is similar to arranging the books on a shelf—each book is equally accessible, and the overall arrangement is not disturbed when a book is temporarily removed from the shelf, photocopied, and then placed back on the shelf (as would happen if a gene or set of genes was transcribed). Next, the fibers are organized into looped domains. This is similar to arranging shelves of books into rows, where they remain organized but occupy less space than if the shelves were scattered about aimlessly. The next step (the formation of heterochromatin) means crossing the line from easy accessibility to maximum storage efficiency. In the library analogy, this is like pressing a group of shelves together with no rows between them—access to the books in the inner shelves is restricted, but the storage capacity of a room grows tremendously. If necessary, one can always push the shelves apart again to allow access to the inner portions, but it is a time-consuming and tiring process.

In the computer example, data is placed on an electromagnetic surface; the data is DNA, and the surface of the hard drive is the protein/RNA scaffold. The data is organized into bytes of information, and a collection of bytes is organized into a data file. This is similar to clustering DNA into histone spools, then twisting them together to form the 30- to 40-nm fibers. Organizing the fibers into loop domains is like organizing files into folders in a computer's operating system: everything has its place, and all files and folders are equally accessible on the hard drive. Moving up to the heterochromatin level is comparable to compressing a set of files to free up storage space on the drive—the data now takes up much less space, but is more difficult to access. The files can be accessed by decompressing them, but this takes time and energy. One interesting difference between this system and the one used for organizing DNA is that electromagnetic data can be distributed anywhere on a hard drive. A computer file can be separated and scattered about, but cells have no such problem. Each "file" of DNA data (a gene, wrapped around a series of histone spools) is always intact, stored as a single linear sequence of DNA. If a piece of a gene is somehow moved to another location in the DNA (i.e., the file is fragmented), the gene no longer works. Cells don't have a defragment program to stay organized.

CONCEPT CHECK ANSWERS #4

The nucleus provides many benefits to a cell. For example, the nucleus allows a cell to develop a specialized chemical environment that is tailored to the care, maintenance, and use of DNA. Second, the double membrane that defines the border of the nucleus helps shield the interior of the nucleus from physical trauma. This protection is enhanced by the nuclear scaffold, which includes lamin proteins. Third, the nuclear membrane helps organize the DNA via the nuclear lamins, which bind chromatin and DNA.

The costs of having a nucleus are not trivial. Considerable cellular energy must be expended to build and maintain the phospholipids and proteins contained in the nucleus. Also, restricting access to the interior of the nucleus via nuclear pore complexes can slow the time it takes for a cell to respond to an external stimulus, because most molecular traffic in and out of the nuclear interior requires complex machinery that consumes energy.

Concerning the question of whether prokaryotes would be better off if they had a nucleus, arguments can be made for and against. Perhaps the best argument for why a prokaryote could benefit from a nucleus is the fact that nucleated cells (i.e., eukaryotes) are capable of forming a multitude of different multicellular organisms, quite possibly because they can control the expression of a much larger number of genes than are found in prokaryotes. Such control might require a specialized compartment like a nucleus. A strong argument against having a nucleus is its high metabolic cost. Plus, the nucleoid in prokaryotes is typically much smaller than the chromosomes in eukaryotes, which suggests that they are more efficient at storing their genetic information than eukaryotes. Their higher efficiency might make having a nucleus unnecessary.

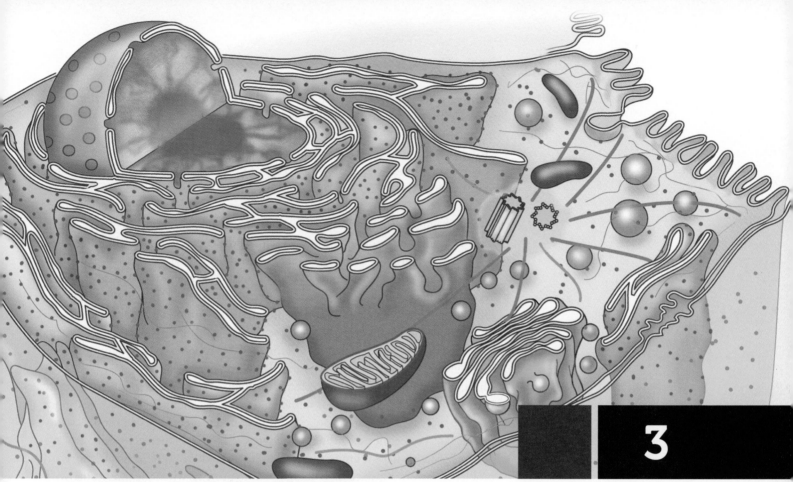

Proteins and Polypeptides

3

3.1 The Big Picture

The goal of this chapter is to introduce the general structure and function of the third type of building block in cells, called proteins. We will begin at the most basic level and work our way up in complexity. Like Chapter 2, it contains four main concepts:

- First, the tremendous diversity of protein structure and function arises from simple polymers of amino acids. This is derived from the relatively simple chemistry used to create these polymers. Similar to our discussions of polysaccharides (Chapter 1), nucleic acids (Chapter 2), and lipids (Chapter 4), our strategy for understanding proteins begins by understanding how their fundamental building blocks are assembled. Learning the structure of amino acids and the peptide bonds holding them together is the primary objective of the first section. Heavy emphasis is placed on memorization of these structures and definitions, because they apply to all proteins.

- Second, proteins are dynamic structures that change shape. There are thousands of different proteins in a typical cell, making it virtually impossible to come up with unique descriptions for every one of these shapes. Instead, the spatial arrangements have been organized into categories, numbered 1 through 4 in increasing complexity, to give us a vocabulary for describing these changes in shape.

- Third, proteins are only useful to cells if they change shape in predictable ways. Here we describe two general ways for inducing these shape changes. Because we cannot describe every type of shape change and function in proteins, we will simply classify proteins into categories according to the types of functions they perform.

- Fourth, proteins have a limited lifetime. Here we address a fairly straightforward problem confronting all cells: What happens to proteins when they wear out? We'll begin our discussion of the answer in this chapter, and pick it up again in later chapters as we learn more about the inner workings of cells.

An important goal of this chapter is to permit students to understand the fundamental structure of all proteins and be able to describe the general spatial arrangement of every protein described in this book. All subsequent chapters assume this knowledge; pay close attention to the new vocabulary introduced here, and practice using it. One easy way to do this is to pick out a few proteins from the later chapters, especially those that appear in figures, and describe their structural organization with the vocabulary introduced in this chapter. At this stage, we'll focus primarily on structure, and address the function of proteins as they appear throughout the text.

3.2 Amino Acids Form Linear Polymers

KEY CONCEPTS

▸ Proteins are composed of polypeptides, which are linear polymers made up of 20 different amino acids.
▸ Amino acids in polypeptides are linked together by peptide bonds.
▸ Amino acids have a characteristic structural polarity that is reflected in the polarity of polypeptides.
▸ In many cases, several polypeptides must join together to form a functional protein.
▸ All proteins exhibit three characteristic traits: (1) they adopt at least two stable three-dimensional shapes, (2) they bind to at least one molecular target, and (3) they perform at least one cellular function.

Amino acids are the building blocks of proteins, analogous to the nucleotide building blocks of nucleic acids. Each amino acid contains a central carbon atom, called the **α-carbon**, attached to four different molecular structures. Two of these are functional groups (an amino

BOX 3-1 TIP

It may be helpful to review the structures of functional groups in Table 1-1 in Chapter 1.

group and a carboxylic acid group, hence the term *amino acid*), and one is a single hydrogen atom (**BOX 3-1**). The fourth structure, often called an amino acid side chain (commonly abbreviated **R** or **R group**), differs in each different amino acid. Proteins are constructed from 20 different amino acids. These 20 amino acids are classified into 3 or 4 groups, according to the chemical nature of their side chains, as shown in Figure 1-14.

The group containing the largest number of amino acids (with nine) is called the *nonpolar* (or *hydrophobic*) group. Note that these amino acid side chains are composed almost entirely of carbon and hydrogen atoms; the sulfur atom in methionine and nitrogen atom in tryptophan do not impart enough polarity to attract water. Six amino acids belong to the *polar* group, and each of the polar side chains contains a polar functional group (hydroxyl, sulfhydryl, or amide). The *ionic* group of five amino acids is sometimes separated into two subgroups: two containing a negatively charged carboxylic acid functional group (acidic subgroup), and three containing a positively charged nitrogen-based functional group (amine or imine—basic subgroup). All five of these side chains contain one charged atom (O^- or H^+).

A Peptide Bond Joins Two Amino Acids Together

Proteins are composed of long, linear sequences of amino acids. These sequences are created by forming a covalent bond between the carboxylic acid group of one amino acid and the amino group of another. One good way to understand how this happens is to practice drawing it, as shown in **FIGURE 3-1**. Start by drawing the α-carbon of one amino acid, and then attach the four functional groups to it. In most drawings such as this, the amino group is placed on the left side, the carboxylic acid group is placed on the right side, a hydrogen atom is placed either above or below, and the side chain is placed opposite the hydrogen. In this case, it doesn't matter which specific amino acid we are drawing, so we can just use the abbreviation R to indicate that a side chain is present. Second, draw another amino acid next to the first, and in the same orientation. Be sure that the carboxylic acid group of the amino acid on the left is near the amino group of the amino acid on the right. This is important, because it makes it easier to see how these two functional groups undergo a chemical reaction to form a covalent bond between the carbon atoms of each group. This reaction, which results in the creation of a water molecule as a byproduct, is called a dehydration reaction (because water has been removed from the initial reactants). The resulting bond is called a peptide bond. Finally, add a box outlining the **amide plane**, as shown.

A Peptide Bond Forms a Rigid, Planar Structure

Peptide bonds have three features that make them especially important in the structure of proteins:

- Because the peptide bond's carbon atom (labeled C_o) is bound to three other atoms (α-carbon [labeled C_α], oxygen, and nitrogen), these atoms are arranged in a triagonal, planar configuration. They are enclosed on the yellow-shaded amide plane in Figure 3-1.
- The bond distance between the C_o carbon and nitrogen atoms is approximately 10% shorter than a typical carbon–nitrogen bond. As a result, this bond has considerable double-bond character, and the nitrogen bonds are also forced into a triagonal, planar configuration. This is also shown in the amide plane in Figure 3-1.

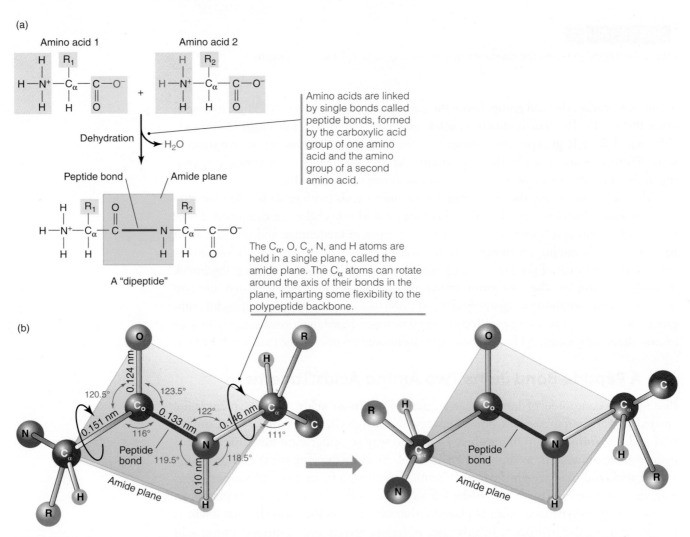

(a)

Amino acid 1   Amino acid 2

Amino acids are linked by single bonds called peptide bonds, formed by the carboxylic acid group of one amino acid and the amino group of a second amino acid.

Dehydration

Peptide bond   Amide plane

A "dipeptide"

The C_α, O, C_o, N, and H atoms are held in a single plane, called the amide plane. The C_α atoms can rotate around the axis of their bonds in the plane, imparting some flexibility to the polypeptide backbone.

(b)

FIGURE 3-1 Creating a peptide bond between two amino acids. (a) A simplified depiction of amino acids to illustrate the formation of a peptide bond. (b) A three-dimensional representation of a peptide bond, demonstrating its planar structure. Adapted from D. Voet and J. G. Voet. Biochemistry, Third edition. John Wiley & Sons, Ltd., 2005. Original figure adapted from R. E. Marsh and J. Donohue, Adv. Protein Chem. 22 (1967): 235–256.

Because these two triagonal configurations overlap, all four affected atoms lie in the amide plane. This means that *the bonds linking amino acids are not flexible.* Keep in mind that each α-carbon is bound to four atoms, so the peptide bonds formed on either side of C_α are arranged in a tetrahedral fashion (see Figure 1-9).

- Although the bonds linking amino acids are not flexible, *the α-carbons can rotate around these bonds.* This is known as **tortional rotation**. This rotation is very important, because it allows a linear chain of amino acids to form many different shapes without compromising the planar structure of the peptide bonds. The drawings at the bottom of Figure 3-1 show two such orientations. Note how different the two shapes are, even though the amide plane is in the exact same orientation. Viewed another way, if one fixed the R groups in place, the planes of the peptide bonds could rotate from one amino acid to the next. The point here is that *the R groups of amino acids can be located in great variety of locations with respect to the peptide bonds, and vice-versa.* This also permits linear sequences of amino acids to adopt characteristic shapes independent of the R groups in the sequences.

- Atoms in peptide bonds are capable of forming hydrogen bonds. Notice that each peptide bond contains an oxygen atom, as well as a hydrogen atom, bound to a nitrogen atom. These atoms can form hydrogen bonds with other molecules, or even with other amino acids in the same protein. This is a very common way to stabilize the structure of proteins. Note that adjacent C=O and H–N groups in the same peptide bond do not form hydrogen bonds with one another, because they are facing in opposite directions. **FIGURE 3-2** shows a number of different hydrogen bonds that can be formed between amino acids.

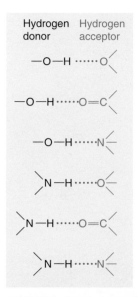

FIGURE 3-2 Some typical hydrogen bonds in proteins. The amino acid residue that supplies the hydrogen atom is designated the donor and the residue that binds the hydrogen atom is designated the acceptor.

▪▪ The Amino Acid Side Chain Does Not Participate in the Formation of a Peptide Bond

All 20 amino acids can form peptide bonds with one another, regardless of the side chain each contains. The reason for this is fairly simple: the peptide bond is formed between an amino group (or imino group: see lysine in Figure 1-14) and a carboxylic acid group (which all 20 amino acids have, in the same orientation), but the side chain is not involved. Because each amino acid has only one amino group and one carboxylic acid group (ignoring the side chains), it can only form one or two peptide bonds, ensuring that the amino acid polymers found in proteins are always organized in a linear fashion. (There are no naturally occurring branching polypeptides, though some have been synthesized in laboratories.) Note that because of the tetrahedral orientation of the bonds formed by the α-carbon, even a straight sequence of amino acids has a characteristic zigzag shape to it.

▪▪ Amino Acids Joined by a Peptide Bond Maintain Structural Polarity

Every amino acid, whether it is part of a peptide bond or not, always has an amino group and carboxylic acid group attached to the α-carbon. This means that amino acids have structural polarity; that is, the functional groups are attached to opposite-facing bonds formed by the α-carbon. Therefore, if several amino acids are linked together by peptide bonds, each of them maintains this polarity. Even the entire polymer has polarity: the amino acid at one end has a free (unbound) amino group, and the amino acid at the other end has a free carboxylic acid group, as shown in **FIGURE 3-3 (BOX 3-2)**. These two ends are called the **amino terminus** and **carboxy terminus** of the polymer, respectively. They are abbreviated in many different ways, such as NH_2 terminus, NH_3^+ terminus, COO^- terminus, COOH terminus, N terminus, and C terminus, or even just N and C.

▪▪ Definitions: Proteins vs. Polypeptides vs. Peptides vs. Subunits

For (mostly) historical reasons, several different names are used to describe amino acids held together by peptide bonds. Each name refers to a specific combination of amino acids. The smallest of these is *two* amino acids held together by a single peptide bond, and it is called a **dipeptide**. Adding additional amino acids to the dipeptide results in the formation of a tripeptide (three amino acids), tetrapeptide (four amino acids), and so on, as shown in Figure 3-3. Notice that the numerical portion of the name (di-, tri-, tetra-, etc.) refers to the number of amino acids in the structure, *not* to the number of peptide bonds.

At some point, scientists stop counting the number of amino acids and simply use the word **oligopeptide** to designate a few amino acids held together by peptide bonds. There is no strict convention for what defines the structure. (Think of the word *group* here.) At yet another somewhat arbitrary point (typically 10 amino acids or more), the structure is called

a **polypeptide**. Regardless of its length, the repeating $\underset{\text{H}}{\text{N}} - \underset{\text{H}}{\text{C}} - \overset{\overset{\text{O}}{\|}}{\text{C}}$ sequence of atoms linking amino acids is called the **backbone**. When discussing proteins, the term *polypeptide* is most appropriate, because most proteins contain more than 20 amino acids (the smallest known human protein has 44), and some proteins contain over 1,500 amino acids.

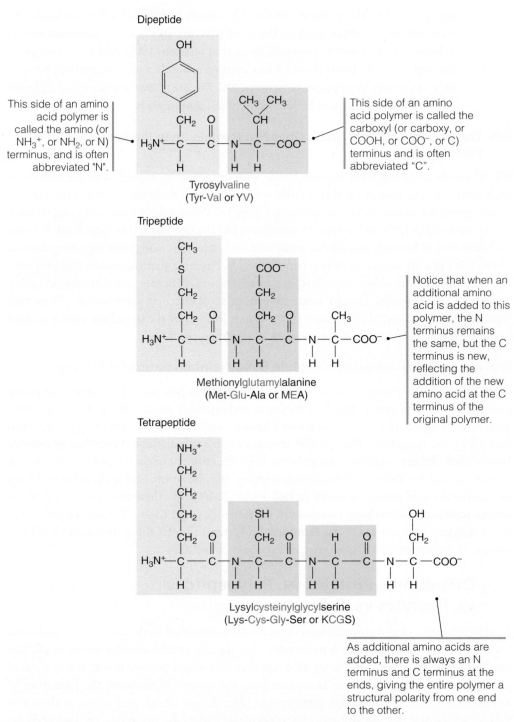

FIGURE 3-3 Conventions for drawing peptides. By convention, the amino acid terminus (N terminus) is on the left and the carboxyl terminus (C terminus) is on the right. Peptides are named as derivatives of the carboxyl terminal amino acid.

■■ Proteins Are Polymers of Amino Acids That Possess Three Important Traits

In many cases, a **protein** consists of a single polypeptide. But this is not always the case, so the two words are *not* interchangeable. When a protein is composed of two or more polypeptides, each polypeptide is called a **subunit**; the term *protein* applies only to the collection of all of the subunits when they become functional. To earn the name protein, any group of amino acids has to have three important traits, as illustrated in **FIGURE 3-4:**

Why are there so many different ways to draw a protein? It is common practice for scientists to draw proteins with only the resolution necessary to illustrate whatever point they are making. In addition to the molecular diagrams used in Figure 3-3, there are at least six other ways to draw proteins, and each has its own particular advantage. The simplest drawings, which look like simple blobs, are used when one is discussing the overall shape of a protein without considering its actual three-dimensional shape. Typically, these blobs are used to illustrate that a protein changes shape under different chemical conditions or when it binds a target, without detailing exactly what kind of shape change occurs. Figure 3-16 is a good example of a blob. Simple lines, as in Figure 3-4, represent the polypeptide backbone and add a bit more resolution by showing roughly where each amino acid in a polypeptide is located in space, but these are usually not very precise. Ribbon drawings (see Figure 3-7C) add additional detail to a line drawing by showing where secondary structures form in a polypeptide, using the standard convention of a cylinder to represent a α-helix, and a flat ribbon (sometimes shaped like an arrow pointing from amino terminus to carboxy terminus) to represent a strand of a β-sheet. Stick diagrams are highly detailed, and show where each atom of a polypeptide is located in three-dimensional space, plus the position of the covalent bonds holding them together; Figure 3-1 is a good example of a stick diagram. Space-filling diagrams (sometimes referred to as van der Waals diagrams) expand on stick diagrams by showing the *volume* occupied by each atom in space (roughly equivalent to the van der Waals radius), as in Figure 3-7A. Finally, sometimes proteins are drawn as cartoons when their three-dimensional structure is well known (e.g., microtubules, showing the microtubule spindle during mitosis). Because molecular diagrams, blobs, and line drawings do not represent the actual structure of proteins, they can be used when the true structure is either not known or does not matter. All of the other drawings require some knowledge of a protein's actual structure, and are therefore much more limited in use since the real three-dimensional structure of most proteins is not known.

■ **PROTEIN TRAIT #1: All proteins adopt at least two stable three-dimensional shapes**. It is obvious that *any* group of amino acids must have at least one shape, and that the greater the number of amino acids it contains, the larger number of possible shapes it can adopt (think of adding links to a chain). What distinguishes proteins from any other amino acid polymers is that proteins organize their amino acids into *stable* shapes. Put another way, each protein adopts a small number of shapes, and tends to use *only* these shapes during its lifetime. Furthermore, every protein composed of the same amino acid sequence in its polypeptide(s) adopts the same small number of shapes. The protein hemoglobin, which we discussed in Chapter 2 (Figure 2-7), is an excellent example: every healthy hemoglobin molecule that has the same amino acid sequence uses the same small number of shapes, and it switches between these shapes depending on whether it is bound to oxygen or not; only mutant forms of hemoglobin form the polymeric fibers that cause sickle-cell disease.

■ **PROTEIN TRAIT #2: All proteins bind to at least one molecular target**. As we discussed in Chapter 2 (*Messenger RNAs Are Translated into Proteins*), proteins are the molecular machines that translate DNA information into some form of cellular activity. While the number of cellular activities performed by proteins is far too large to count, they all share one common property: they take place only if the proteins bind to some other molecule. Students are likely already familiar with one class of proteins where this property is easy to see: **enzymes**. The function of enzymes is to facilitate chemical reactions, and they do so by binding to the reactant(s) via an **active site** on the protein, where the chemical reaction takes place and the product(s) is/are released. An easy way to experience this is to put a starchy food in our mouth and leave it there for a minute without chewing. The enzymes present in saliva bind to the starch and break it down into smaller sugars, partially liquefying the food and making the starch much easier to digest. Note that these

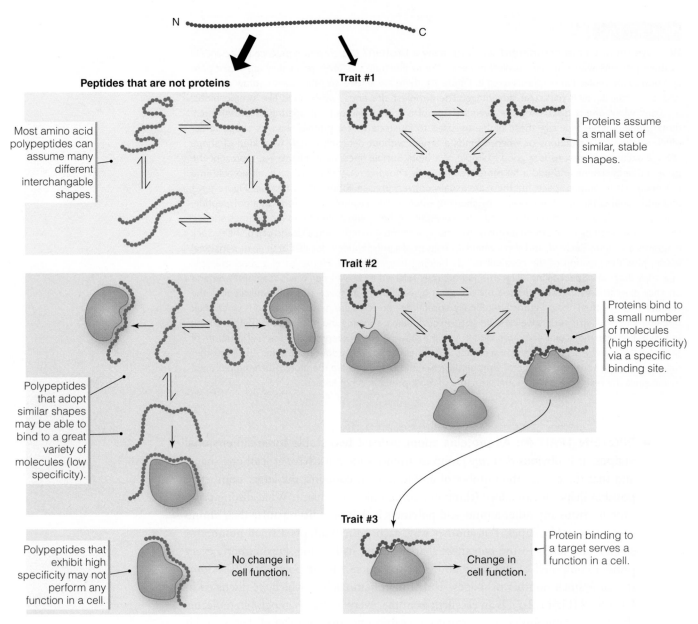

Peptides that are not proteins

Most amino acid polypeptides can assume many different interchangable shapes.

Trait #1

Proteins assume a small set of similar, stable shapes.

Polypeptides that adopt similar shapes may be able to bind to a great variety of molecules (low specificity).

Trait #2

Proteins bind to a small number of molecules (high specificity) via a specific binding site.

Polypeptides that exhibit high specificity may not perform any function in a cell.

No change in cell function.

Trait #3

Change in cell function.

Protein binding to a target serves a function in a cell.

FIGURE 3-4 The three traits of proteins.

enzymes do not digest *everything* they come into contact with: proteins that either bind to a wide variety of molecules (low specificity) or to molecules not found in cells (excessive specificity) are of little use to cells.

■ **PROTEIN TRAIT #3: All proteins perform at least one cellular function.** As we will see in Chapter 8, cells expend a tremendous amount of their metabolic energy on the synthesis of proteins. For this strategy to survive evolution by natural selection, there has to be a substantial return on such a large investment. Cells benefit from making proteins because proteins that adopt useful shapes (sometimes called the native state) perform or control nearly every chemical and physical activity that takes place in cells. It is important to understand that each protein is a *specialist,* performing or assisting with just a few of the countless tasks necessary for a cell to survive and divide. Thousands of different proteins are necessary to maintain a cell in a healthy state.

Anthropomorphism in analogies. Throughout the remainder of this chapter and in the rest of the chapters in this book, there are many analogies for cellular activity in boxes such as this one. Because proteins participate in so many different cellular activities, it is often easy to compare these to everyday activities performed by people. My experience is that students find these anthropomorphized analogies helpful in visualizing what is happening at the molecular level, but it is *very important* not to overinterpret the analogies. Molecules, including proteins, aren't people—they can't see, don't think, don't want, and express no emotions. For these types of analogies to be effective, it is critical to see the underlying biology and chemistry they are helping to explain. As these analogies come up, be sure to be able to trace them back to actual science. If one is unable to, that's a good sign that it's time to leave the analogy and go back to examining the real cells obeying the laws of nature.

Figure 3-4 also shows how these three traits are related; to perform a function, a native protein has to be able to adopt one stable shape when it is *bound* to its target, and another shape when it is *unbound*. More simply, a protein has to be flexible enough to bind *and* release its target. The rearrangement of atomic forces that accompanies the binding and release of a target require this flexibility. Other stable intermediate shapes (sometimes called *states*) may also be necessary for a protein to function properly. There is a balance between stability and flexibility in protein structure.

This is why the first protein trait is so important. Every time a cell synthesizes a polypeptide, there has to be a reasonable probability that the polypeptide will function like all of the other polypeptides of the same sequence (e.g., a newly synthesized hemoglobin will function like all of the other hemoglobins before it). If a polypeptide could adopt a different shape every time it was made, cells would have no way of controlling what it will do, and such a randomly organized polypeptide would not perform *any* cellular function well. Stated more simply, unstable proteins equal chaos, and chaos cannot survive billions of years of natural selection. If there was ever a protein that behaved in a nonpredictable fashion, it likely no longer exists in cells.

This is also why the second protein trait is important. For a protein to participate in any chemical reaction or physical activity, it has to come into contact with its designated target and bind to it in a predictable fashion. Because of trait #1, proteins can form finely tuned regions in their structure (typically called **binding sites**) that only bind to their relevant target molecules. We saw a few examples of binding specificity in Chapter 2, like when we discussed proteins that bind to and modify histones (see *DNA Is "Silenced" in Heterochromatin*). If a protein could bind to *any* molecule, it is highly unlikely it would ever be efficient enough to be of any real use to a cell. How proteins bind and interact with their targets is of such central importance to biology that an entire subdiscipline (**biochemistry**) is devoted to understanding these interactions (**BOX 3-3**).

Subunits Are Polypeptides That Assemble Together to Form a Single Protein

To discriminate between proteins made up of a single polypeptide and those composed of two or more subunits (**FIGURE 3-5**), four additional terms are used. A protein containing a single polypeptide is called a **monomer** (adjective, **monomeric**), while a protein containing two subunits

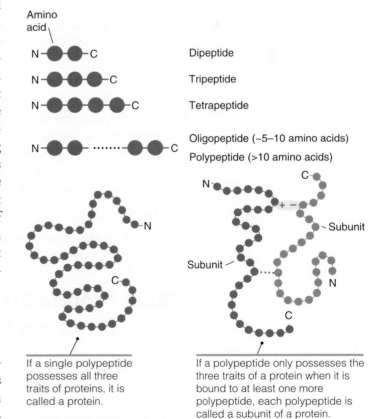

Amino acid

Dipeptide

Tripeptide

Tetrapeptide

Oligopeptide (~5–10 amino acids)
Polypeptide (>10 amino acids)

Subunit

Subunit

If a single polypeptide possesses all three traits of proteins, it is called a protein.

If a polypeptide only possesses the three traits of a protein when it is bound to at least one more polypeptide, each polypeptide is called a subunit of a protein.

FIGURE 3-5 Polypeptides may be proteins or protein subunits.

Understanding protein names. One of the most intimidating aspects of cell biology, especially for students new to the field, is the tremendous number of new words one has to learn to understand the subject matter. This is especially true for protein names. Let's have a look at where all these names come from. The generic names (dimer, oligomer, polymer) are borrowed from organic chemistry and apply to any molecule made up of many subunits (PVC, the material used to make many different types of pipes, is a *polymer* of vinyl chloride). The specific names for individual proteins don't follow any formal rules, but there are some trends that can help. Many protein names end with the suffix *–in* (e.g., actin, tubulin, lamin). This won't help to determine what the protein does, but it will hint that a new word may be a protein name. A second common suffix is *–ase,* and this is applied to *enzymes.* Often, the name includes the substrate for the enzyme (e.g., collagen*ase,* enol*ase,* protein*ase*). A word we will see often is *kinase,* and this means "enzyme that attaches a phosphate to . . .". Kinases are often named for their substrates (the molecules that receive the phosphate), such as protein kinase.

One can name a protein any way they like. Traditionally, the person or group that first identifies and describes a new protein chooses a name for it and uses this name in their publications and lectures, until it is eventually adopted by the greater scientific community. When two or more groups discover the same protein at about the same time, a name contest exists until everyone agrees to adopt one name. Protein names fall into three general categories. The oldest of these relies upon the use of Latin and Greek prefixes and suffixes to describe the known or suspected function of the protein at the time it was discovered (e.g., laminin, fibronectin, hexokinase). Many of these names also include reference to the organic chemistry in its structure or in the molecule it binds to (e.g., transglutaminase, superoxide dismutase); one of the longest of these names has over 1,100 letters and is virtually unpronounceable. A second category that appeared in the middle of the 20th century uses English words in much the same fashion, but introduces acronyms (e.g., Transforming Growth Factor: TGF, Heat Shock Protein: HSP). A third naming convention, which became popular beginning in the 1980s, involves a sense of humor, borrows the acronym strategy, and pokes fun at the stodgy tone of the first category of names. One of the first proteins named this way was Sonic Hedgehog, after the videogame of the same name, and this was followed by another protein named Tiggywinkle Hedgehog, after a character in a children's book by Beatrix Potter. Other protein names in this class include Bad, Boo, CARDIAK, Casper, CLAP, DEDD, MADD, SODD, TANK, TRAMP, TRANCE, and TWEAK.

When encountering a new protein name, don't be intimidated (see Box 1-3). These days, remembering the name of a protein is far less important than understanding its three traits. If the name of a protein gets in the way of understanding its function, call it whatever name is easiest. Then, once it is understood, memorize the real name.

---

is called a **dimer** (adjective, **dimeric**). In some cases, two polypeptides may associate with one another long after they have been synthesized; when this happens, polypeptides are said to dimerize (for example, two subunits of a *dimeric* protein can *dimerize*). If a protein contains three or more subunits, it can be named in a variety of ways: by number of subunits (trimer, quadramer, pentamer, etc.), or by the more general terms (multimer, multisubunit, oligomer, or polymer) (**BOX 3-4**).

CONCEPT CHECK #1

Are the three traits universal? Briefly review the structure of nucleic acids (Chapter 2) and build an argument to answer this question: Do the three traits of proteins also apply to nucleic acids?

▮ 3.3 ▮ Protein Structure Is Classified into Four Categories

KEY CONCEPTS

▸ A vocabulary has been devised for describing types, or classes, of distinctive patterns in protein structure.

▸ The primary structure of any polypeptide is the order (sequence) of its amino acids, from amino terminus to carboxy terminus.

- The secondary structure of a polypeptide describes the three-dimensional arrangement of its primary sequence. The three categories of secondary structure are α-helix, β-sheet, and random coil. Some combinations of secondary structure are called motifs.
- The tertiary structure of a polypeptide is the three-dimensional arrangement of secondary structures. Certain three-dimensional arrangements are shared by several different proteins, and these are called domains.
- Quaternary structure describes the three-dimensional arrangement of polypeptides in a multi-subunit protein.
- Five classes of chemical bonds stabilize polypeptide structure.

Despite the myriad functions performed by proteins, all proteins share enough structural similarities that scientists have developed a vocabulary for describing these features. We can also use this vocabulary to compare different proteins and reach some general conclusions about how the structure of a protein can contribute to its function. There are four categories of protein structure, beginning with the simplest and ending with the most complex, as shown in **FIGURE 3-6**.

Primary Structure Is Defined by the Linear Sequence of Amino Acids

As stated above, all naturally occurring proteins contain at least one polypeptide, which is composed of a linear sequence of amino acids. The first category of a protein's structure (also called its **1°** [**primary**] **structure**) is therefore simply the sequence of amino acids in its polypeptide(s). Remember that polypeptides are synthesized from the amino terminus to the carboxy terminus by ribosomes (see Chapter 2, **Messenger RNAs Are Translated into Proteins**, and Chapter 8, *Proteins Are Synthesized by Rimosomes from an mRNA Template*) by adding one amino acid at a time to the carboxy terminus of a growing polypeptide. Every polypeptide synthesized by a ribosome from a different mRNA has its own unique sequence of amino acids and corresponding 1° structure. When written out in conventional text, the primary sequence always reads left to right, with the amino terminus at the start and the carboxy terminus at the end.

Secondary Structure Is Defined by Regions of Repetitive, Predictable Organization in the Primary Structure

The orientation of 1° structure in three-dimensional space is called the **2°** (**secondary**) **structure** of a protein. As we stated earlier, the tortional rotation of the α-carbon atoms in peptide bonds permits great rotational freedom along the polypeptide backbone. However, somewhat surprisingly, only *two* characteristic orientations appear in proteins, and they are named the **α-helix** and **β-sheet**. A third type of 2° structure, called the **random coil**, consists of all of the rest of the configurations adopted by linear sequences of amino acids (i.e., anything that is not α-helix or β-sheet).

An Alpha Helix Is Shaped Like a Coil

The α-helix is the most common 2° structure, and is shaped like a coil or spring. Because there are 3.6 amino acids per complete turn of the α-helix, the backbone atoms in every fourth amino acid come in relatively close contact, and form hydrogen bonds (N–H•••••O=C) with one another. These hydrogen bonds stabilize the α-helix, and because they can form between any amino acids, there is no theoretical limit to how long an α-helix can be. Notice that the R groups project *outward* from the long axis of the helix. This has special importance for some membrane proteins, as we will discuss in Chapter 4 (see *Membrane Proteins Associate with Membrane in Three Different Ways*). In the so-called ribbon model diagrams of protein structure, α-helices are abbreviated by a corkscrew pattern, as shown in **FIGURE 3-7**.

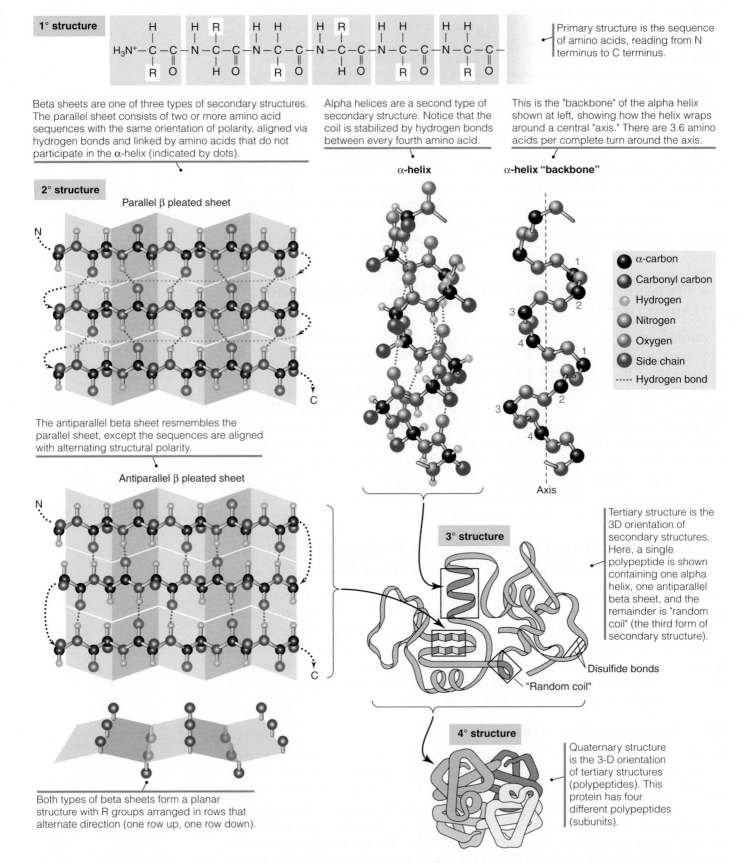

1° structure

Primary structure is the sequence of amino acids, reading from N terminus to C terminus.

Beta sheets are one of three types of secondary structures. The parallel sheet consists of two or more amino acid sequences with the same orientation of polarity, aligned via hydrogen bonds and linked by amino acids that do not participate in the α-helix (indicated by dots).

Alpha helices are a second type of secondary structure. Notice that the coil is stabilized by hydrogen bonds between every fourth amino acid.

This is the "backbone" of the alpha helix shown at left, showing how the helix wraps around a central "axis." There are 3.6 amino acids per complete turn around the axis.

α-helix

α-helix "backbone"

2° structure

Parallel β pleated sheet

- α-carbon
- Carbonyl carbon
- Hydrogen
- Nitrogen
- Oxygen
- Side chain
- ⋯ Hydrogen bond

The antiparallel beta sheet resmembles the parallel sheet, except the sequences are aligned with alternating structural polarity.

Antiparallel β pleated sheet

Axis

3° structure

Tertiary structure is the 3D orientation of secondary structures. Here, a single polypeptide is shown containing one alpha helix, one antiparallel beta sheet, and the remainder is "random coil" (the third form of secondary structure).

Disulfide bonds

"Random coil"

Both types of beta sheets form a planar structure with R groups arranged in rows that alternate direction (one row up, one row down).

4° structure

Quaternary structure is the 3-D orientation of tertiary structures (polypeptides). This protein has four different polypeptides (subunits).

FIGURE 3-6 The four categories of protein structural organization. Adapted from B. E. Tropp. Biochemistry: Concepts and Applications, First edition. Brooks/Cole Publishing Company, 1997.

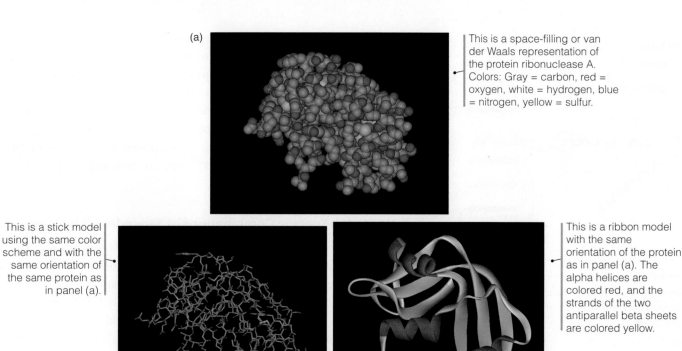

(a)

This is a space-filling or van der Waals representation of the protein ribonuclease A. Colors: Gray = carbon, red = oxygen, white = hydrogen, blue = nitrogen, yellow = sulfur.

This is a stick model using the same color scheme and with the same orientation of the same protein as in panel (a).

This is a ribbon model with the same orientation of the protein as in panel (a). The alpha helices are colored red, and the strands of the two antiparallel beta sheets are colored yellow.

(b)   (c)

FIGURE 3-7 Three ways of representing the structure of a protein. Structures from Protein Data Bank 1JVT. L. Vitagliano, et al., Proteins 46 (2002): 97-104. Prepared by B. E. Tropp.

▌▌ A Beta Sheet Forms a Plane

A β-sheet is the second stable arrangement of an amino acid sequence. As in an α-helix, hydrogen bonds within the polypeptide backbone stabilize the β-sheet structure, but these hydrogen bonds do not form between every fourth amino acid; instead, they form between a row of "downstream" amino acids (they are often called *strands*) that align with the first row. There are two forms of β-sheet: a *parallel* β-sheet forms between amino acid sequences with the *same* structural polarity (i.e., the amino and carboxy terminal ends of each strand of the β-sheet are aligned), and an *antiparallel* β-sheet contains strands with *alternating* polarities, as shown in Figure 3-6.

In both types of β-sheet, notice that the zigzag pattern in the backbone of each strand aligns, giving the entire β-sheet the appearance of a plane, or sheet, that has been folded back and forth (i.e., pleated). These structures are often called **β-pleated sheets** for this reason. Notice too that the folds occur at the α-carbons, such that the R groups project outward from the sheets in rows, and that these rows of R groups alternate from one side of the sheet to the other at each fold. Also, like the α-helix, there is no theoretical limit to how many strands a β-sheet can have.

β-sheets can adopt a number of different shapes. The larger the β-sheet is, the more different shapes it can adopt—in one remarkable example, a β-sheet can be so large that it can be folded into a tube-like shape, forming what is known as a **β-barrel**. Other characteristic shapes of β-sheets exist as well, and we'll see some examples in later chapters. In many models of protein structure, the strands of a β-sheet are illustrated by ribbon-like shapes, as seen in **FIGURE 3-7**. In some cases, the ribbons include an arrow pointed toward the carboxy terminus of the polypeptide to assist with orientation, as shown in **FIGURE 3-8**.

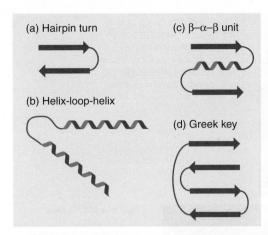

FIGURE 3-8 Examples of common motifs in proteins. Alpha helices are represented by coiled ribbons, strands of beta sheets are represented by arrows pointed from N terminus to C terminus of the polypeptide.

(a) Hairpin turn

(b) Helix-loop-helix

(c) β–α–β unit

(d) Greek key

Another important difference between the α-helix and the β-sheet is that the α-helix is formed by an uninterrupted sequence of amino acids, while the β-sheet, by necessity, is formed by strands of amino acids separated by sequences long enough to form the loops necessary for aligning the strands. The shortest of these loops contains only two amino acids, and is called a *hairpin loop* between strands of an antiparallel β-sheet. For parallel β-sheets, the loops must be even longer than the strands to allow for proper strand orientation. In some cases, these loops are extremely long, and may even contain one or more α-helices.

■■ Motifs Are Specific Combinations of Secondary Structures

The fact that there is no theoretical limit to the size of α-helices or β-sheets raises an important question: Why aren't all proteins made up entirely of these structures? There is no definitive answer to this question yet, but one possibility is that such a protein would be *too stable* to be useful. Because proper protein function requires a balance between stability and flexibility, it makes sense that proteins would include α-helices and β-sheets, but would mix them in with more flexible structures. These non-α-helix or non-β-sheet structures are of course highly variable in shape, but they have a common name: random coil. Though there is no specific definition for a random coil, most biologists consider it a type of 2° structure, as every part of a polypeptide must belong to one of these categories.

In many cases, α-helices, β-sheets, and random coils form characteristic substructures within a protein. These structures are referred to as **motifs**, and Figure 3-8 shows examples of some common motifs found in proteins. Some of these are used to fit into the major groove and minor groove of the DNA double helix (see Figure 2-3). Keep in mind that motifs are *not* proteins; they are simply parts of a protein that have a characteristic shape and contribute to the function of the protein.

■■■ Tertiary Structure Is Defined by the Arrangement of the Secondary Structures in Three Dimensions

The orientation of 2° structures in three-dimensional space is the **3° [tertiary] structure** of a protein. For proteins composed of a single polypeptide, tertiary structure is essentially the overall shape of the entire protein. For multi-subunit proteins, the term 3° structure applies to each subunit, but not to the overall protein. Determining the actual tertiary structure of a protein is still quite difficult, as most protein tertiary structures are determined by growing crystals of proteins and using a mathematical technique, known as Fourier transformation, to analyze a pattern of X-rays deflected off of the crystals. However, this requires scientists to grow a large crystal of a protein in the lab; proteins do not crystallize under normal physiological conditions, so it is extremely difficult to find chemical conditions that will promote the formation of a crystal stable enough to withstand X-ray bombardment. As a result, relatively few proteins have been analyzed in this amount of detail. The 3° structure of relatively small polypeptides (<40 kilodaltons) can be determined by a technique called Nuclear Magnetic Resonance (NMR) spectroscopy, and portions of larger proteins have been analyzed with this instrument.

■■ Most Proteins Require Assistance to Fold into Their Proper Tertiary Structure

Under normal cellular conditions, the 1° structure of a polypeptide will spontaneously organize itself into a stable 2° structure. Alpha helices form quite easily, while β-sheets typically

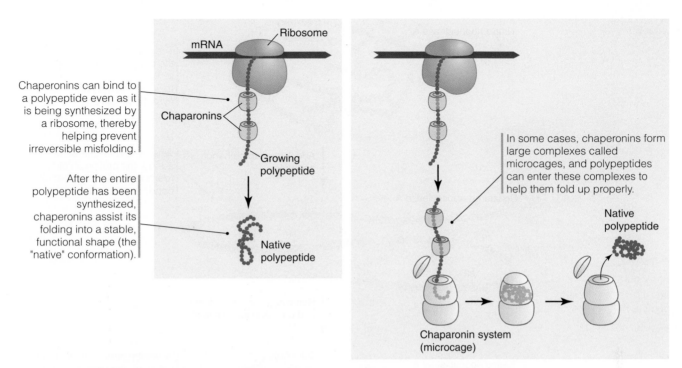

Chaperonins can bind to a polypeptide even as it is being synthesized by a ribosome, thereby helping prevent irreversible misfolding.

After the entire polypeptide has been synthesized, chaperonins assist its folding into a stable, functional shape (the "native" conformation).

mRNA

Ribosome

Chaparonins

Growing polypeptide

Native polypeptide

In some cases, chaperonins form large complexes called microcages, and polypeptides can enter these complexes to help them fold up properly.

Native polypeptide

Chaparonin system (microcage)

FIGURE 3-9 Chaperonins are proteins that assist in the proper folding of newly synthesized polypeptides. Adapted from Young, J. C., et al., Nat. Rev. Mol. Cell Biol. 5 (2004): 781-791.

take more time to find a stable configuration; it is likely that motifs require even longer to organize, but no metabolic energy is required for these structures to form.

Moving up from 2° structure to 3° structure is a different matter, however. Poly-peptides become functional when they reach their proper 3° structure; as we discussed above, to be useful to a cell, a polypeptide/protein needs to adopt a three-dimensional shape that meets the requirements of both being stable enough to perform the same function throughout its lifetime and being flexible enough to bind and release its target(s). Typically, a polypeptide could adopt a number of a different stable shapes that would not be useful to a cell. In effect, a newly synthesized polypeptide has to sort through a number of possible configurations until it finds a useful one, and to find the right shape, most proteins get help from a group of proteins specialized to assist in this sorting, as shown in **FIGURE 3-9** (also **BOX 3-5**). These proteins are known as **chaperonins**, with one very well-known family of chaperonins called **heat shock proteins**, or **HSPs**. Chaperonins usually use ATP energy to chaperone a polypeptide as it folds into the proper 3° structure, and some even help a properly folded polypeptide retain its shape in the face of a denaturing agent (such as elevated temperature).

This observation raises an important question: If most proteins require already-folded chaperonins to fold properly, how do the chaperonins form? One can invoke yet another type of protein to help the chaperonins, but this only moves the question from one protein to another. What helped the first protein to fold up? Early studies on the chemistry of protein

---

BOX 3-5

The personal trainer analogy. If we think of a newly synthesized protein as a person who is performing a set of stretching exercises, then a chaperonin would be like a personal trainer, assisting in the stretching and bending so that the person doesn't hurt themselves. By restricting movement to only those shapes that are healthy, a chaperonin helps ensure that a protein assumes a functional shape after the folding is complete.

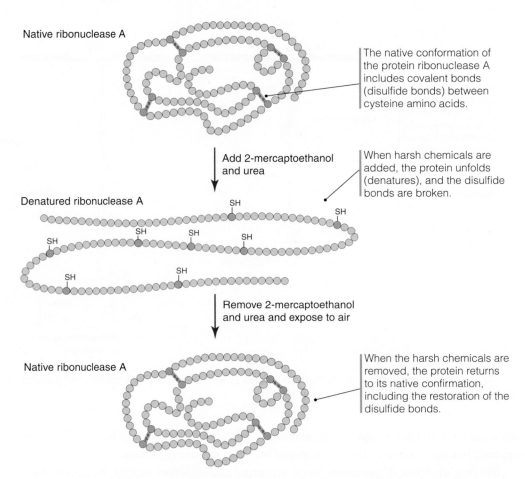

Native ribonuclease A

The native conformation of the protein ribonuclease A includes covalent bonds (disulfide bonds) between cysteine amino acids.

Add 2-mercaptoethanol and urea

When harsh chemicals are added, the protein unfolds (denatures), and the disulfide bonds are broken.

Denatured ribonuclease A

SH SH SH SH SH SH SH SH

Remove 2-mercaptoethanol and urea and expose to air

Native ribonuclease A

When the harsh chemicals are removed, the protein returns to its native confirmation, including the restoration of the disulfide bonds.

FIGURE 3-10 Dr. Anfinsen's experiment demonstrating spontaneous refolding of a denatured protein. Adapted from L. A. Moran and K. G. Scrimgeour. Biochemistry, Second edition. Prentice Hall, 1994.

folding demonstrated that not every protein requires help to fold properly. In the 1950s, Christian B. Anfinsen showed that a protein called RNase would denature (unfold) in the presence of harsh chemicals (urea and 2-mercaptoethanol), and then spontaneously refold once the chemicals were removed, as shown in **FIGURE 3-10**. After a great deal of work, he concluded that "the native [functional] conformation is determined by the totality of inter-atomic interactions and hence by the amino acid sequence, in a given environment."[1] He was awarded the Nobel Prize for this work in 1972, and a great deal of what we now know about protein structure is based on his work. His observations form the foundation for most of the material in this chapter.

■■ A Domain Is a Portion of a Protein That Adopts a Characteristic Shape

As a protein folds into its proper shape, it may form distinct substructures that participate in a subset of the protein's activities. These structures take on shapes independent of the rest of the protein, and many function semiautonomously. These substructures, called domains, exist in a multitude of different proteins, suggesting they are especially beneficial to proteins (**BOX 3-6**). One common usage of domain refers to "that portion or region of a protein necessary to perform a specific task." **FIGURE 3-11** shows some examples of domains found in proteins.

[1] www.nobelprize.org/nobel_prizes/.../laureates/1972/anfinsen-lecture.

▓▓ Many Proteins Are Classified into One of Three Tertiary Structure Categories

While the tens of thousands of proteins typically expressed by a cell can adopt tens of thousands of different 3° structures, most of them conveniently fall into three general categories, as shown in **FIGURE 3-12**. **Globular proteins** adopt a condensed, rounded shape. These proteins typically contain hydrophobic amino acids in their interior, surrounded by a shell of hydrophilic amino acids that interact with the water and other polar molecules making up the greatest percentage of a typical cell's mass. Globular proteins, including the histones and histone-modifying proteins mentioned in Chapter 2, perform a tremendous number of functions in cells, including replication and repair of DNA and generation of metabolic energy from food molecules. **Fibrous proteins** are rich in α-helices and β-sheets and often form filamentous complexes that are exceptionally resistant to mechanical force; the nuclear lamins mentioned in Chapter 2 are good examples of fibrous proteins. Transmembrane proteins span both sides of a cellular membrane and play a critical role in the transmission of information and mechanical forces across membranes. Cell surface receptors are good examples of transmembrane proteins.

(a)

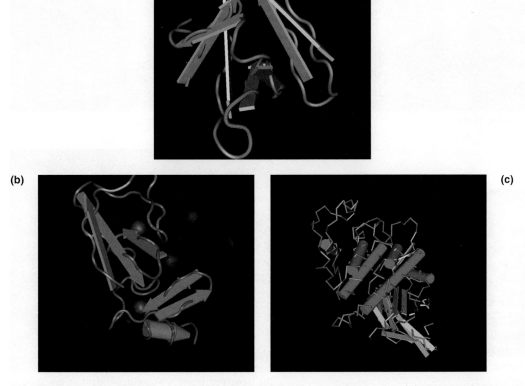

(b)

(c)

FIGURE 3-11 (a) SH3 domain, (b) LIM domain, (c) Kinase domain of cyclin dependent protein kinase I. Examples of domains in proteins. Note that domains contain all three types of secondary structures.

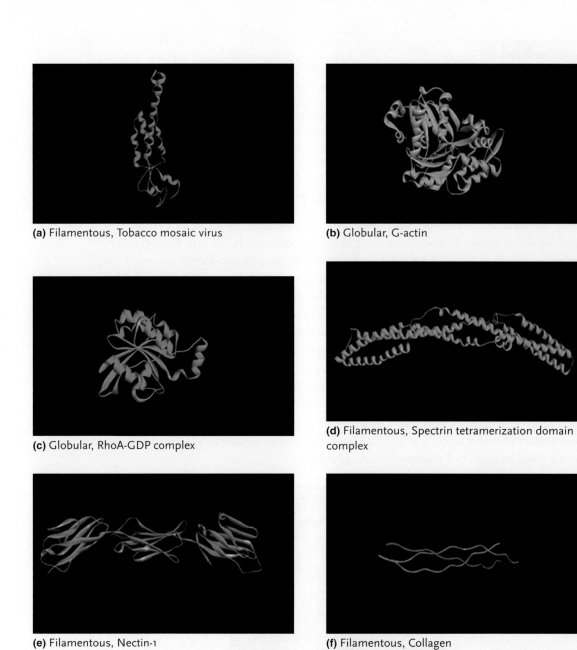

(a) Filamentous, Tobacco mosaic virus

(b) Globular, G-actin

(c) Globular, RhoA-GDP complex

(d) Filamentous, Spectrin tetramerization domain complex

(e) Filamentous, Nectin-1

(f) Filamentous, Collagen

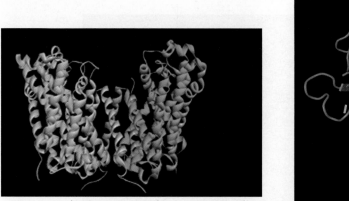

(g) Transmembrane, Transmembrane Amino Acid TransporterT

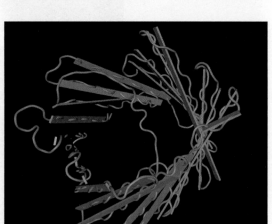

(h) Transmembrane, voltage-dependent anion channel

FIGURE 3-12 Examples of the three structural classes of proteins.

Quaternary Structure Is Defined by the Three-Dimensional Arrangement of Polypeptide Subunits in a Multimeric Protein

The fourth category of protein structure, called **4°** (**quaternary**) **structure**, is reserved for proteins that contain multiple polypeptide subunits. Quaternary structure is used to describe the spatial arrangement of each subunit in these proteins (see Figure 3-6). Different arrangements of the subunits can have a significant influence on the shape and function of these proteins, a concept known as **cooperative interaction**. An excellent example is the protein used by the mammalian immune system to fight disease, the antibody. An antibody protein is made up of four subunits (two copies of "heavy chains" and two copies of "light chains") as shown in **FIGURE 3-13**; the light and heavy chains cooperate to form binding sites for foreign molecules at the tips of the "arms" in the antibody.

Five Classes of Chemical Bonds Stabilize Protein Structure

Trait #1 of proteins is that they adopt *stable* three-dimensional shapes. What makes a particular shape stable? The laws of thermodynamics state that all molecules possess energy (the sum of kinetic, potential, and heat energy), that this energy can be converted from one form into another, and that each molecule seeks an energy state with

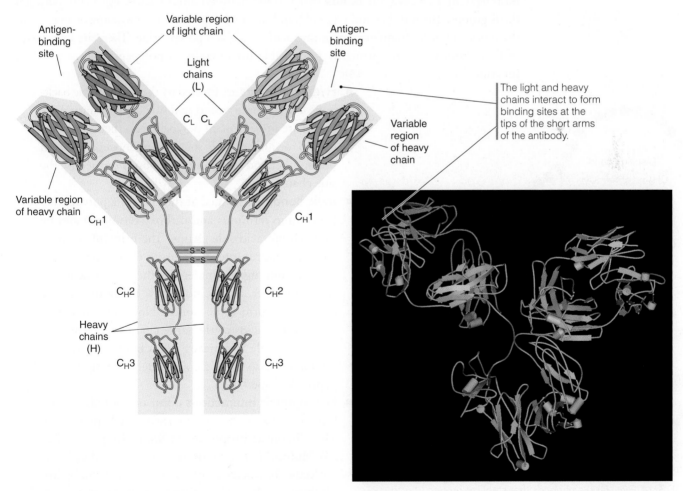

FIGURE 3-13 Two views of quarternary structure in an antibody. This protein is characterized by two large subunits (also known as heavy chains) and two small subunits (also known as light chains) that form a Y-shaped molecule. Protein Data Bank ID: 1IGY. Harris, L. J., Skaletsky, E., and McPherson, A., J. Mol. Biol. 275 (1998): 861–872.

the greatest degree of **entropy** (approximately the number of distinct energy states it can adopt). Each shape of a protein represents a certain energy level, so some shapes are more favorable than others. Moving from one stable shape to another will likely decrease the entropy of the protein; depending on how much the entropy decreases, shape changes like these are considered either unlikely or, in extreme cases, nearly impossible. Plus, when a protein adopts a shape with high entropy, it tends to remain in that shape unless new energy (e.g., heat) is added.

This is why boiling most proteins permanently inactivates them. The additional heat energy permits them to adopt almost any shape possible, regardless of how stable these shapes are. Once the proteins cool, they adopt the highest entropy form that is easiest to reach, and most of these stable forms are not functional. Even adding more energy to the system (reheating) will not cause most of these proteins to revert to a functional state, because there may be several other nonfunctional shapes with nearly the same energy level.

It therefore seems highly unlikely that a polypeptide would adopt a stable, functional shape when so many other stable (but not functional) shapes are available. Yet we know that evolution has selected for polypeptides that *do* prefer a functional shape. How do we reconcile this? Here is where Dr. Anfinsen's conclusions about what determines protein configuration (i.e., shape) can help. He said that "the totality of interatomic interactions and hence . . . the amino acid sequence" play a critical role in controlling protein shape. The key here is the term *interatomic interactions*. The kind of interatomic interactions he was referring to are the **noncovalent** and **covalent bonds** that formed between amino acids, especially between the R groups. The number and types of bonds formed depend, at least to some degree, upon the types and order (sequence) of amino acids in a given polypeptide. There are five classes of these bonds that are primarily responsible for directing a polypeptide into a stable, functional shape (**FIGURE 3-14**):

- **Hydrogen bonds** are formed by the polypeptide backbone and by some amino acid side chains. Since these bonds are present in all polypeptides, their contribution to forming a stable shape is largely dependent on the number and location of the bonds formed by the side chains.

- **Ionic bonds** are formed at the ends of a polypeptide (the amino terminus and carboxy terminus) and by ionic amino acid side chains. The principle is quite straightforward: ions of opposite charge attract one another, pulling the molecules they are attached to closer together. Again, note how many of the side chains can form ionic bonds. Note, too, that two ionic amino acids of the same charge will repel one another. While technically not a bond, this type of interaction also plays a role in determining a polypeptide's shape.

- **Hydrophobic interactions** occur in nonpolar amino acid side chains. As we discussed in Chapter 1 (see **The Chemical Properties of Water Impact Nearly All Molecular Interactions in Cells**), hydrophobic molecules have a tendency to move away from hydrophilic molecules such as water. For most proteins surrounded by water (as in the cytosol or extracellular space), the hydrophobic side chains cluster together

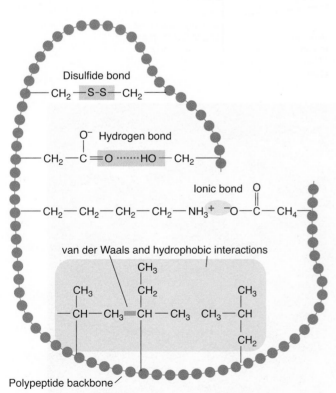

FIGURE 3-14 Five classes of bonds that stabilize protein structure.

CHAPTER 3 Proteins and Polypeptides

in the interior of a protein. Some multi-subunit proteins use hydrophobic interactions to bond their subunits together, and transmembrane proteins use hydrophobic interactions to anchor themselves in the membrane.

- **van der Waals forces** are responsible for a group of attractive interactions between molecules. One clear example is very similar to hydrogen bonding: two atoms that contain some charge polarity (but are not fully ionized) can attract one another if the positive side of one is closest to the negative side of the other (hydrogen need not be present in this case). Likewise, these atoms can repel one another if their positive sides or negative sides are closest. The result is that these atoms reach an optimal distance from one another where the attractive and repulsive forces are equal. (Other types of interactions included in this group are based on quantum physics and are beyond our scope here.) Because they have the potential to occur between nearly all atoms in a polypeptide, they are not selective for any one protein shape, but they do help stabilize a shape once it is formed.

- **Disulfide bonds** are formed by the side chain of cysteine, which contains a sulfhydryl group. When the side chains of two cysteines are close together, they have the potential to form a covalent bond between them, called a disulfide bond because the bond is between two sulfur atoms.

Note that the first four of these five bond types are noncovalent, so their bond energies are 10- to 20-fold weaker than the disulfide bond, which is covalent. However, noncovalent bonds are much more abundant in proteins than disulfide bonds, and therefore typically contribute more to the overall shape of a polypeptide. A final note here is that the number and types of bonds formed depends on the amino acids in a polypeptide and the order in which they are arranged (i.e., the 1° structure). If, for example, a polypeptide contains one large stretch of hydrophobic amino acids, they will tend to cluster together in one area. If the same number of hydrophobic amino acids is scattered throughout the 1° structure, the same hydrophobic interactions will drive them together, but they may form many small clusters rather than a single large cluster. This is what Dr. Anfinsen meant when he mentioned the importance of amino acid sequence in determining a protein's shape.

CONCEPT CHECK #2

The 1°–4° structure vocabulary was designed to help scientists describe the structure–function relationship in proteins. Recall our discussion of the sickle-cell mutation in Chapter 2, and use the vocabulary in this section to describe how the sickle-cell mutation affects the structure and function of hemoglobin.

▇3.4▇ Changing Protein Shape and Protein Function

KEY CONCEPTS

▸ To be functional, proteins must be able to be activated and inactivated. These two functional states are reflected by at least two subtly different shapes. Many proteins adopt several intermediate shapes as well.

▸ To induce and control the switch from one protein shape to another, cells chemically modify the amino acids in a protein. The type and duration of these modifications varies considerably for different proteins.

The "cell as a busy city" analogy, revisited. In Chapter 1, Box 1-2, we introduced the busy city analogy for examining the big picture of a cell's activities. The analogy applies here as well, in that a city populated by thousands of people does not need *all* of its inhabitants to be actively working every minute of the day. In fact, such a condition would be potentially catastrophic: if mechanics tried to repair and maintain taxicabs while they were simultaneously being used by taxi drivers to deliver passengers, it is likely that neither job would be successfully completed, and great harm could result. Some workers must stop so that others can work, or neither set of workers will be completely successful. The exact same principle applies to proteins in cells: if every protein was functionally active at the same time, very little productive work could be done. So, there must be a mechanism for controlling the activity of these proteins, and in fact it is very simple: every protein can be switched on or off by changing its shape. One good example of a situation in which proteins never stop working is cancer.

▸ The formation of new covalent bonds in a polypeptide is a common method for inducing long-term changes in protein structure and function. Most of these bonds are formed between the amino acid side chains and functional groups in other molecules.

▸ Noncovalent binding of additional molecules, via the same kinds of bonds that maintain stability in protein structure, is a means of altering protein structure and function for short lengths of time (<1 second).

▸ Proteins are classified into nine categories according to their function. Some categories of proteins share common structural features as well.

Now that we have established how and why proteins adopt stable shapes, let's focus on how and why these stable shapes *change* during a protein's lifetime. The reason every protein changes shape is simple: even though every protein has a function, it does not necessarily have to perform that function all of the time. It is important to understand that *inactivity* is just as important as *activity* for a protein to be useful (**BOX 3-7**).

All Proteins Adopt at Least Two Different Shapes

A key to cell survival is being able to establish and maintain an internal environment that is different from the environment outside the cell. Most of the machinery responsible for doing this consists of proteins. Maintaining a constant internal environment requires careful coordination of a multitude of chemical reactions, and this requires the ability to activate and inactivate these reactions as needed. Proteins must therefore have at least one inactive shape to complement its active shape. Both of these shapes have to be stable and interchangeable, or the protein will be of no use to a cell.

A central tenet of biochemistry and cell biology, derived from the three traits of proteins, is that all proteins bind to something; when they do bind, they change their shape; and when they change their shape, they change their function. Based on what we have already covered, it is clear that not all shape changes apply here—only those that impact the functional state of a protein. Discriminating useful (functional) shape changes from the rest, and determining the exact functional state of each, are some of the most important challenges in cell biology. As we will see below, cells have devised very clever ways to change the shapes of their proteins, which permits them to regulate their activity with great precision.

Proteins Change Shape in Response to Changes in Their Environment

Because noncovalent bonds play such an important role in determining the 3° and 4° structure of proteins, changes in the environment that affect these bonds have a tremendous impact on the shape and function of proteins. Not all bond types are equally sensitive to these changes, and scientists take advantage of this to selectively disrupt bonds in proteins. For example, hydrogen bonds are quite stable and subject to interference or inhibition by only a few common chemicals (e.g., Dr. Anfinsen used urea to disrupt hydrogen bonds in RNase.) The same is true for van der Waals forces and hydrophobic interactions. When scientists want to isolate proteins in cells, they commonly use a solution containing **detergents** that can disrupt hydrophobic bonds to dissolve the membranes and release the membrane proteins. One common environmental change that affects these bonds is change in temperature—as a cell heats up, the extra energy causes the protein to vibrate more, which may separate the atoms forming the bond. Conversely, cooling a cell will decrease protein vibration, stabilizing it.

The same applies for ionic bonds, but they are sensitive to far more environmental changes than just temperature fluctuation; this is because most molecules in cells are ionic, and cells can change the concentration and location of these ions quite easily. Therefore, an increase in the concentration of an ion increases the strength of the electrostatic interactions between them (also known as *ionic strength*), and these ions may compete for and displace the amino acid ions forming the bond. A simple example is increasing the salt (Na^+ and Cl^-) concentration around a protein: the Na^+ ions may displace positively charged ions on an amino acid (e.g., lysine), while the Cl^- ions may substitute for a negatively charged ion on a different amino acid (e.g., aspartic acid). Either one of these events could break an ionic bond between two charged amino acids, thereby altering the shape of a protein. Keep in mind that a change in pH can have the same effect, due to the presence of protons (H^+) and hydroxyl ions (OH^-) in water.

Cells Chemically Modify Proteins to Control Their Shape and Function

In addition to changing the ionic strength of the solution surrounding a protein, cells use a number of other mechanisms to alter protein shape. In most cases, these modifications require metabolic energy and are therefore very tightly controlled. They can be conveniently broken into two classes, based on the duration of the modification.

Covalent Modifications Are Relatively Long Lasting

Because covalent bonds are stronger than noncovalent bonds, it is much more difficult to break them. Increases in temperature, changes in salt concentration, and most other chemical changes have no impact on these bonds; therefore, cells use covalent bonds to modify proteins when the effect needs to be relatively long-lasting (i.e., at least a few seconds, and as long as several months). Many proteins have more than one covalent modification at any time. **FIGURE 3-15** illustrates the seven most common types of molecules covalently attached to proteins: protein subunits, small proteins, sugars, lipids, phosphate groups, methyl groups, and acetyl groups.

Protein subunits are sometimes joined via **disulfide bonds** between cysteine amino acids, primarily in proteins at the cell surface and in the extracellular space. Disulfide bonds can also form between two cysteines in the same polypeptide, and are formed when the sulfhydryl groups on the cysteine amino acids are *oxidized*. Breaking these bonds, therefore, requires a corresponding *reduction* reaction with an electron donor molecule. Very few regions of a cell contain enough electron donors to break these bonds, so disulfide bonds are effectively permanent. For this reason, they are rare in most proteins, because proteins must preserve some flexibility in order to adopt their two required shapes.

Small proteins (<100 amino acids) can be attached to larger proteins to modify their function or target them for destruction. When the protein **ubiquitin** is attached to lysine or glutamic acid side chains of cytosolic and nuclear proteins, these proteins are destined

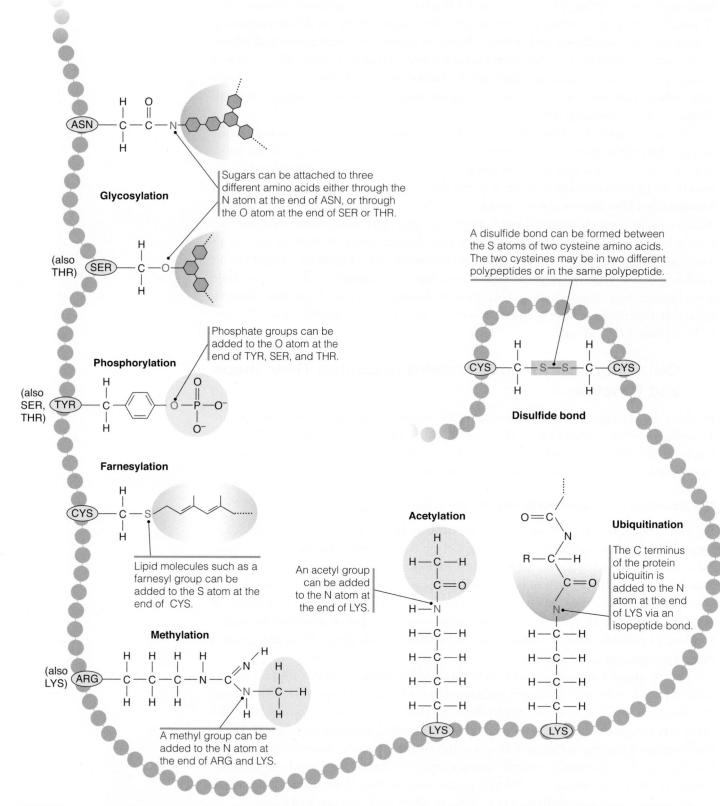

Glycosylation

Sugars can be attached to three different amino acids either through the N atom at the end of ASN, or through the O atom at the end of SER or THR.

A disulfide bond can be formed between the S atoms of two cysteine amino acids. The two cysteines may be in two different polypeptides or in the same polypeptide.

Phosphorylation

Phosphate groups can be added to the O atom at the end of TYR, SER, and THR.

Disulfide bond

Farnesylation

Lipid molecules such as a farnesyl group can be added to the S atom at the end of CYS.

Acetylation

An acetyl group can be added to the N atom at the end of LYS.

Ubiquitination

The C terminus of the protein ubiquitin is added to the N atom at the end of LYS via an isopeptide bond.

Methylation

A methyl group can be added to the N atom at the end of ARG and LYS.

FIGURE 3-15 Seven different types of molecules can be covalently attached to amino acid side chains. Three-letter abbreviations for amino acids are shown, plus the side chain of each.

to be destroyed (i.e., cleaved into very small peptides), as we will discuss shortly. Ubiquitin was so-named because it is found in all cells (it is *ubiquitous*). Proteins known as **S**mall **U**biquitin-like **M**odifiers (or **SUMOs**) are attached to the lysine amino acids of some proteins, including those involved in controlling gene transcription. Curiously, these proteins are attached via an **isopeptide bond**, which is a peptide bond formed between the side chain of one polypeptide and the backbone of another.

Sugars are covalently attached to some asparagine, serine, and threonine amino acids, found in proteins primarily located on the cell surface and in the extracellular space. Because they are highly polar and contain hydroxyl groups, sugars readily form hydrogen bonds with each other, with water, and with amino acid side chains (**BOX 3-8**). In many cases, oligosaccharides containing several sugars are attached to proteins, and because of their size, they have the potential to form and break numerous hydrogen bonds between amino acids. This often results in a dramatic change in the shape of a protein. In many cases, addition of sugars (a process called **glycosylation**) is actually *necessary* for a protein to adopt a fully functional shape. Most glycosylated proteins remain glycosylated throughout their lifetime.

Lipids are covalently attached to the sulfhydryl group of cysteine amino acids on some membrane proteins (**BOX 3-9**). Two specific lipids, the farnesyl group and the geranylgeranyl group, are attached this way. A phospholipid can also be attached to the carboxy terminus of some membrane proteins that are sorted in the Golgi apparatus (see Chapter 9). Due to their hydrophobic nature, lipids achieve their lowest energy state (are most stable) when they interact with the lipid tails of membrane phospholipids. For this reason, when lipids are attached to cysteines or carboxy termini on the exterior of a cytosolic protein, they can serve as anchors that tether cytosolic or extracellular proteins to membranes. This is one of the many mechanisms cells use to attach proteins to membranes, and we will see many more in Chapter 4.

Phosphate groups are covalently attached to the hydroxyl groups of serine, threonine, and tyrosine side chains in a wide variety of cellular proteins. Phosphate groups are the smallest molecules commonly bound to proteins via covalent bonds, and they are negatively charged. Some proteins can have three or more phosphate groups attached to them at the same time. The negative charges have the potential to disrupt existing ionic bonds within a protein, as well as form new ionic bonds with any positively charged amino acids near them. In this way, cells can use phosphate groups to fine-tune the shape of a protein. In some cases, the addition of phosphate activates a protein, and in other cases this inhibits it. These phosphates are readily removed by a class of enzymes called phosphatases, so proteins are often phosphorylated for brief periods. This rapid on-off strategy is commonly used in signaling proteins.

Methyl groups are added to a great variety of proteins, usually on arginine and lysine amino acid side chains. In Chapter 2, we briefly discussed methylation of histones as a means of controlling gene expression (see **Nucleosome Structure Can Be Modified by Cells**).

BOX 3-8 TIP

Take a look at Figure 1-13, to review the general structure of oligosaccharides.

BOX 3-9 TIP

Look at Figure 1-10, to review the general structure of lipids.

Additional proteins that are methylated include those involved in signal transduction, cell growth, and programmed cell death, and we will cover those topics in later chapters.

Acetyl groups are added to lysine amino acids in a number of different proteins, especially the class of proteins known as transcription factors. These proteins regulate gene transcription in response to extracellular stimuli, and therefore control the phenotype of a cell. Other proteins that are commonly acetylated are responsible for cell growth and determination of cell shape; thus, acetylation of proteins has widespread and long-lasting effects on cells.

Despite their relative stability and long duration, all covalent modifications to proteins are reversible. Specialized enzymes are responsible for forming the covalent bonds, while others are responsible for breaking these bonds; in some cases, the enzymes that remove these modifications attract special attention because they are associated with diseases. For example, the enzymes that remove acetyl groups from lysine amino acids in histones (often called histone deacetylases, abbreviated as HDAC, see Chapter 2) have aroused great interest in cancer researchers, because drugs that inhibit them show great promise for treating some forms of cancer.

■■■ NonCovalent Modifications Are Relatively Short Lived

Noncovalent modifications capitalize on the fact that the association between a protein and these additional molecules is weak, in that this allows proteins to bind and release these molecules extremely quickly (i.e., fractions of a second). These short-term interactions are often used to vary the activity of enzymes; they are also quite costly, in that cells expend a great deal of energy to maintain a steady supply of them.

Many proteins contain more than one binding site, as shown in **FIGURE 3-16**. One of these is dedicated to binding the protein target (e.g., an active site in an enzyme that binds a substrate), and the other is available for binding to these short-duration modifiers. Because this extra binding does not directly compete for access to the first binding site, the binding at the second site is often referred to as **allosteric** (from Greek, meaning "other space") binding. Allosteric binding is a common means for controlling the shape of the active site of enzymes without interfering with the actual binding of the substrates.

Short-term binding, cleavage, and release of nucleotides are common ways to regulate protein activity. In addition to serving as sources of metabolic energy and as building blocks for RNA synthesis, ATP and GTP can bind proteins to alter their behavior. Two examples are the binding of GTP by the protein tubulin and the binding of ATP by the protein actin; in both cases, the binding promotes the polymerization of these proteins. In addition, most proteins that bind nucleotide *tri*phosphates can cleave them to form nucleotide *di*phosphates (e.g., ADP, GDP); the diphosphate form has the opposite effect on tubulin and actin, as it promotes the depolymerization of these proteins (see Chapter 5). One class of protein is so dependent on binding of GTP/GDP to function properly that its members are called **G proteins**. **FIGURE 3-17**

Common allosteric factors include Ca^{2+}, ATP, and GTP. These bind to the allosteric site for varying amounts of time and induce a shape change in the protein that also changes the shape of the primary binding site. Once the allosteric factor is released, it may be cleaved (e.g., ATP → ADP + P$_i$) or recycled (e.g., Ca^{2+}).

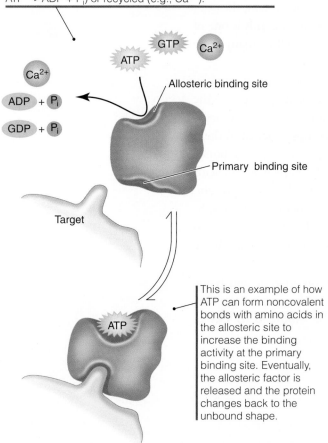

This is an example of how ATP can form noncovalent bonds with amino acids in the allosteric site to increase the binding activity at the primary binding site. Eventually, the allosteric factor is released and the protein changes back to the unbound shape.

FIGURE 3-16 Many proteins have more than one binding site. Allosteric binding sites help control the shape and function of a protein.

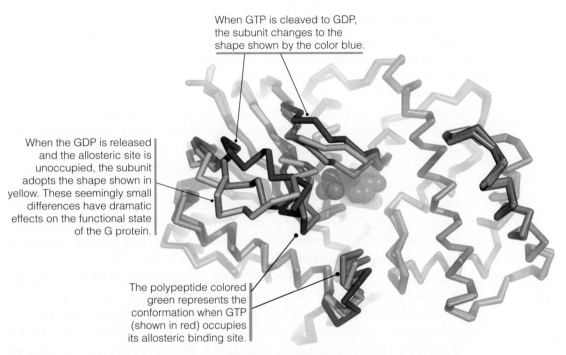

When GTP is cleaved to GDP, the subunit changes to the shape shown by the color blue.

When the GDP is released and the allosteric site is unoccupied, the subunit adopts the shape shown in yellow. These seemingly small differences have dramatic effects on the functional state of the G protein.

The polypeptide colored green represents the conformation when GTP (shown in red) occupies its allosteric binding site.

FIGURE 3-17 Occupancy of an allosteric GTP binding site controls a G protein subunit's shape. Three different conformations of the subunit are overlaid for comparison. Photo courtesy of Heidi E. Hamm and Will Oldham, Vanderbilt University Medical Center.

shows how the shape of a G-protein subunit changes when its allosteric site is bound to GTP, GDP, or no nucleotide. (Regulation of G proteins is covered in much greater detail in Chapter 11.)

Binding and release of calcium ions also regulates the activity of many proteins. Calcium ions are among the smallest known allosteric modifiers, and because many proteins actually *require* calcium ion binding to function properly, removal of calcium (by lowering the concentration of these ions in the surrounding environment, for example) is an efficient means of rapidly shifting these proteins into a nonfunctional shape. We will see many examples of this in later chapters, especially in our discussions of cell signaling, nerve conduction, and muscle contraction.

■■ Proteins Are Classified into Nine Categories According to Their Function

The number of different proteins that exist in nature is not known, because there are simply too many to count. Even human cells contain more than we can count easily—the estimates range in the tens of thousands (**BOX 3-10**). How do we make sense of this dizzying number of proteins? One easy method for seeing the big picture of protein structure and

---

BOX 3-10

The "cell as a busy city" analogy, continued. One does not need to know the exact identity of each citizen of a city to describe how people function within the city. One can use any number of different descriptors to classify workers: manual laborers, administrators, teachers, bureaucrats, and so on. The nine categories listed here for protein functional groups are not the only way to describe groups of proteins—they simply give us a decent idea of how labor is divided into specializations in cells.

function is to organize them into categories according to the functions they perform. This strategy permits us to generalize a bit about the structural characteristics of each category and to highlight important differences between the categories.

The nine categories of protein function are:

- **Enzymes** are proteins that catalyze chemical reactions inside cells; in multicellular organisms, they also facilitate reactions in the extracellular spaces. The primary binding targets for these proteins are their substrates, which are then converted into products. Enzymes exist in every compartment of cells, and are also present in cell membranes. Due to their great diversity of functions, it is difficult to generalize their structural characteristics.

- **Regulatory proteins** control the function of enzymes by binding to them and subtly changing their shapes (e.g., the enzymes responsible for converting DNA sequences into RNA sequences during transcription are controlled by a family of proteins known as transcription factors). It is likely that at least one regulatory protein binds to every enzyme in a cell. Due to the diversity of their binding targets, these proteins do not share any distinctive structural characteristics.

- **Structural proteins** form stable complexes inside and outside of cells that provide structural support to cells and tissues. One common structural characteristic is that they bind quite tightly to one another, forming large complexes that are quite strong. Most structural proteins form fibers or link fibers together, and therefore resemble flexible scaffolding materials. Many of these are multi-subunit proteins that have 4° structure. They do *not* facilitate chemical reactions.

- **Motility proteins** interact with structural proteins to generate motion inside a cell. This motion can be calibrated so that a cell can selectively move small parts (individual proteins, membrane vesicles), organelles (mitochondria, endoplasmic reticulum), chromosomes (during mitosis), or much larger structures (cilia and flagella). These proteins can also move entire cells, such as muscle cells during contraction or skin cells that crawl into a wound during healing. Because this movement requires energy, many proteins in this category bind to and cleave ATP, in addition to binding their primary targets. The complexity of orchestrating these movements usually requires multi-subunit proteins that have 4° structure.

- **Transport proteins** control the movement of small molecules across membranes. They are essential for maintaining cellular homeostasis, and also help store some metabolic energy in the form of ion gradients across membranes (a form of potential energy). All transport proteins are transmembrane proteins, so they share a common trait: they possess at least one stretch of hydrophobic amino acids to stabilize them in the hydrophobic interior of a membrane.

- **Hormonal proteins** are used as chemical signals between cells in a multicellular organism. Typically, cells release hormonal proteins into a circulatory system, which carries them to target cells far (at least 10 cell diameters) away. These proteins are all relatively small (some are even referred to as *peptides* rather than as proteins) so they diffuse easily, and they all bind to cell surface receptor proteins on the target cells. These proteins also share other structural characteristics: they do not form large complexes like structural proteins and do not consume metabolic energy to function.

- **Receptor proteins** bind to signaling molecules, including hormonal proteins. Receptor proteins also belong to a group of proteins called signaling proteins. Most chemical signals inside cells originate from receptor proteins, and are then transmitted throughout the cell by a host of other signaling proteins, many of which are enzymes. Receptors can be found in membranes and in the cytosol, but they do not share many common structural features. Their distinguishing property is that once they bind to a signal, receptors adopt a different shape that targets a different molecule; in other words, receptors have at least two binding targets, and the binding of one target molecule is dependent on binding of the other. This is how receptors pass a signal to a downstream molecule.

- **Defensive proteins** are a specialized class of proteins that target potentially harmful molecules and neutralize them. A good example is the group of antibodies that circulates in vertebrate animals and binds to infectious agents (such as bacteria and viruses). Plants synthesize a different group of proteins to ward off the effects of pollutants, infectious organisms, and other environmental stresses. These proteins are quite diverse in structure.

- **Storage proteins** are used by cells to bind to and protect energy sources from degradation. Plants, insects, and vertebrate animals all contain proteins whose primary function is to bind to energy-rich molecules (typically fats) and facilitate their long-term storage. Examples of the work of storage proteins include the yolk of a bird egg (a storehouse of energy for the chick as it grows inside the eggshell) and plant seeds, which contain proteins that serve as sources of amino acids for the germinating seed.

CONCEPT CHECK #3

Recall from Chapter 2 that DNA undergoes some of the same chemical modifications as proteins. Would cells benefit by using *all* of their protein modification strategies on DNA as well? Why or why not?

3.5 Where Do Proteins Go to Die?

KEY CONCEPTS

▸ An important method for recycling misfolded or unused proteins, called protein degradation, reduces proteins to individual amino acids for use in building new proteins.

▸ Cells use three different mechanisms to degrade proteins, each confined to a specific compartment of the cell.

▸ Proteasomes degrade proteins in the cytoplasm and in the nucleus; these proteins are tagged for destruction by the covalent attachment of many copies of a small protein called ubiquitin.

▸ Lysosomes degrade proteins in the endomembrane system, including those captured by endocytosis.

▸ Metalloproteinases degrade proteins in the extracellular space, allowing remodeling of this space in multicellular organisms.

Nothing in nature is permanent. Just as cells and organisms eventually die, proteins stop functioning. In most cases, a cell outlives its proteins, so when the proteins stop functioning (often by folding into one or more stable, but useless, shapes), the cell can recycle many of the amino acids in the polypeptide(s) making up the protein. It is important to note the difference between a protein that is simply in an inactive conformation and one that can no longer be activated, as the latter are most often the targets of protein degradation, though inactive but otherwise healthy proteins are sometimes degraded, too (we will see examples of this in Chapter 13). Disassembling a protein must be done carefully; simply breaking random bonds in the polypeptide will likely not yield any useful amino acids. Every cell, therefore, has specialized proteins—called **proteolytic enzymes**—that can disassemble proteins into pieces small enough that metabolic enzymes can convert them into new amino acids, sugars, and so on.

Consider for a moment the challenge this presents to a cell. We have devoted this entire chapter to the importance of proteins, and are now confronted with a class of enzyme that breaks proteins apart. How can a cell keep such an enzymatic protein? Wouldn't this protein simply digest every other protein in the cell, thereby killing the cell? In theory, yes. (Some proteins are actually used by cells to commit suicide, but that is a very special use of proteolytic enzymes, and is covered in Chapter 13.) How, then, can cells solve this problem? The answer is surprisingly simple: cells create proteolytic enzymes, but then sequester them away from the rest of the functional proteins. In effect, they build barriers between these proteins and the rest of the cell, so they can control the types of polypeptides that get digested. This is an excellent example of the cell compartmentalization/specialization strategy we first discussed in Chapter 1. There are three compartments where these proteolytic enzymes are allowed to function: the proteasome, the lysosome, and the extracellular space.

Proteins in the Cytosol and Nucleus Are Broken Down in the Proteasome

Proteins in the cytosol are, by definition, not protected by a membrane-bound organelle. If a cell synthesizes a proteolytic enzyme in the cytosol, there is a chance it could destroy some healthy proteins in its vicinity if it is not carefully controlled. Cells have developed an elegant three-part solution to save their healthy cytosolic proteins, as illustrated in **FIGURE 3-18**:

- First, the hydrolytic enzymes are large, multi-subunit proteins. This means that each subunit is inactive until the entire complex is assembled, greatly reducing the chance that random proteolysis of a healthy protein will occur. In fact, the fully assembled complex, called a **proteasome**, looks something like a barrel or can with

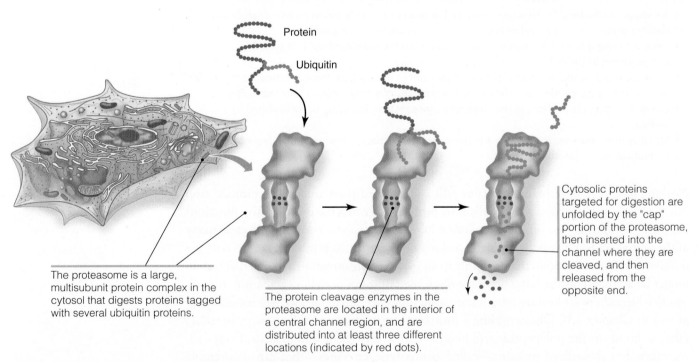

Protein

Ubiquitin

The proteasome is a large, multisubunit protein complex in the cytosol that digests proteins tagged with several ubiquitin proteins.

The protein cleavage enzymes in the proteasome are located in the interior of a central channel region, and are distributed into at least three different locations (indicated by red dots).

Cytosolic proteins targeted for digestion are unfolded by the "cap" portion of the proteasome, then inserted into the channel where they are cleaved, and then released from the opposite end.

FIGURE 3-18 Proteasomes digest proteins in the cytosol.

very thick walls (called the 20S proteasome) and a lid (called the 19S cap) at each end. Together, they form a small channel (~5 nm in diameter) where the polypeptides are broken down by *hydrolyzing* the peptide bonds that hold the amino acids together (recall that peptide bonds are formed by a dehydration reaction; these enzymes simply add water back and break the bond).

- Second, the active sites in this complex are located on the walls of the channel, far away from other cytosolic/nuclear proteins. The active sites (there are at least three in eukaryotic cells) are separate from one another, so each can specialize in breaking down a polypeptide in a different place. These sites break peptide bonds formed by positively charged, negatively charged, and hydrophobic amino acids.

- Third, proteins need to be chemically modified before they can even enter the channel. The chemical modification is **ubiquitination**, the covalent attachment of the protein ubiquitin (discussed earlier in this chapter, see Figure 3-15). Typically, several ubiquitins are added in a series, and the product is called a **polyubiquitin chain**. It appears that one of the most important functions of ubiquitin is to target proteins for destruction in this fashion (additional functions of ubiquitin are discussed in Chapter 13). The steps required to attach ubiquitin to a protein are extremely complex, involving three additional protein complexes. It is therefore almost impossible for a protein to be hydrolyzed in a proteasome by accident.

Proteasomes are found in the nucleus as well as in the cytosol. As we will see in Chapters 8 and 11, most proteins that reside in the nucleus are synthesized in the cytosol and transported into the nucleus after they are properly folded. Maintaining proteasomes in the nucleus permits cells to degrade unwanted nuclear proteins without having to first ship them back into the cytosol.

Proteins in Organelles Are Digested in Lysosomes

Proteasomes hydrolyze nonfunctional proteins in the cytosol. So, what about proteins in organelles? Misfolded (oddly shaped) and nonfunctional proteins in the endoplasmic reticulum are transported out into the cytosol, and then hydrolyzed by proteasomes. But proteins in the Golgi complex, trans-Golgi network (TGN), and other membrane-bound compartments have a different fate. Some vesicles that leave the TGN contain proteins that possess a unique modification: they have a phosphorylated sugar called mannose-6 phosphate (also called M6P) covalently attached to an asparagine amino acid. This M6P tag targets these proteins to an organelle called the lysosome, shown in **FIGURE 3-19**. The lysosome is aptly named: it is a "body" (-*some*) that "lyses" (cleaves) molecules; it is the membrane-bound complement to the proteasome, and its proteolytic enzymes cleave membrane and engulfed proteins that fuse with the lysosome. The lysosome also contains enzymes that digest fats and sugars. We will discuss the complex membrane shuttling mechanisms that control lysosome formation and function in Chapter 9 (**BOX 3-11**).

Proteins in the Extracellular Space Are Digested by Proteinases

The third compartment where proteins are digested is in the extracellular space. The plasma membrane does a good job of protecting the cell interior from external damage, so cells can afford to dump proteolytic enzymes (called **proteinases**) into the fluid that surrounds

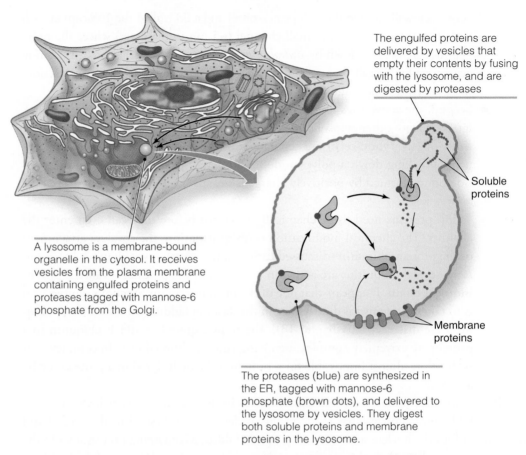

The engulfed proteins are delivered by vesicles that empty their contents by fusing with the lysosome, and are digested by proteases

Soluble proteins

A lysosome is a membrane-bound organelle in the cytosol. It receives vesicles from the plasma membrane containing engulfed proteins and proteases tagged with mannose-6 phosphate from the Golgi.

Membrane proteins

The proteases (blue) are synthesized in the ER, tagged with mannose-6 phosphate (brown dots), and delivered to the lysosome by vesicles. They digest both soluble proteins and membrane proteins in the lysosome.

FIGURE 3-19 Most membrane proteins and engulfed proteins are digested by lysosomes.

the cells if proteins in the extracellular space are damaged and/or need to be replaced. An excellent example of this occurs during wound healing, when cells digest the extracellular proteins and dead cells in the wound, then build new cells and extracellular matrix proteins to replace them. Cells also secrete proteins that specifically target the proteinases and inhibit them, setting up a competition between the proteinases and their inhibitors—this is how cells can control the rate and duration of digestion that takes place outside the plasma membrane, as shown in **FIGURE 3-20**.

CONCEPT CHECK #4

Cells also make proteins that digest DNA, called nucleases. How would you predict that a cell treats these proteins to protect against self-digestion of its genome?

BOX 3-11

The incinerator analogy. Proteasomes and lysosomes observe a simple rule when it comes to cellular proteins: whatever goes in *never* comes back out. A polypeptide that enters either structure is always broken down and digested forever. The enzymes that digest these polypeptides are like the heating elements in an incinerator, in that any foreign material that comes into contact with them gets broken down irreversibly. Note that some viruses have learned how to use this principle to their advantage in lysosomes, and they have developed methods for surviving the proteolytic ordeal.

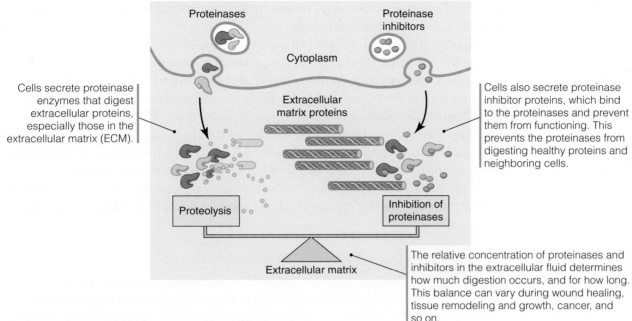

Cells secrete proteinase enzymes that digest extracellular proteins, especially those in the extracellular matrix (ECM).

Cells also secrete proteinase inhibitor proteins, which bind to the proteinases and prevent them from functioning. This prevents the proteinases from digesting healthy proteins and neighboring cells.

The relative concentration of proteinases and inhibitors in the extracellular fluid determines how much digestion occurs, and for how long. This balance can vary during wound healing, tissue remodeling and growth, cancer, and so on.

FIGURE 3-20 Extracellular proteins are digested by proteinases.

3.6 Chapter Summary

Proteins are the actors that convert most information in DNA sequences into cellular activity. Like DNA, they are composed of linear sequences of simple building blocks (amino acids), but they form much more elaborate and complex shapes. This happens because proteins contain 20 different subunits, rather than the 4 subunits (or nucleotides) that compose DNA. In addition, the side chain of each of the 20 amino acids is structurally unique, and this further increases the complexity of protein shapes.

All amino acids share some common structural features, including an amino group and a carboxylic acid group. Polymers of amino acids are held together by a covalent bond formed between the amino group of one and the carboxylic acid group of the next; this bond is called a peptide bond, and long polymers of amino acids are called polypeptides for this reason. Most proteins are composed of a single polypeptide, but some contain several polypeptides bound to one another.

A nomenclature has been developed to describe trends in protein shape. The primary structure is simply the linear sequence of amino acids in a polypeptide, while the secondary structure describes how these linear sequences are arranged in space as α-helices, β-sheets, or random coils. The tertiary structure describes how the secondary structures are arranged in space, and is highly variable in proteins. Three common tertiary structures in proteins are globular structures, filamentous structures, and transmembrane structures. Quaternary structure is reserved only for those proteins that are composed of more than one polypeptide, and it describes how these polypeptides are oriented in space with respect to one another.

All proteins share three central traits that distinguish them from most other molecules in cells: (1) they adopt stable, yet flexible, three-dimensional shapes, (2) they all bind to at least one target molecule with enough specificity to enable them to become specialists in cells, and (3) they all perform at least one function in cells. These three traits have arisen as a product of evolution by natural selection, such that polypeptides not sharing these three traits are not synthesized by cells. Cells can control when a protein is functional by covalently attaching a number of different molecules to it and/or by allosteric (noncovalent) binding of regulatory molecules. Proteins are further classified into nine groups, according to the functions they perform in cells.

Proteins are recycled by cells. All proteins have a lifetime, and will eventually adopt a dysfunctional shape. At that point, the proteins are destroyed by one of three different mechanisms, each targeting a different set of cellular and extracellular proteins. Proteasomes digest proteins in the cytosol, lysosomes digest proteins in the endomembrane system, and metalloproteinases digest proteins in the extracellular space. All three methods share two important features: they break amino acids by hydrolyzing peptide bonds, and they do this in isolated compartments so as not to destroy functional proteins by accident. The products of this destruction are small peptides than can be reused to make new proteins.

Collectively, these features illustrate how proteins perform their functions in cells, and provide the foundation for explaining cellular activities at a molecular level in later chapters.

CONCEPT CHECK ANSWERS #1

We can start by rewriting the three traits. Trait #1 becomes: All *nucleic acids* adopt stable three-dimensional shapes. Is this statement supported by evidence? We know that DNA *always* forms a double helix in cells (and does so spontaneously), so the statement appears to be true. But we also know that DNA can be completely unwound and pulled into single strands (e.g., during DNA replication or gene transcription), then folded back up by proteins, and still remain functional. While some proteins are cyclically disassembled/unfolded and reassembled/folded while functioning in cells, most are not. It therefore appears that remaining in a stable three-dimensional shape (a double helix) is *not* essential for nucleic acids to be functional.

The rewritten trait #2 becomes: All *nucleic acids* bind to something else. We know that functional DNA must be bound by proteins so that the information stored in DNA can be transmitted throughout the cell. However, in most cells, not all DNA is functional, and the non-functional DNA does not likely bind anything in a meaningful way. Therefore, trait #2 does *not* apply to nucleic acids, either.

Revised trait #3 becomes: All *nucleic acids* perform at least one cellular function. Because most cells contain at least some nonfunctional DNA, this statement is not true. The three traits are therefore not universal. Why can DNA escape these requirements? Non-functional DNA is carefully packaged in cells so it does not interfere with normal cellular activities, while nonfunctional proteins would not be sequestered as easily, and would likely disrupt at least some normal activities in cells. Packaging makes the difference.

CONCEPT CHECK ANSWERS #2

The sickle-cell mutation is a point mutation, meaning that one amino acid (glutamic acid) is replaced with a different amino acid (valine). This affects the 1° structure of hemoglobin because it changes the sequence of amino acids in the polypeptide. Because 2° structures can theoretically form anywhere in a polypeptide, it is not possible to predict which 2° structure will be affected by this mutation. We know that the 4° structure of the protein changes dramatically, from a tetramer to a long fibrous polymer. Because the 4° structure changes so dramatically, it is safe to assume that the 3° structure (and hence, the 2° structure) is altered too. This reflects a reorganization of the bonds (perhaps all five classes) that determine the stability of the protein shape. Because an ionic amino acid (glutamic acid) is replaced with a hydrophobic amino acid (valine), we can assume that ionic and/or hydrophobic interactions are disrupted and lead to the subsequent change in higher-order shape. While the tetrameric form of hemoglobin is functional, the fibrous polymer is not (it is an excellent example of a stable shape that is *not* functional). Thus, a change in the 1° structure of a polypeptide can propagate up to 4° structure and ultimately impact the function of a protein.

CONCEPT CHECK ANSWERS #3

The biggest difference between the types of modifications in proteins is the *rate of turnover* for these modifications. Some proteins must make rapid changes in shape to be useful—consider the contractile proteins that drive muscle contractions during a reflex reaction, or those that keep us balanced as we walk. Also, there is usually a correlation between how quickly a modification can be made and how long it needs to last: rapid contraction of muscles also means rapid relaxation in most cases, so the turnover is very high. But the rate of turnover for DNA modifications is comparatively slow. Recall that to be useful, DNA needs to bind to a protein that then changes its function, and that the majority of these DNA-associated functions concern gene expression, which takes at least several minutes to complete. Therefore, one only sees long-term modifications of DNA, and all of them are via covalent attachment of other molecules. Short-term modifications, most of which are noncovalent associations, do not occur for DNA.

CONCEPT CHECK ANSWERS #4

Proteins are found everywhere in cells, but DNA is not, so digestive enzymes have a much greater chance of damaging healthy proteins than damaging healthy DNA. Let's review the strategies cells use to control proteolytic enzymes and look for similarities for proteins that target DNA. First, proteolytic enzymes are *sequestered* in a cell, either in a proteasome or in a lysosome. We would therefore predict that the same was true for nucleases, and it is: some of these enzymes are also sequestered in lysosomes, as we will see in Chapter 9. Second, proteins need to be either chemically modified (by ubiquitination, to enter proteasomes) or actively shuttled (by vesicles, to enter lysosomes) to allow them access to the sequestered enzymes. We therefore predict that target DNA or RNA would have to be either modified or shuttled to be destroyed. This, too, is generally true: many nucleases target single-stranded forms of DNA or double-stranded RNA, and both of these are unusual in healthy cells. Also, entire engulfed cells can be targeted to the lysosomes, where their DNA is digested. Third, cells make inhibitors for proteolytic enzymes to control their activity. We would therefore predict that cells would make inhibitors of nucleases, and, in fact, nuclease inhibitors have been found in some strains of yeast. How nucleic acids are digested is a separate field of study and will not be considered further here.

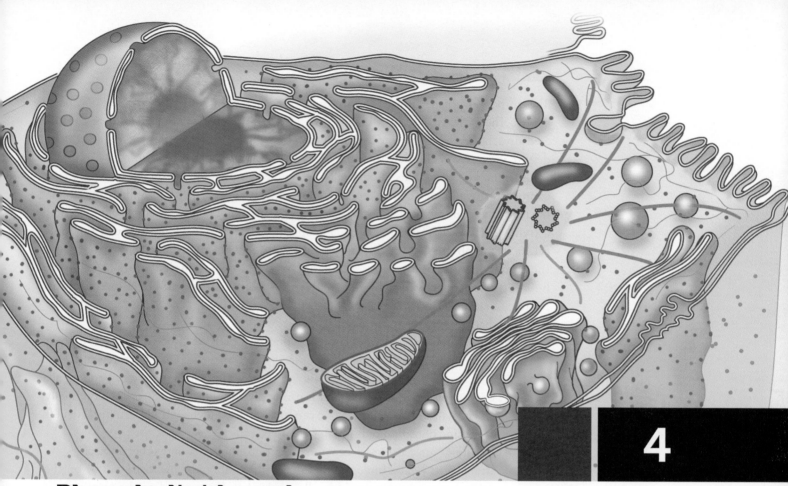

Phospholipids and Membrane Structure

4

4.1 The Big Picture

The purpose of this chapter is to introduce the general structure and assembly of membranes. The presence of at least one membrane is required for any organism to be considered alive, but this does not mean that all membranes are alike. Recall from Chapter 1 that, despite its relatively simple structure, the carbon atom is capable of forming a multitude of different molecules; this is especially evident when one considers the molecules that compose a membrane. Unlike nucleic acids and proteins, which are built from simple nucleotide and amino acid subunits, membranes are much larger structures composed of a tremendous variety of molecules. Phospholipids are some of most abundant molecules in membranes, so in this chapter we will focus on them; our goal is to discuss the common features shared by *all* membranes, and thus complete our survey of the building blocks of cells. We will reserve the details concerning membrane specialization for later chapters.

This chapter contains three major sections:

- Unlike nucleic acids and proteins, phospholipids are not composed of linear polymers of a small group of subunits; instead, they are branched molecules, and each branch is often unique, making it nearly impossible to memorize them all. Because of this, our strategy for learning phospholipid structure will rely heavily on generalizations, such that we can more easily understand the structural details of specific phospholipids in later chapters.
- The fluid–mosaic model explains how phospholipids and other molecules are arranged to yield a biologically functional membrane. Our focus will be on the general concepts that apply to all membranes. Understanding the implications of these concepts will be crucial when we begin discussing specific membranes in later chapters, so this idea will likely require more attention than the first.
- Phospholipid synthesis and membrane formation occur in an assembly-line-like fashion. This concerns the molecular mechanisms cells use to build phospholipids and organize them into functional membranes. Notice that the first two ideas are focused on *what* these structures are, while this idea addresses *how* they are made. It is important to understand the sequence of events, and the reasons why each step is necessary. Fully understanding this information will require solid comprehension of the basic chemistry we discussed in Chapter 1, because we will be discussing enzymatic reactions that give rise to the structurally complex molecules, including phospholipids, that compose membranes.

4.2 Phospholipids Are the Basic Building Blocks of Cellular Membranes

KEY CONCEPTS

- ▸ Of the four major classes of molecules that serve as the building blocks for cells (sugars, nucleic acids, proteins, phospholipids), phospholipids are the most structurally complex.
- ▸ Phosphoglycerides are the most common phospholipids; they are composed of two fatty acyl groups and a polar head group attached to a glycerol backbone.
- ▸ The length and degree of saturation of the fatty acyl tails can vary between different phospholipids.
- ▸ The primary function of phospholipids is to form the lipid bilayer that is characteristic of all cellular membranes.
- ▸ Phospholipid bilayers are selectively permeable to solutes.

Phospholipids are structurally and functionally different from the other cellular building blocks we have discussed in Chapters 1, 2, and 3. Before we begin our detailed examination of phospholipids, let's compare their structural properties to these other basic building blocks. In Chapter 2, we learned that nucleic acids are, structurally, some of the simplest biomolecules in cells; they are composed of only four different nucleotide subunits, joined into linear polymers by one type of chemical bond (the phosphoester bond), and stabilized as linear pairs by hydrogen bonds. The complexity of nucleic acids arises not from their basic structure, but from their *tremendous size*—the average lengths of prokaryotic and eukaryotic genes are approximately 900 and 1,300 base pairs, respectively. Structures of this size can adopt a wide variety of shapes, giving rise to the different folded forms of nucleoid and chromatin (see Chapter 2, *DNA Packaging Is Hierarchical*).

The same pattern is evident in proteins. Polypeptides are composed of 20 different amino acids, joined into linear polymers by one type of chemical bond (the peptide bond), and stabilized by five different types of chemical bonds (see Chapter 3, *Five Classes of Chemical Bonds Stabilize Protein Structure*). The average length of a protein ranges from ~250 to ~300 to ~470 amino acids in archaea, prokaryotes, and eukaryotes, respectively. Notice that over the course of evolution from archaea to eukaryotes, the length of an average protein nearly doubled.

The pattern continues for sugars. Oligosaccharides and polysaccharides are some of the most diverse types of biomolecules, yet most of them are composed of fewer than 20 monosaccharide subunits joined by a small number of bonds (principally the glycosidic bond, a form of ether bond). But size alone does not account for the tremendous diversity of oligosaccharides and polysaccharides. Cells employ *two additional strategies* for increasing variation in these structures: branching and chemical modification. In branching, sugar subunits in these polymers form polygonal, ring-like structures, and the carbon atoms that form most of the vertices in these polygons contain functional groups capable of forming covalent bonds with similar groups on other sugars. This means that a single sugar subunit (a monosaccharide) can form bonds with two or more neighboring sugars, forming a branched network (see Figure 1-14). This way, the same number of monosaccharides can be combined into multiple combinations, increasing the structural variation. This strategy does not apply to nucleic acids, and appears only rarely in proteins (in the form of disulfide bonds formed between cysteine amino acids in two different polypeptides).

The second additional strategy involves modifying the functional groups attached to carbon atoms in sugars (e.g., hydroxyl groups) by adding or substituting different functional groups (e.g., acetyl groups, phosphate groups). Each modification results in the formation of a different type of sugar subunit, greatly increasing the potential diversity of sugar polymers. We have seen this strategy employed before, in both nucleic acids (e.g., methylation, see Figure 2-18) and proteins (e.g., acetylation and phosphorylation, see Figure 3-15). This strategy is commonly used in cells.

Phospholipids also acquire their structural diversity through branching and chemical modification. Branching is so important for phospholipids that *all* phospholipids are branched (hence the complexity), and phospholipids use a special kind of chemical modification not seen in any other classes of cellular building blocks (unsaturation of hydrocarbons); this is where the bulk of the structural variation arises. Unlike oligo/polysaccharides, nucleic acids, and proteins, phospholipids are not polymers of a simple structure, and are considerably less diverse as a result. Of these four major classes of molecules, phospholipids are by far the smallest in size. The molecular weight of most phospholipids is less than 1,000 daltons, while most DNA and proteins are at least 10 to 20 times this size. The size of sugar polymers in cells varies considerably, from disaccharides (<500 daltons) to enormous polysaccharide complexes (>1 million daltons).

It is estimated that eukaryotic cells contain ~1000 different lipids (approximately one-tenth the number of different genes and proteins), including only about 20 different phospholipids. The reason for this relatively small number is that phospholipids perform far fewer unique functions than the other cellular building blocks. This does not mean that these functions are less important: phospholipids are absolutely essential ingredients in any cellular membrane. Instead of focusing on the tremendous number of possible different structures (as we did for proteins in Chapter 3), let's pay attention to the subtle changes we find between different phospholipids. The reason these changes are important is that they ultimately impact the overall structure of the membrane that contains them. And recall from Chapter 1 that the structure–function relationship predicts that when these structures adopt different shapes, they perform different functions in cells.

Phospholipids Contain Four Structural Elements

Like nucleic acids and proteins, phospholipids have a backbone, which serves as an attachment site for different functional groups, as shown in **FIGURE 4-1**. What makes this backbone different, however, is that it does not form covalent bonds with other phospholipid molecules (i.e., make polymers). Each phospholipid is a separate molecule in a membrane.

Glycerol Is a Three-Carbon Sugar-Alcohol

Two molecules serve as backbones for phospholipids (**BOX 4-1**). The more common backbone is a three-carbon sugar-alcohol known as **glycerol**. Glycerol contains three carbon atoms linked together by single bonds, and each carbon atom is also attached to two hydrogen atoms and a hydroxyl group, allowing each carbon atom to arrange the four bonds in a tetrahedral arrangement (see Figure 1-9). Phospholipids constructed from a glycerol backbone are called **phosphoglycerides**, to distinguish them from a phospholipid called **sphingomyelin**, which contains a different backbone molecule. Because they are the most abundant type of phospholipids, we will first focus our attention on the phosphoglycerides.

The Lipid Portion of a Phospholipid Can Vary Widely in Structure

Phospholipids are named as such because they contain lipids. Recall from Chapter 1 that lipids are long, linear polymers made up almost entirely of carbon and hydrogen, which makes them very hydrophobic. Phospholipids contain two lipids called **fatty acids** because they contain a carboxylic acid group at one end. (The remaining portion of the fatty acid is often called the tail.) The ester bond is formed by a dehydration reaction involving a hydroxyl group on glycerol and the hydroxyl group on the carboxylic acid of a fatty acid. The formation of these bonds is never random: one fatty acid attaches to a terminal carbon of glycerol, and the second attaches to the middle carbon atom, as shown in Figure 4-1. The two fatty acids that attach to a glycerol backbone do not need to be identical.

Because there are two fatty acids in each phospholipid, the number of combinations found in different phospholipids is simply twice the number of different fatty acids found in cells. Most cells contain about 10 fatty acids, and they are classified into 2 structural groups, shown in **FIGURE 4-2**. One group contains fatty acids with only single covalent

BOX 4-1 TIP

Beware of falling into the jargon trap. The rest of this chapter introduces a lot of words that may be new to most beginning cell biology students, especially the names of specific molecules. This can sometimes give the illusion that the concepts behind these names are terribly complicated, but that is typically not true. If the name of a molecule causes confusion, rename it and move on. After getting comfortable with how these molecules function, go back and learn their proper names.

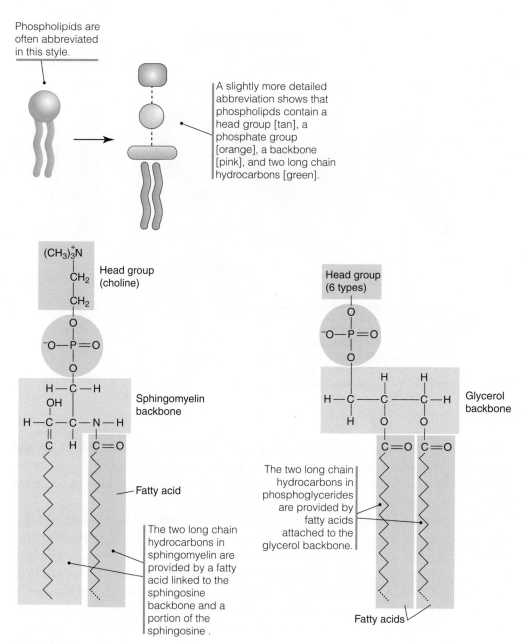

Phospholipids are often abbreviated in this style.

A slightly more detailed abbreviation shows that phospholipds contain a head group [tan], a phosphate group [orange], a backbone [pink], and two long chain hydrocarbons [green].

$(CH_3)_3^+N$

Head group (choline)

CH_2

CH_2

Head group (6 types)

O

$^-O—P=O$

O

$^-O—P=O$

O

H—C—H

OH

Sphingomyelin backbone

H—C—C—N—H

C H C=O

Fatty acid

The two long chain hydrocarbons in sphingomyelin are provided by a fatty acid linked to the sphingosine backbone and a portion of the sphingosine .

H—C—C—C—H

H O O

Glycerol backbone

C=O C=O

The two long chain hydrocarbons in phosphoglycerides are provided by fatty acids attached to the glycerol backbone.

Fatty acids

FIGURE 4-1 Phospholipid structure.

bonds (no double bonds) between the carbons in the tail; this means that as many hydrogen atoms as possible have been attached to the carbon atoms. A common name for members of this group is **saturated** (i.e., they are "fully stocked" with hydrogens). Saturated fatty acids only differ with respect to how many alkyl ($–CH_2–$) groups they contain in their tails. Due to the nature of the chemical reactions cells used to synthesize them, these fatty acids nearly always have an even number of carbons, and the three most common saturated fatty acids used to make phospholipids contain 16, 18, or 20 carbons. The tail end always contains a methyl group ($–CH_3$), and the opposite end contains the carboxylic acid group we discussed previously.

The second, more numerous class of fatty acids *does* contain double bonds between the carbon atoms in the tail. These also usually contain even numbers (16–20) of carbon atoms, but due to the varied placement of the double bonds, many more structures can be made from any given number of carbon atoms in the tail. Because of the double bonds,

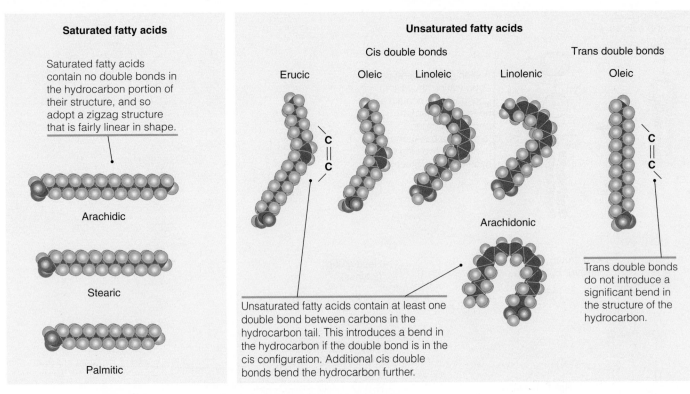

Saturated fatty acids

Saturated fatty acids contain no double bonds in the hydrocarbon portion of their structure, and so adopt a zigzag structure that is fairly linear in shape.

Arachidic

Stearic

Palmitic

Unsaturated fatty acids

Cis double bonds

Trans double bonds

Erucic   Oleic   Linoleic   Linolenic   Oleic

C=C

Arachidonic

Unsaturated fatty acids contain at least one double bond between carbons in the hydrocarbon tail. This introduces a bend in the hydrocarbon if the double bond is in the cis configuration. Additional cis double bonds bend the hydrocarbon further.

C=C

Trans double bonds do not introduce a significant bend in the structure of the hydrocarbon.

FIGURE 4-2 Two classes of fatty acids in cells.

these molecules have fewer hydrogen atoms per carbon than the saturated fatty acids, and so are called **unsaturated** fatty acids. They are further classified according to how many double bonds they contain (monounsaturated vs. polyunsaturated). Most unsaturated fatty acids in membranes contain one or two double bonds.

These double bonds are very important. Recall from Chapter 1 (see Figure 1-9) that when carbon forms four covalent bonds, these bonds are arranged as a tetrahedron around the atomic nucleus, but when it forms three bonds (i.e., combining two single bonds to form one double bond), the bonds adopt a planar, triangular orientation. This switch from a tetrahedral shape to a flat triangular one has a significant impact on the overall structure of the fatty acid tail. Whereas saturated fatty acids have zigzag tails because of the tetrahedral orientation at each carbon, two different orientations are possible for each double bond in an unsaturated fatty acid. If the upstream and downstream carbon atoms that lie on either side of the double bond are positioned on *opposite* sides of the double bond, this is called a **trans configuration**, and the fatty acid is called a **trans fatty acid**. Notice in Figure 4-2 that a fatty acid tail containing one or more trans bonds assumes a somewhat linear configuration, fairly similar to that for a saturated fatty acid. But, if the upstream and downstream carbon atoms lie on the *same* side of the double bond (called the **cis configuration**), the fatty acid assumes a very different shape. These **cis fatty acids** have "kinks," or bends, in their tails. As the number of cis double bonds increases, so does the degree of bending; note that arachadonic acid, which contains four cis double bonds, actually forms a U shape.

■■ Polar Head Groups Confer Additional Specificity on the Structure of Phospholipids

The third structural element in phospholipids is known as the **head group**. The six head groups attached to phosphoglycerides in eukaryotic cells are shown in **FIGURE 4-3**. Note that three of these are ionic (PC, PE, PS), and three (PI, PG, the *bis*-glycerol portion of CL) are clearly polar, due to the high number of hydroxyl groups they contain. All six are linked to the glycerol backbone by a **phosphate group**. The presence of a polar (charged) head group makes this portion of the phospholipid hydrophilic; as a result, these head groups attract water, while the hydrophobic lipid tails repel water.

Because phosphoglycerides are composed of three different functional groups in addition to the glycerol backbone, their formal names are often quite long. For example, a phosphoglyceride built from one stearic acid, one oleic acid, a choline head group, a phosphate group, and a glycerol backbone is called 1-stearoyl-2-oleoyl-phosphatidylcholine. (Note that even a name this long does not mention how many double bonds the oleic acid has, or whether the double bonds are arranged in a cis or trans configuration.) To avoid being bogged down by these long names, most scientists use only the latter portion of the name (e.g., phosphatidylcholine), or even just a simple acronym (e.g., phosphatidylinositol is often abbreviated PI, as shown in Figure 4-3). As a result, the precise lipid composition of individual phospholipids is often not considered in discussions of membrane structure.

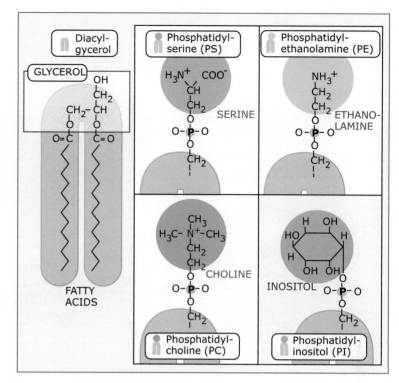

FIGURE 4-3 Six phosphoglycerides in eukaryotes.

The combination of six different head groups and variations in fatty acid structures gives rise to abou t 1,000 distinct phosphoglycerides in eukaryotic membranes.

The Amphipathic Nature of Phospholipids Allows Them to Form Lipid Bilayers in Aqueous Solution

Molecules like phospholipids that both attract and repel water are called **amphipathic**. Amphipathic molecules have characteristic properties when they are added to water, the most noticeable being that they spontaneously aggregate into groups or clusters; this occurs because clustering is energetically favorable. All molecules added to liquid water, regardless of their chemical makeup, have to interact with the nearest water molecules, and this interaction imparts some degree of structural organization on these water molecules (this is sometimes referred to as the **water shell** or **hydration shell**). Recall from Chapter 3 that the laws of thermodynamics state that all molecules, including water, seek an energy state with the highest degree of entropy. The water molecules in the hydration shell have low entropy, so the fewer water molecules necessary to form a hydration shell, the better. The hydrophobic portions of phospholipids aggregate because this reduces the number of water molecules in the hydration shell surrounding them.

When purified phospholipids are added to liquid water, they form three characteristic structures, as shown in **FIGURE 4-4**. The smallest of these is simply a ball, called a **micelle**. The hydrophobic tails of the phospholipids aggregate in the center of the micelle, and the hydrophilic head groups spread out to form the surface facing the water. The next larger structure is called a **liposome**; it resembles a micelle that has formed a shell around an inner compartment of water. Because the phospholipids interact with water at both the

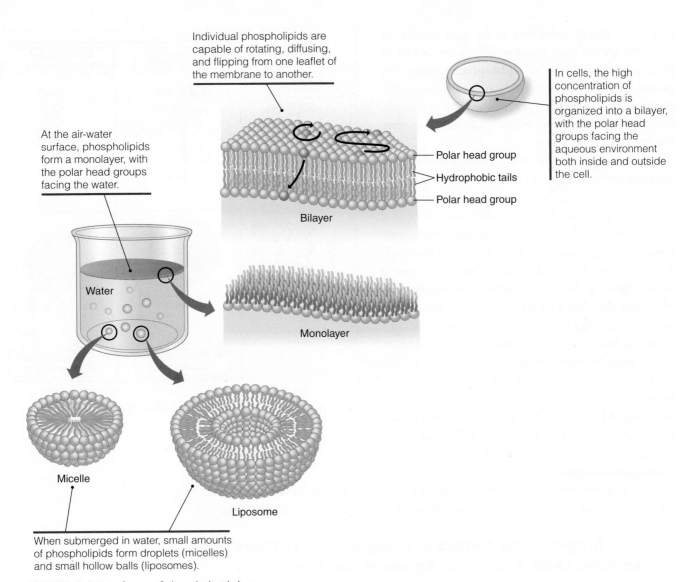

Individual phospholipids are capable of rotating, diffusing, and flipping from one leaflet of the membrane to another.

At the air-water surface, phospholipids form a monolayer, with the polar head groups facing the water.

In cells, the high concentration of phospholipids is organized into a bilayer, with the polar head groups facing the aqueous environment both inside and outside the cell.

Water

Polar head group
Hydrophobic tails
Polar head group

Bilayer

Monolayer

Micelle

Liposome

When submerged in water, small amounts of phospholipids form droplets (micelles) and small hollow balls (liposomes).

FIGURE 4-4 Four forms of phospholipid clusters.

inner and outer surfaces, they reorganize to form two layers. Notice that the hydrophobic tails of both layers cluster together, away from the water.

Micelles and liposomes do not typically form in most cells. However, because cellular membranes enclose aqueous compartments just as liposomes do, the phospholipids in membranes are oriented in essentially the same fashion. The largest structure formed by phospholipids interacting with water is a **monolayer**, formed at the air-water surface. Cells contain water on both sides of their membranes, so every cellular membrane contains phospholipids oriented in a **bilayer** (composed of two **leaflets**), with a single hydrophobic interior and two hydrophilic surfaces that interact with water, shown in Figure 4-4. This bilayer forms spontaneously, requiring no energy input or assistance from cells.

▦ Individual Phospholipids Can Diffuse Freely within a Single Layer

Because phospholipids are not covalently attached to one another, they can diffuse within each leaflet, as seen in Figure 4-4. Note that the hydrophobic tails in the membrane interior repel the polar head groups, discouraging spontaneous flipping of phospholipids from one leaflet to the other. In effect, membranes serve as diffusion barriers for their own phospholipids, making it difficult for them to flip from one side of the hydrophobic interior to the other. Phospholipid bilayers are sometimes described as two-dimensional fluids because of this restricted diffusion (**BOX 4-2**).

Lipid Bilayers Are Asymmetrical

Because phospholipids do not typically spontaneously flip from one side of a membrane bilayer to the other, the relative concentration of individual phospholipids on two sides of the same membrane can differ. For example, phosphatidylcholine (PC) is concentrated in the external leaflet of the plasma membrane, and phosphatidylethanolamine (PE), phosphatidylserine (PS), and phosphatidylinositol (PI) in the cytoplasmic leaflet. The functional consequences of this asymmetry are not well understood, but the loss of asymmetry is usually a property of a damaged or dying cell. The distribution of phospholipids in the two leaflets of a membrane is regulated by three classes of proteins that we will discuss later in this chapter.

Phospholipid Bilayers Are Semipermeable Barriers

Because the hydrophobic tails of phospholipids are concentrated in the interior of a membrane, they repel most hydrophilic molecules and form a barrier to diffusion from one side of the membranes to the other. This is the structural foundation for the selective barrier we discussed at the beginning of Chapter 1. Using artificial membranes composed entirely of phospholipids (i.e., no proteins or other cellular materials), scientists have measured the permeability of phospholipid bilayers to a variety of chemicals that cells commonly encounter, and they observe three simple trends, illustrated in **FIGURE 4-5**:

- **Small molecules are more permeable than large molecules.** Water (molecular weight: 10 daltons), molecular oxygen (O_2; 16 daltons), and carbon monoxide (CO; 14 daltons) are among the smallest molecules in cells, and they diffuse through lipid bilayers quite easily. Larger molecules (sugars, amino acids, nucleotides) cannot diffuse through lipid bilayers.

- **Nonpolar molecules are more permeable than polar molecules.** Remember that the hydrophobic tails of the phospholipids are the primary barriers to diffusion. Therefore, the more nonpolar a molecule is, the more easily it can pass through a membrane. Nitrous oxide (NO), O_2, and CO_2 have little to no electrical polarity and diffuse quite easily. Methane (CH_4), ethanol (CH_3–CH_2–OH), and propane (C_3H_8) are very hydrophobic, and therefore also readily diffuse through the hydrophobic region of a phospholipid bilayer.

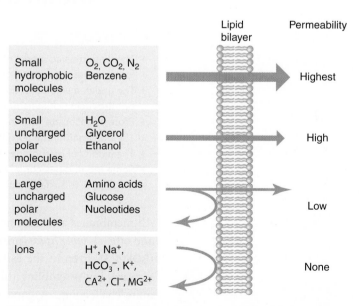

FIGURE 4-5 Phospholipid bilayers have varying permeability to solutes.

- **Charged molecules do not diffuse**. Any charged molecule, regardless of size, is repelled by the hydrophobic portion of a lipid bilayer and therefore cannot diffuse. This includes ions (H^+, Ca^{+2}, Na^+, K^+, Cl^-, HCO_3^-, PO_4^{-2}, etc.). This is one of the most useful properties of phospholipid bilayers, because it allows cells to build gradients of ions and other charged molecules on one side of a membrane, a topic we will cover in Chapter 10.

One shorthand means of estimating the membrane permeability of a given molecule is called the partition coefficient (abbreviated D). The formal definition of this value is the ratio of concentrations of a compound in the two phases of a mixture of two immiscible solvents at equilibrium. In organic chemistry, the value reflects the relative solubility of a molecule in a hydrophobic solvent (e.g., a liquid hydrocarbon such as octanol) versus its solubility in liquid water. In general, the higher the D value, the more membrane permeable a compound is.

■■ Maintaining a Chemical Imbalance across a Membrane Is Essential for Life

Recall from Chapter 1 that for a cell to remain alive, it must *never* be in chemical equilibrium with its surrounding environment. One of the easiest ways to remain out of equilibrium is to build a gradient of impermeable molecules on one side of the plasma membrane. All cells do this routinely by pumping ions from one side of the plasma membrane to the other. Prokaryotes use specialized proteins to pump H^+ or Na^+ ions from their cytosol into the extracellular environment, while eukaryotes use a protein that creates two gradients simultaneously: it pumps K^+ ions into the cytosol while pumping Na^+ ions out of the cytosol. We will have a much closer look at ion pump proteins in Chapters 10 and 11.

CONCEPT CHECK #1

Nonpolar amino acids contain a hydrophobic side chain attached to a polar backbone, similar to the structure of phospholipids. Why do polymers of amino acids (i.e., proteins) form such a wide variety of shapes, while clusters of phospholipids do not?

4.3 The Fluid–Mosaic Model Explains How Phospholipids and Proteins Interact within a Cellular Membrane

KEY CONCEPTS

▸ The fluid–mosiac model of membrane structure, proposed in the 1970s, forms the foundation for our current model of membrane structure.
▸ The updated version of the fluid–mosaic model emphasizes the mosaic property of membranes by introducing clusters of phospholipids, proteins, and cholesterol called lipid rafts.
▸ Proteins associate with membranes in three different ways.
▸ The transmembrane portion of most membrane-spanning proteins form an α-helix.
▸ The fluidity of membranes is dynamic and sensitive to four different variables.
▸ Some membranes contain immobilized protein complexes that permit stable attachment of the membrane to surfaces such as other cells or extracellular molecules.

All membranes are made up of three principal ingredients: *phospholipids* (phosphoglycerides and others), *other lipids* (primarily cholesterol and compounds called glycolipids), and *proteins* (**BOX 4-3**). Even in the early 20th century, identifying these ingredients was fairly straightforward, but assembling this information to create an accurate picture of a membrane was quite challenging. A good example is the bilayer organization of phospholipids: virtually nothing else in nature is arranged as a bilayer, so it was not at all obvious that membrane

BOX 4-3 TIP

Refer to Figure 1-10 for an illustration of lipid structures, including cholesterol.

phospholipids are arranged in this fashion. In 1972, Drs. S. Jonathan Singer and Garth Nicholson assembled the findings of decades of studies and proposed the **fluid–mosaic model** for membrane organization, shown at the top of **FIGURE 4-6**, which represented a tremendous advance in our understanding of how cells are built. The central tenet of this model is that a membrane resembles a mosaic pattern, wherein the phospholipid bilayer acts like a planar fluid, with membrane proteins floating randomly within it, supported by hydrophobic interactions between phospholipid tails and hydrophobic amino acids. (Cholesterol molecules were proposed to float in the phospholipid fluid, acting as "spacers" to maintain proper fluidity of the phospholipids.) Initially, the proteins were thought to be relatively scarce relative to the phospholipids, and have minimal interactions with them.

Recent evidence suggests that the picture is far more complicated, so changes have been made to the model, as shown at the bottom of Figure 4-6:

- We now know that most membrane proteins are extremely large, dwarfing the surrounding phospholipids, and that they can cluster to form large patches (sometimes called **lipid rafts**) that are chemically and physically distinct from the surrounding membrane. These patches contain proteins, phospholipids, and other lipids (especially cholesterol) bound together tightly enough to resist chemical disruption, though they are not covalently linked. One scientist, David M. Engelman, summarized this finding as, "membranes are more mosaic than fluid."[2]

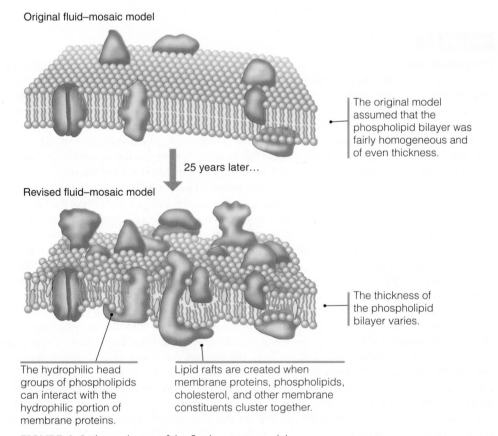

Original fluid–mosaic model

The original model assumed that the phospholipid bilayer was fairly homogeneous and of even thickness.

25 years later…

Revised fluid–mosaic model

The thickness of the phospholipid bilayer varies.

The hydrophilic head groups of phospholipids can interact with the hydrophilic portion of membrane proteins.

Lipid rafts are created when membrane proteins, phospholipids, cholesterol, and other membrane constituents cluster together.

FIGURE 4-6 The evolution of the fluid–mosaic model.

[2] Membranes are more mosaic than fluid. *Nature* 438, 578-580 (1 December 2005) | doi:10.1038/nature04394.

- Hydrophilic portions of phospholipids also associate with membrane proteins. Due to their relatively large size, membrane proteins can spread out over the surface of a membrane, allowing the hydrophilic head groups of phospholipids to form noncovalent bonds with hydrophilic regions of the proteins. This may be one means for stabilizing lipid rafts.

- The interaction of both hydrophilic and hydrophobic portions of phospholipids with membrane proteins distorts phospholipids from the idealized planar arrangement suggested by the original fluid–mosaic model. The implication of this is that due to these differences in phospholipid arrangement, there are variations in the thickness of a membrane.

- The formation of lipid rafts and distortions of the phospholipid bilayer by membrane proteins change over time—a membrane is *dynamic*. How cells control membrane fluidity and the formation or dissolution of lipid rafts is still not very clear (**BOX 4-4**).

Membrane Proteins Associate with Membranes in Three Different Ways

Membrane proteins are organized into three classes, according to how they associate with the lipid bilayer, as shown in **FIGURE 4-7**:

1. **Integral membrane proteins** are embedded into the lipid bilayer. They can either bend a portion of their structure into the bilayer (these are called **monotonic proteins**), or span the bilayer entirely (these are called **membrane-spanning**, or

---

BOX 4-4

The pool analogy. One way to visualize the *traditional* fluid–mosaic model is to imagine pouring enough ping-pong balls into a swimming pool to cover almost the entire surface. If we watched them long enough, we'd see that that the balls do not stay in one place. They jostle about and appear to flow over the water surface. This is a good approximation of how the head groups of phospholipids move in a fluid membrane (the fatty acid tails would be underneath the balls, in the water). To represent proteins in the membrane, we could throw in some buoyant objects (rubber ducks, beach balls, empty soda bottles, etc.). They too would flow over the surface, but not as quickly. Cholesterol molecules could be represented by blocks of wood: they float on the surface of the pool, but don't project above the ping-pong balls (i.e., they would be mostly submerged).

To incorporate the recent modifications of the fluid–mosaic model, we could make the following changes: (1) our protein objects would have to be much larger—imagine replacing some of the rubber ducks, and beach balls with much larger objects, such as air mattresses or floating pool lounge chairs, and increasing their numbers so they become a much greater percentage of the objects in the pool; and (2) we'd have to add some sort of sticky substance to the protein objects that could capture some of the ping-pong balls and wood blocks and permit the proteins to stick together, even if for a short time. If a group of floating chairs, ping-pong balls, and wood blocks clustered together because of the sticky tape, that complex would represent a lipid raft.

Note that our revised pool surface would be much less uniform than the one we began with. Floating complexes / lipid rafts would be mostly immobile, while individual balls, blocks, and protein objects would flow around them. And, because the proteins have such diverse shapes and are moving, the ping-pong balls and wood blocks around them would not be arranged in an even, flat plane. Some might be slightly submerged for a while, and others could be projecting out of the pool by a small amount. Finally, imagine that if we waited long enough, we would see these clusters would break up and reform; no single "picture" of this pool would represent what is happening, because everything is moving.

Also, don't forget that a membrane is a bilayer, so this scene plays out on both sides of the membrane.

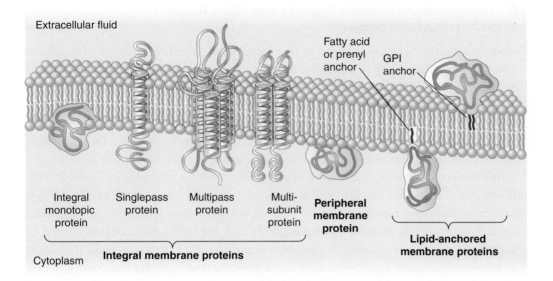

Extracellular fluid

Fatty acid
or prenyl
anchor

GPI
anchor

Integral
monotopic
protein

Singlepass
protein

Multipass
protein

Multi-
subunit
protein

**Peripheral
membrane
protein**

**Lipid-anchored
membrane proteins**

Cytoplasm

Integral membrane proteins

FIGURE 4-7 Three classes of membrane proteins. Each class is represented by a different color protein.

transmembrane proteins). The portion of these proteins that lies within the plane of the bilayer must contain enough hydrophobic amino acids to stabilize its interactions with the phospholipid tails. Membrane-spanning proteins may have one (called **singlepass**) or more (**multipass**) regions that extend entirely through the membrane bilayer. We will discuss how transmembrane proteins achieve these orientations in Chapter 8. In some cases, two or more transmembrane polypeptides bind together to form a multi-subunit transmembrane protein (**BOX 4-5**).

2. **Lipid-anchored membrane proteins** do not actually penetrate the lipid bilayer at all. Instead, they have a lipid covalently attached to the end of a cysteine amino acid (see Chapter 3, *Covalent Modifications Are Relatively Long Lasting*). Two types of lipids, called **prenyl groups**, are attached to the cysteine amino acids of some proteins. Notice that both the farnesyl and geranylgeranyl lipids contain long, trans-polyunsaturated regions that project outward from the cysteine side chain. Similarly, a modified phospholipid called **glycophosphatidylinositol** (GPI) can be attached to the carboxy terminus of some membrane proteins. These long hydrocarbon chains penetrate the lipid bilayer to associate with the hydrophobic interior of a membrane, thus serving as anchors to keep a protein associated with a membrane.

3. Some proteins are classified as membrane proteins even if they do not come into direct contact with the lipid bilayer. These **peripheral membrane proteins** bind to integral membrane proteins, and this binding is stable enough to effectively immobilize them at or near the surface of a membrane. These proteins are classified as membrane proteins because they typically remain associated with isolated membranes and/or membrane proteins when a cell is lysed (broken apart).

BOX 4-5 TIP

For an example of a dimeric membrane spanning protein, see Figure 6-18, (an integrin).

▪▪■ Transmembrane Proteins Typically Use Alpha Helices to Cross the Lipid Bilayer

Despite the multitude of possible shapes that proteins can adopt, virtually all of those that are known to completely span a membrane use the same types of secondary structures in their membrane-spanning regions. As we discussed in Chapter 3 (see *Secondary Structure Is Defined by Regions of Repetitive, Predictable Organization in the Primary Structure*), α-helices and β-sheets are two of the most common secondary structures found in proteins. Of those transmembrane proteins whose secondary structures are known, the most common secondary structure formed by the amino acids in the interior of a membrane is an α-helix.

By taking a look at the orientation of an α-helix within a phospholipid bilayer (**FIGURE 4-8**), we can see why this is true. Remember that the polypeptide backbone for all proteins is polar, due to the presence of the oxygen and nitrogen atoms. These two atoms stabilize secondary structures by forming hydrogen bonds. But the hydrophobic region created by phospholipid tails in a membrane is nonpolar, and discourages hydrogen bond formation. We therefore encounter a problem: how can a protein, which always has polar atoms in its backbone, remain stable in the hydrophobic interior of a membrane? The answer is

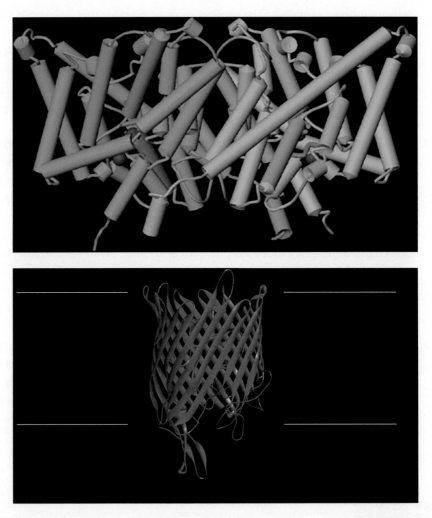

FIGURE 4-8 The orientation of alpha helices and beta sheets in transmembrane portions of integral membrane proteins.

CHAPTER 4 Phospholipids and Membrane Structure

elegantly simple. By forming an α-helix, a protein is able to completely surround the poly-peptide backbone with hydrophobic amino acid side chains while preserving the hydrogen bonds formed along the backbone. Because the side chains always project outward from an α-helix (see Figure 3-6), these side chains can interact with the phospholipid tails, effectively stabilizing the amino acids forming the α-helix in the membrane interior. This results in two forms of complementary stabilization: the hydrogen bonds in the backbone stabilize the α-helix, while the hydrophobic side chains anchor the helix in the lipid por-tion of the membrane.

Therefore, the membrane-spanning region of a singlepass membrane protein is typi-cally a single α-helix composed of mostly hydrophobic amino acids. If an integral mem-brane protein contains more than one transmembrane region, they may cluster together; in this case, the portion of each α-helix that remains in contact with the phospholipid tails must be hydrophobic, but the side that binds to the other α-helices can vary con-siderably, and may even contain a large number of polar and/or ionic amino acids, as illustrated in Figure 4-8. We will see examples of multispanning proteins like these in Chapters 10, 11, and 14.

Beta sheets, usually arranged in an antiparallel fashion, are also found in membrane-spanning regions of integral membrane proteins, though they are less common. When these sheets are large enough, they bend to create a cylindrical structure called a β-barrel, as shown in Figure 4-8. Some of these barrels serve as membrane channels. Many channel-like barrels have alternating hydrophobic and hydrophilic amino acids in each strand. This distri-bution of amino acids creates both a hydrophobic face and a hydrophilic face on the β-sheets. The hydrophobic face projects outward into the phospholipid tails, while the hydrophilic side of the barrel faces inward, facilitating transport of hydrophilic solutes and water through the membrane.

Cellular Membranes Are Both Fluid and Static

Every membrane in a cell has its own distinct structure and function. Some are quite porous, such as the outer membrane of gram-negative bacteria, while others, such as the inner mem-brane of mitochondria and chloroplasts, are permeable to only a small number of molecules. Likewise, membranes can vary in their degree of fluidity: the endoplasmic reticulum and Golgi membranes in eukaryotic cells permit a great deal of diffusion of phospholipids and membrane proteins, whereas in most cells, some regions of the plasma membrane are much more static and heterogeneous. Because each membrane is composed of a unique combina-tion of different phospholipids and proteins, plus varying amounts of other lipids, cells can customize their membranes to optimize the functions they perform. We will see numerous examples of specialized membranes in later chapters.

Membrane Fluidity Is Sensitive to at Least Four Different Variables

The balance point between membrane fluidity and structural stability is determined by a number of different factors. In most experiments concerning membrane fluidity, small probe molecules are added to a membrane, and then their movements are tracked over time with a microscope (note that the fluidity of the phospholipids themselves is not typically measured). The farther the probes move over time, the greater the fluidity of the membrane. Because membrane phospholipids are smaller than membrane proteins and act independently from one another, their chemical composition contributes a great deal to the overall fluidity of most membranes. Specifically, the length and saturation of lipid tails affects membrane fluidity, according to two very simple trends:

- Phospholipids containing short (≤16 carbon atoms) lipid tails increase membrane fluidity, relative to those containing long lipid tails (>16 carbon atoms).
- Cis unsaturated fatty acid tails have kinks induced by the cis double bonds, and therefore occupy a greater volume of space in a membrane than a saturated fatty acid tail of the same length. This increased volume decreases the density of phospholipids in a membrane, and ultimately leads to an increase in membrane fluidity.

Other variables also impact membrane fluidity:

- The amount of cholesterol, a common lipid component of most membranes, present makes a big difference. Other than a single hydroxyl group that interacts with water, cholesterol is composed entirely of carbon and hydrogen, and is therefore very hydrophobic. It inserts spontaneously into membranes, and interacts primarily with the tails of membrane phospholipids. The concentration of cholesterol varies considerably in different cellular membranes (**TABLE 4-1**). At low concentrations (approximately 10–30% of the total number of molecules in a membrane), its rigid, planar structure allows it to act as a spacer between phospholipids, thereby increasing their motion and the associated fluidity of a membrane. At higher concentrations (around 50%, as in most eukaryotic plasma membranes), cholesterol can actually have the opposite effect, making a membrane more rigid (**BOX 4-6**).
- Temperature and atmospheric pressure affect the motion of membrane constituents. A rise in temperature increases the Brownian motion (random movement) of every component of a membrane, and this increased motion translates to increased fluidity. In addition, significant changes in pressure, such as when a marine mammal such as a whale dives thousands of meters below the ocean surface, will literally compress the constituents of cellular membranes into smaller volumes, thereby hindering diffusion and decreasing fluidity.

▆▆■ Membrane Components Can Form Large Molecular Complexes with Little or No Mobility

Membranes or membrane patches do not always need to move to be useful. For example, multicellular organisms use proteins embedded in their plasma membrane to adhere to one another, as shown in **FIGURE 4-9**. To be stable, these so-called **cell–cell junctions** have to be held in place by clusters of proteins in the cell interior. All of this binding between cytosolic

TABLE 4-1 Major lipid components of selected biomembranes.				
	COMPOSITION (MOL %)			
Source/Location	PC	PE + PS	SM	Cholesterol
Plasma membrane (human erythrocytes)	21	29	21	26
Myelin membrane (human neurons)	16	37	13	34
Plasma membrane (E. coli)	0	85	0	0
Endoplasmic reticulum membrane (rat)	54	26	5	7
Golgi membrane (rat)	45	20	13	13
Inner mitochondrial membrane (rat)	45	45	2	7
Outer mitochondrial membrane (rat)	34	46	2	11
Primary leaflet location	Exoplasmic	Cytosolic	Exoplasmic	Both

PC = phosphatidylcholine; PE = phosphatidylethanolamine; PS = phosphatidylserine; SM = sphingomyelin.

The dance floor analogy. To visualize the relationship between fatty acid saturation, cholesterol concentration, and membrane fluidity, let's imagine the dynamics on a crowded dance floor, with the dancers playing the role of the phospholipids. If everyone dances with their arms flat against their side and their legs almost perfectly straight (i.e., mostly jumping up and down), they take up relatively little space, so more dancers can fit on the floor. This is analogous to a membrane composed entirely of fully saturated phospholipids. To create more room for the dancers, one can add a cholesterol analog: nondancing elements, such as security staff (bouncers). The dancers have to make room for the bouncers, so they increase their movement (fluidity) as the bouncers walk across the floor (i.e., moving bouncers always leave a wake). This is how a low concentration of cholesterol can *increase* membrane fluidity—the cholesterol disrupts the packing of the phospholipids, and creates spaces where more movement can take place. If one had approximately equal numbers of dancers and bouncers on the floor (a 1:1 ratio, as in some regions of the plasma membrane), the dancers would have a much harder time moving about; they might even be boxed in by some of the bouncers. This is why high concentrations of cholesterol *decrease* membrane fluidity.

Now, imagine the same scenario, except we will introduce unsaturated fatty acids. Let the legs of the dancers represent the fatty acyl tails of the fatty acids. Saturated and trans unsaturated fatty acids are relatively straight, but cis unsaturated fatty acids are bent. So, if we imagine that one of the two fatty acids on the phospholipids is cis unsaturated, this is represented by having some of the dancers bend one of their legs at the knee. Now, when these "cis unsaturated" dancers move, they occupy much more space, due to their lower legs sticking outward. A dance floor containing a combination of saturated and cis unsaturated dancers is more fluid than one containing only saturated dancers; cis unsaturated phospholipids increase membrane fluidity, by creating space to move.

and membrane proteins results in the formation of some very large molecular complexes. Similar clustering of proteins in the plasma membrane occurs when cells bind to elements of the extracellular matrix, forming **cell–matrix junctions**. Regardless of the type of fatty acid chains in the phospholipids, or the amount of cholesterol in a membrane, large molecular complexes like these cannot diffuse easily. In fact, it is rather useful for these complexes to remain immobile, because they can partition a membrane into regions. We will see several examples of these types of membrane complexes in Chapters 5 and 6.

CONCEPT CHECK #2

One of the most significant additions to the original fluid–mosaic model is the concept of variation in phospholipid distribution in membranes. Based on our discussion of molecular teamwork in Chapter 1, what impact would you expect organized clusters of phospholipids to have on the function of membranes?

4.4 The Smooth Endoplasmic Reticulum and Golgi Apparatus Build Most Eukaryotic Cellular Membrane Components

KEY CONCEPTS

▸ The lipid components of membranes are built from precursor molecules in the cytosol.
▸ In humans, most cells receive lipids, including cholesterol, by engulfing lipoprotein complexes that are synthesized by the liver and circulate in the bloodstream.
▸ Phosphoglycerides are assembled in the smooth endoplasmic reticulum.
▸ Six different phosphoglycerides, each containing a different head group, are synthesized from a common precursor, phosphatidic acid.

- Most sphingolipids are assembled in the Golgi apparatus from precursors made in the endoplasmic reticulum.
- Proper assembly and delivery of complete membranes requires at least nine distinct steps following the synthesis of membrane components.
- Flippases, floppases, and scramblases are enzymes that move phospholipids from one leaflet of a membrane to the other, thereby generating an asymmetrical distribution of membrane components that is necessary for proper membrane function.
- Lipid carrier proteins are capable of transporting lipids from one membrane to another in a cell.

The final major topic of this chapter explores how biological membranes are assembled. Historically, our understanding of the biochemistry responsible for membrane synthesis, especially for phospholipids, has lagged far behind that for the other basic cellular building blocks (sugars, nucleic acids, and proteins). But considerable progress has been made in the last 15 years, such that we can now examine the mechanisms responsible for membrane synthesis with a level of detail comparable to that for DNA and protein synthesis. Our first step is to examine how and where each of the major membrane components are built, and then we will examine how they are organized to form functional membranes.

Glycerol and Fatty Acids Are Synthesized in the Cytosol

Glycerol, which forms the backbone of all phosphoglycerides, is synthesized in the cytosol. For most cells, the precursor to glycerol is a closely related molecule called **glycerol-3 phosphate**. Glycerol-3 phosphate is typically quite abundant in cells because it is an intermediate molecule formed during the digestion of food molecules (sugars, fats, proteins) to make ATP. We will visit this topic again in much more detail in Chapter 10.

The fatty acids used to build the tails of phosphoglycerides and other lipids are synthesized in the cytosol by an enzyme called **fatty acid synthase**. In humans, most cells don't have to synthesize their own fatty acids; instead, they absorb fatty acids from molecular complexes called **lipoproteins**. Lipoproteins are made by cells in the liver that secrete them into the bloodstream, which carries them to the remainder of the cells in the body. The remaining ingredients needed to synthesize membranes (phosphate groups, water, etc.) are in such great abundance in cells that we can take them for granted.

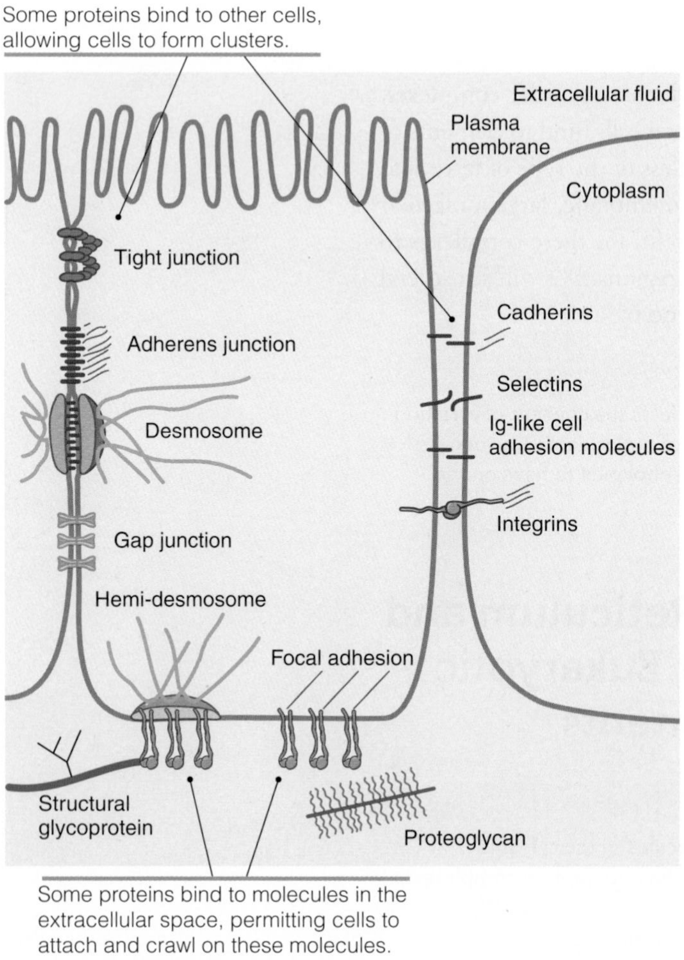

The Synthesis of Phosphoglycerides Begins at the Cytosolic Face of the SER Membrane

All phosphoglyceride synthesis begins with the covalent attachment of two fatty acids to glycerol by transmembrane enzymes called **acyl transferases** in the smooth endoplasmic reticulum (SER). The resulting product, called phosphatidic acid, spontaneously inserts itself into the cytoplasmic leaflet of the smooth endoplasmic reticulum membrane. Six different molecules, called head

Some proteins bind to other cells, allowing cells to form clusters.

Tight junction

Adherens junction

Desmosome

Gap junction

Hemi-desmosome

Structural glycoprotein

Focal adhesion

Proteoglycan

Extracellular fluid

Plasma membrane

Cytoplasm

Cadherins

Selectins

Ig-like cell adhesion molecules

Integrins

Some proteins bind to molecules in the extracellular space, permitting cells to attach and crawl on these molecules.

FIGURE 4-9 Plasma membrane proteins form clusters that permit cell adhesion.

groups, are added to the phosphate group to generate six different phospholipids. The standard convention for naming most phospholipids is "Phosphatidyl-X" where X is the name of the head group. Often, we simply abbreviate them as PC, PS, PE, PI, PG, and bisPG (a derivative of PG also called cardiolipin, or CL).

Each of these phospholipids plays an important role in regulating cellular activities. They are far more than space-filling molecules in the "fluid" portion of a membrane. For example, when PI is modified by the addition of two more sugars to the inositol head group, the resulting glycophosphatidylinositol (GPI) can act as the anchor for some lipid-anchored membrane proteins. GPI synthesis begins on the cytosolic face of the endoplasmic reticulum (ER), and is completed in the ER interior, as shown in **FIGURE 4-10**. The relative concentrations of these phospholipids can vary considerably between different cell types and among different membranes in the same cell. In later chapters, we will refer back to these phospholipids to help explain how specific cellular functions are performed.

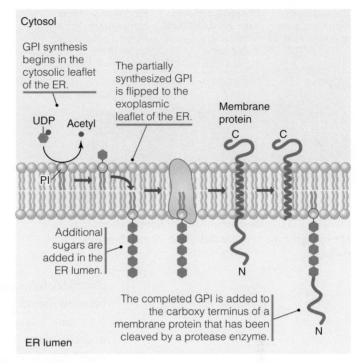

FIGURE 4-10 GPI is synthesized and attached to membrane proteins in the ER.

Additional Membrane Lipids Are Synthesized in the Endoplasmic Reticulum and Golgi Apparatus

Phosphoglycerides are by far the most abundant lipids found in membranes, but other membrane lipids are also noteworthy. For example, the **farnesyl** and **geranylgeranyl lipids** that are attached to lipid-anchored membrane proteins are precursors of cholesterol, and are synthesized in the ER membrane from the same building blocks used to make fatty acids. The enzymatic steps for synthesizing cholesterol are different than those used to make fatty acids, and are so complicated that elucidation of the entire synthesis pathway took over two decades to complete and led to the Nobel Prize in 1964. As with fatty acids, most cholesterol synthesis in humans takes place in the liver.

Most Membrane Assembly Begins in the SER and Is Completed in the Target Organelle

Once the phosphoglycerides and other lipids have been synthesized at the cytosolic face of the SER, they are joined by additional membrane proteins synthesized in the rough endoplasmic reticulum (RER) to form a fluid–mosaic structure that closely resembles a mature cellular membrane. Simply collecting all of these molecules in the same location (the ER), however, is not enough to ensure the functional stability of all cellular membranes. Eight tasks remain before membrane synthesis is complete. These tasks, in rough chronological order (some occur simultaneously, or at multiple times in multiple locations), are:

1. **Membrane proteins in the endomembrane system are inserted in the ER.** All cellular membranes contain proteins. Recall from Chapter 1 (see *The Plasma Membrane, Endoplasmic Reticulum, Golgi Apparatus, Endosomes, Lysosomes, and Peroxisomes Form a Protein-Trafficking Network Called the Endomembrane System*) that the ER is linked to many other organelles by a vesicle-shuttling system. These vesicles are membrane-bound compartments that bud from one organelle and fuse

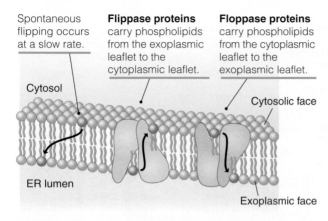

Spontaneous flipping occurs at a slow rate.

Flippase proteins carry phospholipids from the exoplasmic leaflet to the cytoplasmic leaflet.

Floppase proteins carry phospholipids from the cytoplasmic leaflet to the exoplasmic leaflet.

Cytosol

Cytosolic face

ER lumen

Exoplasmic face

FIGURE 4-11 Three mechanisms for translocation of phospholipids from one leaflet of the ER membrane to the other.

with another. The membrane proteins in the endomembrane system originate in the endoplasmic reticulum.

2. **Nonspecific flippases and floppases transport phosphoglycerides from one membrane leaflet to the other in the ER.** Earlier in this chapter we mentioned that membrane phosphoglycerides are not equally distributed in both layers of a membrane. For example, PC is enriched in the external face of the plasma membrane, sometimes called the **exoplasmic** or **ectoplasmic leaflet**, relative to the internal face, called the **cytoplasmic leaflet**. Some of this asymmetry actually originates in the ER. All phosphoglycerides are synthesized on the cytoplasmic leaflet of the ER, but a membrane consists of *two* leaflets: How do the phosphoglycerides get to the other side of the phospholipid bilayer? There are two mechanisms, shown in **FIGURE 4-11**. First, phosphoglycerides have the ability to spontaneously flip-flop from one side of a membrane to the other. This is not a rapid process—the half-life for this flip-flopping (i.e., time it takes for half of them to flip sides at least once) ranges from several hours to days. The second mechanism is much more rapid: a group of transmembrane proteins called **floppases** use ATP energy to carry phospholipids, including phosphoglycerides, to the exoplasmic leaflet of the ER. Note that this is a one-way ride; floppases do *not* move phospholipids back to the cytoplasmic leaflet. The return journey takes place on proteins conveniently named **flippases**, which also use ATP to selectively power the transport in the reverse direction only. The half-life for transport by both types of enzymes is less than one minute.

3. **Membrane vesicles transport new phosphoglycerides and membrane proteins from the endoplasmic reticulum to other organelles in the endomembrane system.** Despite both spontaneous and enzyme-assisted flipping and flopping, the absolute number and relative concentration of phosphoglycerides in both leaflets of the ER are *not* identical. When membrane-bound vesicles shuttle between organelles in the endomembrane system, they preserve the membrane asymmetry of their parent organelles; a vesicle budding from the ER and fusing with the Golgi apparatus will, in effect, donate a piece of ER membrane to the Golgi apparatus, such that whatever asymmetry was present in the ER is passed to the Golgi. Vesicles budding from the Golgi apparatus and fusing with the plasma membrane will likewise preserve the asymmetry of the Golgi, thereby replicating the unequal distribution of membrane components that began in the ER. (Note that these vesicles also contain the proteins synthesized in the ER. We will discuss these proteins in the next section.) This membrane trafficking will be covered in much greater detail in Chapter 9.

4. **Glycolipid synthesis is completed on the exoplasmic face of the Golgi apparatus.** Some lipid precursors synthesized in the ER arrive in the Golgi and are converted into glycolipids by enzymes in the lumen of the Golgi apparatus. Because all of the completed glycolipids are present in the *exoplasmic* face of the Golgi membranes, this adds to the overall membrane asymmetry. This is the final stage of membrane component synthesis. All subsequent steps involve modifying and/or moving components to their proper destination(s). The mechanisms governing proper sorting of membrane components to different destinations are quite complex, and are discussed in detail in Chapter 9.

5. **Final orientation of plasma membrane lipids occurs *in situ*.** Aside from the ER, most attention on membrane asymmetry has focused on the plasma membrane, which is probably the most asymmetric membrane in most cells. The plasma membrane of most cells contains a variety of phospholipid transport proteins, including specialized flippases and floppases that are much more selective for specific phospholipids than their ER counterparts. As a result, one (PC) is highly enriched in the exoplasmic leaflet, while the remainder of the phosphoglycerides are relatively enriched in the cytoplasmic leaflet, as shown in **FIGURE 4-12**. Because this reorganization takes place directly in the

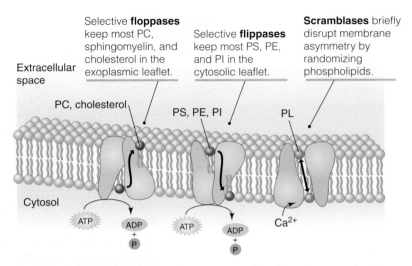

FIGURE 4-12 Establishment and maintenance of lipid asymmetry in the plasma membrane.

plasma membrane, rather than in the ER or Golgi compartments previously traversed by these molecules, it is said to occur *in situ* (from Latin, literally meaning "in place"). The asymmetric organization of the plasma membrane plays an important role in many cellular functions, such as controlled cell death (also called **apoptosis**) and cell signaling, which we will discuss in Chapters 11 and 13. It is perhaps not surprising, then, that a third class of phospholipid transporters, called **scramblases**, can have a profound impact on cellular function. These enzymes are capable of both flipping *and* flopping phospholipids in the plasma membrane, and require no metabolic energy (e.g., ATP) to do so. Short-term activation of scramblases triggers cellular signaling pathways, but promotes apoptosis if the scrambling is not repaired.

6. **Membrane proteins in other organelles are inserted *in situ*.** Most proteins synthesized by free ribosomes in the cytosol (i.e., those not attached to the RER) remain in the cytosol for the duration of their existence. A subset of these, however, contain amino acid sequences (tags) that target them to specific organelles, shown in **FIGURE 4-13**. In addition, cytosolic proteins containing the proper sequences can attach to, or insert themselves into, the membranes of mitochondria, chloroplasts, and peroxisomes. There is some evidence that vesicles from the ER can fuse with and donate membrane components to these organelles in at least some cells, but most membrane proteins in these organelles do not arrive via this route. The mechanisms governing targeting of cytosolic proteins to these organelles will be discussed in Chapter 8.

7. **Fatty acid binding proteins carry phospholipids to peroxisomes, mitochondria, and chloroplasts.** Because most phospholipid synthesis occurs in the ER, those organelles that receive little or no

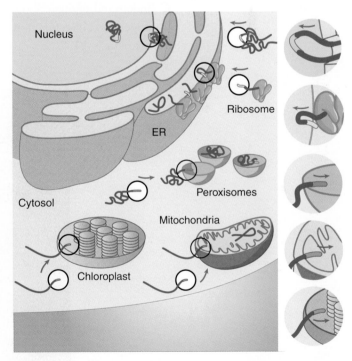

FIGURE 4-13 Protein "tags" direct some cytosolic proteins to enter organelles.

vesicle traffic from the ER must have an alternative means for building their own phospholipids. While the details of this type of transport have yet to be uncovered, at least five families of proteins are involved (**BOX 4-7**); collectively, they are known as **lipid-binding proteins**. The best characterized of these is sterol carrier protein-2, which delivers both fatty acids and cholesterol to peroxisomes, mitochondria, and chloroplasts, as shown in **FIGURE 4-14**.

8. **Cholesterol is delivered to target membranes**. Cholesterol is synthesized in the ER, but most of it is not carried to the plasma membrane via ER-derived vesicles, as is true for nearly all other components of the plasma membrane. Direct observation of cholesterol transport with a microscope shows that it can make the journey from the ER to the plasma membrane much faster than any membrane protein. This observation helps explain why the concentration of cholesterol in the ER and Golgi apparatus is much lower than that found in the plasma membrane.

Collectively, all of these steps are required to build and maintain healthy membranes in cells, as summarized in **FIGURE 4-15**. Most of the molecular details of the mechanisms governing such a complicated web of activities have yet to be discovered; nonetheless, we now understand enough about how the membrane constituents interact to accomplish a wide range of cellular functions, and these subjects will be addressed in much greater detail in Chapters 5–14.

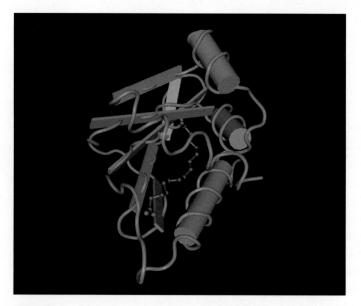

FIGURE 4-14 Sterol carrier protein-2 structure. Note the fatty acid bound to it.

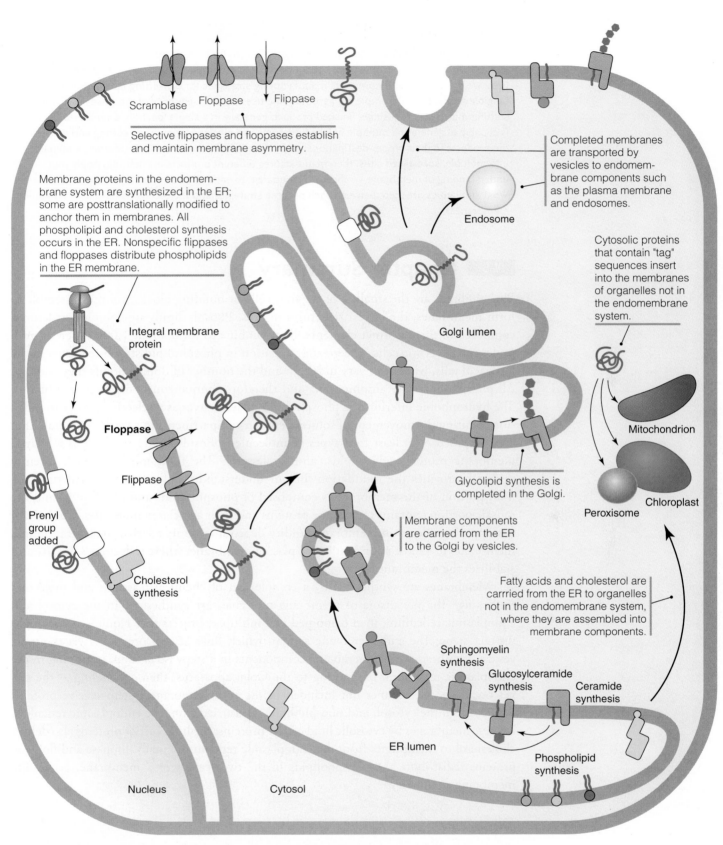

Scramblase   Floppase   Flippase

Selective flippases and floppases establish and maintain membrane asymmetry.

Membrane proteins in the endomembrane system are synthesized in the ER; some are posttranslationally modified to anchor them in membranes. All phospholipid and cholesterol synthesis occurs in the ER. Nonspecific flippases and floppases distribute phospholipids in the ER membrane.

Integral membrane protein

Floppase

Flippase

Prenyl group added

Cholesterol synthesis

Endosome

Completed membranes are transported by vesicles to endomembrane components such as the plasma membrane and endosomes.

Cytosolic proteins that contain "tag" sequences insert into the membranes of organelles not in the endomembrane system.

Golgi lumen

Mitochondrion

Peroxisome

Chloroplast

Glycolipid synthesis is completed in the Golgi.

Membrane components are carried from the ER to the Golgi by vesicles.

Fatty acids and cholesterol are carrried from the ER to organelles not in the endomembrane system, where they are assembled into membrane components.

Sphingomyelin synthesis

Glucosylceramide synthesis

Ceramide synthesis

ER lumen

Phospholipid synthesis

Nucleus   Cytosol

FIGURE 4-15 Summary of membrane synthesis mechanisms in eukaryotes.

The steps outlined in this section describe a mechanism for manufacturing specialized membranes in a cell. This is an excellent example of how cellular information is put to work. In our everyday lives, we are familiar with numerous manufacturing strategies for converting information into tangible products. For example, an assembly line uses a linear sequence of processes to change starting materials into a single finished product, generally in a single location; a centralized network is capable of generating multiple finished products in a single location, then sorting and delivering the products to their proper destinations; and a distributed network sorts the starting materials into multiple, specialized sites, then manufactures different products at each site. Apply your understanding of membrane synthesis to compare it to each of these manufacturing strategies. What similarities are there between each of these strategies and the mechanisms cells use to build membranes?

4.5 Chapter Summary

Phospholipids are the smallest of the four cellular building blocks, yet they assemble to form membranes, the largest structures in cells. Phospholipids are complex structures, composed of three distinct elements, that combine to form about 1,000 different molecules in eukaryotic cells. The greatest variation in phospholipid structure occurs in the fatty acid tails, which can vary in length and the number of double bonds they contain. All phospholipids are amphipathic, and therefore spontaneously organize as a bilayer. The hydrophobic interior of a phospholipid bilayer serves as a selective barrier to diffusion, limiting the movement of solutes between compartments in a cell. Cell membranes are composed of at least four types of molecules: phospholipids arranged as a bilayer, membrane proteins, glycolipids, and cholesterol. The fluid–mosaic model, recently updated, provides the foundation for our understanding of membrane structure; the fluid portion of the membrane is composed of phospholipids and glycolipids, and the mosaic portion contains membrane proteins and their associated molecules. Cholesterol has opposing effects on membrane fluidity depending on its relative concentration in a membrane. It is most abundant in the plasma membrane, where it decreases fluidity and stabilizes the membrane.

Membranes are synthesized by a complex set of chemical reactions and molecular trafficking. The precursors of membrane molecules are synthesized in the cytosol and endoplasmic reticulum, then combined in a multipart process that requires at least eight distinct steps. The endomembrane system, which links several organelles via transport vesicles, synthesizes most membrane components in a stepwise fashion, beginning in the endoplasmic reticulum, progressing to the Golgi apparatus, then concluding at the target organelle. Membranes not included in the endomembrane system receive proteins directly from the cytosol, and phospholipids are carried from the endoplasmic reticulum to these membranes by cytosolic lipid carrier proteins; the lipid carrier proteins also deliver cholesterol to membranes from the endoplasmic reticulum. Finally, flippase and floppase proteins redistribute the phospholipids in the two leaflets of a membrane, generating membrane asymmetry.

The key difference between polypeptides and membranes is that the amino acids in polypeptides are covalently linked to one another, such that a change in shape of one portion of a polypeptide can impact the structure of the entire polypeptide. In contrast, phospholipids are never covalently linked to one another, so each acts independently. If one region of the membrane changes shape (e.g., forms a lipid raft), the phospholipids surrounding this region are largely unaffected.

Variations in membrane fluidity allow for the formation of clustered teams of molecules in phospholipids. This selective clustering permits greater specialization of membranes, enhancing the division of labor that is key to cell survival. For example, clusters of molecules in a membrane can subdivide the membrane into distinct compartments that perform different functions. In general, the more heterogeneous a membrane is, the greater the number of distinct tasks it can perform.

Assembly line: Cells use enzymes to manufacture the precursor molecules of membrane components in one location, somewhat like assembly lines. For example, glycerol and fatty acids are synthesized via a sequence of steps in the cytosol, while cholesterol and ceramide are likewise synthesized in the ER. Another assembly-line-like strategy is the modification of phospholipids (e.g., addition of sugars to glycolipids) and membrane proteins (e.g., addition of GPI) that takes place in the Golgi apparatus after their synthesis in the ER.

Centralized network: The ER resembles a central assembly point in a centralized network, in that both phosphoglyceride and sphingomyelin synthesis is completed in the ER before these molecules are sent to the Golgi apparatus for sorting to their final destinations in the endomembrane system.

Distributed network: The final orientation of phospholipids in a membrane is determined *in situ* by the flippases and floppases for each membrane. Similarly, cholesterol and other lipids are delivered by lipid-binding proteins to organelles without having to pass through the Golgi apparatus.

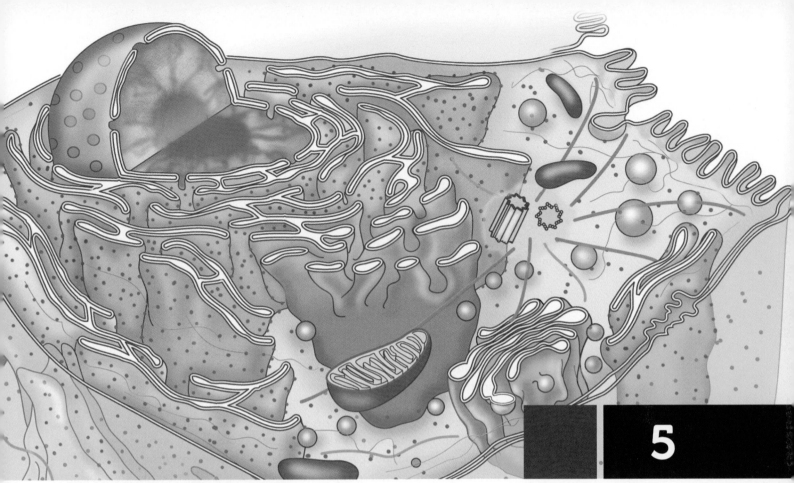

5

The Cytoskeleton and Cellular Architecture

■5.1■ The Big Picture

The first four chapters of this text have focused on describing the molecular constituents found in all cells. So far, the emphasis has been on the general traits of the four classes of essential cellular molecules (sugars, nucleic acids, proteins, and lipids) that allow them to perform their functions in cells. Now it's time to put this information to work, by applying the concepts in the first four chapters to specific molecules in cells. We will start by examining proteins known as cytoskeletal proteins, which collectively constitute the cytoskeleton of a cell.

We are embarking on a detailed examination of cellular structure and function, so let's have a look at what we expect to find. Using the cytoskeleton as an example, we are going to start amassing enough information to build cell biology principle #1: Cells are always in motion.

At first, this principle might be surprising, because we rarely think of ourselves as being in constant motion. Most cellular motion is not the standard type we experience in our everyday lives; instead, it means the cells in our bodies (which we can't see), as well as every other living cell, are undergoing some kind of movement *to stay alive*. We're referring to types of motion that may be quite unfamiliar at this point, which is why it takes a whole chapter to explain this principle (**BOX 5-1**).

There are five major ideas in this chapter:

- All cytoskeletal proteins share two important features. One is that they form polymers inside cells; this means that, just as many amino acids polymerize to form a functioning polypeptide, cytoskeletal proteins polymerize to form a functioning *skeleton* inside cells. Unlike polypeptides, however, the individual units in the skeleton are not always held together by covalent bonds. Instead, many bind via noncovalent bonds, and this means these parts of the skeleton can be disassembled and reassembled relatively easily. The second important feature of cytoskeletal proteins is that they function in much the same way as our own bony skeleton, in that they help provide the mechanical strength necessary to resist physical forces, hold cells in place, and determine the shape and motion of individual cells. Whereas our bony skeleton is primarily composed of one material (bone), the cytoskeleton is made up of three very different classes of proteins, each specializing in a subset of these functions.

- The class of cytoskeletal proteins called the *intermediate filaments* is primarily associated with the mechanical strength function in cells. What does this have to do with cellular motion? Any moving object has to have enough strength to both generate the motion *and* resist the resulting forces without tearing apart, and intermediate filaments help cells resist external forces. Relative to the other two cytoskeletal protein types covered in this chapter, intermediate filaments are specialized to perform a small number of tasks, based on their simple yet distinctive shapes. Pay particular attention to how the method of assembly helps to make these proteins especially tough, and the costs/benefits of this strategy. These features will be handy when comparing the three types of cytoskeletal proteins.

BOX 5-1 TIP ■

My goal in each chapter of this section of the book is for students to understand one principle of cell biology at a level that is suitable for their background and interests. This means that, depending on the degree of detail they are seeking, it may be unnecessary to read an entire chapter, or they may have to seek more advanced texts. As in Chapters 1, 2, 3, and 4, it is important to ask the course instructor what level of sophistication is appropriate for their course.

- The class of cytoskeletal proteins called *microtubules* is primarily associated with the trafficking function in cells. Trafficking is a word that is rarely used in everyday conversation, but cell biologists use it because it describes the kind of remodeling of a cell's contents that takes place continuously throughout a cell's lifetime. Given the structure–function relationship that forms the foundation of cell biology, we can reasonably expect that because these proteins perform a different *function* from intermediate filaments, they will also have a significantly different *structure*. As we discuss the polymerization of microtubules, pay attention to the differences between this strategy and that used for intermediate filament polymerization.

- Another class of cytoskeletal proteins, called *actin*, is associated with large-scale movement of cells (keep in mind that because cells are so small, moving a few millionths of a meter is considered large scale for them). Actin is by far the most complex of the three types of cytoskeletal proteins because it forms so many different shapes. Therefore, comparing its polymerization mechanism with intermediate filaments and microtubules will not be sufficient to explain how it functions; instead, we will address some of the basic functions of actin in this chapter, and examine more specific examples in later chapters.

- Eukaryotic cytoskeletal proteins have a long evolutionary history, evolving from similar proteins that first appeared in prokaryotic cells. If all modern cells move all the time, the earliest prokaryotes likely did, too. A brief look at these prokaryotic cytoskeletal proteins suggests a great deal about the roles they played in the earliest forms of life.

5.2 The Cytoskeleton Is Represented by Three Functional Classes of Proteins

KEY CONCEPTS

▸ The cytoskeleton is a complex mixture of three different types of proteins that are responsible for providing mechanical strength to cells and supporting movement of cellular contents.

▸ The most visible form of cytoskeletal proteins are long filaments found in the cytosol, but these proteins also form smaller shapes that are equally important for cellular function.

▸ The structural differences between the three cytoskeletal protein types underscores their four different functions in cells.

The term *cytoskeleton* is used to describe a network of filamentous proteins that occupy a large portion of the cytosol in most cells, where they appear to link most of the organelles to each other and to the plasma membrane, as shown in **FIGURE 5-1**. This network plays a significant role in determining the distribution of cellular contents, and even the overall shape of the cell. If we consider the function of our own bony skeleton, the analogy makes sense. Our skeleton defines most of our body shape (our height, the length of our limbs, the shape of our hands and feet, etc.), and permits movement by providing the attachment sites for muscles. As we will see, the cytoskeleton contributes greatly to the motion of cells as well.

If we continue the skeleton analogy, we can see additional similarities: Just as our skeleton is built from a relatively small number of materials (mostly bones, ligaments, and cartilage), the cytoskeleton is composed largely of three different types of proteins. Bones can vary widely in size and shape despite their similar molecular makeup, and this same idea applies to cytoskeletal proteins. Damage to our bones and ligaments can have devastating

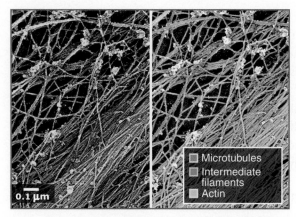

FIGURE 5-1 The cytoskeleton forms an interconnected network of filaments in the cytosol of animal cells. Reprinted from J. Struct. Biol., vol. 115, M. T. Svitkina, A. B. Verkhovsky, and G. G. Borisy, Improved Procedures for Electron Microsopic Visualization..., pp. 290-303, Copyright (1995) with permission from Elsevier [http://www.sciencedirect.com/science/journal/10478477]. Photos courtesy of Tatyana Svitkina, University of Pennsylvania.

effects on our ability to function, while damage of the cytoskeleton can likewise be fatal to a cell.

If we apply the three traits of proteins from Chapter 3 to the cytoskeletal proteins, we see even more similarities. By necessity, all of them bind to a target; in many cases, the target is simply another copy of the same type of protein. When they bind, they change shape and form polymers that are typically the largest protein-based structures in cells. Intermediate filament proteins bind to other intermediate filament proteins to form long, filamentous polymers named intermediate filaments. Likewise, tubulin proteins bind to other tubulin proteins to form long, tube-shaped polymers called microtubules, and actin proteins polymerize to form actin filaments, sometimes known as microfilaments. Note that they do not form mixed polymers: each protein type will only polymerize with others of the same type. As we will see below, these cytoskeletal proteins can bind to many different proteins as well; these proteins can connect the polymers to organelles or the plasma membrane, for example. Finally, different filament types can be interconnected by bridging proteins to form a diverse network.

Only when we consider the third trait of proteins do we see obvious differences between these filaments. Each filament performs a distinct set of functions in cells. When one considers the how many different proteins each cytoskeletal polymer can bind to, the number of combined functions increases tremendously. How do we keep track of all of these functions? By clustering similar functions together into a small, manageable set. Each of the cytoskeletal protein types will be classified into a different functional category, and our mission is to find the information that will justify classifying them in this way.

5.3 Intermediate Filaments Are the Strongest, Most Stable Elements of the Cytoskeleton

KEY CONCEPTS

▸ Intermediate filaments are highly stable polymers that have great mechanical strength.
▸ Intermediate filament polymers are composed of tetramers of individual intermediate filament proteins.
▸ Several different genes encode intermediate filament proteins, and their expression is often cell and tissue specific.
▸ Intermediate filament assembly and disassembly are controlled by posttranslational modification of individual intermediate filament proteins.
▸ Specialized intermediate-filament-containing structures protect the nucleus, support strong adhesion by epithelial cells, and provide muscle cells with great mechanical strength.

One striking similarity between our bony skeleton and the cytoskeleton is that each is strong and durable. Without bones, we would not be able to resist the force of gravity enough to allow us to stand, walk, or even sit upright. Likewise, without a cytoskeleton, our cells cannot resist the mechanical stresses that we mostly ignore in everyday life, such as the frictional forces created when our clothing rubs against us. One of the major functions of the cytoskeleton, therefore, is to provide the mechanical strength necessary for cells to resist these forces. The intermediate filament portion of the cytoskeleton, shown in **FIGURE 5-2**, is primarily responsible for supplying this strength to cells, as illustrated in **FIGURE 5-3**. Whenever we discuss intermediate filaments, we will map them to this function (**BOX 5-2**).

Intermediate Filaments Are Formed from a Family of Related Proteins

Different cells in a multicellular organism are exposed to different types and degrees of mechanical stress (defined as the amount of force applied to a given surface area). For example, the cells lining our blood vessels are exposed to varying degrees of fluid sheer stress caused by the constant flow of blood over their surface, while our skeletal muscle cells experience stress when they contract. Cells have the ability to construct different intermediate filaments suited to resist the types of force they are exposed to. Pathologists often look for the specific makeup of these "custom" intermediate filaments as an easy way to distinguish one cell type in a tissue from another.

The Intermediate Filament Genes in Humans Are Classified into Six Groups

Each type of intermediate filament protein is encoded by a separate gene. Humans contain 70 intermediate filament genes that form approximately 75 different filament proteins (the number of proteins exceeds the number of genes due to alternative splicing of some of these genes; see Chapter 8, *In Eukaryotes, Messenger RNAs Undergo Processing Prior to Leaving the Nucleus*). These 70 genes are classified into 6 groups, named type I, type II, and so on, as shown in **TABLE 5-1**. Clusters of closely related genes (those that share the most sequence homology) are given a single name (e.g., type I and type II keratins, type IV neurofilaments, type V lamins). In humans, the type I and type II keratins are the most numerous intermediate filament proteins.

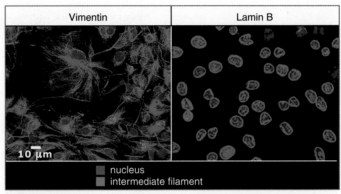

FIGURE 5-2 Two types of intermediate filaments. Fluorescence microscopy was used to label a type III intermediate filament protein (vimentin, left panel) and a type V intermediate filament protein (lamin, right panel) in cultured fibroblast cells. Photos courtesy of John Common and Birgit Lane, Institute of Medical Biology, Singapore.

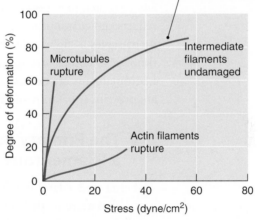

FIGURE 5-3 Intermediate filaments have the most mechanical strength of the cytoskeletal proteins, as measured by the strain (deformation) of the filaments in response to application of sheer stress (force/area). Note that each of three cytoskeletal filaments has a different response to the application of stress. Adapted from P. A. Janmey, et al., J. Cell Biol. 113 (1991): 155-160.

TABLE 5-1	The major classes of intermediate filaments in mammals.		
Class	Protein	Distribution	Proposed Function
I	Acidic keratins	Epithelial cells	Tissue strength and integrity
II	Basic keratins	Epithelial cells	Tissue strength and integrity
III	Desmin, GFAP,vimentin, periphevin	Muscle, glial cells, mesenchymal cells, perphevin neurons	Sarcomere organization, integrity
IV	Neurofilaments (NFL, NFM, and NFH)	Neurons	Axon organization
V	Lamins	Nucleus	Nuclear structure and organization
VI	Nestin	Neurons	Axon growth

■■ Intermediate Filament Expression Is Largely Cell and Tissue Specific

The expression pattern for each of the six groups of intermediate filament proteins is usually restricted to a specific cellular location, cell, or tissue type, as shown in Table 5-1. This reflects the customization property of intermediate filaments, in that each different filament serves a slightly different purpose. For example, the type V intermediate filaments (known as lamins) are found exclusively in the nucleus (see Chapter 2 and Chapter 8), neurofilaments are expressed only in nerve cells, and the expression of type VI intermediate proteins (phakinin and filensin) is restricted to the eye lens.

■■ The Primary Building Block of Intermediate Filaments Is a Filamentous Subunit

To better understand how intermediate filaments provide cells with mechanical strength, we need to examine their molecular structure. The first notable feature is that *all intermediate filament proteins are only effective when they form polymers.* While there is some turnover of intermediate filaments in cells (e.g., during cell division), these polymers are the most stable of the three types of cytoskeletal filaments.

Let's use the fact that intermediate filaments must be mechanically strong to understand how they are made. What property of proteins can we exploit to make them especially strong? Let's start at the simplest level, the 1° structure. Recall from Chapter 3 that all proteins are linear polymers of amino acids linked together by the same covalent (peptide) bonds. This means that the backbone of all proteins has essentially the same strength, and suggests that changes in only the 1° structure will not be sufficient to provide the extra strength needed in an intermediate filament.

■■ A Central Alpha-Helical Domain Confers Tremendous Tensile Strength to Intermediate Filament Subunits

If we look at the 2° structure of proteins, we see some clues. Recall that there are three 2° structures: α-helices, β-sheets, and random coils. Are any of these stronger than the others? Both α-helices and β-sheets are stabilized by hydrogen bonds, but random coils may not be; this suggests that either of the first two may be especially abundant in intermediate filament proteins. When we look at their general structure, as shown in **FIGURE 5-4**, we do indeed see an abundance of α-helices in the central rod domain (in fact, this is where α-helices were first discovered), flanked on either side by small, globular head and tail domains at the amino and carboxy ends, respectively. This makes sense, because α-helices are long and thin: remember that we are trying to make a long, thin polymer from these protein subunits. It is easy to imagine that if one tugs on the head and tail ends of this structure, the abundance of hydrogen bonds stabilizing the α-helices will prevent the protein from stretching out and/or collapsing. This simple property contributes a great deal to the function of these proteins.

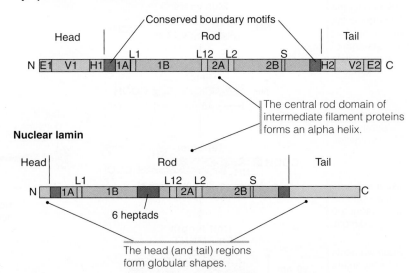

FIGURE 5-4 The general structure of two different intermediate filament proteins.

Intermediate Filament Subunits Form Coiled-Coil Dimers

Now let's look at the 3° structure of intermediate filaments. **FIGURE 5-5** illustrates how complete intermediate filaments are assembled. Notice that two intermediate filament proteins (also called *monomers* or *subunits* in this context) wrap around one another in parallel to form a dimer (**BOX 5-3**). Some intermediate filament dimers only form when the two subunit types are different (e.g., keratins and neurofilaments make heterodimers), some can assemble from identical or different subunits (e.g., vimentin and nestin make homo- and heterodimers), and the rest make only homodimers. Because it is a coil formed by two α-helical polypeptides, this structure is called a **coiled coil** (the redundancy is intentional). Coiled coils are some of the strongest structures found in proteins, because they maximize the amount of surface contact between the two polypeptide chains; the close proximity of the amino acid side chains permits hydrophobic bonds to form between the two chains. Because many of these bonds form between leucine amino acids, this structure is sometimes referred to as a leucine-zipper motif.

Heterodimers Overlap to Form Filamentous Tetramers

The next level of intermediate filament assembly is an antiparallel staggered tetramer, as seen in Figure 5-5. Notice that at this stage, even the coiled coils line up and interact with one another, further increasing the number of bonds stabilizing the structure. These tetramers are the smallest stable form of intermediate filaments in cells, and usually represent the "replacement pool" on hand to replace damaged tetramers in the polymers. It is important to note that these tetramers have no structural polarity; the amino termini (tails) of each dimer project outward at both ends, and the carboxy termini (heads) interact with the rod domains on the adjacent subunit. This means that the two ends of the tetramer are identical; there is no front or back to this structure. This feature is important because we will see later that the other two types of cytoskeletal filaments *do* have structural polarity, and this difference helps explain why these filaments play different functional roles in cells (**BOX 5-4**).

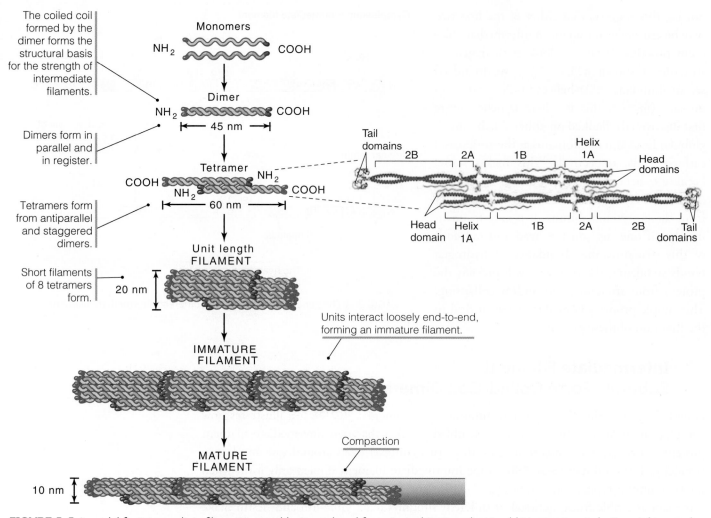

The coiled coil formed by the dimer forms the structural basis for the strength of intermediate filaments.

Monomers

NH$_2$ ⟿⟿⟿ COOH

Dimers form in parallel and in register.

Dimer

NH$_2$ ⟿⟿⟿ COOH

↔ 45 nm ↔

Tetramers form from antiparallel and staggered dimers.

Tetramer

COOH ⟿⟿⟿ NH$_2$
NH$_2$ ⟿⟿⟿ COOH

↔ 60 nm ↔

Tail domains 2B 2A 1B Helix 1A Head domains

Head domain Helix 1A 1B 2A 2B Tail domains

Short filaments of 8 tetramers form.

Unit length FILAMENT

20 nm

Units interact loosely end-to-end, forming an immature filament.

IMMATURE FILAMENT

Compaction

MATURE FILAMENT

10 nm

FIGURE 5-5 A model for intermediate filament assembly. Reproduced from Annual Reviews by Harald Herrmann, et al., Copyright 2004 by Annual Reviews, Inc. Reproduced with permission of Annual Reviews, Inc. in the format of Textbook via Copyright Clearance Center. Photo courtesy of Harald Herrmann, German Cancer Research Center.

BOX 5-3 TIP

Revisiting the protein/subunit issue. In Chapter 3, we discussed the use of different words to describe the functions of polypeptides. In most cases, the word "subunit" cannot be used interchangeably with "protein." But, be careful: when we talk about the cytoskeleton, the rules change. A cytoskeletal protein is very often called a subunit, because these proteins are frequently considered functional only when they form polymers. This applies to all three cytoskeletal filament types. Because they can form functional polymers of different sizes, there is no rigorous standard to define when a "subunit" becomes part of a functional polymer. As frustrating as this may be for new cell biology students, it's simply the way things are, and no real effort has been made to fix it. We all just have to get used to this exception.

BOX 5-4

The Tetris analogy. If the nonpolarized tetramer is the basic building block of intermediate filaments, let's use a popular videogame to illustrate it. In the game Tetris®, the goal is to arrange different geometric shapes into a continuous horizontal line. One of the shapes, called the S-unit, resembles the outline shape of an intermediate filament tetramer. In this analogy, a fully formed intermediate filament is like an enormous Tetris game containing only S-units, and all of them fit together perfectly to make many complete horizontal lines. Considering how difficult that would be in the game, it is amazing that cells do this routinely.

Assembly of a Mature Intermediate Filament from Tetramers Occurs in Three Stages

Intermediate filament assembly has been studied mostly *in vitro,* using purified proteins. Under conditions where the assembly can be slowed down enough to watch it progress, it occurs in three stages, illustrated in Figure 5-5. First, eight tetramers associate into a loosely packed unit-length filament approximately 17 nm in diameter. Next, these unit-length filaments align end-to-end to form an immature filament, still 20 nm in diameter. Finally, the filament compacts to form a dense, mature, 10-nm-thick filament. Note that no additional proteins are required to form intermediate filaments this way, and assembly is spontaneous. It has been very difficult to verify that the filament assembly occurs in the exact same steps in cells, but the evidence accumulated thus far suggests that the assembly process is at least very similar.

Posttranslational Modifications Control the Shape of Intermediate Filaments

Recall from Chapter 3 (see *Cells Chemically Modify Proteins to Control Their Shape and Function*) that proteins form both covalent and noncovalent bonds with other molecules. Intermediate filaments undergo many posttranslational modifications, including phosphorylation, glycosylation, farnesylation, and transglutamination of the head and tail domains on the individual intermediate filament subunits. While the function of most of these modifications is only now being truly appreciated, it is clear that the phosphorylation/dephosphorylation of domains in intermediate filament proteins helps control the assembly of the subunits into mature filaments. This is especially evident when intermediate filaments disassemble during cell division: the nuclear envelope dissolves (requiring dissolution of the lamin intermediate filaments) and the contents of the cell (including the cytosolic intermediate filament proteins) are partitioned into roughly equal halves during cytokinesis (see Chapter 7, *Cytokinesis Completes Mitosis by Partitioning the Cytoplasm to Form Two New Daughter Cells*). In general, phosphorylation of the globular domains dissolves intermediate filaments; upon removal of the phosphates by protein phosphatases (see Chapter 3, *Covalent Modifications are Relatively Long Lasting*), the subunits spontaneously reassemble into filaments.

Intermediate Filaments Form Specialized Structures

The restricted cell and tissue distribution of intermediate filament proteins further emphasizes the importance of cell specialization in multicellular animals, and allows us to examine how the mechanical strength function is applied to different cell and tissue types in greater detail.

Lamins Form a Strong Cage inside the Nucleus

Lamins (not to be confused with *laminins,* see Chapter 6, *Laminins Provide an Adhesive Substrate for Cells*) are found exclusively in the nucleus of cells, and are structurally quite different from all of the other intermediate filament proteins. Three different lamin genes in mammals undergo alternative splicing (see Chapter 8, *The Spliceosome Controls RNA Splicing*) to yield six different lamin subunits, called lamins A, B1, B2, B3, C1, and C2. Each cell contains at least one version of all three (A, B, C) types, resulting in three different homodimers. All of these proteins contain a longer helical domain than cytoplasmic intermediate filaments; this difference helps ensure that lamins do not copolymerize with the shorter cytoplasmic intermediate filament subunits.

Lamins form strong filaments that line the inner surface of the nucleus, protecting chromatin from potential damage caused by mechanical trauma (see Figure 2-26). More recent evidence suggests that the lamins are linked to the chromatin by other structural proteins and may play a role in controlling gene expression by influencing the degree of chromatin compaction. When the nuclear envelope dissolves during prometaphase in cell division, lamins are phosphorylated in the head and tail globular domains, and the resulting shape change of the proteins causes them to dissolve into individual tetramers. It is not yet known whether this phosphorylation causes nuclear envelope disassembly, or merely coincides with it.

■■ Epithelial Intermediate Filaments Form Strong Attachment Sites at the Cell Surface

The largest and most diverse group of intermediate filament proteins is the keratins expressed in epithelial cells (see Chapter 14, *Epithelial Tissues form Protective, Semipermeable Barriers Between Compartments*). Humans have 54 different keratin genes (28 type I and 26 type II) that encode proteins, and these are named according the convention "K#," where # is a number between 1 and 86 (the details of these numbers are not important for us here). Keratins are considered **obligate heterodimers**, in that the dimers forming their filaments must contain one type I and one type II subunit. Different combinations of keratin pairs are expressed in different stages of development, and in different kinds of epithelial tissues, as shown in **FIGURE 5-6**.

Keratins are always located in the cytosol of epithelial cells, and in some cases they are the most abundant proteins these cells express (keratins can represent up to 80% of proteins in the cytosol). Keratins are essential components of two important cell adhesion complexes in epithelial cells: desmosomes and hemidesmosomes (see Chapter 6, *The Hemidesmosome Is a Specialized Junction Formed Between Cells and the Basal Lamina, and Desmosomes Are Intermediate Filament-Based Cell Adhesion Complexes*). The primary function of these complexes is to ensure strong adhesion between neighboring epithelial cells and between epithelial cells and their underlying extracellular matrix.

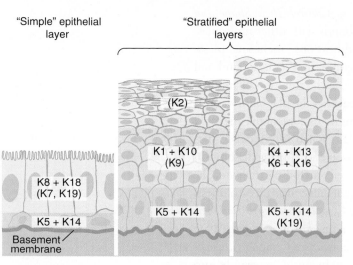

FIGURE 5-6 Keratin expression patterns vary in different epithelial tissues. Note that K5/K14 is expressed in all three types of epithelia shown, but K8/K18 is restricted to simple epithelia.

■■ The Protein Desmin Is Essential for Muscle Function

Desmin is a type III intermediate filament protein that forms homodimers and is found in all three types of muscle tissue (skeletal, cardiac, smooth; see Chapter 14 for a more detailed description of these tissues). The distribution of desmin differs in each type of muscle cell, but desmin is always part of the contractile apparatus. In smooth muscle cells, desmin is found in the dense bodies that connect to the actin microfilament network (see *Actin-Binding Motor Proteins Exert Force on Actin Filaments to Induce Cell Movement* later in this chapter), while in skeletal and cardiac muscle cells, desmin is concentrated in the Z-lines that border the sarcomeres and intercalated discs. Desmin is somewhat unusual in that it does not form the long filamentous structures that we typically see formed by most intermediate filament proteins; instead, it appears to function as part of molecular clusters that link disparate parts of muscle cells together. One good example is a structure called a

costamere, illustrated in **FIGURE 5-7**. Costameres link the contractile apparatus of muscle cells with receptor protein complexes in the plasma membrane, which in turn anchor the cells to the extracellular matrix. Mutations in costamere proteins, including desmin, lead to a variety of muscular diseases, including muscular dystrophies. Experiments in three different animal model systems (developing chicken embryos, mice, and zebrafish) confirm that desmin is essential for proper muscular function. When desmin is mutated or deleted in these animals, muscle cells will continue to form, but their ability to contract is severely compromised.

CONCEPT CHECK #1

Compare the properties of intermediate filaments with six other materials that have great mechanical strength: rope, chain, steel cable, steel rebar, silk, and Kevlar. In what ways are intermediate filaments similar to / dissimilar from these structures? Using evolution by natural selection as a driving force for optimizing intermediate filament structure and function, suggest an argument for why the current properties of intermediate filaments are the best means for providing mechanical strength inside cells (for example, why don't cells build a structure that more closely resembles steel chains?).

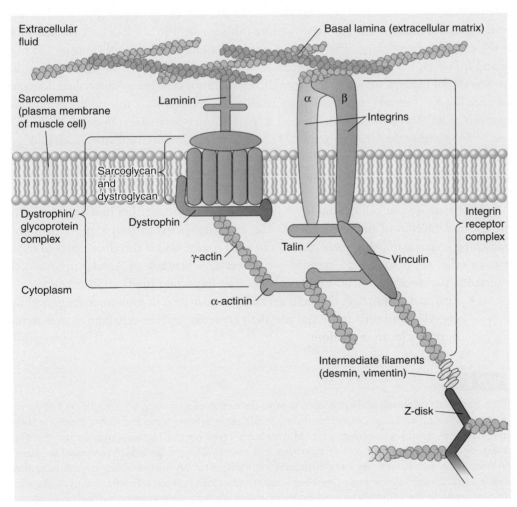

FIGURE 5-7 Costameres link the contractile apparatus of muscle cells to the plasma membrane and extracellular matrix.

5.4 Microtubules Organize Movement inside a Cell

KEY CONCEPTS

▸ Microtubules are hollow, tube-shaped polymers composed of proteins called tubulins.

▸ Microtubules serve as "roads" or tracks that guide the intracellular movement of cellular contents.

▸ Microtubule formation is initiated at specific sites in the cytosol called microtubule-organizing centers. The basic building block of a microtubule is a dimer of two different tubulin proteins.

▸ Microtubules have structural polarity, which determines the direction of the molecular transport they support. This polarity is caused by the binding orientation of the proteins in the tubulin dimer.

▸ The stability of microtubules is determined, at least in part, by the type of guanine nucleotides bound by the tubulin dimers within it.

▸ Dynamic instability is caused by the rapid growth and shrinkage of microtubules at one end, which permits cells to rapidly reorganize their microtubules.

▸ Microtubule-binding proteins play numerous roles in controlling the location, stability, and function of microtubules.

▸ Dyneins and kinesins are the motor proteins that use ATP energy to transport molecular "cargo" along microtubules.

▸ Cilia and flagella are specialized microtubule-based structures responsible for motility in some cells.

Microtubules are cytoskeletal proteins that guide the trafficking of proteins, vesicles, and even organelles inside a cell. Before we pursue this idea any further, let's have a closer look at the concept of trafficking. The Oxford English Dictionary defines trafficking as "To pass to and fro upon; to frequent (a road, etc.); to traverse." When we discuss trafficking in the context of cell biology, we are referring to the mechanisms cells use to move their contents around. We have already encountered an excellent example of how this occurs during the assembly of cellular membranes in Chapter 4 (see *Most Membrane Assembly Begins in the SER and Is Completed in the Target Organelle*). This illustrates an important theme in cell biology: the function of a molecule in a cell is often determined by its location. This means that one good way of regulating cellular functions is to permit movement of molecules in cells, then regulate this movement very closely.

Why don't we just call this molecular movement? The answer lies in a single word in the above definition of trafficking: road. The difference between simple movement and trafficking is that trafficking takes place on defined routes, and is therefore easier to regulate. In cells, the microtubule cytoskeleton serves as a network of "roads" upon which molecules "pass to and fro" (**BOX 5-5**). This has four important implications:

- First, it demonstrates that cells are far more than bags of molecules. Even stationary cells constantly move and sort their contents, and the distribution of material in cells is far from random.

BOX 5-5

The highway network analogy. Let's expand the notion of trafficking on a road a bit further to highlight a few other features of microtubules. Roads come in many sizes and forms, from one-lane dirt roads to multilane superhighways. Microtubules resemble the highways more than the dirt roads because they permit two-way trafficking and several different "vehicles" can travel on them at once. Also, sometimes very large structures are transported on microtubules, analogous to the "wide load" convoys that are used to move large structures on highways. Finally, microtubules are, in proportion to the cargo that travels on them, extremely long. Cargo can be carried tremendous distances (up to hundreds of millimeters in very long cells) on a single microtubule. One important difference between microtubules and highways is that cargo can attach to and detach from a microtubule anywhere along its length.

- Second, microtubule-based movement of molecules limits where some of these molecules can go; if a region of a cell (e.g., the interior of the nucleus or the matrix of a mitochondrion) lacks microtubules, molecules that require microtubules for movement will not be transported there. Conversely, if a region of a cell contains many microtubules (e.g., the mitotic spindle), molecular traffic in that region can be tightly controlled.

- Third, the transport of microtubule-dependent cargo depends upon the structural stability of the microtubules it rides on. Molecular traffic in cells can easily be redirected by simply dissolving some microtubules in one region of the cell and building new microtubules in another.

- Fourth, transport along microtubule "roads" requires energy. Just like the roads in our everyday lives, these roads don't move by themselves; rather, we have to generate some sort of force to push or pull our cargo. The same holds true for cells. We will encounter proteins specialized to convert metabolic energy into forces that move cargo along microtubule roads later in this chapter.

The distribution of this microtubule network varies greatly from one cell type to another (e.g., in some yeast cells, the microtubules actually penetrate the nucleus), but a good example of such a network is shown in **FIGURE 5-8**. Notice that it is most dense in the center of the cell, and appears to radiate outward toward the cell periphery. These are

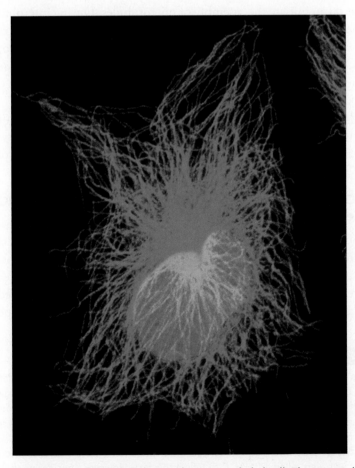

FIGURE 5-8 The distribution of microtubules in a human epithelial cell. The microtubules are stained green in this fluorescence micrograph, and the DNA is stained red to show the location of the nucleus. Photo courtesy of Holger Lorenz, Zentrum für Molekulare Biologie der Universität Heidelberg, Germany.

the "roads" upon which many molecules travel as they make their way through the cytosol. When we discuss microtubules, we'll map them to the *intracellular transport* function.

■■ Microtubule Assembly Begins at a Microtubule-Organizing Center

Because of their different functions in cells, it makes sense that the assembly and structure of microtubules are different than that for intermediate filaments. One striking difference between the two is that microtubule assembly typically requires a large complex of proteins, whereas all intermediate filaments spontaneously assemble. An important similarity between them is that both are made up of two protein subunits; the subunits that form microtubules belong to a protein family called the **tubulins**.

There is an advantage to investing so many resources on controlling microtubule assembly. In Chapter 4 we discussed how microtubule-based transport of membrane vesicles is crucial for building membranes. Recall from Chapter 4, *Concept Check #3* and *Concept Check Answers 33* that at least three different strategies exist for organizing the steps required to complete a complex process, such as the synthesis and sorting of membrane components. Microtubules are an excellent example of two of these strategies: the **centralized network** and the distributed network.

The centralized network appears in cells that contain **microtubule-organizing centers** (**MTOCs**) called **centrosomes**. While many animal cells contain centrosomes, this feature is not are found in all eukaryotes (e.g., plant cells) (**BOX 5-6**). This MTOC was so-named because it can initiate the assembly of microtubules. During interphase, it typically lies close to the nucleus, as shown in Figure 5-8. During the S phase of the cell replication cycle (see *The Cell Cycle Is Divided into Five Phases* in Chapter 13), it is copied, and the two new MTOCs form the poles of the mitotic spindle (see *Mitosis and Cytokinesis Occur in M Phase* in Chapter 13). At the completion of the cell cycle, each pole of the spindle becomes the centrosome for the daughter cells.

The distributed network is represented by cells that do not always contain an MTOC. During the interphase stage of the cell cycle, microtubules in these cells arise instead from clusters of proteins distributed throughout the cytosol. Some of these proteins are also found in MTOCs, as we will discuss below.

■■ The MTOC Contains the Gamma Tubulin Ring Complex (γTuRC) That Nucleates Microtubule Formation

FIGURE 5-9 shows two electron micrographs of the centrosome. Notice that two structures are clearly visible in this MTOC; these are called the centrioles. Each centrosome contains two centrioles, arranged perpendicular to one another. Intact centrioles are very difficult to isolate from most cells. So far, we know that these structures are composed primarily of tubulins, plus at least 100 other protein types. How they achieve their perpendicular ori-

---

BOX 5-6 TIP ■

Centrosomes vs. MTOCs. The history of these terms is somewhat complicated, so let's clarify what each refers to. The word centrosome was coined first, by microscopists who literally saw a "body" (in Greek, *soma*) in the "center" of a cell. They didn't know what its function was, so they decided to leave the name at that. But, at the time, use of Latin and Greek words to derive biological terms was still very popular—recall the history of naming proteins in Chapter 1—so "central body" was converted to *centrosome*. Much later, when it became clear that the function of the centrosome is to nucleate microtubules, it was also called an MTOC to reflect this fact. However, considering that other cellular structures also nucleate microtubules, the term MTOC is now used as a blanket term for them all (like "organelle"), and each structure has its own name. So we come full circle, and the centrosome is still the name of the MTOC near the nucleus, even though we now know what it does.

CHAPTER 5   The Cytoskeleton and Cellular Architecture

entation, and what significance this orientation may have, are still not well understood, and the function of these centrioles is not known.

Figure 5-9 also shows the cluster of centrosome proteins that surrounds the centrioles and appears as a dense cloud when viewed with an electron microscope. This cloud is called the **pericentriolar material**, and like the centrioles it is composed of many different proteins. Among these are a set of proteins that forms a scaffold, or lattice, around the centrioles. One protein in the lattice, called **gamma- (γ) tubulin**, is organized into a helical or ring shape, somewhat like a single coil of a spring. Two additional proteins help form the coil, and many more attach to one face of this coil, forming a structure called the **gamma-tubulin ring complex (γTuRC)**. Because microtubules grow outward from the γTuRCs, most cytosolic microtubules appear to originate from a single place in these cells. Cells lacking interphase MTOCs also contain γ-tubulin in many of the protein clusters that promote microtubule growth.

Most microtubules that grow from the centrosome remain attached to it, but microtubule formation and anchoring are not performed by the same proteins. The function of γ-tubulin is to initiate the formation of microtubules in cells, but it does not hold the resulting microtubules in place. At least some of the other proteins in the γTuRC play this role, though the exact mechanism is still not known. Gamma-tubulin rings have been found in many different locations in cells, strongly suggesting that these other sites initiate microtubule formation as well. Many of these sites contain additional proteins that could anchor the microtubules.

■■ The Primary Building Block of Microtubules Is an Alpha-Beta Tubulin Dimer

Regardless of where they form, microtubules always assemble the same way. It is easiest to observe this assembly in a test tube (*in vitro*) with purified proteins, and that is how we gained most of our understanding of this process. Microtubules are composed of only two proteins, called α- and β-tubulin. These two proteins bind to one another so strongly that they form a very stable dimer, shown in **FIGURE 5-10**. Both are similarly shaped, globular proteins, and they align front-to-back to form a distinctive shape that looks roughly like a peanut or a figure eight.

A second important observation is that each α- and β-tubulin protein binds to a GTP molecule. The α-tubulin never releases its GTP, but the β-tubulin is capable of cleaving off the terminal phosphate to create GDP (a process called **GTP hydrolysis**), and has the ability to release this GDP and bind a new GTP to take its place. This swapping of GDP for GTP is crucial for controlling the assembly of α-β tubulin dimers into microtubules, as we will discuss below.

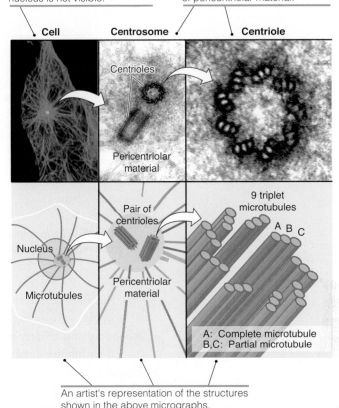

In this fluorescence micrograph, the microtubules are green and the centrosome is yellow. The nucleus is not visible.

Electron micrographs (low and high magnification) show that the centrosome is composed of two centrioles surrounded by a cloud of pericentriolar material.

An artist's representation of the structures shown in the above micrographs.

FIGURE 5-9 The structure and location of the centrosome. Photos courtesy of Lynne Cassimeris, Lehigh University.

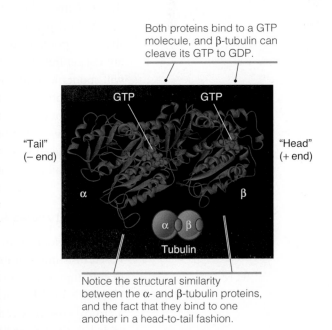

Both proteins bind to a GTP molecule, and β-tubulin can cleave its GTP to GDP.

Notice the structural similarity between the α- and β-tubulin proteins, and the fact that they bind to one another in a head-to-tail fashion.

FIGURE 5-10 A three-dimensional model of the dimer formed by α- and β-tubulin. Structure from Protein Data Bank 1TUB. E. Nogales, S. G. Wolf, and K. H. Downing, Nature 391 (1998): 199–203.

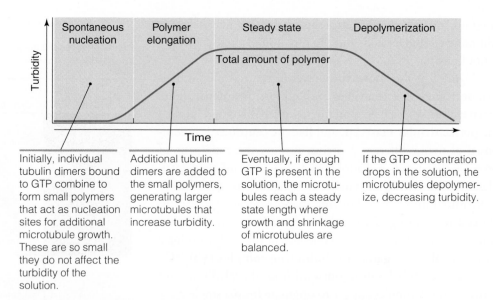

Initially, individual tubulin dimers bound to GTP combine to form small polymers that act as nucleation sites for additional microtubule growth. These are so small they do not affect the turbidity of the solution.

Additional tubulin dimers are added to the small polymers, generating larger microtubules that increase turbidity.

Eventually, if enough GTP is present in the solution, the microtubules reach a steady state length where growth and shrinkage of microtubules are balanced.

If the GTP concentration drops in the solution, the microtubules depolymerize, decreasing turbidity.

FIGURE 5-11 *In vitro* assembly of microtubules is spontaneous and GTP dependent. The graph represents the turbidity of a solution of α-β tubulin dimers over time.

The difference in the way α-tubulin and β-tubulin bind GTP is an excellent illustration of how the positioning of the nucleotide-binding pocket affects the relationship between secondary and tertiary structures of these proteins. The GTP that is bound to α-tubulin is buried within the dimer (resulting in stabilization), whereas the GTP bound to β-tubulin is exposed. If the exposed α-β dimer on the plus end of the microtubule undergoes a conformational change, it is hypothesized that the nucleotide in the β-tubulin is hydrolyzed and this alters the shape of the tubulin dimer—from straight to curved—which then promotes disassembly, at least in part, by disrupting lateral interactions between adjacent dimers. We'll examine the impact of this conformational change more closely when we discuss dynamic instability later.

A third property of microtubule assembly that we can easily observe in a test tube is that it is spontaneous, rapid, and reversible. If purified α-β tubulin dimers bound to GTP are concentrated enough (i.e., they reach what is called the **critical concentration**), they will spontaneously form microtubules, as shown in **FIGURE 5-11**. This is easy to measure: as the microtubules elongate, they deflect more light passing through the test tube (making it appear cloudy, a property called *turbidity*), so we can simply shine light through the tube and watch what happens over time. Later, if no additional GTP is added and the initial GTP is cleaved to GDP by β-tubulin, the microtubules will spontaneously disassemble, and the turbidity will drop. This switch from growth to shrinkage can take place in seconds.

■■ Microtubules Are Hollow "Tubes" Composed of 13 Protofilaments

Let's have a look at the molecular mechanisms behind this rapid assembly and disassembly. Based on studies of purified solutions of α-β tubulin dimers, we know that the first step in building a microtubule is the assembly of small polymers of these dimers, shown in **FIGURE 5-12**. Most of these polymers are unstable, and quickly dissociate, but some reach a critical size (between 6 and 12 dimers), and these begin to grow. Additional dimers are added to the tips and sides of the polymer, until a sheet composed of 13 **protofilaments** is formed. This sheet then folds into a tube shape, giving rise to the final microtubule structure, approximately 25 nm in diameter. These tubes, also called nuclei (not to be confused with the organelles of the same name), are rather short, but can quickly increase in

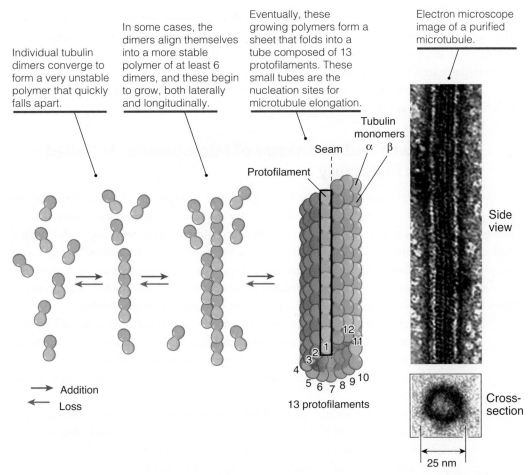

Individual tubulin dimers converge to form a very unstable polymer that quickly falls apart.

In some cases, the dimers align themselves into a more stable polymer of at least 6 dimers, and these begin to grow, both laterally and longitudinally.

Eventually, these growing polymers form a sheet that folds into a tube composed of 13 protofilaments. These small tubes are the nucleation sites for microtubule elongation.

Electron microscope image of a purified microtubule.

Seam

Protofilament

Tubulin monomers
α β

Side view

→ Addition
← Loss

12
11
1
2
3
4
5 6 7 8 9 10

13 protofilaments

Cross-section

25 nm

FIGURE 5-12 A simple model of microtubule assembly. Micrograph provided by Harold Erickson, Duke University School of Medicine. Top photo courtesy of Lynne Cassimeris, Lehigh University.

length by addition of more dimers to both ends. This association is not permanent, however; eventually, the dimers fall off the ends of the microtubule. At steady state, the rate of dimer addition will equal that of dimer dissociation, keeping the average length of the microtubules constant.

■■ GTP Binding and Hydrolysis Regulate Microtubule Polymerization and Disassembly

To help explain how microtubules can grow and shrink at the same time, let's use two of the three traits of proteins (see Chapter 3, *Proteins Are Polymers of Amino Acids That Possess Three Important Iraits*). First, tubulin dimers adopt at least two different stable shapes (trait #1). One appears when both the α- and β-tubulin proteins are bound to GTP. Dimers in this condition favor polymerization, and therefore readily attach themselves to the ends of microtubules (trait #2). The second shape appears when the β-tubulin in the dimer cleaves its GTP to GDP (recall that the α-tubulin never cleaves its GTP, so we will ignore it). This cleavage is a spontaneous process. Because the β-tubulin is now bound to a different molecule (GDP has a different shape than GTP), it changes shape. This change causes dimers to adopt a shape that promotes depolymerization. Thus, at steady state, some dimers are in the GTP-bound state, while others are in the GDP-bound state, and the resulting shapes of these dimers causes some dimers to add to a microtubule, while other dimers will not.

Because the α- and β-tubulin proteins are aligned front-to-back in the dimer, one end of the dimer is structurally different from the other. This means that the dimer, as well as the microtubule it forms, has structural polarity. By convention, one end is called the "plus end" and the other the "minus end." These names were chosen because microtubules do not grow at the same rate from both ends; *in vitro,* α-β dimers in the GTP-bound state preferentially attach to the plus end, so the plus end grows faster than the minus end. Note that this is quite different from intermediate filaments, which do not possess structural polarity when they are polymers.

■■ The Growth and Shrinkage of Microtubules Is Called Dynamic Instability

Studying microtubule assembly *in vitro* is a great way to grasp the fundamentals of the process, but it is clearly not as realistic as studying microtubule dynamics *in vivo* (i.e., in living cells). *In vivo* research is typically much more difficult than working with purified molecules, but it provides an opportunity to tackle questions that cannot be otherwise addressed. An excellent example concerns the complex mechanisms cells use to control the location, stability, and growth rate of microtubules. To help us get started, let's mention three important differences between *in vitro* and *in vivo* studies of microtubules.

- First, microtubules are seldom "naked" in cells and therefore do not all behave the same way. Microtubules are known to bind to many different molecules in cells—so many that one cannot easily keep track of them all at once. Thus, we can only estimate what the combined effect all these molecules have on any given microtubule.

- Second, every microtubule in a cell lies in its own unique environment. *In vitro,* one can suspend all of the microtubules in a single solution and observe their collective behavior in that one environment, but the interior of a cell is so heterogeneous that no two microtubules are ever in the exact same state.

- Third, living cells don't "hold still." Regardless of what our wishes are, cells have their own agenda: to maintain their disequilibrium with the surrounding environment, and thereby stay alive. This means that the internal state of a cell changes over time, so the microtubules we see at the end of an experiment may have been influenced by factors we cannot (or did not) control during the experiment.

Despite these limitations, researchers have made considerable progress toward understating how cells control their microtubules. One of the most important advances is our understanding of how microtubule assembly, elongation, and stabilization occur. The centrosome and other MTOCs in cells play a central role in this.

■■ Some Microtubules Rapidly Grow and Shrink in Cells

Using specially prepared light microscopes, one can see that the length and position of microtubules changes rapidly in most cells, as shown in **FIGURE 5-13**. This growing/shrinking behavior is collectively called **dynamic instability**. The growth of microtubules begins when γ-tubulin rings nucleate polymerization of the α-β tubulin dimers, such as in MTOCs like the centrosome. Because the centrosome is the most heavily studied MTOC, we use it to illustrate how dynamic instability takes place.

Microtubule growth begins when enough GTP-bound α-β tubulin dimers bind to the γ-tubulin ring (located on the surface of the centrosome) to form a microtubule nucleus, as we discussed above. These nuclei always have the same

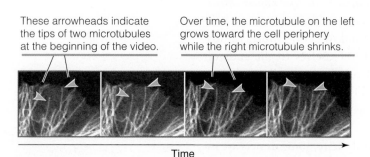

These arrowheads indicate the tips of two microtubules at the beginning of the video.

Over time, the microtubule on the left grows toward the cell periphery while the right microtubule shrinks.

Time

FIGURE 5-13 Growth and shrinkage of microtubules in a living cell. The microtubules have been tagged with a fluorescent molecule, and recorded by video over time. Photos courtesy of Lynne Cassimeris, Lehigh University.

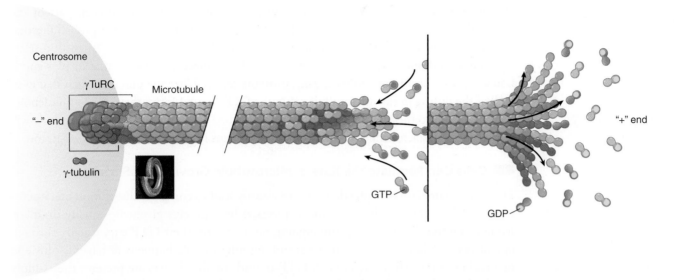

FIGURE 5-14 The growth of microtubules begins at the gamma tubulin ring and continues as long as the plus end contains GTP-bound tubulin dimers. Photo courtesy of Lynne Cassimeris, Lehigh University

orientation: their minus ends are nearest the γ-tubulin, and their plus ends face away from the γ-tubulin, into the cytosol. At some point, the minus end of a growing microtubule releases from the γ-tubulin, but remains attached to the centrosome by additional proteins in the pericentriolar material (most of these have yet to be identified). If additional GTP-bound dimers are nearby, they will add to the nucleus, but mostly at the plus end, as shown in **FIGURE 5-14**. This causes the microtubule to elongate, and it will continue as long as enough GTP-bound tubulin dimers are present at the plus end (this is often referred to as a GTP cap). We see this in the microscope as the emergence of a microtubule from the centrosome and its trajectory through the cytosol. Where a microtubule ends up is determined by a host of factors, including several proteins in the cytosol that attach to it as it elongates.

Eventually, the supply of GTP-bound tubulin dimers runs out, and the microtubule stops growing. At this point, two things can happen. The most common outcome, illustrated in **FIGURE 5-15**, is that the microtubule begins to depolymerize at the plus end (and perhaps also at the minus end), and the microtubule literally falls apart, somewhat like a rope uncoiling. Cell biologists who study this process gave the action the dramatic name **catastrophe** to emphasize the rapidity of this collapse. What causes catastrophe? Recall that all α-β tubulin dimers bind to GTP molecules, and the β-tubulin cleaves its GTP to GDP, switching it from a pro-polymerization shape to a shape that does not favor polymerization. When the last of the GTP-bound tubulin dimers is added to the end of a microtubule, they eventually convert the GTP bound to the β-tubulin into GDP, and this drives them to drop off the end of the microtubule. Once they fall off, the dimers that were previously added to the microtubule now face the same

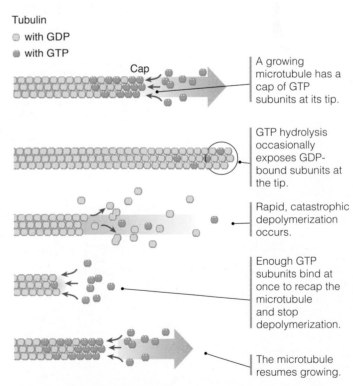

Tubulin
- ◯ with GDP
- ◉ with GTP

A growing microtubule has a cap of GTP subunits at its tip.

GTP hydrolysis occasionally exposes GDP-bound subunits at the tip.

Rapid, catastrophic depolymerization occurs.

Enough GTP subunits bind at once to recap the microtubule and stop depolymerization.

The microtubule resumes growing.

FIGURE 5-15 Two fates of the plus ends of microtubules.

choice: to remain with the microtubule, or fall off. Because they have been part of the microtubule even longer than those at the very tip, there is an even greater probability that they are now in the GDP-bound form, so they too fall off.

The second outcome for a microtubule that has stopped growing, is that its plus end is bound by a set of **microtubule capping proteins** that stabilize the plus end, such that even if all of the dimers in the microtubule are in the GDP-bound form, they still do not depolymerize. As long as the capping proteins remain attached to the plus end, the microtubule will remain intact. Attachment of the pericentriolar proteins to the minus end stabilizes it, too.

■■ Cells Can Regulate the Rate of Microtubule Growth and Shrinkage

Despite its name, catastrophe does not necessarily lead to complete dissolution of a microtubule. Often, the depolymerization is reversed by a process given the equally dramatic name of **rescue**. Microtubules undergoing rescue regain their GTP cap, usually through one of two mechanisms. First, if the shrinking microtubule happens to have its plus end in a region of the cell where enough GTP-bound tubulin dimers are present, they simply attach themselves to the shrinking plus end, thereby stopping further dissolution. The second mechanism is a bit more complicated, because it requires additional proteins to rescue the microtubule; these proteins attach themselves to the shrinking plus end and promote the attachment of new GTP-bound dimers. When viewed under a microscope, these proteins appear to "track" with the plus end of the regrowing microtubules, though researchers do not yet fully understand the mechanisms underlying these behaviors.

■■ Some Microtubules Exhibit Treadmilling

In cases where neither end of a microtubule is stabilized, tubulin dimers are added to the plus end and lost from the minus end. The overall length of these microtubules remains fairly constant, but the dimers are always in flux, as shown in **FIGURE 5-16**. This process

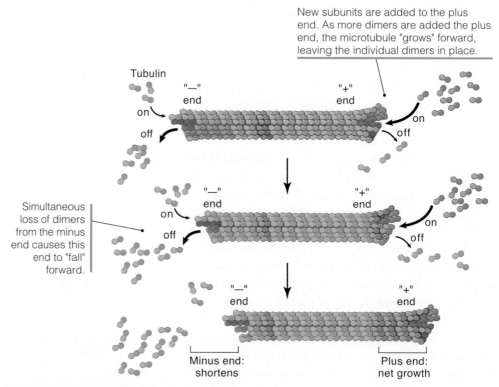

FIGURE 5-16 Treadmilling in microtubules.

CHAPTER 5   The Cytoskeleton and Cellular Architecture

is called **treadmilling** because it somewhat resembles the motion of an exercise treadmill. But be careful about over interpreting this: an exercise treadmill *actually moves* (rotates), but a treadmilling microtubule doesn't move in the same way. To understand this concept, it is important to keep in mind that the *relative* position of a given dimer in a microtubule changes over time, even if the *absolute* position does not change; that is, when it is first added to the microtubule, a dimer lies at the extreme plus end, but over time, as more dimers are added on top of it, the dimer is now somewhere in the middle on the microtubule. Eventually, after enough other dimers have fallen off the minus end, the same dimer is now found at the minus end. Assuming that enough time has passed between its incorporation into the plus end and its appearance at the minus end, it switches from a GTP-bound state to a GDP-bound state, and eventually falls off. Though the name treadmilling has stuck, it might be more useful to consider the motion of a tread on a tractor as it moves: once a given part of the tread contacts the ground, it remains there, stationary, until it is pulled back up by the passing wheels, as shown in Figure 5-16.

■■ Dynamic Instability Allows Cells to Explore Their Cytosol

How does a microtubule undergoing dynamic stability benefit cells? If it won't hold still, how can it serve as any kind of useful "road" for molecular traffic? The answer, simply put, is that a moving microtubule isn't supposed to support trafficking. Instead, dynamic instability is a clever way to allow cells to do something we can't do with a conventional road: move it from location to location, according to where it is needed most; once it is stabilized in its new location, trafficking resumes. In some special cases, the stabilized road can be pushed or pulled, even as cargo is moving on it. This is very useful for cells that are undergoing major shape changes, or for cells attempting to convert their interphase microtubules into a mitotic spindle at the beginning of mitosis (**BOX 5-7**). In effect, it allows a cell to explore its cytosol as it changes position, or to redirect molecular traffic to a different region of the cell as necessary.

Another benefit of dynamic instability is that it allows cells to exert force. If a molecule can remain attached to the plus end of a moving microtubule, it can be transported through the cell as the microtubule shrinks or grows, as shown in **FIGURE 5-17**. The structural basis for this force is provided by extensive, non-covalent, intramolecular bonding in microtubules illustrated in **FIGURE 5-18**. Because the 13 protofilaments in a microtubule are stabilized by both longitudinal bonds (between successive dimers in the same protofilament) and lateral bonds (between dimers

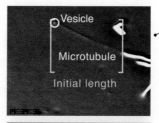

In this figure, a shortening microtubule has a small vesicle attached at its end. As the microtubule shortens, the vesicle remains attached and is moved toward the nucleation site.

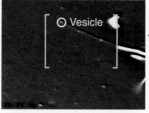

This was an *in vitro* experiment using purified tubulin added to diluted cytoplasm. *In vivo*, microtubule depolymerization may help move kinetochores during mitosis.

FIGURE 5-17 Microtubules exert enough force to move cargo by dynamic instability. Photos courtesy of Lynne Cassimeris, Lehigh University.

BOX 5-7 ■

The fishing analogy. One way of understanding the value of repeatedly building and disassembling microtubules at the centrosome is to consider why someone using a fishing pole does something similar. Imagine fishing from a boat, by casting a hook at the end of a fishing line into the water. You might wait for a few seconds or minutes, and if no fish bites the hook, you reel the hook in and try again, perhaps in a different part of the water. This is similar to the behavior of the centrosome, in that microtubules are extended and then left to drift for a few seconds or minutes, and if a capping protein does not capture the plus end (i.e., bite the hook), the microtubule is disassembled (rather than "reeled in") and recycled so a new microtubule can emerge from the centrosome, perhaps in a different region of the cytosol.

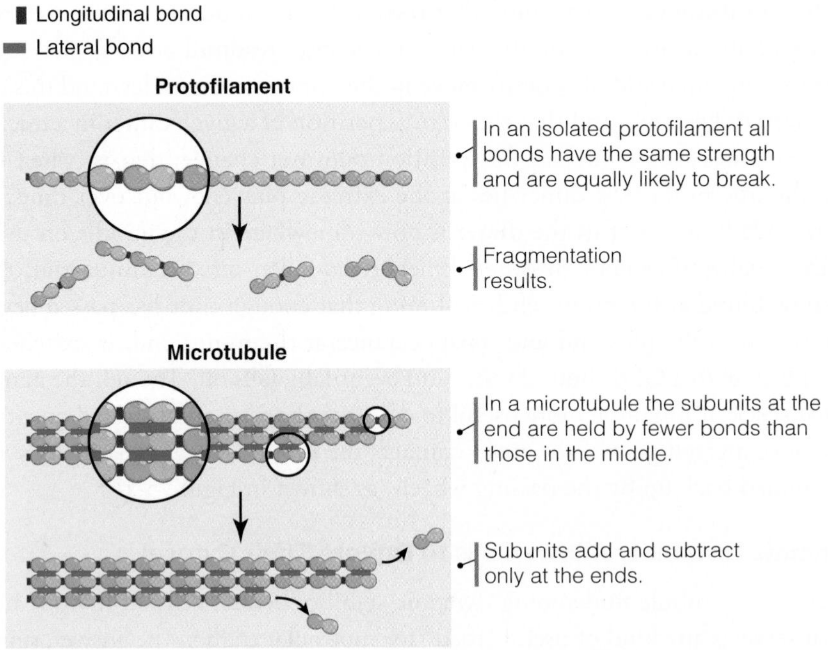

■ Longitudinal bond

■ Lateral bond

Protofilament

In an isolated protofilament all bonds have the same strength and are equally likely to break.

Fragmentation results.

Microtubule

In a microtubule the subunits at the end are held by fewer bonds than those in the middle.

Subunits add and subtract only at the ends.

FIGURE 5-18 Longitudinal and lateral bonds make microtubules strong.

in adjacent protofilaments), microtubules are strong enough to transport even relatively large items, such as the vesicle shown in Figure 5-17.

Microtubule-Associated Proteins Regulate the Stability and Function of Microtubules

The capping proteins, rescue-associated proteins, and proteins that govern the motion in Figure 5-17 are examples of a large group of proteins called **microtubule-associated proteins (MAPs)**. Other examples of MAPs include proteins that crosslink, bundle, or sever microtubules; proteins that control microtubule stability; and proteins that link microtubules to other elements of the cytoskeleton. Some of these proteins bind only to the tips of microtubules, while others can bind along the entire length of a microtubule.

Microtubule-Based Motor Proteins Transport Organelles and Vesicles

Some of the most important MAPs are those that push/drag cargo along the length of a microtubule; called motor proteins, these are the proteins that perform the actual trafficking of molecules in a cell. In our analogy of microtubules serving as "roads," motor proteins often play the role of flatbed trucks. Actual cargo is attached to one side of these proteins, while the other side is responsible for determining the speed and direction of the movement. In some cases, this cargo may be a vesicle, such as those we discussed in Chapter 4, or, entire organelles such as the endoplasmic reticulum, Golgi apparatus, and mitochondria can be transported this way. During mitosis, condensed chromosomes are transported along the length of microtubules in the mitotic spindle to ensure their proper segregation to the resulting daughter cells; in other cases, the microtubule motor proteins remain stationary and move the microtubule. There is no easy example of this in our road analogy—it would be the equivalent of the trucks being held in place, spinning their wheels and moving the road beneath them.

The two families of microtubule motor proteins are called dyneins and kinesins, shown in **FIGURE 5-19**. Most dyneins move in the same direction, toward the minus end of a microtubule; most kinesins move toward the plus end of a microtubule, but a few move in the opposite direction. To accomplish this movement, the proteins use ATP energy, but more specifically, they adopt at least four different shapes, depending on their ATP/ADP and target-binding status, as shown for kinesins in **FIGURE 5-20**. The portion of each protein that binds to microtubules is called a head domain, and that is where ATP is used to generate the motion. Notice that the two heads swing past one another, very similar to the way a person walks. This is one of the rare instances where a cellular event (motor protein movement) very closely resembles an equivalent activity in our own lives (walking along a road). Also, notice that all movement is relative: if a microtubule is anchored (e.g., to a centrosome), the motor protein will move along it, but if the motor protein is anchored (we will see examples of this in later chapters), then the microtubule will move.

Microtubule-based transport in cultured cells is illustrated in **FIGURE 5-21**. In this figure, pigment granules are concentrated in the center of a pigment cell in response to a chemical stimulant. Notice that all of the granules move in the same direction with respect to the polarized organization of the microtubule network: they move toward the

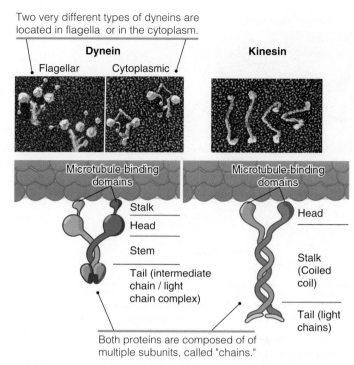

FIGURE 5-19 The structure of dynein and kinesin, the two most common motor proteins that bind to microtubules. Photos courtesy of John Heuser, Washington University School of Medicine.

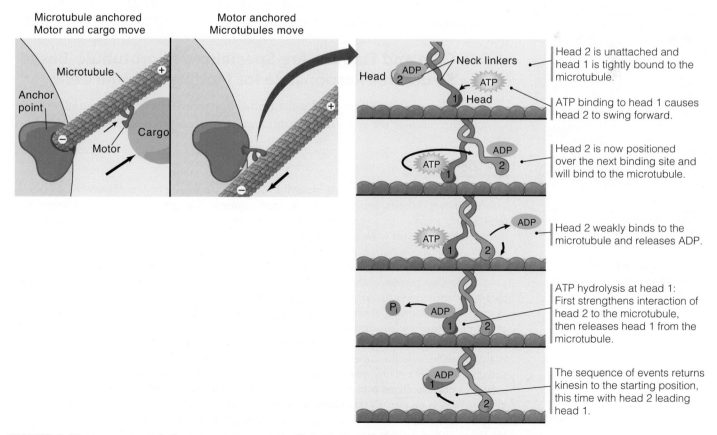

FIGURE 5-20 How a microtubule motor protein moves along a microtubule.

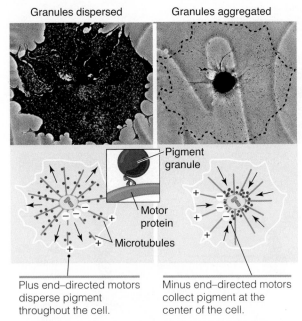

Granules dispersed   Granules aggregated

Pigment granule

Motor protein

Microtubules

Plus end–directed motors disperse pigment throughout the cell.

Minus end–directed motors collect pigment at the center of the cell.

FIGURE 5-21 Microtubule motor proteins transport pigment granules in a pigment cell. Photos courtesy of Vladimir Rodionov, University of Connecticut Health Center.

One motor protein binds both the cargo and the microtubule and actively transports in one direction, while the other does not bind the microtubule.

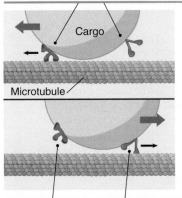

Cargo

Microtubule

Switching between different motor proteins determines the direction that the cargo moves.

FIGURE 5-22 A model for biodirectional movement on a microtube.

minus ends of the microtubules, which are attached to the centrosome. This illustrates another important trait of microtubule trafficking: individual motor proteins "walk" along microtubules in only one direction. When the stimulant was added to the pigment cell in Figure 5-19, only those motor proteins that walk toward the minus end were activated. If any plus-end motor proteins were attached to the pigment granules, they were inactive, and were therefore simply dragged along as excess baggage, as shown in **FIGURE 5-22**. Once the stimulant is removed (not shown in Figure 5-19), the plus-end motors are active again, and the pigment granules are dragged in the opposite direction, dispersing them throughout the cytosol. This is how some animals can rapidly change their coloring: when they condense pigment granules in their pigment cells, the cells adopt the color of the pigment, and dispersal of the granules changes the color of the cells back to their original state. We'll have a much closer look at the microtubule motor proteins in Chapters 7 and 9.

This microtubule trafficking network generates and maintains structural polarity in cells. No cell is perfectly symmetrical: an unbalanced distribution of cellular contents is often essential to cell function (consider how useful a skeletal muscle cell would be if it spread its contractile proteins randomly throughout the cytosol). This distribution does not happen by accident. Rather, it represents a complex and carefully coordinated program of trafficking. This is especially evident in highly polarized cells such as epithelial cells and neurons (see Chapter 14 for more discussion of these cell types). It is speculated that the centrioles, which lie in the middle of the centrosome and are always oriented perpendicular to each other, may also contribute to establishing this polarized organization (**BOX 5-8**).

Cilia and Flagella Are Specialized Microtubule-Based Structures Responsible for Motility in Some Cells

In addition to coordinating molecular trafficking, some cells use microtubules to generate force *outside* the cell. The microtubules are clustered into special structures called

---

BOX 5-8 TIP

Different definitions of the word "polarity." So far in this text, we have used the word polarity in several different contexts, and each has a different meaning. With reference to chemical bonds, the term polarity refers to an unequal distribution of electrons, such as are found between oxygen and hydrogen in water. This is *electrical polarity*. *Structural polarity* has several different meanings: it can mean an unequal distribution of atoms (the 5' end of a nucleic acid has a different chemical structure than the 3' end, for example), or it can be applied to larger structures, such as the arrangement of polypeptides in a protein, or the organization of proteins in a polymer. In this chapter, the structural polarity we are referring to concerns the three-dimensional arrangement of proteins in a polymer, such as intermediate filaments. A handy way to determine if a structure has this kind of polarity is to simply ask: Is one end of the structure the same as the other? For example, a single intermediate filament protein is polarized because it has an amino and a carboxy end; the dimer formed by these proteins is also polar, because the dimer has different ends. When the dimer forms a tetramer, however, the polarity disappears because now both ends of the tetramer have the same structure. Finally, the word *polarity* can be associated with entire cells, such as epithelial cells or neurons. These cells have very different regions on their plasma membranes, so one can always distinguish one side from the other. Skeletal muscle cells are less polarized, because they pull (contract) from both ends with the same structures. In fact, just about every cell is polarized at least to some degree.

Are prokaryotic cilia and flagella the same as those in eukaryotes? No. First, bacteria only form structures called flagella, not cilia. Also, despite the fact that prokaryotic and eukaryotic flagella appear to perform the same functions, they are completely different structures: prokaryotic flagella are composed of a protein called flagellin and contain no microtubules at all.

cilia and flagella (singular: cilium, flagellum) (**BOX 5-9**). In single-celled organisms or specialized motile cells in multicellular organisms (e.g., sperm cells), these structures propel the cell through the aqueous environment, a motion often referred to as swimming; in stationary cells, the force is used to move fluid over the surface of the cell. Cilia and flagella are fairly similar in structure, and both share the same structure, called an **axoneme**, that organizes their microtubules (**FIGURE 5-23**). Axonemes are composed of a pair of central microtubules that run the length of the cilium, surrounded by nine pairs of microtubules that form a "cage" around the central pair. This is commonly referred to as a 9+2 arrangement. Both cilia and flagella move in a whip-like fashion in liquid, shown in **FIGURE 5-24**, and this is caused by the dynein-driven sliding of the peripheral microtubules past one another in a coordinated fashion. The extensive crosslinking of microtubules within the axoneme provides the strength necessary to resist the bending forces induced by the dynein motor proteins. The minus ends of the microtubules are anchored in an MTOC called the **basal body** that lies at the cytoplasmic end of these structures.

In mammals, a modified nonmotile cilium called the **primary cilium** forms on almost every cell. While it cannot "whip" like normal cilia (it lacks the central pair of microtubules), it is thought to play a central role in sensing extracellular stimuli, including photons of light, chemical "odorants," and shear forces in the extracellular space caused by fluid flowing past a cell.

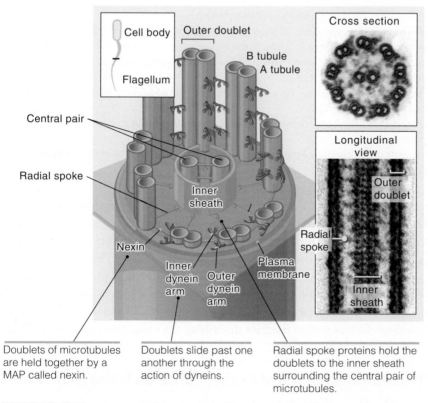

Doublets of microtubules are held together by a MAP called nexin.

Doublets slide past one another through the action of dyneins.

Radial spoke proteins hold the doublets to the inner sheath surrounding the central pair of microtubules.

FIGURE 5-23 The structure of an axoneme. Photos courtesy of Dr. Gerald Rupp, Institute for Science and Health.

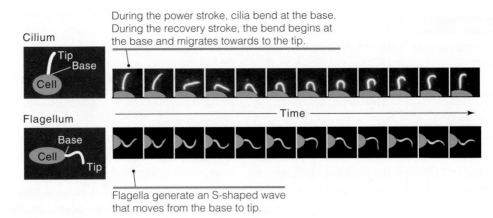

During the power stroke, cilia bend at the base. During the recovery stroke, the bend begins at the base and migrates towards to the tip.

Cilium

Tip
Base
Cell

Time

Flagellum

Base
Cell
Tip

Flagella generate an S-shaped wave that moves from the base to tip.

FIGURE 5-24 The coordinated motion of a cilium and a flagellum. Photos courtesy of D. R. Mitchell, SUNY Upstate Medical University.

CONCEPT CHECK #2

When microtubules are grown *in vitro* from purified tubulin proteins and poured onto a glass slide that has been coated with dynein or kinesin proteins, the microtubules appear to "surf" across the slide if ATP and GTP are also present. How can you explain this behavior? What would happen if you then added a large amount of ADP or GDP instead of ATP or GTP?

■ 5.5 ■ Actin Filaments Control the Movement of Cells

KEY CONCEPTS

▸ Actin filaments are thin polymers of actin proteins.
▸ Actin filaments are responsible for large-scale changes in cell shape, including most cell movement.
▸ Actin filament polymerization is initiated at numerous sites in the cytosol by actin-nucleating proteins.
▸ Actin filaments have structural polarity, which determines the direction that force is exerted on them by myosin motor proteins.
▸ The stability of actin filaments is determined by the type of adenine nucleotides bound by the actin proteins within them.
▸ Actin-binding proteins play numerous roles in controlling the location, stability, and function of actin filaments.
▸ Cell migration is a complex process, requiring assembly and disassembly of different types of actin filament networks.

The actin cytoskeleton is, in many ways, similar to the microtubule cytoskeleton. For example, both are composed of filaments made from globular protein subunits, and both are acted upon by motor proteins to induce cellular movement, but the kind of movement induced by the actin cytoskeleton is dramatically different from that associated with microtubules. Actin and its associated proteins have a much greater impact on the overall shape of a cell than microtubule-based movement, so we see them dominate in cells where *large-scale movement* is especially important, such as in muscle cells. In this section, we will follow the same format as that used for describing the other two cytoskeletal filaments, to help compare and contrast them.

■■ The Building Block of Actin Filaments Is the Actin Monomer

The simplest building block of actin filaments is a monomeric actin protein. There are as many as six different isoforms of the actin protein (named α, β, γ, δ, etc.) expressed in a given cell, each encoded by a different (but highly homologous) gene. Actin monomers bind together to form an actin filament, also called a microfilament because it has the smallest diameter—7 nm—of the three cytoskeletal filament types. A microfilament is actually composed of two strands that are intertwined, shown in **FIGURE 5-25**, somewhat like the coiled coils formed by intermediate filament subunits. Both lateral and longitudinal bonds hold the actin monomers together; this suggests that these filaments share an important functional feature with intermediate filaments: they have great tensile strength, and they can resist pulling forces much better than microtubules can. But like microtubules, these filaments have structural polarity: the plus end is often called the barbed end, and the minus end is known as the pointed end. These names were coined by microscopists who used an electron microscope to examine actin filaments bound to another protein called myosin—the complex these two proteins form looks somewhat like an arrowhead, as shown in **FIGURE 5-26**.

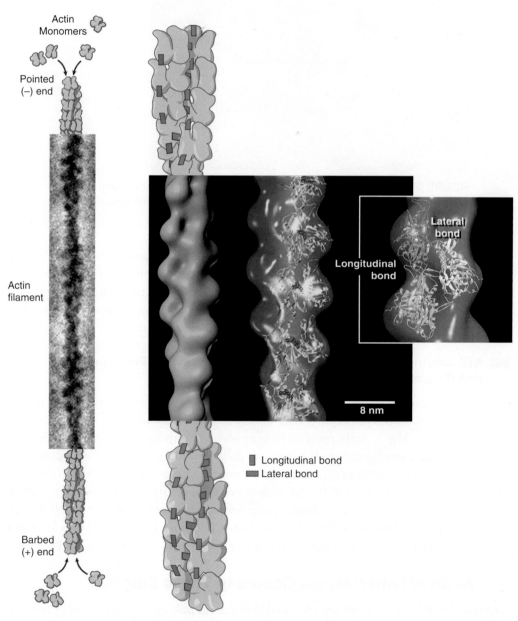

FIGURE 5-25 The general structure of an actin filament. The lateral and longitudinal bonds holding actin monomers together are indicated on the right. Photos courtesy of Ueli Aebi, University of Basel.

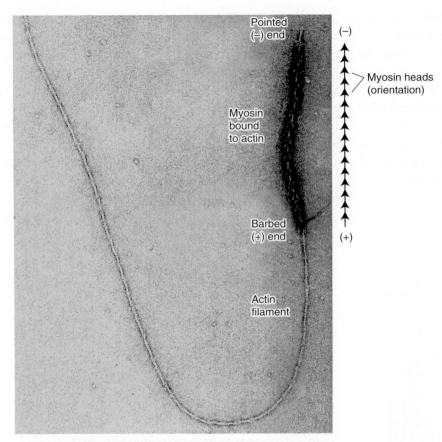

FIGURE 5-26 An electron micrograph of an actin filament partially coated with myosin proteins. Photo courtesy of Marschall Runge, John Hopkins School of Medicine and Thomas Pollard, Yale University.

As a rule, actin filaments are found wherever large-scale movement and/or tensile strength are required in a cell. Since these are common requirements, actin filaments are found in a wide variety of locations and in a myriad of configurations; some of these are shown in **FIGURE 5-27**. We will have a closer look at many of these structures in later chapters, when we can focus on how they help solve specific problems cells encounter.

◼◼ ATP Binding and Hydrolysis Regulate Actin Filament Polymerization and Disassembly

Like tubulin proteins, actin monomers adopt at least two different shapes, and each is determined by the nucleotide phosphate bound to the monomer, plus a single divalent cation (e.g., Ca^{+2}, Mg^{+2}). Actin proteins bind to adenosine nucleotides: when they bind to ATP, they adopt a configuration that *promotes* polymerization, and when they cleave the terminal phosphate to convert ATP to ADP, the protein changes to a shape that *discourages* polymerization. When the ADP-bound protein encounters a fresh ATP molecule, it drops the ADP and picks up the ATP. In short, actin monomers behave very similar to α-β tubulin dimers, except they bind to ATP/ADP instead of GTP/GDP. This ATP/ADP is held in a cleft in the middle of the monomer, as shown in **FIGURE 5-28** and discussed in **BOX 5-10**.

◼◼ Actin Polymerization Occurs in Three Stages

Like microtubules, much has been learned about actin polymerization by using purified proteins *in vitro*. The polymerization mechanisms of these two cytoskeletal proteins are strikingly similar, so we can capitalize on our earlier discussion of microtubule assembly

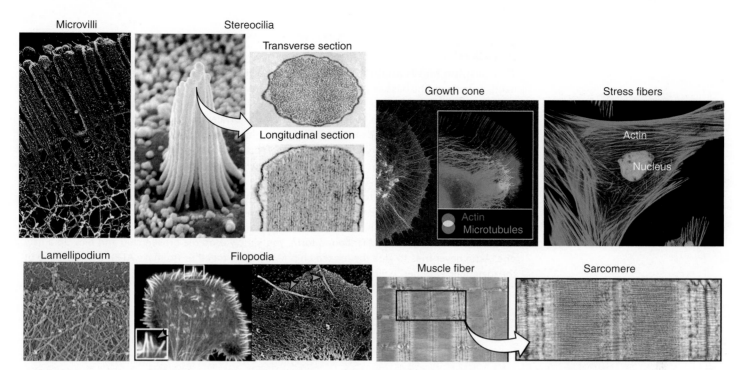

FIGURE 5-27 A number of different actin filament–based structures in cells. Neuronal growth cone photos © Schaefer, Kabir, and Forscher, 2002. Originally published in The Journal of Cell Biology, 158: 139–152. Used with permission of Rockefller University Press. Photos courtesy of Paul Forscher, Yale University. Neuronal growth cone photos © Schaefer, Kabir, and Forscher, 2002. Originally published in The Journal of Cell Biology, 158: 139–152. Used with permission of Rockefller University Press. Photos courtesy of Paul Forscher, Yale University. Photo of cell with stress fibers courtesy of Michael W. Davidson and Florida State University Research Foundation. Lamellipodium photo courtesy of Tatyana M. Svitkina, University of Pennsylvania. Filopodia photos reprinted from Cell, vol. 118, M. R. Mejiliano, et al., Lamellipodial versus filopodial mode of the actin..., pp. 363–373, Copyright (2004) with permission from Elsevier [http://www.sciencedirect.com/science/journal/00928674]. Photo courtesy of Tatyana M. Svitkina, University of Pennsylvania. Filopodia photos reprinted from Cell, vol. 118, M. R. Mejiliano, et al., Lamellipodial versus filopodial mode of the actin..., pp. 363–373, Copyright (2004) with permission from Elsevier [http://www.sciencedirect.com/science/journal/00928674]. Photo courtesy of Tatyana M. Svitkina, University of Pennsylvania./science/journal/00928674]. Photo courtesy of Tatyana M. Svitkina, University of Pennsylvania. Electron micrograph of microvilli © Hirokawa, et al., 1982. Originally published in The Journal of Cell Biology, 94: 425–443. Used with permission of Rockefeller University Press. Photos courtesy of John E. Heuser, Washington University in St. Louis. Scanning micrograph of stereocilia reproduced from A. J. Hudspeth and R. Jacobs, Proc. Natl. Acad. Sci. USA 76 (1979): 1506–1509. Photo courtesy of A. J. Hudspeth and R. A. Jacobs. Electron micrographs of stereocilia © Tilney, DeRosier, and Mulroy, 1980. Originally published in The Journal of Cell Biology, 86: 244–259. Used with permission of Rockefeller University Press. Photos courtesy of Michael J. Mulroy, Medical College of Georgia.

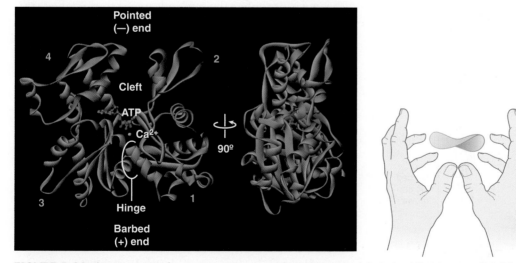

FIGURE 5-28 The structure of an actin monomer. Left, a ribbon model, derived from a crystalized form of the protein. Right is an analogy of an actin monomer as a pair of hands. Structures from Protein Data Bank 1ATN. W. Kabsch, et al., Nature 347 (1990): 21–22.

The clasping hands analogy. One simple way to visualize the structure of actin is to press your hands together at the base of your thumbs so that they form a clam-shaped structure that can open and close by pivoting at the base (see Figure 5-28). Next, pick up a small potato chip with your two hands and have it settle into your palm: this is the equivalent of an actin protein bound to ATP. Holding the chip in your hands keeps your palm slightly open and keeps your hand in a shape that somewhat resembles two parentheses: (). This is the shape of actin when it prefers polymerization, which would be roughly equivalent to having your two hands grasp the base of someone else's hand held the same way.

Now, squeeze your palms together to break the chip (no need to crush it, just break it). This is the equivalent of cleaving the ATP to ADP. Notice that your hands are now flatter, more like two parallel lines: ||. Your hands are less inclined to grab someone else's, because there is less space between them. To switch back to the ATP-bound form, you simply drop the broken chip and pick up a new one. Actin never tries to glue the broken chip (ADP) back together (to make ATP); that's the job of the metabolic enzymes, which we will discuss in Chapter 7.

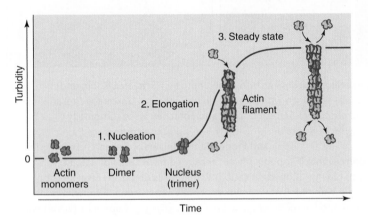

FIGURE 5-29 The three stages of actin filament assembly *in vitro.*

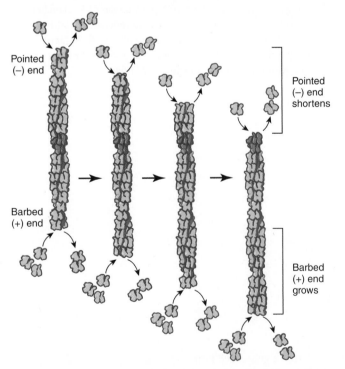

FIGURE 5-30 Treadmilling in actin filaments. Note the similarity of this treadmilling with that shown for microtubules in Figure 5-16.

to help us learn how actin behaves. For example, actin filaments form spontaneously once a critical concentration is reached, and filament assembly occurs in the same three stages we saw for microtubules, as demonstrated in **FIGURE 5-29**. First, nuclei form from the binding of three actin proteins, and then additional monomers are added to the nuclei (called the elongation stage) until a steady state is reached, where the rate of monomer addition is equal to the rate of disassembly.

Actin Filaments Have Structural Polarity

In a growing actin filament, the ATP-bound monomers are preferentially added to the barbed (plus) end. They align in a front-to-back orientation with the monomers at the end of the filament, and this gives the entire filament the same kind of structural polarity we see in microtubules. This polarity confers the same advantage to microfilaments that we saw in microtubules: motor proteins that bind to actin filaments can easily determine which direction to pull on them. We will examine the motor proteins in greater detail below.

Actin Filaments Undergo Treadmilling

Yet another similarity between actin filaments and microtubules is that both undergo treadmilling. Notice how closely the treadmilling shown in **FIGURE 5-30** resembles that shown in Figure 5-16. As you might expect, the ATP-bound monomers added to the plus end eventually hydrolyse the ATP to form ADP, and once these monomers reach the minus end, they tend to fall off. One important difference between actin filaments and microtubules is that actin filaments are not generally nucleated by a single structure in cells, and so there is no equivalent to an MTOC for actin. As a result, many more actin filaments have exposed minus ends, and treadmilling is therefore more likely for actin filaments than for microtubules.

The abundance of free minus ends in the actin cytoskeleton requires us to look more closely at how they behave in cells. Technically speaking, each end of an actin filament has its own critical concentration. Because ATP–actin monomers prefer to add to the plus end, we can conclude that the critical concentration for the plus end is lower than that for the minus end—it takes a lower concentration of ATP–actin to reach the polymerization point at the plus end. If we increase the ATP–actin concentration enough, we will reach the critical concentration for the minus end as well, and actin monomers will begin adding to both ends. Notice that treadmilling of actin therefore only happens when the concentration of ATP–actin monomers falls between the two critical concentrations.

This matters because it offers cells a tool to help control the organization of their cytoskeleton. If they require rapid assembly of a strong cytoskeletal filament in a new region of the cytosol, they can trigger the disassembly of existing actin filaments by simply reducing the amount of free ATP–actin monomers. This can be done in several ways, including binding of free monomers by actin-binding proteins. Recall that the disassembly of the other strong filaments, the intermediate filaments, requires phosphorylation of the subunits. The strategy for removing actin filaments is much simpler and faster. The corollary to this is that assembly of new actin filaments is likewise much faster, because they can initially grow from both ends when the pool of ATP-actin is highest, then level off once steady state is reached.

The main point here is simply that treadmilling is an especially important trait for actin filaments, and we will see examples of it being put to use in many different instances. When we discuss actin filaments in other chapters, keep in mind that treadmilling will often be invoked to help explain how these filaments function.

Six Classes of Proteins Bind to Actin to Control Its Polymerization and Organization

Given how dynamic actin filaments can be, for them to be of any use, cells must be able to control when they form, where they form, how fast they form, their orientation with respect to one another (and other cellular structures), and so on. That's a tall order for a filament made out of only one protein. Recall from Chapter 3 that posttranslational modification is a common method for controlling protein shape and function. Actin proteins are modified by phosphorylation, alkylation, and formation of intrachain disulfide bonds, but because actin filaments are used in so many different cellular structures (recall Figure 5-27), even these modifications are not enough to fully control actin filaments. A tremendous number (close to 300, by some estimates) of **actin-binding proteins** are expressed by cells to assist. We cannot possibly discuss them all, but we can have a look at some representatives. The actin-binding proteins are grouped into six classes, according to the effects they have on actin filaments.

Monomer-Binding Proteins Regulate Actin Polymerization

Perhaps the simplest way to regulate actin filament behavior is to simply control how many actin monomers are available for polymerization; this is the role played by **monomer-binding proteins**. In metazoan cells, the two most abundant proteins in this class are thymosin β4 and profilin. Thymosin β4 binds to actin monomers (in a 1:1 ratio) and prevents the monomers from polymerizing, even if the monomer is also bound to ATP. This is an excellent example of how the function of one protein is to bind to and control the function of another. Profilin plays a slightly different role. It too binds 1:1 with actin monomers, but

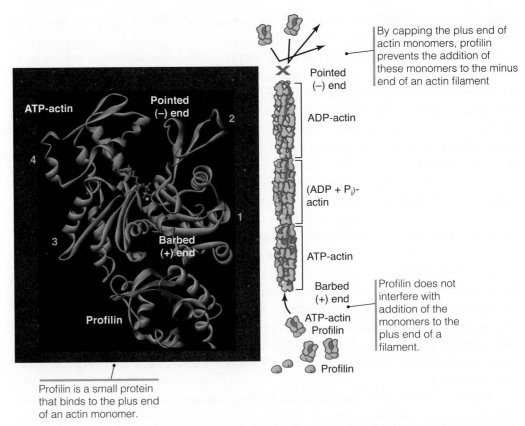

By capping the plus end of actin monomers, profilin prevents the addition of these monomers to the minus end of an actin filament

Pointed (−) end

ATP-actin

ADP-actin

(ADP + P$_i$)-actin

ATP-actin

Barbed (+) end

ATP-actin Profilin

Profilin does not interfere with addition of the monomers to the plus end of a filament.

Profilin

Profilin is a small protein that binds to the plus end of an actin monomer.

FIGURE 5-31 The structure and function of profilin, an actin monomer-binding protein. Structure from Protein Data Bank 2ITF. J. C. Grigg, et al., Mol. Microbiol. 63 (2007): 139-149.

only prevents the monomer from binding to the minus end of an actin filament. Because it still has a free minus end, a monomer bound to profilin can still add to the plus end, as shown in **FIGURE 5-31**. This is a good example of how a cell can drive actin filaments to grow in a preferred direction.

▪▪ Nucleating Proteins Regulate Actin Polymerization

For all three types of cytoskeletal filaments, the rate-limiting step is always the initial nucleation process. Small clusters of subunits are quite unstable, and easily fall apart. The role of actin **nucleating proteins** is to stabilize these clusters to facilitate the formation of actin nuclei. Some of the best-known nucleating proteins belong to a multi-subunit cluster called the ARP2/3 complex, so named because it contains *A*ctin *R*elated *P*roteins 2 and 3, plus five other proteins. The ARP2/3 complex, in conjunction with several other regulatory proteins, initiates the formation of actin nuclei. Interestingly, it does this by binding to the side of an existing actin filament, and therefore promotes the formation of branched actin filaments networks, as shown in **FIGURE 5-32**. Other nucleating proteins can initiate the formation of new linear filaments as well.

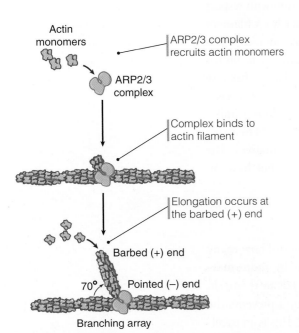

Actin monomers

ARP2/3 complex recruits actin monomers

ARP2/3 complex

Complex binds to actin filament

Elongation occurs at the barbed (+) end

Barbed (+) end

70° Pointed (−) end

Branching array

FIGURE 5-32 ARP2/3 nucleates the formation of a new actin filament off the side of an existing filament.

▪▪ Capping Proteins Affect the Length and Stability of Actin Filaments

Capping proteins inhibit actin filament elongation by binding to either the plus or minus ends of the filaments. They can also prevent the shortening of filaments by inhibiting loss of monomers from the ends. This is important because it prevents the runaway polymerization/depolymerization of actin, and permits cells to build up a high concentration of monomers without risking their random attachment to existing filaments. In short, it uncouples actin filament growth from the concentration of monomers in the cytosol. When used with the monomer-binding proteins, these proteins give cells considerable latitude for deciding when and where to build new actin filaments. Examples of capping proteins include CapZ and gelsolin, and we will see how muscle cells use CapZ to control the length of actin filaments in striated muscle cells in Chapter 14.

■■ Severing and Depolymerizing Proteins Control Actin Filament Disassembly

As important as actin filaments are, there are times when their presence is a problem for a cell. For example, when a cell divides, it must eventually split into two, and the presence of long actin filaments in the cytosol of the dividing cell can interfere if these filaments are longer than the diameter of the resulting daughter cells. Likewise, when a cell is actively migrating, it literally has to pick itself up and crawl forward; any long actin filaments that remain in place once the cell begins to move must be disassembled or the cell runs the risk of ripping itself apart. For these reasons, cells usually keep a small pool of **depolymerizing proteins** on hand to facilitate the rapid breakdown of the filaments. Cofilin and other members of the actin depolymerizing factor (ADF) family are a widely used group of depolymerizing proteins. **Severing proteins** are aptly named—they break actin filaments all along the length of the filaments (rather than at the ends), and when combined with the depolymerizing proteins, can dissolve a network of actin filaments in seconds.

■■ Cross-Linking Proteins Organize Actin Filaments into Bundles and Networks

Thus far, we've focused our attention on actin-binding proteins that target one filament at a time. The last class of actin-binding proteins contains those that work with *groups* of actin filaments. These proteins are called **cross-linking proteins**, because their function is to form linkages between actin filaments. It is important to point out that these links are *not* the covalent bonds that sometimes cross link proteins. In this case, the cross links are the actin-binding proteins themselves, and the bonds holding the filaments to the cross-linking proteins are noncovalent. Actin cross-linking proteins are classified into the three groups shown in **FIGURE 5-33**. The cross-linked filaments can take the form of tight parallel bundles, loose bundles, or branched networks, as **FIGURE 5-34** shows.

Actin-Binding Motor Proteins Exert Force on Actin Filaments to Induce Cell Movement

To help distinguish actin from the other two cytoskeletal networks in cells, we'll again emphasize that actin and its associated proteins are responsible for the large-scale movement of cells. To move an entire cell is an enormous undertaking, especially when one considers how carefully

Group	Protein	Molecular Weight (kda)	Location
I	Fascin	55	• Acrosomal process • Filopodia • Lamellipodia • Microvilli • Stress fibers
	Scruin	102	• Acrosomal process
II	Villin	92	• Intestinal and kidney brush border microvilli
III Calponin homology-domain superfamily	Fimbrin	68	• Adhesion plaques • Microvilli • Stereocilia • Yeast actin cables
	Dystrophin	427	• Muscle cortical networks
	Abp120 (Dimer)	92	• Pseudopodia
	α - actinin (Dimer)	102	• Adhesion plaques • Filopodia • Lamellipodia • Stress fibers
	Filamin (Dimer)	280	• Filopodia • Pseudopodia
	Spectrin (Tetramer)	α 280 β 246–275	• Cortical networks

FIGURE 5-33 Actin cross-linking proteins are organized into three groups, based on the way they bind actin filaments.

Actin bundles

Tightly packed bundle

Fimbrin

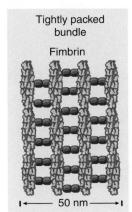

Loose bundle

α-Actinin

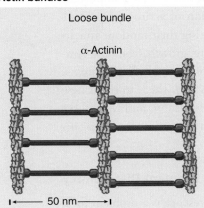

Actin network

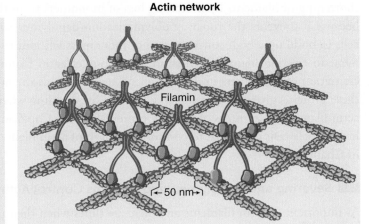

Filamin

FIGURE 5-34 Three forms of cross-linked actin filaments created by different cross-linking proteins

organized it is. The exact mechanisms cells use to control their movements are likely as varied as the cells themselves, but we can nonetheless discuss some common features shared by almost all moving cells.

Central to all directed movement is the need for some form of propulsive force. Actin-binding motor proteins provide this force by converting the metabolic (potential) energy stored in ATP into kinetic energy (actual motion). These proteins are so special that they are not generally grouped with the six classes of actin-binding proteins we've just discussed. This special status is based on a simple fact: without these motor proteins, most actin networks (regardless of how they were organized) would be practically useless. The one feature all of these motor proteins possess is the ability to reversibly bind to actin filaments and to exert force on them during the period when they are bound. This results in a bind-move-release-shift cycle for these proteins, not unlike the way a person tugs on a rope. Also, these motors can cooperate, like tug-of-war teams. The more motor proteins that are used, the stronger the force a cell can exert on its actin network.

Motor-based motion serves a tremendous number of purposes for cells. During development, when a multicellular organism is still only a disorganized cluster of cells, actin-based movement is responsible for reorganizing the cluster into a primitive body plan. Organisms deprived of this early cell-migratory ability die well before maturity; in mammals, this known as an embryonic lethal phenotype. Later in development, cells must be able to move away from the central mass to create distinct organs, limbs, and the networks (nervous system, circulatory system, lymphatic system) that connect them. Even in adulthood, cell migration is absolutely essential—it permits some cells to heal a wound, and others to maintain an active immune system. Even perfectly normal tissues require cell movement to remain healthy; for example, the cells lining our gastrointestinal (GI) tract are constantly being replaced by cells migrating from deeper layers in the GI lining. Some highly specialized motor proteins even control molecular trafficking by moving cargo along actin filaments, analogous to the function of microtubule motor proteins.

■■■ Cell Migration Is a Complex, Dynamic Reorganization of an Entire Cell

Rather than attempt to describe how every cell moves, let's envision a generic, eukaryotic animal cell and use it to integrate some well-known aspects of cell migration into a model of how most real cells move. Keep in mind that the goal here is not to comprehensively understand how any given cell moves, but simply to put what we've learned about the cytoskeleton to work and see how simple concepts can be combined to reveal a portion of a cell's elegant inner workings.

We can start by deciding what kind of cell we want to move. Let's make it the simplest cell possible, by throwing out any specialized structures that aren't necessary, such as cilia or flagella, special sensors for environmental stimuli (e.g., light-sensing structures in the visual system), or highly specialized cell shapes (e.g., the elongated shape of many neurons, the cell–cell junctions that hold epithelial cells together). What we're left with is a slightly oval-shaped cell that can move in any direction it pleases, with only the most basic structural organization.

Now, consider what it will take to move this cell. Keep in mind that this cannot be a rash decision; we already know from previous chapters that almost every part of a cell is connected to some other part, so randomly moving one part of the cell forward could have drastic consequences for the rest. Cell migration is therefore carefully orchestrated, ensuring that every part of a cell remains as functional as possible. Let's use an analogy: if each of us was roughly the size of an average protein in our generic cell, that cell would be at least as big as any major sports stadium in the United States. Imagine attending a game at such a stadium, then being told that the entire stadium would have to be moved two miles (or kilometers) in one direction while the game continued. Seems nearly impossible, doesn't it?

In one sense, it really is impossible, as stadiums aren't designed to be moved. Let's imagine how we would construct a building of that size that could be moved at will. Most of our building materials should possess at least one of these traits:

- **Modular Construction**. The structure should be almost as easy to take apart as it is to assemble. Let's make almost nothing permanent, so we would stay away from concrete as a building material, for example.
- **Lightweight and strong**. Concrete is strong, but it is used primarily to resist compressive forces (pushing). We want our materials to resist shear forces, tensile strain (pulling), *and* compression. Not only is concrete heavy, but pulling forces can easily crack it.
- **Elasticity.** Elastic materials can absorb kinetic energy by deforming without breaking, as well as spontaneously return to their original shape once the kinetic energy is removed. Elasticity would give us some latitude when we start moving large sections of our stadium/cell.

Recall that almost all of the cellular molecules we have described so far in this book possess at least one of these three traits. We can briefly describe how our imaginary stadium could be moved: we would first partially disassemble the largest stadium structures, and then apply some motive force to all of the pieces to move them forward. Remember that we can't push or pull the structure from the outside, because cells can't do that, either; all of the motion has to be generated *inside* the plasma membrane. And the plasma membrane has to remain intact throughout the entire process, of course, or the cell would rupture and die. If this sounds a bit daunting, it should, because it's a tougher problem than most building contractors ever have to face.

■■ Migrating Cells Produce Three Characteristic Forms of Actin Filaments: Filopodia, Lamellopodia, and Contractile Filaments

Let's leave the analogy for a while, and return to the cellular solution to this problem. How does a cell generate enough force to move it a long distance without help from an outside (pushing/pulling) force? Fortunately, the answer is elegantly simple: cells use actin filaments and their associated motor proteins to both push and pull themselves forward. To explain this, we will need actin-binding proteins and actin-based motor proteins to help control the location, shape, and size of actin filament networks.

Cell migration begins when a cell pushes a portion of its plasma membrane outward. This pushing force comes from the polymerization of actin filaments near the plasma membrane, and it requires the coordinated action of several different actin-binding proteins: nucleating proteins initiate filament formation, capping proteins cap the minus ends to prevent filament breakdown, severing and disassembly proteins dissolve existing actin filaments to provide new monomers, monomer-binding proteins release their monomers to add to the plus ends, and cross-linking proteins organize the filaments into networks. As the filaments grow, they distort the plasma membrane; the elasticity provided by the hydrophobic interactions between the membrane phospholipids helps ensure that the membrane does not rip when the actin filaments push against it.

One type of actin cluster that pushes the plasma membrane is called a **filopodium** (plural, **filopodia**), which means "thread-like foot" (**BOX 5-11**). These structures are thin, parallel bundles of filaments that all have the same polarity, with the minus ends pointing toward the cytosol, and the plus ends pointing toward the plasma membrane, as shown in **FIGURE 5-35**. A second type of actin cluster is called a **lamellopodium** (plural, **lamellopodia**), which means "sheet-like foot." Lamellopodia contain branched networks of filaments, held together by different cross-linking proteins than those found in filopodia. Lamellopodia cause broad, relatively shallow distortions of the plasma membrane.

Once the plasma membrane has been distended, the next step is to secure it in its new location. This requires a different set of membrane-associated proteins that we will discuss in detail in Chapter 6, but for now, we will simply call them **integrin complexes**. Filopodia and lamellopodia are called "feet" because they are analogous to the feet of a walking animal: they are the first structures to make contact with the new surface the cell is walking on, and they must be able to generate some traction if the cell is going to move forward. The adhesive proteins help keep the plasma membrane in place, like the treads on the sole of a shoe.

One important feature of these integrin complexes is that they can bind to actin filaments. The first filaments they bind are usually those making up the filopodia and lamellopodia, but later, once several adhesive proteins have been positioned in these "feet," a new type of actin bundle attaches to them. These bundles resemble filopodia in some ways in that they are parallel bundles with identical polarity, but rather than growing outward toward the periphery of the cell, the plus ends of these actin filaments grow inward, deeper into the cytosol. Whether the integrin complexes actually nucleate actin filament

FIGURE 5-35 Different forms of actin in stationary and migrating cells. Reproduced from A. Hall, Science 279 (1998): 509-514 [http://www.sciencemag.org]. Reprinted with permission from AAAS. Photos courtesy of Alan Hall, Memorial Sloan Kettering Cancer Center, New York.

growth, or simply capture neighboring clusters, is still not clear. The thickest of these bundles are called **stress fibers**. Stress fibers also contain motor proteins that cause individual actin proteins in the fibers to slide past one another, shortening the fibers.

Keep in mind that the modularity of actin filaments makes their assembly and disassembly easy to control, and a host of actin-binding proteins are responsible for keeping this cycle properly balanced. Actin is also the smallest of the cytoskeletal filaments, and the lateral and longitudinal bonds made between monomers make actin filaments very strong. This helps explain why actin is the protein of choice for moving entire cells.

◼◼ Myosins Are a Family of Actin-Binding Motor Proteins

Having assembled a network of actin filaments in our generic cell, we now have to explain how a cell pulls itself forward. To help us, let's briefly introduce a new analogy. Imagine we are watching a rock climber scale a cliff face. As she climbs up, we see a repeating pattern: she extends her arms, clutches a stable portion of the cliff with her hands, and then pulls herself upward to a new location, where the process repeats itself (we must ignore her legs pushing upward, because cells cannot do this). Like the rock climber, a cell must *contract* to actually crawl forward.

The contractile force in a cell is provided by the motor proteins we mentioned previously. The most prominent actin-associated motors are proteins called myosins. There are a multitude of different myosin proteins, and they are classified into 18 different families (named myosin I, myosin II, and so on, up to myosin XVIII). The most widely expressed myosins in humans are different forms of myosin II, and these are classified into muscle and nonmuscle types. Many cell biologists call these conventional myosins, and they are by far the best understood.

Despite the great variability between myosins, they all share some common properties that explain how they make cells move. First, myosins are multi-subunit proteins organized into three structural domains, each of which performs a distinct job (**FIGURE 5-36**). The different subunits are called **light and heavy chains**, based on their relative size. The motor domain, which is formed by the heavy chain, binds to both the actin filament and

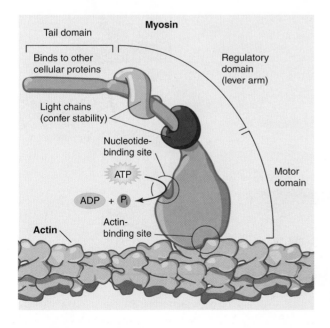
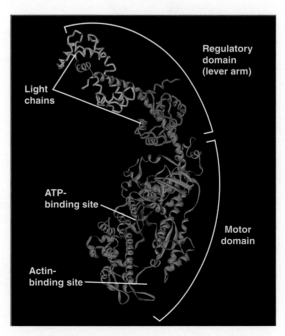

FIGURE 5-36 Myosin proteins contain three funtional domains. Structure from Protein Data Bank 2MYS. I. Rayment, et al., Science 261 (1993): 50-58.

to ATP. It is also responsible for cleaving the ATP to ADP to generate the contractile force. The **regulatory domain**, composed of both a heavy chain and two light chains, serves as the "lever arm" that swings back and forth as the myosin slides along an actin filament. The tail domain determines which other proteins each myosin protein binds to; some bind to other myosins, enabling cells to form large clusters of myosins that can generate tremendous force.

To move our generic cell forward, we simply introduce myosin II molecules into the integrin complexes and actin stress fibers. The tail domains of some myosin II proteins can bind directly to proteins in integrin complexes, and the head domains bind to the actin stress fibers. Once activated, the myosin begins a **contractile cycle** that causes it to move toward one end of the actin filaments (myosin V crawls toward the minus end, all other myosins crawl toward the plus end). The contractile cycle for myosin requires ATP binding and hydrolysis by the head domain, as shown in **FIGURE 5-37**.

Now we need to have a close look at a very subtle feature that will help put the whole picture together. Notice that when the myosins attached to the integrin complexes at the front of migrating cells attempt to crawl toward the plus end of the actin filaments, they can't move; they are stuck in place by the integrin complexes. This is extremely important. If the myosins can't move, what happens to the force they exert on the actin? Inevitably, the actin filaments must be pulled forward instead, or the filaments will rip apart (this is why actin filaments must be so strong). Notice that this is what we mean when we say that

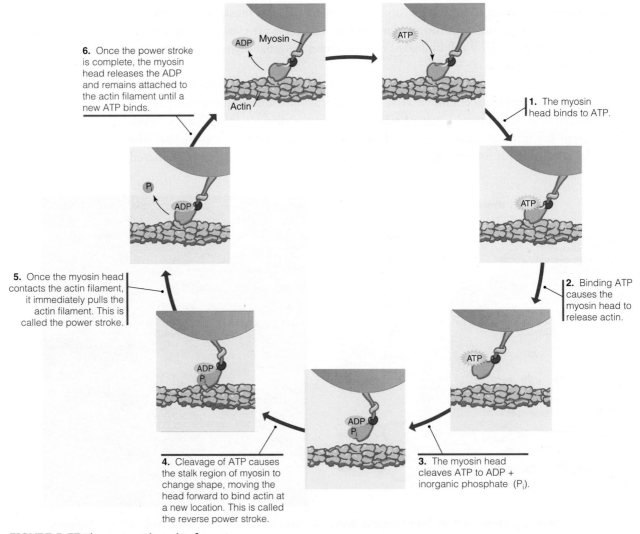

6. Once the power stroke is complete, the myosin head releases the ADP and remains attached to the actin filament until a new ATP binds.

1. The myosin head binds to ATP.

2. Binding ATP causes the myosin head to release actin.

3. The myosin head cleaves ATP to ADP + inorganic phosphate (P$_i$).

4. Cleavage of ATP causes the stalk region of myosin to change shape, moving the head forward to bind actin at a new location. This is called the reverse power stroke.

5. Once the myosin head contacts the actin filament, it immediately pulls the actin filament. This is called the power stroke.

FIGURE 5-37 The contractile cycle of myosin.

a cell pulls itself forward: whatever is attached to the actin filament gets pulled along with it. For example, if other actin-binding proteins are attached to the plasma membrane at the rear of the cell, the force exerted by the myosins is sufficient to drag the attached membrane forward too. If a lot of myosins in stress fibers work together, they can easily tug the rest of the cell, including the organelles, along with the membrane. To make things even easier, some actin filaments form close to and parallel with the plasma membrane at the rear of the cell, creating a **cortical-actin network** (the word *cortical* in this context refers to the outer edge of the cytosol). When myosins pull on these filaments, the cell essentially squeezes itself forward, somewhat like the pastry bags used to decorate cakes. Combined with successive waves of filopodia and lamellopodia, followed by integrin complexes that can pull actin filaments, the cell both "rolls" and "squeezes" itself forward.

This leaves one final question: what happens when the cell is pulled so far forward that the first integrin complexes are now at the rear of the cell? The answer, fortunately, is simple: cells capitalize on the modular nature of the contractile apparatus, and simply disassemble the integrin complex, releasing the rear of the cell from the surface it was attached to. Any remaining actin filaments attached to the integrin complexes are also disassembled and recycled.

Note that this entire operation is under the control of many types of signaling molecules that which orchestrate the timing and location of each molecular event. The molecular mechanisms cells used to pass signals from one location to another in a single cell, such as takes place here, are discussed in Chapter 11.

■■ Not All Cell Movement Is the Same

The above scenario is just one example of one type of cellular movement, and cells move in innumerable ways. For example:

- Cortical actin–myosin complexes can pinch one region of cell to compensate for expansion in another region.
- Cortical actin–myosin complexes can squeeze a dividing cell in the middle so tightly that the two halves break apart to create daughter cells.
- The cells lining the blood vessels can change shape to regulate the amount of fluid they allow to flow between them.

Remember, cells are always in motion, even seemingly stationary cells.

■■ Striated Muscle Contraction Is a Well-Studied Example of Cell Movement

Perhaps the easiest cells to observe contracting are the muscle cells that move our bones and cause our heart to beat. These cells have a very distinctive and highly specialized actin cytoskeleton; in fact, nearly 20% of the total protein in these cells is actin. Skeletal muscle cells are specialized to contract in only one direction, and so their actin and myosin filaments are all arranged in parallel, forming what appears as stripes under a microscope. Cardiac muscle cells have a very similar arrangement (though they form branched rather than entirely linear patterns). For this reason, both cell types are called striated muscle (striated is another word for striped) cells. **FIGURE 5-38** shows the overall structure of a typical skeletal muscle. Notice that the stripes are actually distinct regions of a highly organized structure called a sarcomere. A large skeletal muscle, such as a triceps muscle in

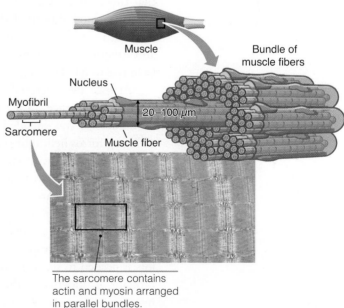

FIGURE 5-38 Striated muscle cells contain parallel bundles of actin and myosin that appear as stripes when viewed with microscopes. Photo courtesy of Clara Franzini-Armstrong, University of Pennsylvania, School of Medicine.

the arm, contains millions of sarcomeres, all capable of pulling in the same direction. Note that smooth muscle cells, the third type of muscle cells, lack sarcomeres because they are capable of contracting in multiple directions at once. We will have a much closer look at the structure and function of all three muscle cells and contractile tissues in Chapter 14.

CONCEPT CHECK #3

Actin and myosin work so closely together that they are sometimes called by a single name, *actomyosin*. These proteins provide the kinetic energy necessary to move cells, including muscle cells, and sometimes the actomyosin cytoskeleton is compared to real muscles, like our pectoral muscles. Actomyosin is sometimes called "the '*muscle element*' that allows *muscle cells* to form *entire muscles*."[3] Is this a fair comparison? Develop an argument that supports, and and another that rejects, this comparison.

5.6 ■ Eukaryotic Cytoskeletal Proteins Arose from Prokaryotic Ancestors

The great benefits provided by cytoskeletal proteins are not reserved for eukaryotic cells only; these proteins evolved from ancestral cytoskeletal proteins that first appeared in prokaryotic cells. Modern prokaryotic cells express a number of cytoskeletal proteins that are homologous to eukaryotic cytoskeletal proteins and play very similar roles in these cells. For example, despite the fact that prokaryotes do not possess true intermediate filaments, the intermediate filament protein vimentin contains binding regions for single-stranded DNA, and is thought to be a descendent of a DNA-binding protein that first appeared in primitive prokaryotes.

Likewise, a protein called FtsZ is a member of a prokaryotic family of proteins closely related to eukaryotic tubulin proteins. It is widely expressed in bacteria, where it binds and cleaves GTP and adopts two different shapes accordingly. FtsZ also forms dimers that generate protofilament polymers, and organizes into a ring around the middle of bacterial cells that contracts during cell division. One interesting observation is that this protein is also found in mitochondria and chloroplasts, suggesting eukaryotic cells may have inherited this protein directly from its own prokaryote-derived organelles. At least four other bacterial proteins are known to bind to and alter the structure and function of FtsZ polymers, and these are likely related to an ancestral protein that gave rise to MAPs in eukaryotes.

At least two bacterial homologs for actin proteins have also been identified, though their gene sequence similarity to eukaryotic actin genes is quite low, suggesting their common ancestor was encoded by one of the earliest genes in cells. The protein MreB binds ATP and forms helical filaments in bacteria that promote cell movement. Despite the weak sequence homology, the three-dimensional structure of this protein is remarkably similar to that of actin. ParM is another filament-forming protein that resembles actin filaments, and is known to participate in the segregation of plasmids in bacteria, again suggesting a role in controlling movement in these cells.

Together, these proteins conjure up an image of a primitive, single-celled organism that gained the ability to specialize by producing proteins that performed a subset of the behaviors it needed to survive. Some of these proteins were the ancestors of modern cytoskeletal proteins. It underscores that protecting DNA, compartmentalizing the cell's contents, and moving through the environment are some of the most important properties of all living organisms.

---

[3] Patel TJ & RL Lieber (1997) Force transmission in skeletal muscle: from actomyosin to external tendons, *Exec Sports Sci Rev*, 25:321-363.

All cells adopt different shapes, reflecting the wide range of functions they perform; however, these shapes do not come for free. Left by itself, a phospholipid bilayer will always form a sphere, the lowest energy state in an aqueous environment. This means that without some active intervention by proteins, all cells would be spherical in shape. The name "cytoskeleton" is given to the proteins discussed in this chapter because they play a role analogous to our own skeleton: they distort the plasma membrane to create different shapes, sort the cell's contents into specific regions of these shapes, and stabilize these shapes by providing resistance to mechanical forces that would otherwise distort (or even destroy) the cell.

Intermediate filaments are the most stable type of cytoskeleton proteins, and they are especially effective at resisting mechanical forces. This arises from the way they are built: they are highly coiled filaments, which permits maximum contact between the individual proteins and increases the number of bonds holding these proteins together. Intermediate filaments are found primarily in the cytosol, but one type forms a cage-like structure in the nucleus that protects the DNA. Different cell types express different intermediate filament proteins, permitting them to customize this part of their cytoskeleton to suit their specific needs.

Microtubules behave quite differently from intermediate filaments. These filaments can be very dynamic in cells; in fact, the term *dynamic instability* was coined to describe their rapid assembly and disassembly in cells. Cells capitalize on this dynamic instability by using microtubule-associated proteins (MAPs) to control when and where microtubules form. Once a microtubule has been stabilized, its primary role is to serve as a track for directing the trafficking of molecules in the cytosol. Examples of this transport include vesicle traffic between the ER and the Golgi apparatus, movement of mitochondria from one region of the cell to another, and segregation of chromosomes during mitosis. The transport of most molecules on microtubules is driven by specialized MAPs called motor proteins. The structure of microtubules makes them quite strong; tubulin proteins form both lateral and longitudinal bonds with a microtubule, and this gives them enough strength to literally push and pull molecules attached to their tips as they undergo dynamic instability. Tubulin proteins are a good example of a common theme in cell biology: they bind to GTP and GDP, and use the difference in shape between these two molecules to influence their own shape and function.

Actin filaments are the most complex of the three cytoskeletal elements. This complexity arises from two features of these structures: they can easily form both linear and branched networks, and they undergo a process called treadmilling that keeps them constantly in flux, even if their overall shape remains constant. Both properties are controlled by a wide variety of actin-binding proteins that are analogous to the MAPs that bind microtubules. Actin filaments are also smaller and simpler than intermediate filaments or microtubules, making them the preferred choice for rapid filament assembly and disassembly in subregions of the cytosol. Motor proteins called myosins pull on actin filaments; when these actin filaments are immobilized on the cell surface, the resulting force causes the plasma membrane to move. Coordinated teams of actin and myosin can cause an entire cell to change its shape or even move to a new location.

Every cell has a cytoskeleton. The importance of cytoskeletal proteins to cellular function is evident when one considers how long they have been present in cells. Evidence from modern prokaryotes suggest that the ancestral proteins that gave rise to modern cytoskeletal networks were some of the earliest proteins to appear in cells. These proteins also illustrate how important cytoskeletal functions are for all organisms.

See the table below for comparison.

	Ropes	Chains	Steel cables	Steel rebar	Silk	Kevlar
Similarities to intermediate filaments	High degree of surface contact between individual strands increases strength; use of coiling. No structural polarity.	Modular structure; each link is somewhat analogous to a tetramer. No structural polarity.	Flexible, uniform shape throughout the length of the structure. No structural polarity.	High degree of interaction with other supporting structures (e.g., concrete). No structural polarity.	Polymeric protein. Highly insoluble in water. Very lightweight. High tensile strength.	In filament shape, it is spun like rope. Also shaped into cables, analogous to cross-linked intermediate filament bundles. Very lightweight.
Differences with intermediate filaments	Chemical bonds holding intermediate filaments together are stronger; ropes are more susceptible to fraying at ends.	No equivalent loop-like structure in intermediate filaments. All subunits in intermediate filaments held together with the same strength, much less variation along their length.	Steel alloy's atomic structure is based on a lattice formed by iron and (most often) carbon atoms. Often heterogeneous in structure. Requires tremendous heat to form.	Poor flexibility.	Much greater elasticity. Contains abundance of β-sheets rather than α-helices. Possesses structural polarity.	Possesses structural polarity.

Evolution by natural selection acts on the heritable portion of an organism's phenotype, and therefore influences gene allele frequency in a population. This means that for organisms to use evolution as a design strategy, the material must be a gene or its products. Since proteins are much stronger than nucleic acids, evolution affects genes that are translated into proteins. Once this is established, the options for developing a strong, flexible, lightweight, insoluble material are limited to those that can be formed by proteins. Theoretically, cells could use proteins to build filaments of a different composition (e.g., sugar polymers, minerals); nothing like steel would ever form, however, because forming the chemical nature of steel from its constituent elements requires high temperatures that far exceed anything cells can survive.

The surfing behavior is caused by the immobilized motor proteins pulling on the soluble microtubules. If enough force is generated in the same direction by one or more motor proteins, the microtubule attached to the motors slides in the direction that the motors pull. The random deposition of the motors on the slide causes the microtubules to change directions as they encounter differently oriented motor proteins. If a large amount of ADP were added to the solution, microtubule sliding would slow or stop altogether, because the excess ADP would occupy the ATP-binding site in the motors instead of the ATP, thereby preventing them from cleaving any ATP to generate kinetic energy. If GDP were added, the microtubules would partially or fully dissolve because the tubulin dimers would bind the excess GDP, shifting them into a form that favors the monomer subunit form over the polymer form.

Here is one way to paraphrase the statement: actomyosin generates the contractile force necessary make a single muscle cell move. When multiple muscle cells work together this way, they can all coordinate their movement to generate enough strength to move an entire muscle, like a pectoral muscle.

An argument in support of this assertion is that an entire muscle is no stronger than the sum of its actomyosin elements, since these proteins are the only ones that make the muscle cells contract. The more actomyosin a muscle has, the stronger it is.

An argument for rejecting this assertion is that it is oversimplified. Consider the costamere we discussed in the intermediate filaments section of this chapter; if proteins that link actomyosin to the plasma membrane and/or the extracellular space fail, the muscle cell cannot function properly. Mutations in these linking proteins cause several forms of muscular dystrophy for this reason.

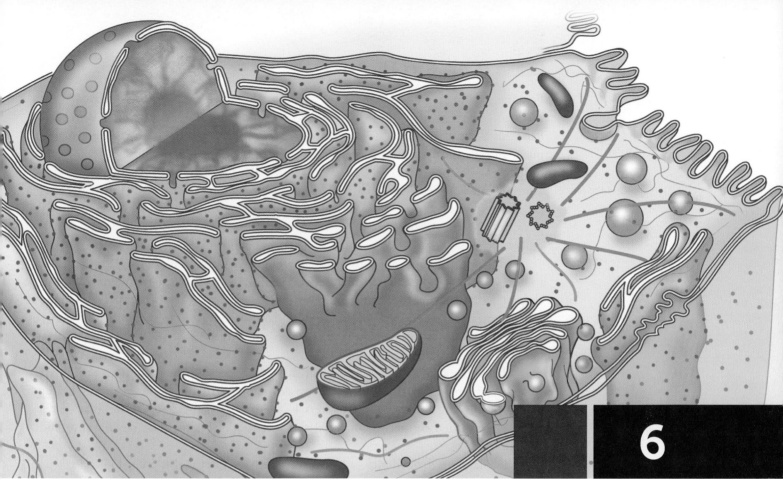

6

The Extracellular Matrix and Cell Junctions

■6.1■ The Big Picture

For many students new to cell biology, one of the more challenging concepts to get used to is how crowded the space is between the cells in multicellular organisms. Perhaps this is because none of us can see an average cell with our naked eye, so our mental image of cells is driven almost entirely by the drawings we use to represent them, and, in many of these drawings, the space between cells is often left blank. This is possibly a product of how cells were first described: the first microscopes capable of visualizing cells were developed over 400 years ago, and by today's standards, they were not very powerful. The intracellular spaces often looked empty, and the first drawings of cells reflected this.

Modern microscopes allow us to see an abundance of molecules in and around cells. In fact, we now see more in our microscopes than we can ever adequately describe. In this chapter, we'll look closely at extracellular spaces, the molecules that occupy them, and the receptor proteins cells use to bind to extracellular molecules and to each other. Doing so will allow us to explore cell biology principle #2: cells within tissues are physically contiguous with their surroundings.

What this means is that cells are not that different from the inanimate world that surrounds them. In fact, cells are composed of many of the same chemicals as the world they live in, as we saw in Chapters 1–4. For most cells, their extracellular spaces are filled with densely packed material that is essential for them to remain alive and functional.

This chapter is divided into two major sections, each devoted to a key topic:

- In the first section, we will examine representative molecules that are commonly found in the space between cells. These molecules, collectively called the extracellular matrix, are highly specialized to perform distinct functions in those extracellular spaces; we will pay a great deal of attention to their molecular structure in this section. When we discuss the receptor proteins that cells use to attach to the extracellular matrix (called cell–extracellular matrix junctions) at the end of this section, we will be able to access some of the information from Chapter 5 because these receptors are often connected to the cytoskeleton.

- The second major section of this chapter is devoted to the molecules that form direct links between cells (called cell–cell junctions). For many cells, direct contact with other cells is absolutely essential for their survival. This section will introduce several different kinds of cell–cell junctions, and will therefore rely heavily on the material we discussed in Chapters 3 (proteins) and 4 (membrane structure). As in the first section, we will also use information in Chapter 5, because many of these junctions are physically connected to elements of the cytoskeleton.

This chapter also provides an opportunity to apply much of the material we've covered in previous chapters. For example, we will consider how the primary, secondary, tertiary, and quaternary structure of the proteins called collagens contribute to their functions; describe the three traits (Chapter 3) of integrin receptors; learn to use the language from Chapter 4 to explain why tight junctions are so important for those cells that assemble them; and use the information in Chapter 1 to explain why tissue hydration is provided mostly by proteoglycans, not glycoproteins. Building these kinds of links between concepts in different chapters is intended to help keep the "big picture" of cell biology in focus.

6.2 The Extracellular Matrix Is a Complex Network of Molecules That Fills the Spaces between Cells in a Multicellular Organism

KEY CONCEPTS

▸ The extracellular matrix is a dense network of molecules that lies between cells in a multicellular organism and is made by the cells within the network.
▸ The principal function of collagen is to provide structural support to tissues.
▸ The principal function of fibronectin is to connect cells to matrices that contain fibrillar collagen.
▸ The principal function of elastin is to impart elasticity to tissues.
▸ The principal function of laminins is to provide an adhesive substrate for cells and to resist tensile forces in tissues.
▸ Proteoglycans consist of a central protein "core" to which long, linear chains of disaccharides, called glycosaminoglycans (GAGs), are attached.
▸ The basal lamina is a thin sheet of extracellular matrix found at the basal surface of epithelial sheets and at neuromuscular junctions and is composed of at least two distinct layers.
▸ Cells express receptors for extracellular matrix molecules. Virtually all animal cells express integrins, which are the most abundant and widely expressed class of extracellular matrix protein receptors.

Multicellular organisms exhibit three essential properties that distinguish them from unicellular organisms:

- First, they form stable bonds between neighboring cells. To function effectively as a group, each cell in a cluster must be able to maintain contact with, and/or close proximity to, other cells in the group for a relatively long time.

- Second, as the number of cells in an organism increases, the material that fills the extracellular space plays an increasingly important role in governing the location and behavior of the cells.

- Third, these cells possess a means for communicating with one another, either directly via cell–cell contact or indirectly via signaling molecules that span the extracellular spaces. This is how group decisions are made by clusters of cells.

The extracellular matrix (ECM) supports all three of these properties. It is a complex network of proteins, sugars, minerals, and fluids that simultaneously provide both stable adhesion sites for cells and transmit signals between them. Together, the ECM, plus water and some minerals and salts, completely fills the spaces between cells. The ECM is organized into fibers, layers, and sheet-like structures, and in some tissues, the ECM is organized into a complex sheet named the **basal lamina**, which is in direct contact with cell layers. In all cases, these molecules impart tremendous strength and flexibility to tissues while also serving as a selective filter to control the flow of particulate (undissolved) material between cells.

Glycoproteins Form Filamentous Networks between Cells

Recall from Chapters 1 and 3 that sugar molecules can be covalently attached to the side chains of some amino acids after a protein has been formed. As a general rule, when the relative abundance of the sugars is low, the molecules are called *glycoproteins*, and when so many sugars are added that they constitute the majority of the molecular mass, the molecules are called *proteoglycans*. The ECM contains several different molecules of each type, and each one has its own function in multicellular organisms. Let's examine some of the most common ECM glycoproteins first.

▪▪■ Collagen Provides Structural Support to Tissues

The family of glycoproteins known as **collagens** have been present in multicellular organisms for at least 500 million years and may have been the first ECM proteins to evolve in animal cells. The collagen family consists of at least 27 proteins that are, collectively, the most abundant proteins in the animal kingdom. Nearly all animal cells synthesize and secrete at least one form of collagen. If one could somehow remove every component of human beings except the collagens, they would still be recognizable as distinct individuals—that means there is *a lot* of collagen in our bodies.

Because they are so abundant, collagens have a major impact on how our bodies are assembled, and the mechanical and chemical properties of our tissues. Collagens provide structural support to tissues and come in a variety of shapes. Their primary functional property is *tremendous strength*. All proteins of the collagen family are bundled together as thin (approximately 1.5 nm diameter), triple helical, coiled coils composed of three collagen protein subunits, held together by both noncovalent and covalent bonds. Like intermediate filament proteins (see *Intermediate Filaments Are the Strongest, Most Stable Elements of the Cytoskeleton* in Chapter 5), this high degree of coiling allows an abundance of these bonds, and this is why collagens are so strong. Because they can be bundled into much larger structures (e.g., tendons and ligaments) than intermediate filaments, they can withstand much greater forces.

The coiled coils form three kinds of collagen structures—called fibrillar, sheet-like, and fibril-associated collagens—as illustrated in **FIGURE 6-1**:

- In fibrillar collagens, the coiled coils are organized into fibrils, or ropes, that provide great strength along a single axis. (This is analogous to the bundling of wires to form strong steel cables.) Parallel bundles of these fibrils, such as in those in tendons, impart tremendous strength capable of resisting the strain imposed by muscles on bones.

- Sheet-like collagens contain coiled coils organized into branched networks that, while less able to resist muscular force, are better able to withstand stretching in multiple directions; these networks are found in skin, for example.

- A third type of collagen, known as fibril-associated, forms coiled coils used to bind fibrillar collagens together. They are analogous to the cross-linking proteins that hold actin filaments together inside cells (see *Cross-Linking Proteins Organize Actin Filaments into Bundles and Networks* in Chapter 5).

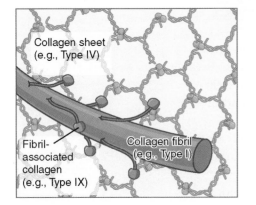

FIGURE 6-1 Collagen subunits are assembled into triple-helical coiled coils that can be organized into fibrils or sheets joined together by other extracellular matrix proteins, including fibril-associated collagens.

Most of the 27 different types of collagens can be grouped into the four classes listed in **FIGURE 6-2**. Each triple-helical structure is given a Roman numeral type designation (I, II, III, etc.). Each collagen subunit is called an α subunit, and the type of subunit is designated by a number (α1, α2, α3, etc.), which is followed by the Roman numeral of the type in which it is found. For example, the principal fibrillar collagen found in rat tails (and other tissues), named type I collagen, consists of two copies of the α1(I) subunit and one copy of the α2(I) subunit; therefore, it is designated $[\alpha1(I)]_2\alpha2(I)$. In most cases, the term *type (number) collagen* is adequate to describe the different collagens.

Class	Example	Location
Fibril-forming (fibrillar)	$[\alpha1(I)]_2\alpha2(I)$	Bone, cornea, internal organs, ligaments, skin, tendons
Fibril-associated	$\alpha1(IX)\alpha2(IX)\alpha3(IX)$	Cartilage
Network-forming	$[\alpha1(IV)]_2\alpha2(IV)$	Basal lamina
Transmembrane	$[\alpha1(XVII)]_3$	Hemidesmosomes

FIGURE 6-2 Collagens are organized into four major classes that vary according to their molecular formula, polymerized form, and tissue distribution. Some classes encompass several types of collagens.

The structure of collagen fibers is shown in **FIGURE 6-3**. Three polypeptide subunits are wrapped together in parallel to form a 300-nm-long coiled coil. Collagens contain a characteristic repeating sequence of amino acids consisting of glycine-X-Y, where X and Y can be any amino acid (but are usually proline and hydroxyproline, respectively; see *Acids Are Carbon-Rich Polymers That Contain an Amino Group and a Carboxylic Acid Group* in Chapter 1). This sequence enables tight packing of the three subunits and facilitates coiled-coil formation. These 300-nm-long units are held together by covalent bonds formed between the N terminus of one unit and the C terminus of an adjacent unit. In fibrillar collagens, the coiled coils are arranged in parallel with small (64–67 nm) gaps between them. These gaps give the fibrils their characteristic striped, or striated, appearance when viewed with an electron microscope.

Fully assembled collagen structures, be they fibrils or sheets, are much larger than the cells that synthesize them; in fact, some fibrils can be several millimeters in length. Collagen subunits are synthesized as coiled coils bound on the N and C termini by globular protein domains that actually interfere with collagen assembly inside the ER and Golgi apparatus, as shown in **FIGURE 6-4**. Portions of these globular domains are eventually cut off by enzymes, and the final steps in assembly occur outside of the cell. The remaining amino and carboxy ends of the collagen subunits, which are not enclosed in the coiled coil, are quite different in each different type of collagen, and these regions are largely responsible for determining which molecules the collagens bind to. Because they are secreted,

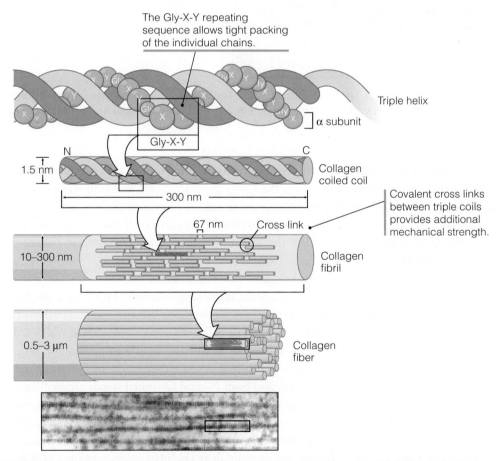

FIGURE 6-3 Schematic diagram of collagen triple-helical coiled coil (top), organization of coiled coils within a fibril (middle), and fibrils in a collagen fiber (bottom). The 67-nm gap between adjacent coiled coils results in a striated appearance in the fibrils that compose the fiber. Photo courtesy of Robert L. Trelstad, Robert Wood Johnson Medical School.

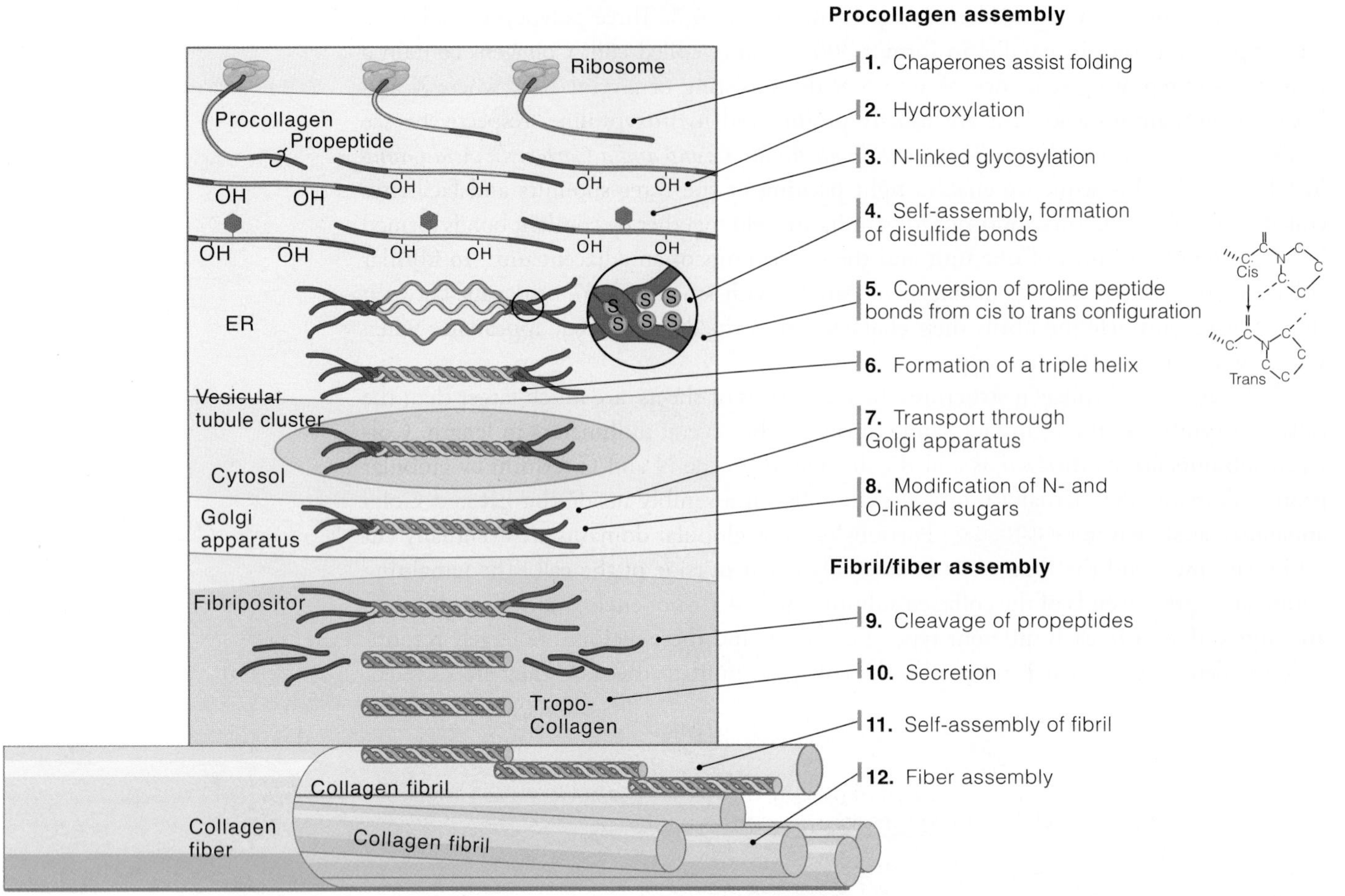

Procollagen assembly

1. Chaperones assist folding
2. Hydroxylation
3. N-linked glycosylation
4. Self-assembly, formation of disulfide bonds
5. Conversion of proline peptide bonds from cis to trans configuration
6. Formation of a triple helix
7. Transport through Golgi apparatus
8. Modification of N- and O-linked sugars

Fibril/fiber assembly

9. Cleavage of propeptides
10. Secretion
11. Self-assembly of fibril
12. Fiber assembly

FIGURE 6-4 Posttranslational modification and assembly of procollagen subunits into triple-helical coiled coils occurs during intracellular trafficking through the secretory pathway. Fibril assembly begins in fibripositors, and is completed in the extracellular space. For simplicity, hydroxyl and sugar groups are not shown in the triple-stranded structures.

collagens follow the same path that other secreted proteins follow, called exocytosis. This begins in the rough ER, moves through the Golgi apparatus, and results in secretion at the plasma membrane. This highlights an important concept in muticellular organisms: cells synthesize and secrete the majority of the material that surrounds them. Exocytosis will be examined in much closer detail in the section titled *Exocytosis Is Vesicular Transport of Molecules to the Plasma Membrane and Extracellular Space* in Chapter 9.

During exocytosis, transport of collagen triple helices from the ER to the Golgi apparatus often presents a challenge for cells. Most trafficking between the ER and the Golgi occurs in small (approximately 60–80 nm diameter) vesicles, but the triple helical portion of most collagens is, on average, 300 nm long, so collagens don't fit into these vesicles. The solution to this problem is to assemble a different membrane-bound compartment, called a vesicular tubular cluster, to carry these long molecules. Once they reach the Golgi apparatus, some collagen chains are O-glycosylated, and both N- and O-linked sugars are modified by Golgi-resident enzymes. For more information on protein glycosylation during exocytosis, see *As Proteins Enter the ER Lumen, They May Be Posttranslationally Modified* in Chapter 8 and *Posttranslational Modification of Proteins in the Golgi Is a Stepwise Process* in Chapter 9.

Collagen fiber assembly begins inside cells near the plasma membrane (**BOX 6-1**). Similar to the vesicular tubular clusters that transport collagen to the Golgi apparatus, elongated membrane-bound compartments called **fibripositors** cluster at or near the plasma membrane. This is where collagen fibrillogenesis begins. One of the first steps is the removal of the N and C terminal propeptides, leaving only the central triple helical rod-like structure. The resulting protein, known as **tropocollagen**, is almost entirely organized as a triple helix, and is the fundamental building block of collagen fibrils. Once freed from their propeptides, tropocollagens begin to spontaneously aggregate in the fibripositors. At some point that is not yet known, the fibripositors fuse with the plasma membrane, releasing the growing tropocollagen aggregates into the extracellular space.

Fibronectins Connect Cells to Collagenous Matrices

Fibronectins (from the Latin, *fibra*, fiber, and *nectere*, to connect) are glycoproteins expressed in nearly all animal connective tissues. In humans, at least 27 different fibronectin proteins can arise from a single fibronectin gene by a process called alternative splicing (for more on alternative splicing, see *The Spliceosome Controls RNA Splicing* in Chapter 8). These variants are classified into two groups: soluble (or plasma) fibronectins, found in a variety of tissue fluids (such as plasma, cerebrospinal fluid, and amniotic fluid); and insoluble (or cellular) fibronectins, which form fibers in the extracellular matrix of virtually all tissues.

Fibronectins attach cells to extracellular matrices in tissues, regulate the shape and cytoskeletal organization of these cells, and help control the behavior of many cells during development and wound healing. Fibronectins are essential for proper development: fetal mice incapable of making fibronectins die soon after the embryos begin to grow. To fulfill their various functions, fibronectins bind to many other molecules in the extracellular matrix. We can see how this is accomplished by looking at the primary and secondary structure of a fibronectin protein. Recall from Chapter 3 (see *A Domain Is a Portion of a Protein That Adopts a Characteristic Shape*) that a **domain** is a subregion of a protein that folds into a stable structure and often plays a distinct role in protein function. Fibronectins contain many such domains, called **fibronectin repeats**. The repeating sequences are classified into three groups, named type I, II, and III, and are numbered consecutively beginning at the amino terminus of the protein. All of this terminology is necessary to describe how each portion of the protein functions, and these functions are shown in **FIGURE 6-5**. Notice that fibronectins can bind to collagens, proteoglycans, and even other fibronectin molecules.

The mature fibronectin protein secreted by cells is always a soluble dimer, held together by two disulfide bonds near the carboxy terminus of each fibronectin molecule, and usually contains two copies of the same splice variant of fibronectin (shown in Figure 6-5). In this way, it somewhat resembles the dimers formed by intermediate filament proteins, which also form strong filamentous structures. Moreover, dimerization of fibronectin is essential for proper formation of the insoluble fibronectin fibers. Assembly of soluble fibronectins into

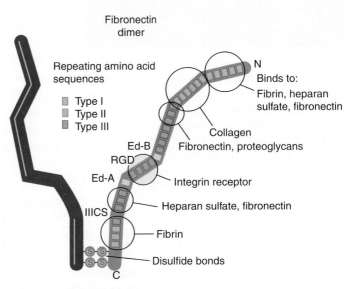

FIGURE 6-5 Two fibronectin polypeptides are covalently linked via disulfide bonds near the carboxyl terminus. Each polypeptide is organized into six domains that consist of small repeating sequences. The major protein binding regions are indicated.

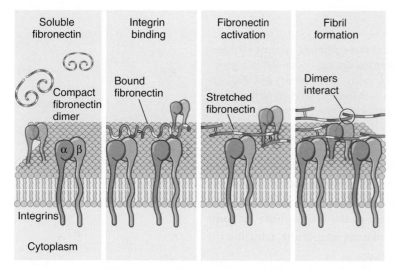

Soluble fibronectin	Integrin binding	Fibronectin activation	Fibril formation

Compact fibronectin dimer

Bound fibronectin

Stretched fibronectin

Dimers interact

Integrins

Cytoplasm

FIGURE 6-6 The fibronectin dimer is secreted in a folded conformation that is stabilized by interactions between fibronectin repeats I1-5, III2-3 and III12-14, and discourages association with other dimers. Upon association with cell surface integrin receptors, the fibronectin dimers are stretched out to reveal binding regions that attract other fibronectin dimers (e.g., I1–5 binds to III1–2 and III12 – 14). The resulting accumulation of fibronectin dimers is organized into a fibril associated with the cell surface.

FIGURE 6-7 Primary mouse dermal fibroblasts preincubated with fluorescently labeled fibronectin (green) for 48 h and counter stained with rhodamine-labeled phallodin to detect actin microfilaments (red). Image courtesy of Melinda Larsen and William Daley, University at Albany, SUNY, and Mary Ann Stepp, George Washington University. Photo courtesy of Melinda Larsen and William Daley, University at Albany, SUNY, and Mary Ann Stepp, George Washington University.

insoluble fibronectin networks requires direct contact with cells, and though the mechanism of fibronectin fiber formation is not yet fully understood, most models suggest that fibronectin dimers first bind to cell surface receptors called **integrins**, as shown in **FIGURE 6-6**. Attachment to integrins triggers assembly of an actin-based cytoskeletal complex on the cytosolic side of the integrin. Pulling of the actin filaments by myosins (see *Myosins Are a Family of Actin-Binding Motor Proteins* in Chapter 5) causes the integrin receptors to pull on the fibronectin, stretching the fibronectin dimers into a more exposed, nearly linear shape. Additional fibronectin dimers are attached to the outstretched dimers, via interactions between specific domains, thus forming a dense network. Also, the force applied by the integrin-actin network elongates each fibronectin molecule. (Notice how very different this assembly mechanism is from that for intermediate filaments: compare Figure 6-6 with Figure 5-5.) The result appears under the microscope as a collection of fibers that line up with the actin fibers inside the cells, as **FIGURE 6-7** shows. Because fibronectin also binds to other ECM molecules, these fibers can weave the matrix into a strong, supportive structure.

■■ Elastic Fibers Impart Flexibility to Tissues

Elastin, as its name implies, is the extracellular matrix protein primarily responsible for imparting elasticity to tissues. The formal definitions of the word *elastic* in the Oxford English Dictionary include one used primarily by biologists and other scientists: "That (which) spontaneously resumes (after a longer or shorter interval) its normal bulk or shape after having been contracted, dilated, or distorted by external force"; or, in plain English, something that is elastic snaps back into place after it has been deformed. Your outer ear is a great example of an elastic object. Elastin allows tissues to stretch and return to their original size without expending any additional energy. It is particularly abundant in tissues such as blood vessels, skin, and lungs, where this flexibility is critical for proper organ function; the flexibility of blood vessels is important for maintaining proper blood pressure, and flexibility in lungs allows proper filling and evacuation of the lungs with each breath.

Elastin is synthesized and secreted by many cell types. These cells also secrete collagens, which resist stretching. By varying the proportion of elastin to collagen in the extracellular matrix, cells can regulate the flexibility and strength of organs.

Elastin is organized into elastic fibers, which consist of a core region enriched in elastin proteins surrounded by a tough coating called a **microfiber** (or **microfibrillar**) **sheath**, as illustrated in **FIGURE 6-8**. The sheath is comprised of at least 10 different proteins, and some of these proteins can bind to cell surface integrin receptors. These fibers are so strong and stable that until recently they were thought to last a lifetime. Elastin is constantly undergoing a slow turnover (i.e., degradation and replacement) in healthy tissues. The elastin in these fibers is also the most insoluble protein in the vertebrate body.

How can elastin be tremendously strong and stable, yet highly flexible? The answer lies in its structural organization, as shown in Figure 6-8. The elastin gene encodes two very different types of sequences: some are hydrophilic, while others are rich in hydrophobic amino acids (**BOX 6-2**). The hydrophobic sequences are interspersed among the hydrophilic regions, giving rise to a large protein with two different properties. The hydrophobic regions impart elasticity by clustering into coils under low stretch conditions, and uncoiling when stretching force is applied, as shown in Figure 6-8. These regions spontaneously coil up again when the stress is removed. Much of the strength in elastic fibers arises from the covalent cross links formed between hydrophilic portions of adjacent elastin proteins, similar to the cross links found in collagens (see Figure 6-4).

Assembling such an insoluble protein poses special problems for a cell. If these proteins spontaneously aggregate before they are secreted, they could interfere with the secretion of other proteins by clogging the exocytosis pathway or by rupturing organelles or the plasma membrane. Cells synthesize and secrete the elastin proteins as monomers, but assemble them into fibers only in the extracellular space, after they have been secreted and pose no threat to the cell interior.

A current model of elastin fibrilogenesis suggests it occurs in seven steps, as shown in **FIGURE 6-9**:

1. Microfiber sheath proteins are secreted into the extracellular space, where they assemble into lattices, strengthened by the covalent cross links formed between them by extracellular enzymes.
2. Elastin monomers (known as **tropoelastin**) are synthesized in the rough endoplasmic reticulum and carried to the plasma membrane by exocytosis. They bind to a chaperone/receptor protein complex in the endoplasmic reticulum. This chaperone complex remains attached to the tropoelastin throughout the exocytosis pathway and prevents aggregation of tropoelastin monomers in the cell.
3. Upon secretion, the tropoelastin is held at the cell surface by elastin-binding proteins. It is cross linked there by extracellular enzymes (very similar to collagen cross linking, see Figure 6-4) to generate small, disorganized clusters.
4. Over time, as additional tropoelastins are secreted and added, the cell surface cluster grows in size, and is further cross linked.
5. At some still undefined point, the aggregate is moved from the external face of the plasma membrane and binds to the microfibrillar sheath.
6. Aggregates on the microfibers condense to form large complexes.
7. Aggregates are further cross linked to form the final structure, which is then covalently linked to the sheath. The term **mature elastin** is reserved for the elastin proteins that have been fully assembled into this polymer.

Elastin is closely integrated into tissues—at least 30 different proteins either bind to elastin fibers or form a portion of the microfibrillar sheath—so alterations in the assembly or function of elastin and elastic fibers can therefore have dramatic consequences. Cutis laxa, a disease involving loss of elastic fibers in the skin and connective tissues, ranges in severity from slightly disrupted fibers and mild skin wrinkling to nearly undetectable amounts of elastin fibers; patients with little or no elastin cannot maintain tissue integrity and die in early childhood.

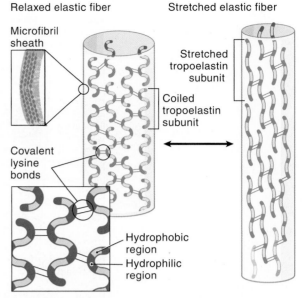

FIGURE 6-8 Schematic representation of relaxed and stretched elastic fibers. Note the dramatic difference in elastin subunit structure in each condition. The exact structure of these subunits is still not known.

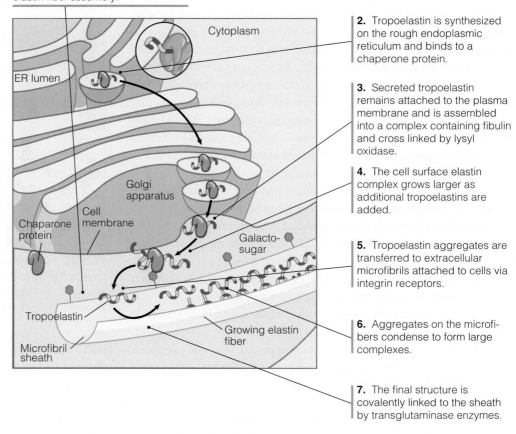

1. Fibrilins and MAGPs are secreted into the extracellular space, and form the microfibrillar lattice that nucleates elastin fiber assembly.

Cytoplasm

ER lumen

Golgi apparatus

Cell membrane

Chaparone protein

Galacto-sugar

Tropoelastin

Growing elastin fiber

Microfibril sheath

2. Tropoelastin is synthesized on the rough endoplasmic reticulum and binds to a chaperone protein.

3. Secreted tropoelastin remains attached to the plasma membrane and is assembled into a complex containing fibulin and cross linked by lysyl oxidase.

4. The cell surface elastin complex grows larger as additional tropoelastins are added.

5. Tropoelastin aggregates are transferred to extracellular microfibrils attached to cells via integrin receptors.

6. Aggregates on the microfibers condense to form large complexes.

7. The final structure is covalently linked to the sheath by transglutaminase enzymes.

FIGURE 6-9 Seven steps of elastin fiber assembly. Adapted from J. E. Wagenseil and R. P. Mecham, Birth Defects Res. C Embryo Today 81 (2008): 229-240.

Patients with Williams syndrome produce truncated forms of elastin that lack some crosslinking domains and are poorly organized into fibers; these patients develop severe narrowing of their large arteries, possibly due to abnormal growth of smooth muscle cells around the arteries to compensate for the loss of the elastic fibers normally found in the artery walls.

■■ Laminins Provide an Adhesive Substrate for Cells

The **laminins** are a diverse family of large extracellular matrix proteins expressed in invertebrate and vertebrate animals; the degree of homology between laminin family members is quite low, suggesting that laminins have a long evolutionary history. Laminin experts believe that the ancestral gene that gave rise to current laminins closely resembles the single laminin gene in the invertebrate *Hydra vulgaris*.

Like collagens, laminins consist of three polypeptide subunits wrapped together to form a triple helical coiled coil. In laminins, the sequence responsible for establishing this coiled coil is seven amino acids long and is found in multiple repeats in each of the three subunits. The coiled coil maximizes the number of noncovalent bonds formed between the subunits and confers structural stability on the completed trimer, again repeating the theme of "coiled-coil strength" found in collagens and intermediate filaments. Once the coiled coil is formed, the subunits are covalently linked via disulfide bonds. Only a portion of each subunit is organized into the coiled coil; each subunit also extends "arms" out from the coil, giving rise to a cross-shaped structure, as illustrated in **FIGURE 6-10**. Like collagens and elastins, laminins are routinely cleaved by enzymes called proteases to generate mature, functional forms of these proteins.

Laminin proteins are heterotrimers: the three subunits contained in a single laminin protein are products of different genes, and are classified into three groups, α, β, and γ. So far, five α, three β, and three γ subunits have been identified. Together, these subunits could theoretically combine to form over 100 different heterotrimer combinations, yet only 16 combinations have been identified so far; still, this allows for a wide range of laminin networks to be constructed in a single organism. The method for naming the laminins identifies the constituent chains in each (e.g., a laminin called laminin-332 consists of α3, β3, and γ2 subunits).

Unlike the other major classes of glycoproteins found in the extracellular matrix, laminins do not form fibrils; rather, they are organized into web-like networks that are able to resist tensile (stretching) forces from many directions at once. The short arms of the laminin heterotrimer, which represent the amino-terminal portion of each subunit, contain domains that associate with other components in the extracellular matrix to form this large web. One excellent example, shown in **FIGURE 6-11**, is a web formed by laminin-111 called the basal lamina, where it interacts with other extracellular matrix proteins and proteoglycans. Some laminin subunits lack these domains, and the mechanism for polymerization of these laminins is not yet known.

Structurally, laminins resemble both collagens (both have a triple-helical, central rod domain) and fibronectins (both have multiple binding sites for cell surface receptors and

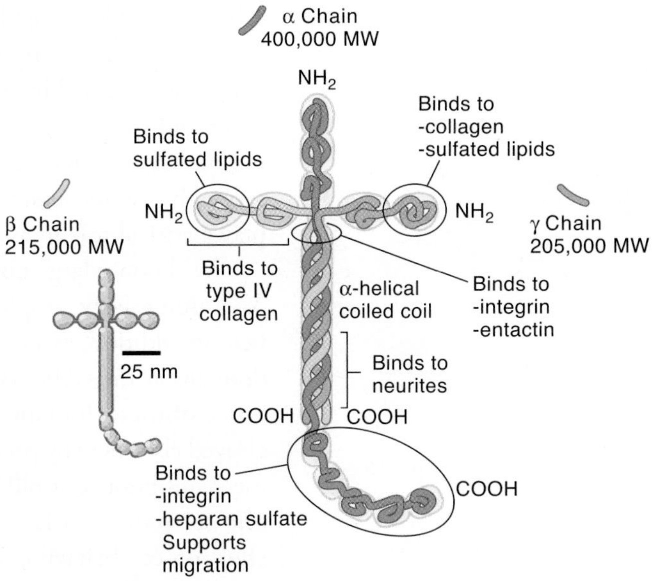

FIGURE 6-10 The three chains of the laminin molecules are wrapped into a central core. The amino-terminal portion of each chain extends from the central core to form a cross-shaped structure. The carboxyl tail of the α chain extends beyond the central core to form up to five globular domains. Important binding regions are indicated.

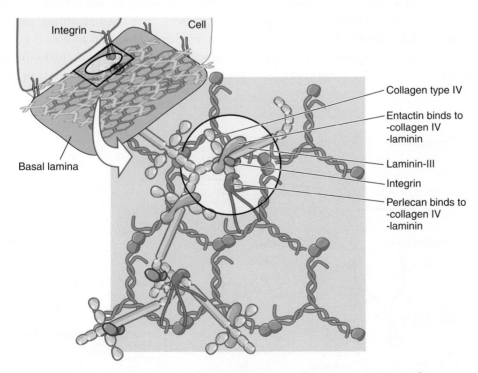

FIGURE 6-11 Laminin associates with at least three other extracellular matrix proteins to form a network within the basal lamina. Laminins also bind to integrin receptors projecting outward from cells attached to the basal lamina.

other components of the ECM), and these similarities are reflected in the functions of laminins as well. Early morphological and biochemical studies showed that laminin-111 is widely expressed in the basal lamina of epithelial tissues and that it supports the attachment and spreading of many epithelial cell types through integrin receptors. Cell spreading on laminins requires that the laminin network be strong enough to resist the tension caused by cytoskeletal remodeling. Many proteins bind to laminin-111, and some of these play a critical role in assembling laminin-111 into networks.

We know a large number of the functions and binding partners associated with regions of laminins. For example, more than 20 different receptors for laminin-111 have been identified; in addition, as shown in Figure 6-10, multiple sites on laminins, such as the globular domains at the carboxyl terminus of the α chains, play a role in regulating cell migration. Some of these domains are "cryptic," meaning they are only exposed after the protein is cleaved elsewhere by protease enzymes. In fact, selective cleavage of laminins is an important mechanism for controlling cellular adhesion and migration in development and cancer. Laminin receptors associate with different elements of the cytoskeleton and can have a distinct effect on cell behavior. The mechanisms responsible for regulating cell response to laminin binding are yet to be elucidated.

Proteoglycans Provide Hydration to Tissues

Proteoglycans are the counterpart to the structural glycoproteins (such as collagen and elastin) in the extracellular matrix. While structural glycoproteins provide tensile strength, proteoglycans ensure that the extracellular matrix is a hydrated gel. This is important for tissues to resist compressive forces, like those generated by the pounding of feet on the ground during a sprint; for this reason, proteoglycans are especially abundant in the cartilage that protects our load-bearing joints such as knees, hips, and spine.

Like other glycoproteins that are abundantly expressed on the cell surface, a proteoglycan is composed of a single polypeptide core (hence the term *proteo*) to which sugars (glycans) are attached. Over 40 different proteoglycan core proteins have been identified, and each contains modular structural domains that can bind to other components of the extracellular matrix, such as carbohydrates, lipids, structural proteins, integrin receptors, and other proteoglycans. **FIGURE 6-12** shows examples of the types of proteoglycans. Most proteoglycans,

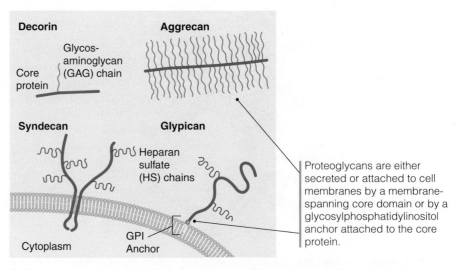

FIGURE 6-12 Summary of proteoglycan structures. All proteoglycans have GAG chains attached to a core protein.

such as decorin and aggrecan, are secreted from the cell, but two types are membrane bound. Members of the syndecan family of glycoproteins contain a transmembrane domain, and the glypicans are anchored to the membrane via a glycosylphosphatidylinositol linkage (see *Membrane Proteins Associate with Membranes in Three Different Ways* in Chapter 4).

Proteoglycans are distinguished from glycoproteins by the type and arrangement of the sugars attached to them. The sugars attached to proteoglycans are termed **glycosaminoglycans (GAGs)**, which are arranged as long, linear chains of repeating disaccharides. These chains may contain hundreds of linked sugars and can reach molecular weights of up to 1,000 kilodaltons (kDa). As **FIGURE 6-13** shows, GAGs are organized into five classes based on the disaccharides they contain, and all but one (hyaluronic acid) can be attached to proteins to form proteoglycans. All GAGs contain acidic sugars and/or sulfated sugars, which gives them a highly negative charge.

The steps involved in building proteoglycans are shown in **FIGURE 6-14**. The core protein is synthesized at the rough endoplasmic reticulum, and most are secreted by exocytosis. Syndecans remain embedded in the plasma membrane after exocytosis is complete. Glypican core proteins are modified by addition of glycosylphosphatidylinositol.

As the core protein progresses through the exocytosis pathway, **glycosyltransferase** enzymes attach sugars to the core protein. Special amino acid sequences within the core protein determine the type and location of the sugars that are attached; these sugars serve as the attachment sites for additional sugars that make up the GAG chains. The GAGs may be modified by still more enzymes that rearrange the structure of the sugars or add sulfate groups to the sugars. Some proteoglycans also have the N- and O-linked oligosaccharides typical of glycoproteins. After the newly synthesized proteoglycans pass through the Golgi apparatus, they are released by exocytosis.

A proteoglycan can have from 1 to over 100 large GAGs attached to it. Because many of the sugars are negatively charged, GAGs repel each other; on proteoglycans containing many GAGs, this forces the core protein into a linear, rod-like shape, with the GAGs projecting outward. The result is that the mature proteoglycan resembles a bristly rod, much like a hairbrush, as seen in Figure 6-12.

This distinctive shape provides proteoglycans with special properties that help define the nature of the extracellular matrix. First, their relatively rigid structure helps them act as structural scaffolds that define the overall shape of the tissues in which they are found. Second, proteoglycans assist the immune system: GAG bristles filter out bacteria and viruses in the extracellular fluid, reducing the chance of infection in tissues. Third, the negative charge of the GAG chains attracts positively charged ions (cations), and these in turn attract water molecules, so proteoglycans are sufficiently hydrated to form a gel. These gels help keep cells hydrated, provide for an aqueous environment that facilitates diffusion of small molecules between cells, and

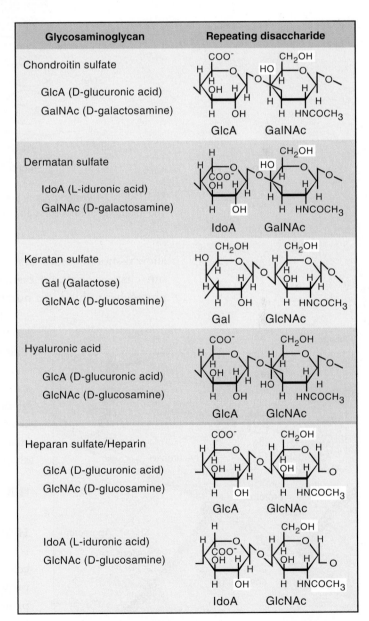

Glycosaminoglycan	Repeating disaccharide
Chondroitin sulfate GlcA (D-glucuronic acid) GalNAc (D-galactosamine)	GlcA GalNAc
Dermatan sulfate IdoA (L-iduronic acid) GalNAc (D-galactosamine)	IdoA GalNAc
Keratan sulfate Gal (Galactose) GlcNAc (D-glucosamine)	Gal GlcNAc
Hyaluronic acid GlcA (D-glucuronic acid) GlcNAc (D-glucosamine)	GlcA GlcNAc
Heparan sulfate/Heparin GlcA (D-glucuronic acid) GlcNAc (D-glucosamine) IdoA (L-iduronic acid) GlcNAc (D-glucosamine)	GlcA GlcNAc IdoA GlcNAc

FIGURE 6-13 GAGs are classified according to the type of repeating disaccharide they contain. Sulfate groups are added at the highlighted positions in the disaccharides. Adapted from K. Prydz and K. T. Dalen, J. Cell Sci. 113 (2000): 193-205.

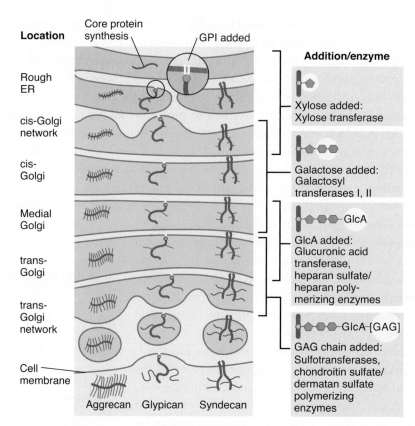

Location

Core protein synthesis

GPI added

Rough ER

cis-Golgi network

cis-Golgi

Medial Golgi

trans-Golgi

trans-Golgi network

Cell membrane

Aggrecan   Glypican   Syndecan

Addition/enzyme

Xylose added: Xylose transferase

Galactose added: Galactosyl transferases I, II

GlcA

GlcA added: Glucuronic acid transferase, heparan sulfate/ heparan poly- merizing enzymes

GlcA–[GAG]

GAG chain added: Sulfotransferases, chondroitin sulfate/ dermatan sulfate polymerizing enzymes

FIGURE 6-14 Proteoglycans are assembled during their transport through the secretory pathway. The locations of important enzymes are indicated.

allow tissues to absorb large pressure changes without significant deformation (these pressure changes occur, for example, during blunt force, injury, or vigorous exercise). Fourth, proteoglycans bind to a number of other proteins and keep them immobilized near cells.

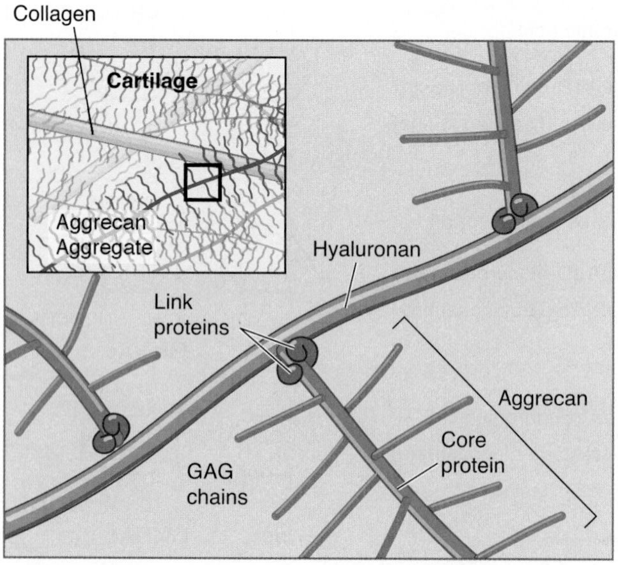

Proteoglycans also bind to and direct the assembly of other extracellular matrix proteins, including collagen. A proteoglycan called aggrecan forms large aggregates with fibers of type II collagen in cartilage, as shown in FIGURE 6-15. To form the aggregates, aggrecan molecules bind to GAGs via linker proteins. Another proteoglycan called decorin acts as a spacer between collagen fibers and controls the fiber diameter as well as the rate of their assembly. Mice that have had their decorin gene knocked out develop irregularly shaped collagen fibrils and have especially fragile skin as a result.

■■ Hyaluronan Is a Glycosaminoglycan Enriched in Connective Tissues

Hyaluronan (HA), also known as hyaluronic acid or hyaluronate, is a glycosaminoglycan. Unlike the other GAGs found in the extracellular matrix, it is not coupled covalently to proteoglycan core proteins; rather, it forms enormous complexes with secreted proteoglycans. One of the most important of these complexes is found in cartilage, where HA molecules secreted by chondrocytes (cartilage-forming cells) bind to as many as 100 copies of the proteoglycan aggrecan (Figure 6-15); the aggregate can be more than 4 mm long and have a molecular mass exceeding 2×10^8 daltons.

Collagen

Cartilage

Aggrecan Aggregate

Hyaluronan

Link proteins

Aggrecan

Core protein

GAG chains

FIGURE 6-15 Proteoglycans such as aggrecan complex with collagen II fibers in cartilage. The aggrecan complexes bind to hyaluronan molecules and attract water, which absorbs compressive forces and acts as a lubricant.

In this way, HA acts to create large, hydrated spaces in the extracellular matrix of cartilage. These spaces are especially important in tissues with a low density of blood vessels because they facilitate diffusion of nutrients and wastes through the extracellular spaces.

The structure of HA is quite simple. Like all GAGs, it is a linear polymer of a disaccharide. HA molecules contain an average of 10,000 (and up to 50,000) of these disaccharides linked by a $\beta(1,4)$ bond, as shown in Figure 6-13. Because these disaccharides are negatively charged, they bind up cations and water. Like proteoglycans, HA increases the stiffness of the extracellular matrix and serves as a lubricant in connective tissues such as joints. The hydrated HA molecules also form a water cushion between cells that enables tissues to absorb compressive forces.

Synthesis of HA is catalyzed by transmembrane enzymes called HA synthases in the plasma membrane. These enzymes are somewhat unusual in that they assemble the HA polymer on the cytosolic face of the plasma membrane and then translocate the assembled polymer across the membrane into the extracellular space. This is entirely different from synthesis of other GAGs, which are synthesized in the Golgi complex and are covalently attached to proteoglycan core proteins as they pass through the secretory pathway.

In addition to its role in tissue hydration, HA binds to specific cell surface receptors that control processes such as cell growth, survival, differentiation, and migration. Virtually all human cells express at least one of these receptors; the principal HA receptor is a protein called CD44, which belongs to a family of related proteins, known as hyladherins, all of which bind to HA.

It is generally thought that HA plays two roles in promoting cell migration. First, by binding to extracellular matrix molecules, it disrupts cell–cell and cell–matrix interactions. Mice that fail to express HA have much smaller spaces between cells and consequently do not develop properly. Because HA has such a large hydrated volume, increased secretion of HA in a tumor may disrupt the integrity of the extracellular matrix and create large spaces through which tumor cells may crawl. Second, HA binding to CD44 receptors may lead to cytoskeletal rearrangements and increased cell migration (see *Cell Migration Is a Complex, Dynamic Reorganization of an Entire Cell* in Chapter 5).

■■ Heparan Sulfate Proteoglycans Are Cell Surface Coreceptors

Heparan sulfate proteoglycans (HSPGs) are defined as those proteoglycan core proteins that are attached to heparan sulfate (HS), a glycosaminoglycan (GAG). Heparan sulfate is mostly found on two families of membrane-bound proteoglycans, the syndecans and the glypicans (see Figure 6-13). Because they remain attached to the cell surface after their assembly, these proteoglycans play a critical role in regulating the adhesion of cells to other components in the extracellular space, including structural glycoproteins, signaling molecules, and other cells. In this section we will first explore the structural diversity of HSPGs and then describe the biochemical and genetic evidence linking HSPGs to a variety of cellular functions.

HSPG synthesis takes place in the Golgi apparatus, and due to its complex structure, there is a tremendous amount of structural variability in HSPGs. In fact, its 32 different disaccharide building blocks give rise to greater structural complexity in HS chains than is found in proteins, which are made up of 20 different amino acids. Because so many different forms of HS can be made for a single HSPG molecule, cells can express multiple different forms of HSPG at the same time, with each form folding into a slightly different shape and, thus, having different binding properties for extracellular proteins.

Not surprisingly then, HSPGs bind specifically to over 70 extracellular proteins and can play many different roles in tissues. Using genetic analysis of model organisms is a powerful tool for elucidating the function of HSPGs in development and disease. One of the best model organisms for genetic analysis is the fruit fly, *Drosophila melanogaster*. Flies with mutations either in the HSPG core protein or in the sugar processing enzymes required for HS synthesis have the same defects as flies with mutations in genes for growth factors or their receptors. Similar studies in mice demonstrate a multitude of functions for HSPGs that are difficult to detect in simpler model systems. For example, knockout mice that fail to express syndecan-1, a major cell surface HSPG, have compromised immune systems and a severely weakened wound-healing response. Knockout of the HSPG perlecan in mice disrupts proper cartilage formation during development, and leads to severe skeletal deformities and early death.

The Basal Lamina Is a Specialized Extracellular Matrix

The term *basal lamina* refers to a thin sheet (or *lamina*) of extracellular matrix that lies immediately adjacent to, and is in contact with, many cell types. The basal lamina is recognized as a distinct form of extracellular matrix because it contains proteins, such as collagen IV and nidogen, found only in this structure, and because it adopts a distinct, sheet-like arrangement. Originally this term applied only to the sheet of extracellular matrix in contact with the basal surface of epithelial cells, where it was first seen with an electron microscope. Now that the major constituents of the basal lamina have been identified, we also apply this term to the sheet that lies between muscle and nerve cells at the neuromuscular junction, because this sheet contains many of the same proteins as the basal laminae underlying epithelial cells.

Over the years many names have been given to this layer of extracellular matrix. When viewed with a scanning electron microscope, the basal lamina appears as a distinct sheet separating two cell layers, and when viewed with a transmission electron microscope, the basal lamina appears as two layers, each approximately 40 to 60 nm wide. The region closest to the epithelial cell plasma membrane appears almost empty and is termed the lamina lucida (from the Latin, meaning bright layer), while the region furthest from the plasma membrane stains darkly with electron-dense dyes and is named the lamina densa (dense layer) (**FIGURE 6-16**). Beyond the lamina densa lies a network of collagen fibers that is sometimes called the reticular lamina; under a light microscope, the basal lamina and reticular lamina appear as a single boundary, often called the basement membrane, as shown in **FIGURE 6-17**. Often, the terms *basal lamina* and *basement membrane* are used interchangeably.

The basal lamina performs four principal functions:

- It serves as the structural foundation underneath epithelial cell layers. Cells attach to laminin and collagen fibers in the basal lamina through specialized structures known as hemidesmosomes, which also connect to the intermediate filament network. In this way, the basal lamina connects the intermediate filament networks of several cells, strengthening the tissue. This is especially prevalent in the skin, which is a very tough organ.
- It is a selectively permeable barrier between epithelial compartments. The proteoglycans in the basal lamina trap particulate matter (dead cells, bacteria, etc.), thereby containing infections and assisting the immune system.
- Proteoglycans in the basal lamina bind, immobilize, and concentrate proteins (such as growth factors) from tissue fluid. This enhances cellular access to these proteins and, in some cases, facilitates binding by their receptors.
- Laminin proteins in the basal lamina serve as a guidance signal to the growth cones of developing neurons (see Chapter 14). This is one of the ways in which the long projections extending from neurons find their cellular targets.

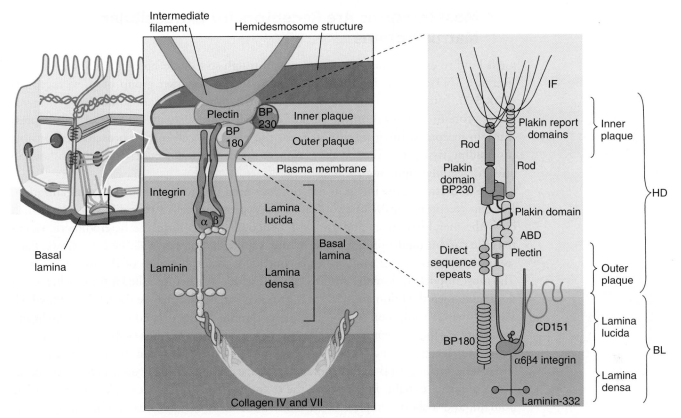

FIGURE 6-16 Hemidesmosomes connect to the basement membrane, which consists of the basal lamina and a network of collagen fibers. The drawing on the right shows some of the molecular components found in the basal lamina (BL) and hemidesmososme (HD). In the basal lamina, BP180 (light green) and α6β4 integrin (purple) bind to laminin-332, which is connected to collagens IV and VII. In the cytosolic portion of the HD, BP180 and α6β4 integrin bind to BP230 (dark green) and plectin (light blue), respectively. The plakin domain, plakin report domain, and actin binding domain (ABD) mediate the interactions between these proteins. Adapted from C. Margadant, et al., Curr. Opin. Cell Biol. 20 (2008): 589-596.

Given this broad range of functions, it is not surprising that the molecular components of the basal lamina can vary in different tissues, or even over time in the same tissue. Nearly 20 different proteins have been identified in the basal lamina, and there are 4 principal components found in nearly all basal laminae; these are type IV collagen, laminin, heparan sulfate proteoglycans, and entactin (also known as nidogen). A current model suggests how these components are woven into a sheet-like configuration that defines the basal lamina.

In this model, shown in Figure 6-11, type IV collagen and laminin polymerize to form networks. These networks are stacked upon one another to form layers and are held together by bridging proteins such as perlecan (a heparan sulfate proteoglycan) and entactin, which are able to bind to both networks. Other components, such as laminin-332 and type VII collagen filaments, are interwoven between the layers. How these additional proteins associate with the principal components is unknown, though there is some evidence that cell contact, via integrin receptors, is responsible for proper assembly of an intact basal lamina, while the presence of entactin is not required. Once assembled, the basal lamina forms a tightly woven, complex web of proteins that provides enough structural stability to support epithelial tissues, yet is porous enough to act as a selective filter of the extracellular fluid.

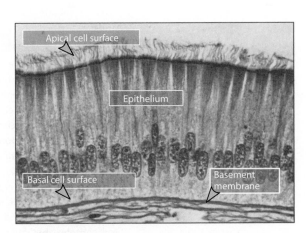

FIGURE 6-17 The basement membrane. Photo reproduced from W. Bloom and D. W. Fawcett. A Textbook in Histology, 1986. Used with permission of Don W. Fawcett, Harvard Medical School.

Most Integrins Are Receptors for Extracellular Matrix Proteins

Cells bind to extracellular matrix proteins through specific receptors. The integrin (pronounced INT-ah-grin) family of proteins is the best-known group of these receptors (**BOX 6-3**). Integrins bind to extracellular matrix proteins and, in some cases, to membrane proteins expressed on the surface of other cells. Virtually all animal cells express integrin receptors. Integrins appear to be the principal cell surface proteins responsible for holding tissues together, as they connect the extracellular matrix to the cytoskeleton and intracellular signaling proteins.

To understand how integrins function, we need to know their structural organization. Integrin receptors consist of two different polypeptides, called α and β subunits, which cross the membrane once and associate noncovalently to form the heterodimeric receptor. The current model of integrin structure is quite complex, as **FIGURE 6-18** shows. Each subunit contains several domains that contribute to the function of the intact receptor. The most important domains in the α chain include a structure called a β-propeller and a domain termed the I domain. Half of the known α subunits include the αI domain, which interacts with an adhesion site (MIDAS) found in the β subunit and determines the ligand specificity for these receptors. Closer to the plasma membrane, the α subunit contains four domains that together make up a "leg" structure, and these are called the thigh, genu, calf1, and calf2 domains. All β chains contain a globular βI domain that contacts the β-propeller domain of the α chain. Finally, both the α and β chains contain a single transmembrane domain, and a short cytoplasmic domain at the C terminus. Because integrin receptors illustrate the three traits of proteins very well we will look at these proteins in great detail.

There are currently 18 α and 8 β subunits known in vertebrates. (Most are numbered consecutively, while some have been given letter names that reflect how they were identified.) These 26 subunits interact to form at least 24 different α–β receptor combinations. In addition, variants of some subunits arise from alternative splicing, which yields further alternatives in subunit composition. Most cells express more than one type of integrin receptor, and the types of receptors expressed can vary during development or in response to specific signals.

Why are there so many integrins? Genetic knockout of some integrin subunits is lethal in developing organisms, while knockout of other integrins appears to have a mild effect, suggesting that some of these receptors may be able to compensate for one another. This ability to compensate is known as functional redundancy.

BOX 6-3 TIP

In this chapter we will encounter several different types of receptors that cells use to adhere to the ECM and to each other, and it is often difficult for beginning students to tell the differences between them. One convenient way to keep track of them is to learn how their names were chosen. Integrins were so named because they *integrate* the tensile forces in the extracellular matrix and cytoskeleton; if a tissue is stretched, for example, the ECM proteins that bind integrins will pull on these receptors, transmitting the force from the extracellular space to the cytoskeleton. Focusing tensile forces onto a small number of receptors gives cells an easy way to gauge the stress they must resist. Linking these receptors to the cytoskeleton also provides cells the opportunity to respond to this stress at the points where the membrane is especially vulnerable to tearing. In contrast, selectins play essentially no role in transmitting force. Instead, they belong to a broader class of proteins called lectins, which bind to polysaccharides. Recall from Chapter 1 that polysaccharides are structurally complex, but far more fragile than glycoproteins. Selectins play very precise roles in circulating immune cells and therefore must be especially *selective* for specific polysaccharides. (The name *selectin* is an intentional pun.) The derivation of the names cadherin and CAM proteins are explained in the sections devoted to these receptors.

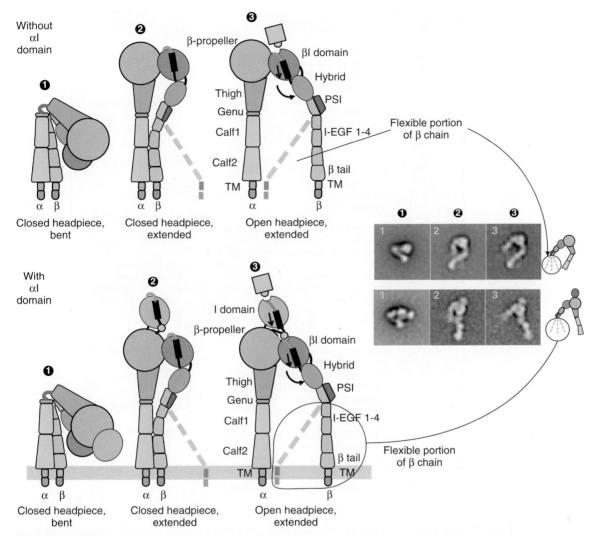

FIGURE 6-18 Model of integrin structure. Top row: three conformations of intergrins lacking the α-I domain. Bottom row: three conformations of integrins containing the αI domain. Note that the juxtamembrane portion of the β chain in each is flexible, and can adopt several different orietntations in the "open" confomations of the integrins. Right: electron micrographs of integrins corresponding to each drawing. Reprinted, with permission, from the Annual Reviews of Immunology, Volume 25 © 2007 by Annual Reviews www.annualreviews.org. Courtesy of Bing-Hao Luo, Loyola University of Chicago.

Integrins are classified into three subfamilies based on the β subunits, as listed in **FIGURE 6-19**. The β₁ integrins bind mostly to extracellular matrix proteins and are by far the most widely expressed group of integrins. The β₂ integrins are expressed only by leukocytes (immune cells) and some of these bind to other cell surface proteins. Some of the β₃ integrins are expressed on platelets and megakaryocytes (platelet precursor cells), and play critical roles during platelet adhesion and blot clotting. Other β₃ integrins are also expressed on cells that line blood vessels (endothelial cells), connective tissue cells (fibroblasts), and some tumor cells. The receptors that include β₄–β₈ subunits are relatively few and very diverse, so they are not classified into any subgroups.

Integrins support cell adhesion by binding directly to an extracellular matrix protein using the extracellular domains of both the α and β chains. Ligand specificity is determined almost exclusively by the αI domain for those integrins that contain it, and by a combination of the extracellular domains of the α chain β chains for those that do not. With the exception

Class		Ligands	Location/Function
β₁	α₁	Collagens, laminin	Extracellular matrix
	α₂	Collagens, laminin	
	α₃	Fibronectins, laminin, thrombospondin	
	α₄	Fibronectin, vascular cell adhesion molecule-1	Cell–cell adhesion
	α₅	Collagen, fibronectin, fibrinogen	Extracellular matrix Blood clotting
	α₆	Laminin	
	α₇	Laminin	
	α₈	Cytotactin/tenasin-C, fibronectin	Extracellular matrix
	α₉	Cytotactin/tenasin-C	
	α₁₀	Collagens	
	α₁₁	Collagens	
β₂	αD	Intercellular adhesion molecule-3, vascular adhesion molecule-1	Cell–cell adhesion
	αL	Intercellular adhesion molecules 1–5	
	αM	C3b	Host defense
		Fibrinogen, factor X, intercellular adhesion molecule-1	Blood clotting Cell–cell adhesion
	αX	Fibrinogen, C3b	Blood clotting Host defense
β₃	αIb	Collagens	Extracellular matrix
	αIIb	Collagens, fibronectin, thrombospondin, vitronectin,	
		fibrinogen, von Willebrand factor, plasminogen, prothrombin	Blood clotting
	αV	Collagen, fibronectin, laminin, osteopontin, thrombospondin, vitronectin,	Extracellular matrix
		disintegrin, fibrinogen, prothrombin, von Willebrand factor,	Blood clotting
		matrix metalloproteinase-2	Protease

FIGURE 6-19 Integrins are organized into subgroups that share β subunits.

of the fibronectin receptor α₅β₁, all integrins can bind to more than one ligand, and each extracellular matrix protein also can bind to more than one integrin. Although it is impossible to predict an integrin-binding site based on the amino acid sequence of the ligand, an acidic amino acid (such as aspartic acid) is common to all known binding sites on extracellular matrix proteins. Many ligands, such as collagen, vitronectin, and fibronectin, contain the sequence arginine-glycine-aspartic acid (RGD).

Cells can adjust the binding of integrin receptors to extracellular matrix proteins, and their subsequent signaling activity, by changing the shape of the integrins. A current model for integrin activation is shown in **FIGURE 6-20**. In the inactive form, the extracellular portion of both the α and β chains is bent backward toward the plasma membrane, and the receptor does not bind to its ligand. Activation of an integrin requires that this extracellular portion of the receptor be straightened out; it is thought that the straighter the receptor, the higher its affinity state for ligand is. At some point in this straightening process, the receptor binds to its ligand, and it is possible that the ligand binding may even assist in the further straightening of the receptor. One conformational change that correlates well with integrin activation is the separation of the α and β cytoplasmic tails. Further changes in the three-dimensional structure of

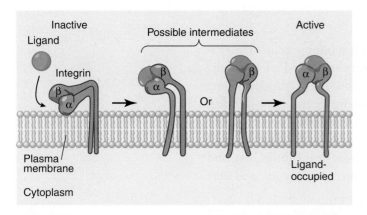

FIGURE 6-20 In this model for integrin activation, the extracellular portion of an inactive receptor is folded in toward the plasma membrane. As the integrin receptor straightens, it binds ligand and becomes active. The fully straightened form has the highest affinity for ligand.

the receptor can be triggered either by altering the local concentration of divalent cations, which bind to the β-propeller region of the receptor, or in response to cellular signaling. One of the most important aspects of this model is that it illustrates how integrins can adopt many different shapes and consequently many different binding states. These shapes are also represented in Figure 6-18. This model helps explain how cells can fine-tune their binding to extracellular matrix proteins.

■■ Specialized Integrin Clusters Play Distinct Roles in Cells

Many receptor proteins form clusters on the cell surface that are specialized to perform a subset of the tasks required to establish and maintain functional cell groups. Receptor clusters that permit adhesion to the ECM, called **cell–ECM junctions**, link the extracellular matrix on the external face of the plasma membrane to the cytoskeleton on the cytosolic surface. Integrin receptor clusters are the best known of these clusters.

When integrin receptors cluster on the cell surface, the cytoplasmic tails of the α and β subunits serve as docking sites for the assembly of a range of different proteins. Based on studies conducted with purified ECM proteins in cell culture, plus *in vivo* studies,

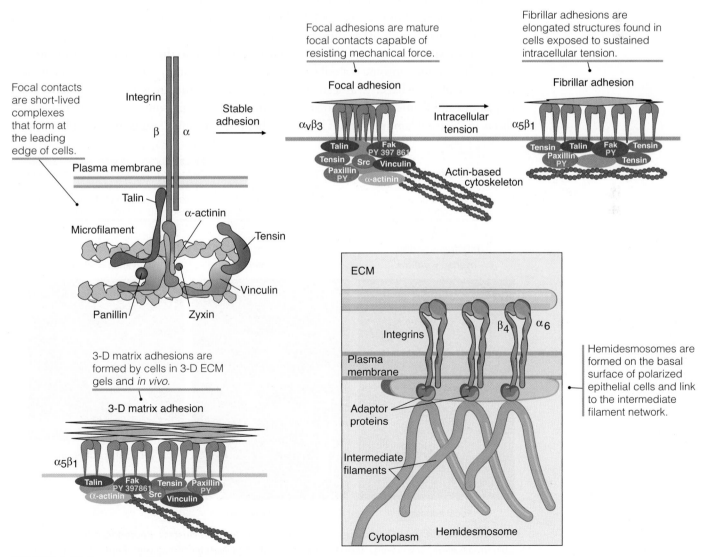

FIGURE 6-21 Five types of integrin clusters.

integrin clusters are classified into five types, as illustrated in **FIGURE 6-21**. Those that contain β_1, β_2, and β_3 integrins form four different linkages with the actin cytoskeleton. **Focal contacts** are the first integrin clusters to form at the leading edge of migrating cells, and are induced by stimulation of actin filament growth near the plasma membrane. If a focal contact generates a stable link between the ECM and actin filaments that is sufficient to resist mechanical force (supplied by myosin pulling on the actin filaments: see *Myosins Are a Family of Actin-Binding Motor Proteins* in Chapter 5), the focal contact increases in size to form a **focal adhesion**. Focal adhesions were initially thought to exist only in cells grown in culture on flat surfaces, but recently they have also been found in cells *in vivo*. Cells cultured in three-dimensional ECM gels form elongated integrin clusters called **three-dimensional matrix adhesions**. These most likely resemble the type of integrin clusters formed by cells *in vivo*. A good name for all of these complexes is simply integrin complexes, because all require clusters of integrins to form.

To date, over 50 proteins have been identified in these complexes, and they are classified in four groups: transmembrane receptors (e.g., growth factor receptors, syndecans), and three types of cytosolic proteins: structural proteins, adaptors, and enzymes. The exact composition of each cluster varies depending on the type(s) of integrins in the cluster, the type of extracellular matrix bound by the integrins, the degree of tensile strain imposed on the cluster, the location of the cluster in the cell, and the type of cell in which the cluster forms, as illustrated in **FIGURE 6-22**. Collectively, these proteins function to control a vast range of cellular functions, as shown in **FIGURE 6-23**.

Hemidesmosomes contain the $\alpha_6\beta_4$ integrin, and link to the intermediate filament network. The hemidesmosome is a cell surface junction found at the basal surface of the plasma membrane of epithelial cells; as seen in **FIGURE 6-24**, this structure is a complex interweaving of plaques and filaments. The primary function of hemidesmosomes is to anchor epithelial sheets to the basal lamina. Lack of functional hemidesmosomes results in severe blistering of

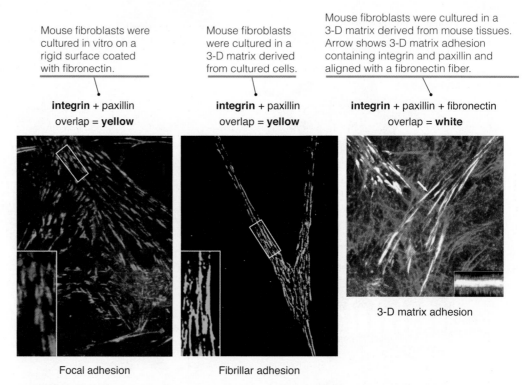

Mouse fibroblasts were cultured in vitro on a rigid surface coated with fibronectin.

integrin + paxillin
overlap = **yellow**

Mouse fibroblasts were cultured in a 3-D matrix derived from cultured cells.

integrin + paxillin
overlap = **yellow**

Mouse fibroblasts were cultured in a 3-D matrix derived from mouse tissues. Arrow shows 3-D matrix adhesion containing integrin and paxillin and aligned with a fibronectin fiber.

integrin + paxillin + fibronectin
overlap = **white**

Focal adhesion

Fibrillar adhesion

3-D matrix adhesion

FIGURE 6-22 Differences in shape and composition in integrin clusters. Reproduced from E. Cukierman, et al., Science 294 (2001): 1708-1712 [http://www.sciencemag.org]. Reprinted with permission from AAAS. Photos courtesy of Kenneth M. Yamada, National Institutes of Health/NIDCR.

CHAPTER 6   The Extracellular Matrix and Cell Junctions

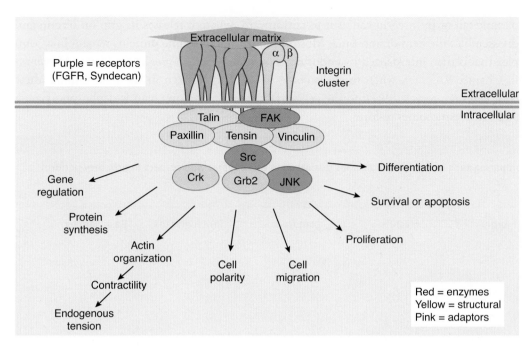

FIGURE 6-23 Summary of integrin cluster components and the cellular activities they control.

many epithelial tissues, including the skin, and these diseases can be fatal.

Figure 6-16 shows the composition of hemidesmosomes. At the cytoplasmic side of a "typical" (type I) hemidesmosome, we see a cluster of intermediate filaments (keratins 5 and 14) attached to an inner plaque (Figure 6-24). This plaque is composed of the adaptor proteins BP230 and plectin, which bind to the intermediate filament proteins. The outer plaque contains two types of transmembrane proteins: the $\alpha_6\beta_4$ integrin receptor and an unusual member of the collagen family. In the extracellular space, anchoring filaments project outward from the plasma membrane, through the lamina lucida and into the lamina densa. The lamina densa contains a variety of basal lamina proteins. Finally, anchoring fibrils of type VII collagen connect the lamina densa to the subbasal lamina densa, which is composed of several extracellular matrix proteins.

Despite their critical importance in maintaining tissue integrity, hemidesmosomes are not static structures; for example, cuts in the skin require that neighboring cells detach from the underlying basal lamina and migrate to the wound area, where the cells divide to repopulate the wound area, and then they reattach to the basal lamina and to each other. This change in phenotype requires that the cells be able to disassemble, and then later reassemble, cell–cell and cell–matrix junctions such as desmosomes and hemidesmosomes, respectively.

One model for the mechanism of disassembly and assembly of hemidesmosomes is that phosphorylation of the cyto-

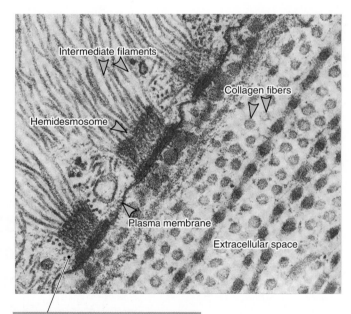

The hemidesmosome is characterized by the collection of filamentous material that terminates in a dense plaque at the cell surface.

FIGURE 6-24 Hemidesmosomes are specialized structures that form at the junction of epithelial cells and the specialized extracellular matrix called the basal lamina. Photo © Kelly, 1966. Originally published in The Journal of Cell Biology, 28: 51-72. Used with permission of Rockefeller University Press. Photo courtesy of Dr. William Bloom and Dr. Don W. Fawcett. A Textbook in Histology, 1986.

plasmic tail of β_4 integrin causes it to curl inward, where it releases its grip on plectin and disassembles the hemidesmosome. Most of the hemidesmosome proteins remain in a complex that is then internalized by endocytosis (see *Endocytosis Begins at the Plasma Membrane* in Chapter 9). Later, when the cells come to rest and reattach to the basal lamina, these complexes are recycled back to the cell surface, the β_4 integrin chain is dephosphorylated, and hemidesmosomes re-form.

CONCEPT CHECK #1

Complete the table below by briefly comparing each of the ECM molecules discussed in this chapter with respect to how they perform these five functions.

Function	Collagens	Fibronectins	Elastins	Laminins	Proteoglycans	Basal lamina
1. Impart mechanical strength to tissue						
2. Provide adhesion sites for cells						
3. Provide elasticity to tissues						

■6.3■ Cells Adhere to One Another via Specialized Proteins and Junctional Complexes

KEY CONCEPTS

▶ Cell–cell junctions are specialized protein complexes that allow neighboring cells to adhere to and communicate with one another.

▶ Tight junctions regulate transport of particles between epithelial cells and preserve epithelial cell polarity by serving as a "fence" that prevents diffusion of plasma membrane proteins between the apical and basal regions.

▶ Adherens junctions are a family of related cell surface domains that link neighboring cells together.

▶ The principal function of desmosomes is to provide structural integrity to sheets of epithelial cells by linking the intermediate filament networks of neighboring cells.

- Hemidesmosomes are found on the basal surface of epithelial cells, where they link the extracellular matrix to the intermediate filament network via transmembrane receptors.
- Gap junctions are protein structures that facilitate direct transfer of small molecules between adjacent cells. They are found in most animal cells.
- Nonjunctional adhesion takes place via receptors that do not form large junctional complexes.
- Cadherins constitute a family of cell surface transmembrane receptor proteins that are organized into eight groups. The best-known group of cadherins, called classical cadherins, plays a role in establishing and maintaining cell–cell adhesion complexes such as the adherens junctions.
- Neural cell adhesion molecules (NCAMs) are expressed only in neural cells and function primarily as homotypic cell–cell adhesion and signaling receptors.
- Selections are cell–cell adhesion receptors expressed exclusively on cells in the circulatory system. They arrest circulating immune cells in blood vessels so that they can crawl out into the surrounding tissue.

Receptors that permit attachment of one cell to another form clusters called cell–cell junctions; sometimes these clusters are long lasting, while other receptors form cell–cell contacts that are by necessity relatively weak and short lived. We will begin by looking at the most stable junctions and end with the least stable.

Tight Junctions Form Selectively Permeable Barriers between Cells

Along the lateral surfaces of adjacent cells in epithelial and endothelial cell layers, three separate cell–cell junctions function as a group called the junctional complex. In vertebrates, these junctions are the **tight junction**, **adherens junction**, and **desmosome**; in invertebrates, the septate junction often acts in place of the tight junction. The relative positions of these junctions are shown in **FIGURE 6-25**. Together, these junctions help to segregate a multicellular organism into discrete, specialized regions and to regulate the transport of molecules between them. These junctions also help protect the cells from physical and chemical damage.

When viewed with a transmission electron microscope, tight junctions appear as a series of small contacts (sometimes referred to as *kisses*) between the opposed lateral membranes of neighboring cells, as **FIGURE 6-26** shows. Proteins on the cytoplasmic face of the membrane adjacent to these contacts are seen as electron-dense clouds. A different technique, called freeze fracture, reveals the protein distribution in the two lipid monolayers separated through the middle of the plasma membrane. The tight junctions appear as a web-like network of thin fibrils (or strands), where the proteins remain embedded in the membrane or as a network of grooves if the proteins have come off during the fracture process.

The molecular composition of tight junctions is complex. Over 24 proteins have been identified in tight junctions, and they are classified into four groups: transmembrane proteins, polarity proteins, cytoskeletal proteins, and signaling proteins. Thus far, three types of transmembrane proteins have been found in the tight junction: **claudins, occludins**, and the **junctional adhesion molecule (JAM)**, as shown in **FIGURE 6-27**. Claudins form the core protein in the tight junction fibrils by clustering their extracellular domains into loops that form selective channels in the fibrils; there are at least 24 different claudin proteins in mammals. Expressing claudin genes in cells that normally do not express them results in formation of tight junctions. Occludins copolymerize laterally with claudins along tight junction fibrils, but their function is unknown.

The three transmembrane proteins attach stably to nine or more structural proteins, including actin. Occludins also bind to the main constituent of gap junctions (a protein called **connexin**), suggesting that tight junctions and gap junctions are both structurally

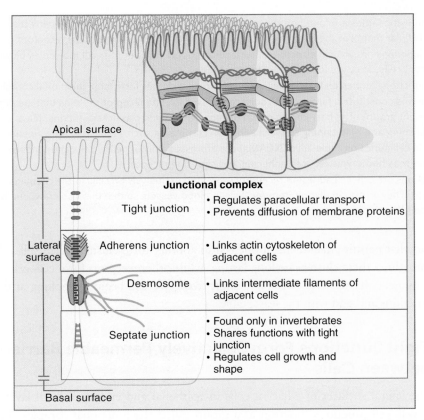

Junctional complex		
	Tight junction	• Regulates paracellular transport • Prevents diffusion of membrane proteins
	Adherens junction	• Links actin cytoskeleton of adjacent cells
	Desmosome	• Links intermediate filaments of adjacent cells
	Septate junction	• Found only in invertebrates • Shares functions with tight junction • Regulates cell growth and shape

Apical surface

Lateral surface

Basal surface

FIGURE 6-25 The junctional complex is composed of at least three distinct cell–cell junctions. It permits epithelial cells to provide structural support and to function as a selective barrier to transport. The septate junction is found only in invertebrates; often, it is present instead of the tight junction.

and functionally related. The transmembrane proteins also bind transiently to over a dozen signaling proteins; this suggests that the tight junction may have an additional role as an organizer for signaling at the cell surface, akin to the role played by focal adhesion complexes at the basal surface of cells.

Tight junctions play two important roles. First, they are the molecular structures responsible for regulating **paracellular transport** (the transport of material in the space between cells) in an epithelial or endothelial cell layer. In this role, tight junctions may be thought of

This freeze fracture electron micrograph shows the web-like fibers that make up the tight junction.

This transmission electron micrograph shows the membrane connections formed in the tight junctions.

FIGURE 6-26 Two views of the tight junction. Photos reproduced from Cell Communications, by Dr. Rody P. Cox, Copyright © 1974. Reprinted with permission of John Wiley & Sons, Inc.

CHAPTER 6   The Extracellular Matrix and Cell Junctions

as a "molecular sieve" through which extracellular molecules are filtered as they cross epithelial and endothelial boundaries. Not all of these sieves are the same, however, since each tissue serves to filter a unique set of diffusing molecules: smoke particles need not be filtered in the kidney, for example. In fact, the cutoff size for free diffusion across tight junctions ranges between 4 and 40 Å, depending on the tissue in which they are found.

The physical barriers for transport of ions and other solutes are quite different; ions are transported instantaneously, but other solutes require minutes or even hours to cross the tight junction. How is this possible? A recent model proposes that the tight junction permeability barrier is composed of rows of charge-selective pores that form the web of fragile strands seen in Figure 6-26. Ions are transported through the pores, but other solutes must wait until the strands break before they can move through the junction. As the strands break and reseal, the solutes move stepwise through the barrier, as illustrated in **FIGURE 6-28**.

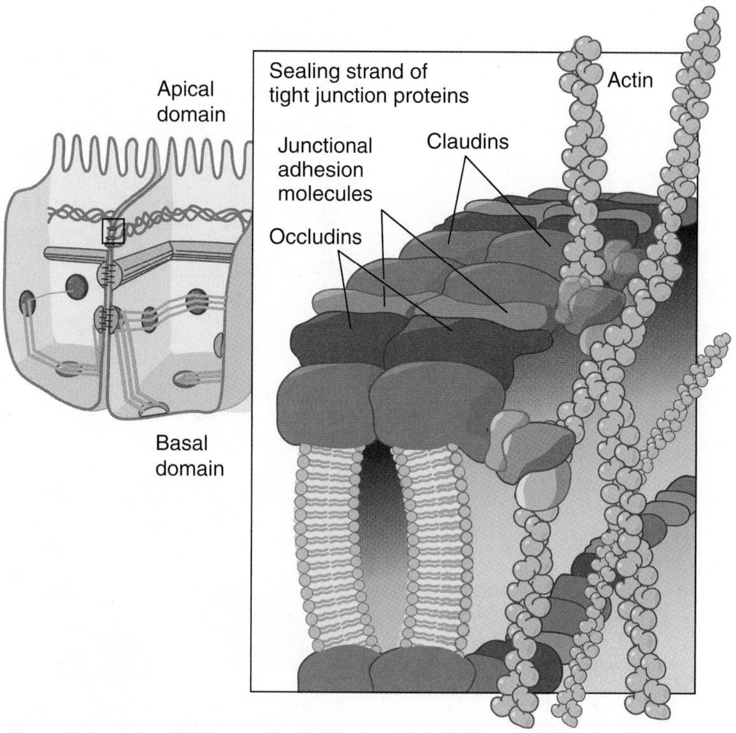

The second role played by tight junctions is to structurally and functionally separate the plasma membrane of polarized cells into two **membrane domains**,

FIGURE 6-27 Tight junctions are held together by occludin, claudin, and junctional adhesion molecules.

as shown in Figure 6-27. This is a critical feature of these cells. The **apical** (from the Greek word *apex*, or top) surface is the portion of the plasma membrane that is oriented toward a cavity or space on one side of the epithelial sheet; the **basal** (or bottom) surface is the region on the opposite side, in contact with the extracellular matrix; and the lateral surface makes up the sides between these two regions. Tight junctions completely encircle epithelial and endothelial cells along the lateral surface at the apical–lateral border, thereby separating the cell into two zones: the apical domain and the basolateral domain. These domains effectively divide the cell surface into "top" and "bottom" regions and play different roles in controlling molecular traffic through the cells.

Recall that membranes are two-dimensional fluids (see *Individual Phospholipids Can Diffuse Freely within a Single Layer* in Chapter 4). Although membrane proteins are able to diffuse in the plane of each domain, they are prevented from diffusing from one domain to the other across the tight junction. In this role, tight junctions serve as a "fence" that maintains the unique molecular composition of the two membrane domains. Three proteins, known as aPKC (atypical protein kinase C), Par (partitioning defective)-3, and Par-6, form a core cluster that plays at least two roles in establishing polarity in the plasma membrane. First, these molecules help stimulate actin polymerization near the plasma membrane, and second, these proteins bind to the occludins and thereby help direct where the tight junctions form along the lateral membrane. Different proteins play a similar role in the septate junction, the functional analog of the tight junction in invertebrate animals, shown in **FIGURE 6-29**. Altering expression of any of these proteins results in a dramatic loss of cell polarity.

◼ Adherens Junctions Link Adjacent Cells

Adherens junctions are components of the junctional complex that hold epithelial and endothelial cells together. In the electron microscope, adherens junctions appear as dark, thick bands that lie near the plasma membranes of adjacent cells, bridged by rod-like structures that project into the intercellular space. The most well-known adherens junction is the

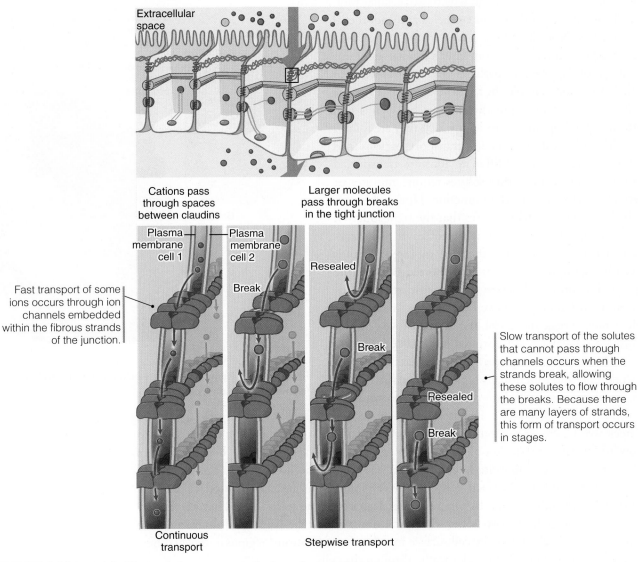

Extracellular space

Cations pass through spaces between claudins

Larger molecules pass through breaks in the tight junction

Plasma membrane cell 1

Plasma membrane cell 2

Resealed

Break

Fast transport of some ions occurs through ion channels embedded within the fibrous strands of the junction.

Break

Slow transport of the solutes that cannot pass through channels occurs when the strands break, allowing these solutes to flow through the breaks. Because there are many layers of strands, this form of transport occurs in stages.

Resealed

Break

Continuous transport

Stepwise transport

FIGURE 6-28 A model of fast and slow transport of solutes through tight junctions.

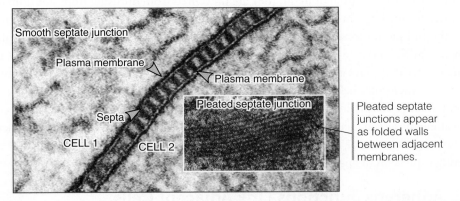

Smooth septate junction

Plasma membrane

Plasma membrane

Pleated septate junction

Septa

CELL 1

CELL 2

Pleated septate junctions appear as folded walls between adjacent membranes.

FIGURE 6-29 Smooth septate junctions appear as linear walls between adjacent cells in invertebrate animals. Reprinted from Tissue Cell, vol. 13, C. R. Green, A clarification of the two types of invertebrate..., pp. 173–188, Copyright (1981) with permission from Elsevier [http://www.sciencedirect.com/science/journal/00408166]. Photo courtesy of Colin R. Green, University of Auckland.

CHAPTER 6   The Extracellular Matrix and Cell Junctions

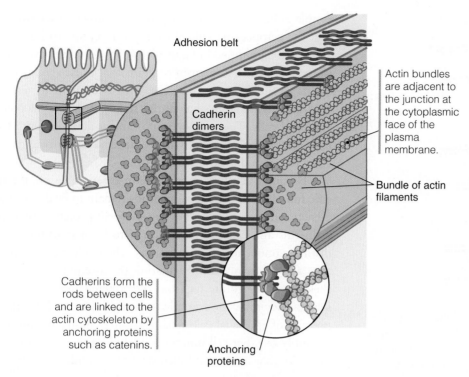

Adhesion belt

Cadherin dimers

Actin bundles are adjacent to the junction at the cytoplasmic face of the plasma membrane.

Bundle of actin filaments

Cadherins form the rods between cells and are linked to the actin cytoskeleton by anchoring proteins such as catenins.

Anchoring proteins

FIGURE 6-30 The zonula adherens is part of the junctional complex.

zonula adherens, shown in **FIGURE 6-30**. It is found just beneath the tight junctions in the junctional complex formed between some epithelial cells (see Figure 6-25). As **FIGURE 6-31** shows, other examples of adherens junctions include the adhesive junctions in the synapses between neurons in the central nervous system, in intercalated disks between adjacent cardiac muscle cells, and in junctions formed between layers of the myelin sheath surrounding peripheral nerves (see Chapter 14).

Regardless of their location, adherens junctions share two properties. First, they contain transmembrane receptor proteins known as **cadherins** that bind to identical cadherins on neighboring cells, as Figure 6-30 shows. Binding of receptors on one cell to the same type of receptor on another cell is called **homophilic binding**. It is thought to play an important role in determining the cellular organization of tissues by helping cells find specific binding partners.

The dimeric cadherin receptors used in adherens junctions contain five extracellular domains that determine exactly how homophilic binding occurs. As shown in **FIGURE 6-32**, three different overlapping arrangements of these domains are possible. The strongest binding occurs when the receptors overlap completely in an antiparallel arrangement, while weaker binding interactions form when the receptors partially overlap. By changing the number of cadherin receptors clustered on their surface, cells can change the strength of their binding to their neighbors.

The second property shared by adherens junctions is that they form adhesions strong enough to allow tissues to change shape and/or resist sheer stress; for example, the

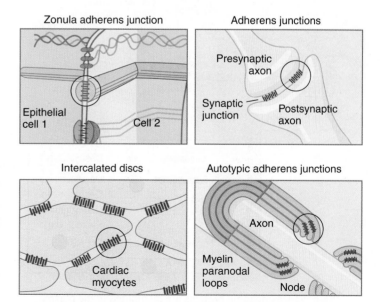

FIGURE 6-31 Each type of adherens junction functions to hold adjacent cells together tightly.

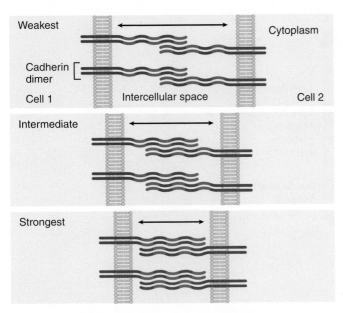

Weakest

Cadherin dimer

Cell 1

Intercellular space

Cell 2

Cytoplasm

Intermediate

Strongest

FIGURE 6-32 Cadherin proteins form dimers that bind to one another. Three different configurations of cadherin interaction are shown. Direct measurement of the strength of adhesion shows that the configuration with the most overlap is also the strongest.

zonula adherens uses anchor proteins, known as **catenins**, to link the cytoplasmic tails of cadherin receptors to bundles of actin, as diagrammed in Figure 6-30. These actin filaments are, in turn, attached to myosin proteins that cause the actin filaments to slide past one another. This is thought to result in contractions that can change the shape of the apical pole of epithelial cells. This may be important in development of the neural tube, for example, when epithelial cells invaginate to close the neural groove (see Figure 6-38).

Beyond the function of cadherin-based adhesion, the specific functions of adherens junctions remain to be determined. Genetic analysis in fruit flies suggests that proteins other than cadherins and catenins are required to form morphologically distinct zonula adherens junctions. These additional proteins may also be involved in regulating the assembly of the cytoskeleton at sites quite far from the adherens junctions themselves. They may, for example, be involved in establishing the polarity of epithelial cells and, thus, may indirectly affect the assembly of other cell junctions, including the tight junction. How this may be accomplished is currently under investigation.

Desmosomes Are Intermediate Filament-Based Cell Adhesion Complexes

The desmosome is a component of the junctional complex in epithelial cells (see Figure 6-25) and is also located in some nonepithelial cells. Three features of desmosomes are immediately apparent in electron micrographs, as **FIGURE 6-33** shows:

- Thick accumulations of fibrils running across a gap (the desmosomal core, about 30 nm wide) between the plasma membranes of two adjacent cells.
- These fibrils appear to terminate in a thick patch of electron-dense material on the cytosolic side of the plasma membrane.
- The electron-dense patches are connected to filaments in the cytosol of each cell.

The accumulation of dense material at the plasma membrane consists of two distinct structures, the **inner dense plaque** and **outer dense plaque**. Each desmosome is rather small (average diameter is about 0.2 μm), and several of them can be seen along the edge of two adjacent cells. The structure looks something like a suspension bridge: cytosolic filaments in neighboring cells are linked together by bridging extracellular filaments connected to supporting anchors on the plasma membrane. For this reason, the structure was given the name desmosome, derived from the Greek words *desmos* (bond, fastening, chain) and *soma* (body). It seems obvious that the purpose of such a structure is to link two cells together.

What function might such a linkage serve in cells? Remember the two main functions of the junctional complex

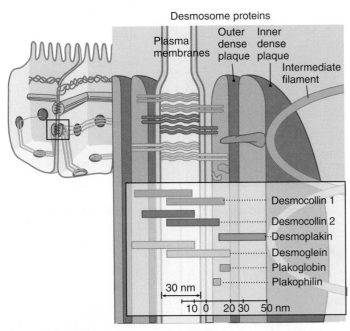

Desmosome proteins

Plasma membranes

Outer dense plaque

Inner dense plaque

Intermediate filament

Desmocollin 1

Desmocollin 2

Desmoplakin

Desmoglein

Plakoglobin

Plakophilin

30 nm

10 0 20 30 50 nm

FIGURE 6-33 Desmosome proteins are distributed in the plasma membrane and a distinctive double plaque arrangement at the cell surface.

shown in Figure 6-25: controlling paracellular transport and resisting physical stresses imposed on the epithelium. Because desmosomes are especially abundant in cells exposed to physical stress such as skin and cardiac muscle, cell biologists thought that they contributed to the latter function. Consequently, the cytoplasmic filaments attached to the dense plaques were called tonofilaments to reflect the supposition that they were under strain (Greek translation, *tonos*). Later, it was determined that these filaments are intermediate filaments, though they are still sometimes called tonofilaments.

In addition to the intermediate filament fibers, at least seven other protein types have been identified in desmosomes, and these are organized into three families. The first are the major transmembrane proteins found in desmosomes, and they are major components of the outer dense plaque, as Figure 6-33 shows. They form the "bridging filaments" that stretch across the intercellular space and serve as binding sites for the second and third families of cytoplasmic proteins. These, in turn, bind to the intermediate filament proteins in the inner dense plaque. The exact makeup of the desmosome, as well as the number of desmosomes formed, varies in different cell types, reflecting the wide variety of stresses that cells must endure.

A common description of the desmosome is that it serves as a "spot weld" between two neighboring cells. The most dramatic proof of desmosome function comes from cases where desmosome structure is compromised; in these cases, epithelial sheets are especially fragile, and the organs they cover are easily damaged. This is especially true of the skin, which is severely blistered. When viewed under a microscope, epithelial cells lacking desmosomes are badly disorganized, lack junctional complexes, and are detached into small clusters (rather than forming a single continuous sheet).

Patients with damaged or missing desmosomes have a wide range of blistering diseases, and these diseases can be lethal. Current treatments for patients with these diseases are focused primarily on protecting the skin and avoiding risky behaviors that might induce blister formation, and the result is a rather poor quality of life. One experimental treatment currently being evaluated is the application of tissue-engineered skin; by substituting the damaged skin with a fresh layer of living, normal skin cells embedded in an engineered extracellular matrix, researchers hope to develop more stable, trauma-resistant skin that forms normal desmosomes.

Gap Junctions Allow Direct Transfer of Molecules between Adjacent Cells

Gap junctions are specialized structures on the cell surface that facilitate the direct transfer of ions and small molecules between adjacent cells. They are found in most vertebrate and invertebrate cell types and are the only known means of **cell-to-cell transport** for animal cells. The gap junctions between cardiac muscle cells also facilitate transmission of electrical signals during muscle contractions.

As shown in **FIGURE 6-34**, these gaps are bridged by **gap junction channels**, which cluster into patches (or plaques) that project out of the plasma membrane, as shown by freeze fracture of the plasma membrane. Gap junctions can contain from a few dozen to many thousand gap junction channels and can extend to several micrometers in diameter on the cell surface. The gap junction channels are made up of two halves called hemichannels or **connexons**, which dock together in the intercellular gap. Each connexon is composed of six protein subunits, called **connexins**

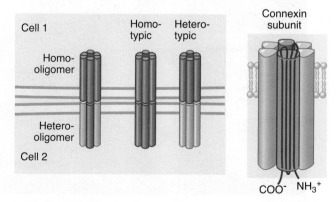

FIGURE 6-34 The principal structural unit of the gap junction is the connexon, which consists of six membrane-spanning connexin subunits. Each connexon is 17 nm long and 7 nm in diameter.

(**BOX 6-4**). A connexon is a 17-nm-long, hydrophilic, cylindrical channel measuring 7 nm in diameter at its widest and about 3 nm in diameter at its narrowest point. Connexins contain four membrane-spanning α helices linked by two extracellular loops. High-resolution structures suggest that the extracellular loops of opposing connexins bind to each other via antiparallel β sheets, thereby forming an α barrel.

Gap junction channels can vary in composition. The human genome sequence suggests that at least 20 different connexin proteins exist in humans, and many cells express more than one connexin type, allowing for the formation of **homo-oligomeric** connexons (consisting of only one subunit type) and **hetero-oligomeric** connexons (containing multiple subunit types). In addition, connexons can dock with connexons of the same (homotypic channels) or different (heterotypic channels) composition. A single gap junction plaque can contain connexons of different connexin composition. Within the plaque, the connexons are either homogenously mixed or are spatially segregated according to their connexin composition, as shown in **FIGURE 6-35**.

In the original experiment to test the hypothesis that cells use channels to directly exchange small molecules, fluorescent molecules were injected into cells growing in culture. The diffusion of the molecules was followed over time by microscopy. These experiments showed that the molecules diffuse between neighboring cells much faster than would be expected if the molecules had to pass through the lipid bilayer of each plasma membrane. This result implicated the presence of a direct channel joining the cytosol of the neighboring cells; these channels were later identified as gap junctions. By using fluorescent molecules of different sizes, it was determined that gap junctions allow passage of molecules up to 1,200 daltons in size (corresponding to a molecule approximately 2 nm in diameter), but exclude molecules larger than 2,000 daltons. This means proteins and nucleic acids do *not* pass through gap junctions, because they are too big. A recent version of the experiment showing exchange of fluorescent molecules between cells expressing connexins is shown in **FIGURE 6-36**.

Experiments such as these showed that ions could pass freely between the cytosolic compartments of cells linked by gap junctions. Other small molecules, including sugars, nucleotides, and second messenger molecules such as cAMP and cGMP, may be exchanged as well. Communication through gap junctions can be critical when rapid, well-coordinated responses are required of a large number of cells; for example, rapid reflex reactions in the brain are mediated by neurons, linked by gap junctions, that allow nearly instantaneous exchange of ions, and the carefully

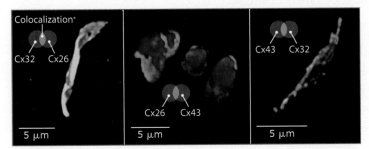

FIGURE 6-35 Double label immunofluorescence staining of connexin (Cx) subunits in gap junction plaques. Cells were transfected with the indicated pairs of connexin genes, then stained with antibodies to the connexins. Cx32 colocalizes with Cx26 but not with Cx43, for example. The cell bodies are not visible. Photos courtesy of Matthias Falk, Lehigh University.

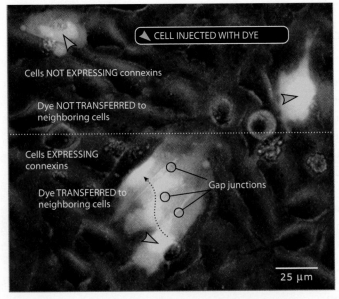

FIGURE 6-36 Fluorescent imaging of dye transfer through gap junctions. Cells were transiently transfected with DNA encoding for a connexin, so not all cells express connexin protein. Photo courtesy of Matthias Falk, Lehigh University.

controlled timing of contraction by heart muscle fibers is mediated by rapid exchange of ions as well.

Gap junction permeability appears to be controlled by channel opening and closing (a process referred to as channel **gating**). There is good evidence that gap junction channel gating is controlled by proteins that phosphorylate connexin subunits, changes in intracellular pH, and alterations in intracellular calcium ion concentration. For example, as calcium concentrations rise from 10^{-7} Molar (M) to 10^{-5} M, gap junction permeability falls; concentrations above 10^{-5} M result in complete closure of the junctions. This may serve as a self-defense mechanism, since cells undergoing apoptosis (sometimes known as *programmed cell death*) typically experience a burst of cytosolic calcium, and closure of gap junctions may protect neighboring cells from initiating their apoptotic programs by accident. Connexins are also rapidly degraded in most cells: the half-life for a connexin protein is estimated to be approximately 15 hours.

Two additional families of gap junction proteins have been discovered. The **innexins** (invertebrate connexins) are found only in invertebrate animals and, despite their name, share no sequence homology with connexins. Nonetheless, they are capable of forming intercellular junctions that look and behave like vertebrate gap junctions. Taking a cue from the "nexin" nomenclature, the second family is called **pannexins** (from the Latin *pan*, meaning *all*). Pannexins are found in both vertebrates and invertebrates and are structurally distinct from both connexins and innexins. Pannexins are found almost entirely in neuronal cells, suggesting that they may have an important role in neuronal development and function, even in organisms with very primitive nervous systems. These observations also suggest that gap junctions arose at least twice during evolution of animals, through completely separate means, a process known as **convergent evolution**.

▪▪ Calcium-Dependent Cadherins Mediate Adhesion between Cells

The **cadherin superfamily** consists of over 70 structurally related transmembrane proteins, all of which share two properties: the extracellular regions of these proteins bind to calcium ions to fold properly (hence *Ca*, for calcium) and these proteins adhere to other proteins (hence, *adherin*). The cadherins are involved in cell–cell adhesion, cell migration, and signal transduction. The first group of cadherins discovered includes those found in the zonula adherens junctions formed between epithelial cells; these are now termed *classical cadherins* to distinguish them from their more distantly related family members.

All classical cadherins are transmembrane receptors with a single membrane-spanning domain, five extracellular domains at the amino end of the protein, and a conserved cytoplasmic C-terminal tail, as diagrammed in Figure 6-32. In vertebrates, the five classical cadherins are termed E-, P-, N-, R-, and VE-cadherins, based on the sites where they were first discovered: epithelium, placenta, nerve, retina, and vascular endothelium, respectively. We now know that expression of each type is more widespread, but the names have stuck.

Classical cadherins function as clusters of dimers on the cell surface. These dimers bind to identical dimers on neighboring cells (this is called homophilic binding); N- and R-cadherin pairs will also bind to each other (**heterophilic binding**). Cells can control their strength of adhesion by varying both the total number of receptors on the cell surface and the lateral diffusion of the receptors within the plasma membrane. Cadherins that are not clustered will not form strong adhesions with neighboring cells. Adhesion occurs in two steps: a weak association is made almost immediately after opposing cadherins come into contact, and then a second, more stable association is formed.

A 56-amino-acid region at the C terminus of the conserved cytoplasmic tail of classical cadherins binds to a cytosolic protein known as **α-catenin**. When attached to cadherins,

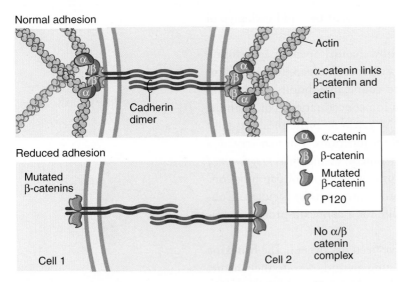

Normal adhesion

Actin

α-catenin links β-catenin and actin

Cadherin dimer

Reduced adhesion

Mutated β-catenins

Cell 1

Cell 2

α	α-catenin
β	β-catenin
	Mutated β-catenin
	P120

No α/β catenin complex

FIGURE 6-37 Cadherin cytoplasmic tails are linked to actin filaments via catenin proteins. Cadherin complexes also contain other linkers and signaling proteins that are not shown. The α-catenin binding domain of β-catenin is required for normal cadherin-mediated cell adhesion.

β-catenin acts as an adaptor molecule by binding to α-catenin, which in turn binds to actin filaments, as shown in **FIGURE 6-37**. (α-catenin and β-catenin are unrelated in sequence.) β-catenin, therefore, is a link in the chain of proteins formed between the cadherin dimers at the cell surface and the actin cytoskeleton.

When not complexed with cadherins, β-catenin plays a role in signal transduction. Cadherins are thought to indirectly play a role in controlling this signaling by sequestering β-catenin in adhesion complexes and limiting its ability to bind other signaling proteins. Other signaling proteins bind to cadherin cytoplasmic tails as well. However, there is no evidence that cadherins are signaling proteins per se; rather, it is likely that because they can associate with the cytoskeleton and with signaling proteins, cadherins and the junctions they form may act as scaffolds that organize signaling molecules to regulate their activity.

Classical cadherins play a significant role during development by controlling the strength of cell–cell adhesion and by providing a mechanism for specific cell–cell recognition. For example, during development, E-cadherins are thought to increase cell–cell adhesion when tight junctions form, and epithelial cells subsequently polarize in the developing embryo; not surprisingly, genetic knockout of E-cadherin genes is lethal early in development.

Functional mutations or knockout of other cadherin family members affect development of a wide variety of organs including the brain, spinal cord, lung, and kidney. An important theme common to all of these developmental events is a process of cellular movement known as **invagination**. For example, the first nervous tissue arises in vertebrates when the cells composing the outermost cell layer (called the ectoderm) form a ridge along the outer surface of the embryo, which then deepens into a cleft and pinches off to form the neural tube. To form this tube, epithelial cells must constrict their apical domains and bend inward, forming a groove, then dissociate and move to new locations to close the tube, as illustrated in **FIGURE 6-38**. Similar movements occur in the formation of many ectodermally derived tissues, and all require variations in the types of cell–cell contacts. Deletion of cadherin genes results in a wide variety of developmental abnormalities, such as poor motor skills due to mistargeted neurons, which also result from errors in epithelial invaginations.

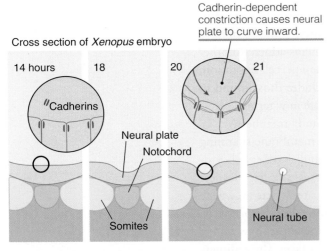

Cross section of *Xenopus* embryo

14 hours   18   20   21

Cadherin-dependent constriction causes neural plate to curve inward.

Cadherins

Neural plate

Notochord

Somites

Neural tube

FIGURE 6-38 As the neural tube is formed, the apical surface of the neural plate cells constricts, causing the neural plate to curve inward.

Calcium-Independent NCAMs Mediate Adhesion between Neural Cells

While some cell adhesion proteins, such as cadherins and integrins, must bind to extracellular calcium ions to promote adhesion, this is not true for all of them. One major class of calcium-independent cell adhesion proteins is the neural cell adhesion molecules (NCAMs).

NCAMs function primarily as cell–cell adhesion receptors, though they can also bind to heparan sulfate proteoglycans. NCAMs are expressed only in neural cells. They are found at the junctions between neighboring cells in both the central nervous system and in the peripheral nervous system, especially in nerve fibers.

Nerve cells express three different NCAM proteins, which are formed by alternative splicing of a single NCAM gene. NCAM-180 and NCAM-140 (the numbers refer to the size of the proteins in kilodaltons) contain a single transmembrane domain and differ in their C-terminal cytoplasmic tails. In contrast, NCAM-120 is anchored to the cell surface by a glycosylphosphatidyl inositol tail.

Following their synthesis in the endoplasmic reticulum, NCAMs undergo many posttranslational modifications in the Golgi apparatus on their way to the cell surface. These modifications include phosphorylation, sulfation, and glycosylation, but perhaps the most significant of these modifications is the addition of long, linear chains of sialic acid sugars, called polysialic acid (PSA), as shown in **FIGURE 6-39**. Addition of the PSA chains significantly changes both the shape and function of NCAMs. Because sialic acid is negatively charged, PSA chains project outward from the NCAM protein, attract cations and bind to water molecules, similar to the glycosaminoglycan chains found on proteoglycans.

The most striking effect of PSA addition to NCAMs is on the adhesive function of the receptors. Membrane-bound NCAMs bind primarily to identical NCAM receptors on neighboring cells. The exact mechanism of this homophilic binding is not known, but it does involve domains at the amino terminus of each receptor. One current model for this interaction suggests that when all of these domains on NCAM receptors overlap, strong, stable adhesion between neighboring cells results. However, PSA-NCAM receptors do not overlap completely, presumably because the large hydration volume and negative charge of the PSA chains repel these domains on the complementary receptors on the neighboring cells, as shown in **FIGURE 6-40**. Consequently, cells expressing PSA-NCAMs bind less strongly to their neighbors than do cells expressing NCAMs that lack PSA.

What advantage is gained by having both strongly and weakly adhering forms of the same receptor? Remember that during development, cells must grow and move about the body to form tissues. While this process is taking place, cells may form and break contacts with each other many times. This is especially true of cells that must form multiple, highly specific cell contacts in the mature organism, such as neurons. Consequently, a low affinity but highly specific interaction, such as occurs between PSA-NCAMs, may be quite useful to developing neurons.

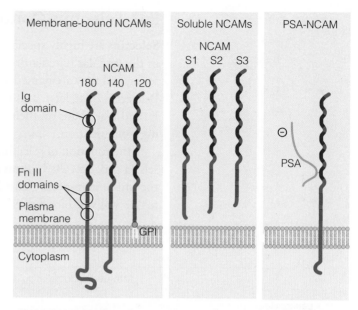

FIGURE 6-39 NCAMs are produced as both membrane-bound and soluble proteins of different sizes. The domain organizations of NCAMs are shown. The extracellular portions of NCAMs can be modified by addition of polysialic acid (PSA), which is attached to asparagine residues during transit through the secretory pathway (shown is a model of the 140 kDa transmembrane form of PSA-NCAM).

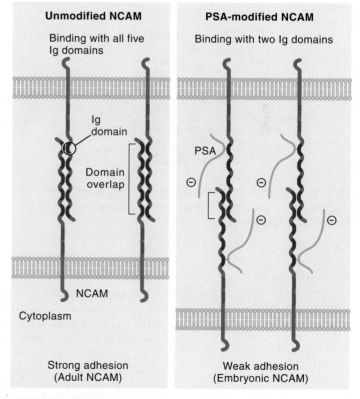

FIGURE 6-40 Unmodified NCAMs can bind to each other with five Ig domains, which results in strong cell–cell adhesion. PSA-modified NCAMs bind with only two Ig domains, which reduces the strength of cell–cell adhesion. The presence of PSA disrupts overlap of selectins, resulting in weak cell–cell adhesion during development. In adults, PSA is not added to selectins, so that firm cell–cell attachment is maintained.

Selectins Control Adhesion of Circulating Immune Cells

Selectins are highly specialized cell surface receptors that are expressed exclusively on cells in the vascular (circulatory) system. Three different types of selectins have been identified so far, and they are named according to the cells that express them: L-selectin (leukocytes), P-selectin (platelets), and E-selectin (endothelial cells). Endothelial cells can express both E- and P-selectins on their surfaces after the cells have become activated by cytokines during inflammation.

The function of selectins is to facilitate the movement of leukocytes out of blood vessels (a process called **extravasation**) and into inflamed tissues, where they contribute to the immune response. This is a difficult task: leukocytes that extravasate must first adhere to the walls of blood vessels despite the sheer forces exerted by the flowing blood. How do leukocytes solve this problem? The answer is elegantly simple: they come to a **rolling stop**, which is mediated by selectins. This way, they are able to gradually reduce their speed in the blood vessel, and as they come to a stop, the leukocytes engage the integrin receptors on endothelial cells. These receptors increase adhesion and assist in the leukocyte's escape from the blood vessel.

How might a cell "roll" to a stop in a blood vessel? The cell must be able to form transient, reversible, adhesive interactions with the endothelial cells lining the vessel. As the leukocyte forms these associations, it drags (or rolls) along the vessel wall until it forms enough associations to come to a complete stop. This is known as **discontinuous cell–cell adhesion**. As shown in **FIGURE 6-41**, leukocytes express selectin ligands such that they will bind only to endothelial cells that express the E- and P-selectins on their cell surface and, thus, will not adhere to vessel walls in noninflamed tissues (**BOX 6-5**).

The key to this selective adhesion is the use of protein–sugar-binding interactions, such as those used by proteoglycans and their receptors. Selectins are so named because the ligand-binding portion of these receptors resembles that found in **lectins**, a group of proteins that bind specifically to cell surface oligosaccharides. The ligand-binding region of selectins lies at the N terminus of the protein, connected

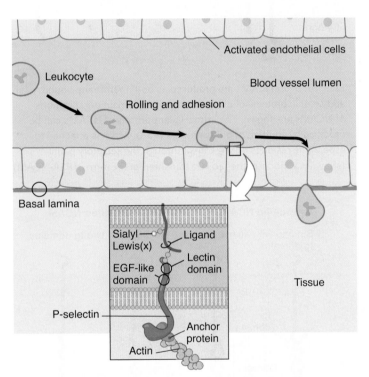

FIGURE 6-41 An illustration of the "rolling stop" function of selectins. Inset: a model of selectin structural organization and binding to a leukocyte ligand.

BOX 6-5

The kayak analogy. One way to visualize how the "rolling stop" occurs is to imagine yourself kayaking down a narrow, fast river; this would represent a leukocyte flowing along in a blood vessel. Then, imagine that you see a group of friends on the bank of the river ahead of you, so you decide to stop and talk with them. However, the river is flowing so quickly that you can't stop by yourself, and you will need some help. Your friends hold out their kayak paddles into the path of your kayak, but the paddles are wet and slippery so you can't get a firm grip on them; as you whiz past, each time you grab one it quickly slips away. But, each time you grasp a paddle, the kayak also slows down a bit, so if this happens many times (you have lots of friends), you will eventually come to a complete stop, and you can then pull your kayak out of the water. The biological equivalent of this is that your arm is a selectin ligand / carrier protein, your hand is a sialyl Lewis(x) sugar structure, and the paddles are selectins expressed by the endothelial cells lining the blood vessel (your friends). Leukocytes make and break connections between their sialyl Lewis(x) and selectins repeatedly until they slow down enough to exit the bloodstream.

to a series of short consensus repeats, followed by a single membrane-spanning domain and a short cytoplasmic domain at the C terminus. Like cadherins and integrin receptors, selectins require extracellular calcium to fold properly and bind to their ligands. Selectins bind to a complex and specific arrangement of sugars, known as **sialyl Lewis(x) (sLex)**, which is attached to "carrier proteins" expressed on the surface of target cells. Selectins can distinguish between subtly different forms of sLex that are attached to different core proteins and, thus, can establish high binding specificity.

CONCEPT CHECK #2

One theme running through this chapter is that cells form specialized protein complexes on their plasma membrane to assist in their direct adhesion to other cells. Taken to the extreme, this suggests that (1) all cells in a multicellular organism should make at least one of these complexes, and (2) the more cell–cell adhesion complexes a cell creates, the better it functions in a multicellular organism. Do you agree with these statements? Why or why not?

6.4 ▌ Chapter Summary

The extracellular matrix fills the space between cells in multicellular organisms, and integrates tissues by providing a common substrate for the adhesion of multiple cells. It consists of hundreds of different molecules that interact in complex and highly organized ways. The two major classes of macromolecules in the extracellular matrix are the structural glycoproteins (e.g., collagens, elastins, fibronectins, and laminins) and the proteoglycans (e.g., heparan sulfate) that impart structural stability and provide a hydrophilic environment in tissues. Each of these molecules consists of structural modules that mediate attachment of cell receptors, growth factors, and other extracellular matrix molecules. These molecules control cell behavior by determining the three-dimensional arrangement of cells in tissues and by providing substrates upon which cells migrate.

At least 100 different proteins combine to form specialized cell surface complexes that mediate attachment to the extracellular matrix and between adjacent cells; the integrin clusters alone interact with over 50 of them. These complexes perform a variety of specialized functions. Tight junctions and septate junctions regulate paracellular transport across epithelial sheets; adherens junctions and focal adhesions link the cell surface to the actin cytoskeleton, thereby controlling cell movement; and desmosomes and hemidesmosomes link the cell surface to the intermediate filament network, thereby providing structural stability and distributing tensile forces across large networks.

CONCEPT CHECK ANSWERS #1

Function	Collagens	Fibronectins	Elastins	Laminins	Proteoglycans	Basal lamina
1. Impart mechanical strength to tissue	Cross-linked triple-helical domain in all collagens; especially prominent in fibrillar collagens. Some are clustered into large cross-linked aggregates (e.g., tendons).	Contain binding sites for other structural glycoproteins (e.g., collagens) to form an interconnected network. Cross-linked dimers clustered into fibers.	Little if any contribution.	Triple-helical domain in all laminins. Contain binding sites for other structural glycoproteins (e.g., collagens) to form an interconnected network.	Negative charge on GAGs attracts water, provides resistance to compression.	Sheet-like collagens form web with laminins and other proteins to resist shearing forces.

Function	Collagens	Fibronectins	Elastins	Laminins	Proteoglycans	Basal lamina
				Some are clustered into branched networks (e.g., basal lamina)		
2. Provide adhesion sites for cells	Contain accessible and cryptic integrin binding sites, including RGD	Contain RGD integrin binding site.	Microfibril sheath contains RGD.	Contain accessible and cryptic integrin binding sites, no RGD. Binds syndecans.	Provides binding sites for non integrin receptors such as CD44, syndecans.	Substrate for hemidesmosomes and other integrin-based adhesion structures.
3. Provide elasticity to tissues	Little to none.	Type III repeats can unfold under stress and refold.	Spontaneous clustering induced by hydrophobic domains is reversible.	Little to none.	Little to none.	None known.

CONCEPT CHECK ANSWERS #2

If one accepts the premise that all cells in a multicellular organism must adhere to one another, statement (1) appears to be correct. However, this premise is false. This chapter discusses several materials that lie between cells, and therefore by definition do not permit direct cell–cell adhesion. Recall from the beginning of the chapter that multicellular organisms possess two traits that distinguish them from unicellular organisms: they can stably adhere to one another, or adhere to an extracellular matrix that lies between them; for example, most fibroblasts do not adhere directly to other cells, but always remain in contact with the extracellular matrix.

Statement (2) might be correct if each cell in a multicellular organism was expected to perform as many different functions as possible, but this is in direct conflict with our understanding of multicellularity. Cells form clusters so that they can become specialized, performing only a subset of functions necessary for the organism. The presumption here is that by focusing on a limited set of tasks, each cell in a multicellular organism can become more proficient in these tasks than their unicellular counterparts. Therefore, cell–cell adhesion complexes represent one type of specialization that cells can use to perform their jobs. Thanks largely due to evolution by natural selection, most cells make only those specialized structures that they must have to perform their functions. As clever as these junctional complexes are, they provide no benefit to cells that can't put them to use. In general, the fewer of these complexes that a cell makes, the more efficient it becomes.

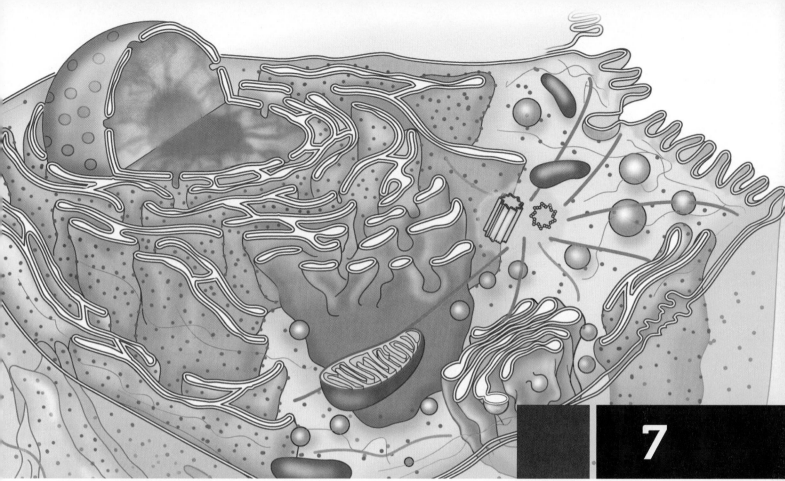

The Nucleus
and DNA Replication

7.1 The Big Picture

Learning any complex subject is itself a complex process. In many college-level courses, subjects are commonly taught and learned in a cumulative, linear fashion, starting with the basic facts in the beginning of the term and culminating with the more complex material at the end. Thus far, Chapters 1–6 have been arranged this way, starting with simple building blocks and culminating with a model of a generic cell. But this strategy has important limitations in the sciences, because most scientific fields are not linear; that is, biological complexity often arises from multiple functional links between simpler units, and thus more closely resembles a web. Applying the linear model to cell biology reveals strands in the web, but fails to capture the interrelationships between strands. One strategy for overcoming these limitations is to simply return to basic subjects again and again and build new strands from them. When enough linear strands are gathered together, they begin to intersect, and finally the real complexity of the subject is revealed.

In this book, we have reached the point where we will start a new strand rather than continue adding to the Chapter 1–6 one. The theme of this new strand is how cellular information is replicated, repaired, and passed from the nucleus to other organelles. Because we are building a new strand, some of the subjects we have discussed in previous chapters will appear again here; for example, we'll revisit the cytoskeleton in this chapter, but this time not in a context that involves cell junctions. Instead we'll view the cytoskeleton from the perspective of how it helps the nucleus and its chromosomes. Our new strand begins with cell biology principle #3: DNA integrity is the top priority for all cells.

Our goal in this chapter is thus to explain how cells preserve DNA integrity. To help us, I've divided the chapter into three major sections:

- The first is quite brief and concerns the anatomy of the nucleus. It will give us an opportunity to organize and expand on the details discussed in previous chapters.
- The second section introduces the protein complexes necessary for building, replicating, and maintaining the structure of DNA. As we mentioned in Chapter 2, this is a subject that is so complex that it is recognized as its own field of biology (molecular biology). It is essential to remember the structure of DNA and its building blocks for this section to make sense, so we will put the information in Chapter 2 to good use here. We will cover just enough molecular biology to illustrate the cell biology principle #3, and leave the rest for other texts. There will be a subtle shift in this section: despite the fact that this chapter (and entire book) is mostly about metazoan (eukaryotic animal) cells, the best information about the details of DNA assembly and repair comes from prokaryotes, primarily bacteria. To make the material easier to learn, we will emphasize the similarities between prokaryotes and eukaryotes here, rather than focus on the differences.
- In the final section of this chapter, we focus on one of the most fascinating behaviors of eukaryotic cells: mitosis. Rather than attempt a comprehensive discussion of mitosis, we will emphasize specific behaviors that help illustrate cell biology principle #3. For example, if DNA integrity is critical, then how do cells ensure this integrity during mitosis? Also, we will capitalize on our knowledge of the cytoskeleton to illustrate just how much effort cells invest in successfully completing mitosis. Look for connections between this chapter and the microtubule section in Chapter 5, because they are closely linked.

7.2 ■ The Nucleus Contains and Protects Most of a Eukaryotic Cell's DNA

KEY CONCEPTS

▸ The nucleus is a highly specialized organelle committed primarily to protecting, copying, and transcribing DNA.

▸ The interior of the nucleus is highly compartmentalized.

▸ DNA copying, plus transcribing and splicing of RNA, are accomplished by large, highly specialized molecular complexes.

The third principle, "DNA integrity is the top priority for all cells," means that given a choice between preserving the structural integrity of DNA (fixing breaks, stitching together fragments, protecting vulnerable ends) and doing anything else (generating energy, crawling, expressing genes, etc.), the first activities *always* take priority. In fact, this is the secret behind many cancer drugs: they disrupt DNA integrity so well that the cells can't fix the damage, and the cells literally die trying to overcome the problem. (It also explains why so many cancer drugs are toxic to healthy cells, too.) If simply staying alive were their first priority, cancer cells could tolerate DNA damage so long as their key survival functions (e.g., maintaining disequilibrium with the external environment, importing nutrients) were operating. But even the most bizarre, mutant cancer cell has largely *intact* DNA, even if the actual order of the genes is scrambled and the genes don't function properly anymore. Without intact DNA, cells don't grow; unless, of course, their cellular machinery for obeying this principle is itself damaged, and these cells grow for only a few generations before they literally fall apart.

■■ The Nuclear Envelope Is a Double Membrane Structure

The nucleus is defined by a double membrane called the **nuclear envelope**, which encloses a space called the nucleoplasm, as shown in **FIGURE 7-1**. Like the other double membrane organelles, cell biologists think that this organelle may have appeared when one prokaryotic cell engulfed another without destroying it, a process called **endosymbiosis** (**FIGURE 7-2**). The inner membrane is much smaller, and only surrounds the nucleoplasm.

■■ The Outer Membrane of the Nuclear Envelope Is Continuous with the Endoplasmic Reticulum

Unlike in other double-membrane organelles, the nuclear outer membrane is highly elaborated (meaning it has grown tremendously during the course of evolution), and in modern eukaryotes it forms an extensive network that can extend throughout the cytosol of most cells. This network is called the endoplasmic reticulum, and it performs a set of functions so different from the nucleus it is considered to be a separate organelle, even though it is physically connected to the nuclear outer membrane (**FIGURE 7-3**). We have already discussed some of the functions of the endoplasmic reticulum in Chapter 4 (see *The Smooth Endoplasmic Reticulum and Golgi Apparatus Build Most Eukaryotic Cellular Membrane Components*), and will examine it in closer detail in Chapters 8 (see *Secreted Proteins and Proteins Targeted to the Endomembrane System Contain an Endoplasmic Reticulum Signal Sequence*) and 9 (see *Exocytosis Begins in the Endoplasmic Reticulum*).

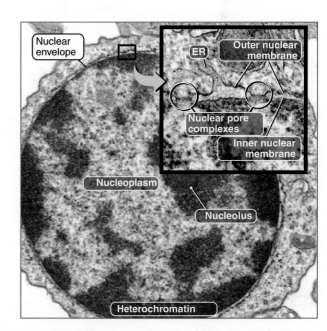

FIGURE 7-1 Electron micrograph of the nucleus of a white blood cell (lymphocyte). Photo courtesy of Terry Allen, University of Manchester.

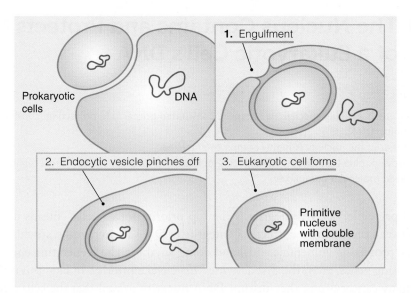

FIGURE 7-2 The nucleus may have arisen by endosymbiosis, a process in which one prokaryotic cell engulfs another cell, which then becomes a primitive organelle, such as the nucleus.

■■ Nuclear Pore Complexes Regulate Molecular Traffic into and out of the Nucleus

We briefly mentioned nuclear pore complexes in Chapter 1 (see *The Nucleus Is the Central Storehouse of Genetic Information*); now let's have a closer look at them. They are by far the most complicated protein structures found in cellular membranes, as **FIGURE 7-4** shows. This elaborate machinery is necessary to help nuclei carefully control which molecules enter and exit the nucleoplasm. The key features of the nuclear pores are several layers of rings stacked on top of one another that span the two nuclear membranes, linked to filamentous protein fibrils that extend into the cytosol, and a complementary set of fibrils that project into the nucleoplasm to form a basket structure. Nuclear pores are formed at areas where the two nuclear membranes fuse together (how this fusion occurs is not yet known), and a typical nucleus usually has dozens of nuclear pore complexes.

The entire structure undergoes complex conformational changes when it transports material into and out of the nucleus. The mechanisms controlling this trafficking are complex, and will be covered during our discussion of protein transport in Chapter 8 (see *The Nuclear Import/Export System Regulates Traffic of Macromolecules through Nuclear Pores*). The point of highlighting these structures again in this chapter is to emphasize how carefully cells control access to their DNA, which in turn helps illustrate that DNA is the most valuable material a cell possesses. *How* the nuclear pores help control access is not important for demonstrating principle #3, so we will leave that for discussion in Chapter 8.

■■ The Interior of the Nucleus Is Highly Organized and Contains Many Subcompartments

In Chapters 1 and 2 we discussed how eukaryotic cells compartmentalize tasks into organelles to increase their efficiency. This spatial organization theme extends even further, to *inside* organelles; the nucleus is a good example. It performs several different functions at once (controlling gene

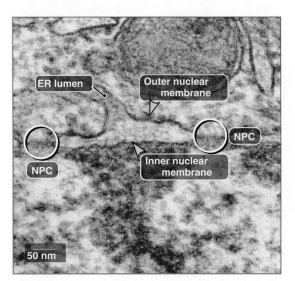

FIGURE 7-3 The nuclear envelope is continuous with the endoplasmic reticulum. Photo courtesy of Terry Allen, University of Manchester.

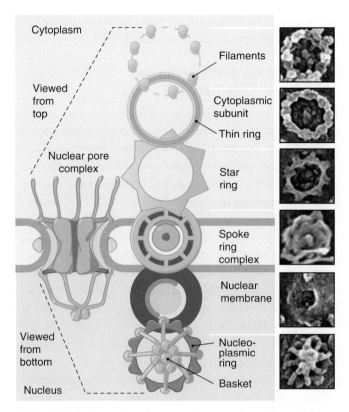

Cytoplasm

Viewed from top

Nuclear pore complex

Viewed from bottom

Nucleus

Filaments

Cytoplasmic subunit

Thin ring

Star ring

Spoke ring complex

Nuclear membrane

Nucleo-plasmic ring

Basket

FIGURE 7-4 The nuclear pore complex appears to be constructed from modular components. Electron micrographs of these components are shown at different stages of nuclear pore complex reassembly following mitosis. Photos courtesy of Terry Allen, University of Manchester.

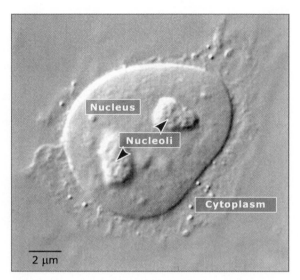

FIGURE 7-5 Light micrograph of a human cancer cell with two nucleoli in its nucleus. Photo courtesy of Zheng'an Wu and Joseph G. Gall, Carnegie Institution.

expression, replicating and repairing DNA, building ribosomes, etc.), and many of these are performed at specialized sites. Keep in mind that these sites are *not* surrounded by their own distinct membrane; they are all part of the nucleus.

The Nucleolus Contains DNA That Encodes Ribosomal RNAs

The largest, and perhaps best-known, substructure in the nucleus is called the nucleolus. All nuclei contain at least one, and sometimes several, of these structures; **FIGURE 7-5** shows two prominent nucleoli in a human cancer cell grown in a cell culture dish. The function of the nucleolus is to assemble the two large particles that form ribosomes in the cytosol. These particles, called **ribosomal subunits**, are composed of proteins (imported from the cytosol through nuclear pore complexes) complexed with large ribosomal RNA molecules (rRNAs) (**BOX 7-1**). This means that nucleoli are sites of high transcriptional activity for rRNA genes. Once the rRNAs are synthesized and joined with the ribosomal proteins in

---

BOX 7-1 TIP

More subunit issues. The word *subunit* is referring here to a structure very different from the "subunits" in proteins. As we found in Chapters 3 and 5, individual polypeptides or proteins can also be called subunits if they only function properly in polymers or clusters. Here, we see that even clusters of proteins can be called subunits if they only function when combined with other clusters. Remember, the word "subunit" refers to different structures in these different contexts, but the term is used repeatedly throughout cell biology because it indicates that a functional structure (regardless of its size) is composed of parts, and the parts are called subunits to reflect this fact. Other similar words (part, portion, etc.) do not have this specific connotation, so we don't use them.

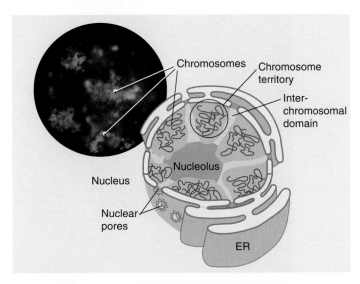

FIGURE 7-6 Individual chromosomes occupy distinct areas of the nucleus called chromosome territories. Photo reproduced with permission from J. Cell Sci., vol. 114 (16): 2891–2893. [http://jcs.biologists.org/cgi/content/full/114/16/2891]. Photo courtesy of Thomas Reid, National Institutes of Health.

the nucleolus, the completed subunits are exported into the cytosol through the nuclear pore complexes.

▮▮ The Nuclear Matrix Helps to Organize Chromosomes

The fact that nucleoli are enriched in rRNA genes implies that the spatial organization of chromosomes in the nucleus is not random. By using different colored fluorescent dyes attached to distinct chromosomes, we can see that chromosomes are compartmentalized into regions in the nucleus, called **chromosome territories** (FIGURE 7-6). In Chapter 2 we discussed how the proteins that form chromatin bind to additional proteins, including the nuclear lamins, to help control the shape of chromosomes and thereby regulate the transition between heterochromatin and euchromatin (see Chapter 2, *DNA Packaging Is Hierarchical*). The official name for this protein network, shown in FIGURE 7-7, is **nuclear matrix**. Note that this function sounds very similar to the role played by cytoskeletal proteins in the cytosol, but the nuclear matrix, though still poorly understood, is not known to contain any cytoskeletal proteins other than the nuclear lamins.

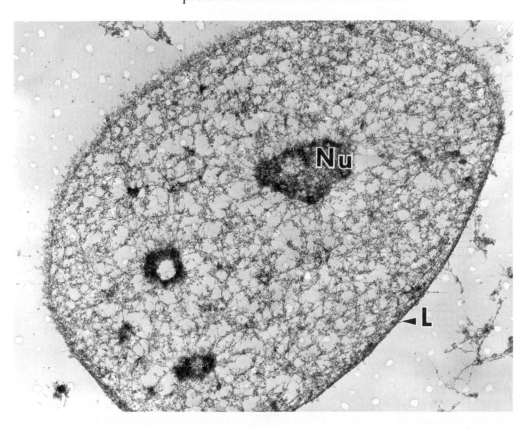

FIGURE 7-7 The nuclear matrix is a network of filaments bound to the nuclear envelope and to DNA. Nickerson, J. a, Krockmalnic, G., Wan, K. M., and Penman, S. (1997). The nuclear matrix revealed by eluting chromatin from a cross-linked nucleus. Proceedings of the National Academy of Sciences of the United States of America 94, 4446–50.

■■ DNA Replication Occurs at Sites Called Replication Factories

Consider for a moment how much work a cell has to do to replicate its DNA during cell division. The millions upon millions of deoxyribonucleotides that make up the DNA of an average cell cannot be copied haphazardly, as random copying here and there would be a terribly inefficient method for achieving replication. And recall our principle #3: the cellular machinery responsible for copying DNA has one of the most important jobs in a cell, because it has to delicately pull apart, copy, and reassemble the DNA strands *without breaking them*. Cells don't trust such a special task to amateurs. The proteins that accomplish this task form large complexes in the nucleus devoted entirely to copying DNA with 100% accuracy and absolutely no breaks. These machines are called **DNA replication factories**, and the smallest functional unit in these factories, which is responsible for copying one segment of DNA, is called a **replisome**. It is relatively easy to see the product of these factories in growing cells by simply labeling newly made DNA with a fluorescent tag and looking at them with a fluorescence microscope. Over time, as the amount of new DNA accumulates, the fluorescent signal increases until bright spots appear in the nucleus, as shown in **FIGURE 7-8**. Directly observing these factories is extremely difficult, however, because they are completely surrounded by clouds of unwound DNA; sample preparations that remove the DNA from a nucleus also remove the factories attached to them.

It is important to note that, no matter how expert these factories are at copying DNA, they do make mistakes: nothing in nature is absolutely perfect. It is these mistakes that give rise to much of the genetic variation between generations that we discussed in Chapter 2 (see *Mutations in DNA Give Rise to Variation in Proteins, Which Are Acted Upon by Natural Selection*). Of course, massive errors are usually fatal to a cell, so natural selection helps ensure that only the most efficient machines are passed from one generation to the next.

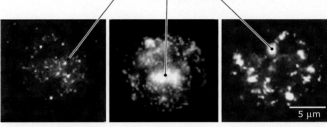

Notice that the fluorescence is concentrated in specific regions of the nucleoplasm, demonstrating that these are areas of active DNA replication.

| 8 hr postmitosis (early S phase) | 14 hr postmitosis (mid S phase) | 16.5 hr postmitosis (late S phase) |

FIGURE 7-8 DNA replication factories appear as bright spots of fluorescently tagged, newly synthesized DNA in these images of an actively growing cell that has entered the S phase of the cell cycle, when DNA is replicated. Reproduced with permission from J. Cell Sci., vol. 107 (8): 2191-2202. [http://jcs.biologists.org/cgi/content/abstract/107/8/2191]. Photos courtesy of Peter R. Cook, University of Oxford.

■■ RNA Polymerase Complexes and Spliceosomes Are Distinct Structures within the Nucleus

The theme of devoting specialized, highly efficient protein complexes to accomplish one or two essential tasks is common in cell biology. Just as replisomes within replication factories are used to copy DNA, additional complexes exist for performing many of the other critical tasks in the nucleus. Two of the most important are RNA polymerase complexes, which are responsible for transcribing the DNA sequence in genes into mRNA, rRNA, tRNA, and other RNAs; and **spliceosomes**, which are responsible for splicing the newly synthesized RNAs into their mature form. In Chapter 8 we will examine transcription (see *RNA Polymerases Transcribe Genes in a "Bubble" of Single-Stranded DNA*) and RNA splicing (see *In Eukaryotes, Messenger RNAs Undergo Processing Prior to Leaving the Nucleus*). Actually observing these structures in the nucleus is as difficult as it is to see replication factories, for similar reasons.

CONCEPT CHECK #1

Which structures do you think evolved first in cells: the molecular machines in the nucleus (replisomes, spliceosomes, etc.) or the nuclear matrix? Based on what you learned about the cytoskeleton in Chapter 5, propose a scenario that can explain the origin of the nuclear matrix.

7.3 ■ DNA Replication Is a Complex, Tightly Regulated Process

KEY CONCEPTS

▸ DNA replication in all organisms is performed by a small number of highly conserved proteins.
▸ Both prokaryotes and eukaryotes express many different forms of DNA polymerase, the enzyme responsible for synthesizing DNA.
▸ DNA replication begins at specific sites called origins of replication.
▸ During replication, double-stranded DNA is unwound and dissociated into single strands that serve as templates for the synthesis of complementary DNA strands.
▸ DNA synthesis occurs only in the 5'-to-3' direction.
▸ DNA polymerases must bind a double-stranded portion of a DNA molecule to begin synthesis. Most often, the double strand consists of the template DNA strand and a short, complementary RNA primer.
▸ DNA ligase connects individual pieces of newly synthesized DNA to form a complete strand.
▸ The enzyme telomerase adds extra DNA to the ends of chromosomes to protect them from degradation.

We now turn our attention to the chemical reactions cells use to replicate their DNA. In line with principle #3, we will focus on cellular strategies for maintaining the structural integrity of chromosomes while also teasing them apart to provide the templates necessary for semiconservative replication. Cells must break and create covalent and noncovalent bonds to accomplish this, and the brief periods when the bonds are broken are the most critical for the cell, because DNA is highly susceptible to irreversible damage and degradation at these times. Cells have evolved highly specialized enzymes that focus on making and breaking these bonds, while preserving DNA structural integrity with nearly 100% efficiency.

■■ DNA Polymerases Are Enzymes That Replicate DNA

The most important enzymes involved in DNA replication are **DNA polymerases**, whose function is to build linear polymers of deoxyribonucleotides from individual deoxyribonucleotide triphosphate subunits (**BOX 7-2**). Despite the fact that the phosphodiester bond linking the sugar-phosphate backbone between each deoxyribonucleotide is identical, several different DNA polymerases exist in both prokaryotes and eukaryotes because the conditions under which these bonds are formed are not all the same. Each polymerase can build DNA strands at different times and in response to different stimulants. This is somewhat analogous to how the telephone and power lines in a city might be built and maintained if some workers specialized in installing new lines and others specialized in repairing damaged, existing lines. Most cells contain several different DNA polymerases.

DNA polymerases share a common structure, shown in **FIGURE 7-9**. Conveniently, the structure looks so much like a person's right hand that the different structural domains in the protein are actually named *palm, thumb,* and *fingers.* To visualize how it works, imagine I hold my right hand out, fingers straight and palm facing upward toward my face. Now imagine someone handing me two chains of different lengths. One is very long (the template DNA strand), so I drape it over my palm and let the ends hang off

---

> **BOX 7-2 TIP** ■
>
> To review the chemical structure of deoxyribonucleotides and DNA, see Chapter 2, *DNA Is Carefully Packaged Inside Cells.* This information is essential for understanding how DNA polymerization occurs.

on either side; the other chain (representing the growing DNA strand) actually ends in my palm. If I clasp both chains with my thumb and use my fingers to help position the two chains such that the last link in the short chain (the 3′ end of the strand) rests in a very specific region of my palm, this represents the region where new deoxyribonucleotides are added to the growing DNA strand.

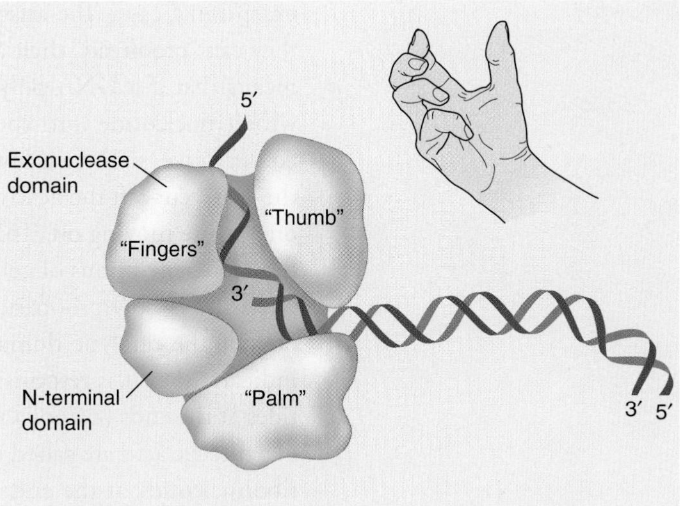

FIGURE 7-9 The common organization of DNA polymerases has a palm that contains the catalytic site, fingers that position the template, a thumb that binds DNA and is important in processivity, an exonuclease domain with its own active site, and an N-terminal domain.

◼◼ DNA Polymerases Add New Deoxyribonucleotides to the 3′ End of a DNA Strand

The actual chemical reactions for adding additional deoxyribonucleotides to the 3′ end are fairly straightforward, and they take place in the **catalytic domain** portion of the palm where the template strand rests. Before we describe the reactions, let's first establish why they have to happen at the 3′ end of the DNA strand. It is critical to know the atomic structure of the reactants, which we described in Chapter 2 (see *DNA Is a Linear Polymer of Deoxyribo-Nucleotides*), for this to make sense. Let's go through this carefully—drawing out the structures as we go along, like we did in Chapter 2, is a good way to see what is happening. Recall that the 3′ end of a DNA strand is the third carbon atom of deoxyribose, connected to four atoms: two other carbons in the sugar, as well as a hydrogen atom and a hydroxyl group. Draw the last deoxyribonucleotide in a DNA strand, and be sure to identify the hydroxyl group on the 3′ carbon. The reaction that adds a new deoxyribonucleotide to this 3′ end involves joining the 5′ end of a deoxyribonucleotide triphosphate to the hydroxyl group of the 3′ carbon.

Now we have an apparent problem. When we drew out a polymer of DNA in Chapter 2 (Figure 2-9), we simply attached deoxyribonucleotide monophosphates together by linking the 5′ phosphate group of one to the 3′ hydroxyl group of another. But it isn't that simple. Recall that the building blocks of DNA are deoxyribonucleotide *triphosphates,* not *monophosphates.* We can't just link a 5′ monophosphate form to the 3′ end.

Why not? To see the answer, let's look very closely at what is happening at the atomic level of this reaction. Due to its high electronegativity, the oxygen in the 3′ hydroxyl group has a partial negative (δ^-) charge (i.e., it is polar). Also, the single-bond oxygen atoms in the phosphate groups of a deoxyribonucleotide have a full negative charge. It is quite unlikely that these two negatively charged groups would form a covalent bond *spontaneously.* To speed the process up, the DNA polymerase enzymes that complete this reaction input extra energy, and here is the key: this extra energy comes from breaking the bond between the first and second phosphate groups in a deoxyribonucleotide *triphosphate.* (Deoxyribonucleotide *monophosphates* can't provide this energy, because they don't have the extra phosphate groups.) This also explains why the backbone of DNA has a single phosphate between each sugar, not two or three.

◼◼ DNA Polymerases Proofread Their Work

Now let's consider a counterargument: Couldn't the same chemical reaction happen between the 5′ triphosphate at the end of a DNA strand and the 3′ hydroxyl of a deoxyribonucleotide triphosphate? In other words, couldn't we use the same reaction to grow DNA in a 3′-to-5′ direction? Theoretically, yes. Then why does this not happen, not even in a single exceptional case? The answer illustrates a second critical point about DNA polymerases: they can "proofread" their activity, and can correct their own errors 99.9% of the time. This means that if a DNA polymerase misreads the template strand or accidentally inserts the wrong nucleotide into the growing DNA strand (remember, DNA replication is semi-conservative, so there is always an intact parental DNA strand to read off of), it can detect the error, cut out the newly added (wrong) deoxynucleotide, and replace it with the correct one before moving on. This is the main reason why DNA sequences do not change much between generations of cells.

Here is how proofreading works. A second functional domain in DNA polymerase, outside the catalytic domain, is called the exonuclease domain. The name "exonuclease" indicates that it is responsible for cleaving nucleotides (*nuclease*), and that it cuts nucleotides at the ends (*exo*) of a nucleic acid. (Enzymes that cut DNA and RNA in the "middle" of a nucleic acid are called **endonucleases**.) If a DNA polymerase inserts the wrong deoxyribonucleotide at the end of a DNA strand, it will not form the proper hydrogen bonds with the corresponding deoxyribonucleotide on the template strand; this results in a change in the shape of the DNA at that site, which is in the "palm" of the enzyme. In keeping with the three traits of all proteins, if the enzyme changes shape, it also changes function, and the shape induced by a mismatched pair of deoxyribonucleotides activates the exonuclease domain, and the mismatched deoxynucleotide at the 3′ end is cut off. What remains attached to the enzyme is *exactly* the same as what was present before the bad linkage was made: a 3′ end of a partially completed DNA strand. Now the DNA polymerase can find a new, correct deoxyribonucleotide triphosphate to add to that growing DNA chain. This is illustrated in **FIGURE 7-10**.

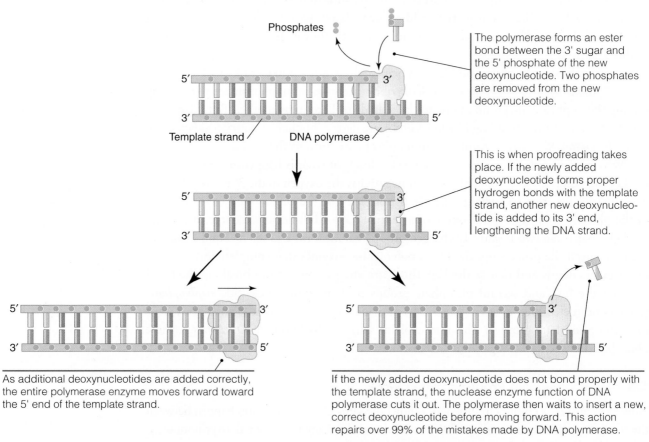

The polymerase forms an ester bond between the 3' sugar and the 5' phosphate of the new deoxynucleotide. Two phosphates are removed from the new deoxynucleotide.

This is when proofreading takes place. If the newly added deoxynucleotide forms proper hydrogen bonds with the template strand, another new deoxynucleotide is added to its 3' end, lengthening the DNA strand.

As additional deoxynucleotides are added correctly, the entire polymerase enzyme moves forward toward the 5' end of the template strand.

If the newly added deoxynucleotide does not bond properly with the template strand, the nuclease enzyme function of DNA polymerase cuts it out. The polymerase then waits to insert a new, correct deoxynucleotide before moving forward. This action repairs over 99% of the mistakes made by DNA polymerase.

FIGURE 7-10 DNA is synthesized in the 5′-to-3′ direction to permit DNA polymerase to proofread the new strand. Most of the data collected to construct this scenario comes from prokaryotic cells, but the same approach is used by all cells.

Before we proceed, it is important to point out two conventions that cell and molecular biologists use when discussing and diagramming the steps of DNA replication and other chemical reactions that take place on DNA. First, the helical structure of DNA makes accurate drawing difficult, and this degree of accuracy usually isn't necessary to get the point across. As such, we often use a very simplistic representation of DNA, like that shown in Figure 7-10. DNA strands are usually drawn as straight lines, with small perpendicular lines to one side to represent the base portion of the nucleotides; the resulting structure looks a lot like a comb. Two antiparallel DNA strands drawn this way look like ladders with small gaps in the perpendicular lines (the "steps") to represent the hydrogen bonds that hold the two strands together. When a protein is shown attached to DNA, usually little effort is made to draw its accurate three-dimensional shape, and instead we use ovals and random shapes, because the precise shape is not that important for the point we are trying to make. (Notice how different this is from the drawings in Chapter 5 of cytoskeletal proteins and the motor proteins that bind them, which are often very accurate.) We will use these types of drawings throughout the rest of this chapter.

The second convention is also a compromise of specific details. For technical reasons, the best information about DNA replication comes from simple, single-celled model organisms like bacteria and yeast, but we extrapolate this information whenever possible to represent universal truths about DNA chemistry. In other words, the best place to learn about how our own cells make and preserve DNA is usually in these simple organisms. That's why Figure 7-10 illustrates how a bacterial DNA polymerase proofreads. Both conventions sacrifice precision for ease of understanding.

Now let's return to why DNA polymerization must occur in only the 5′-to-3′ direction. Because proofreading is an essential feature of DNA polymerases, cells must be able to excise a mismatched nucleotide and be able to add a new one in its place. We have just seen how this happens at the 3′ end of a DNA strand. If a DNA polymerase was capable of adding nucleotides to the 5′ end of the strand, where would that extra energy it needs come from? The 5′ deoxyribonucleotide triphosphate *in the DNA strand*. And once that nucleotide had its two phosphates removed, there is no going back: that energy gets used once, to add one new nucleotide. If that new nucleotide is a mismatch and needs to be replaced, where will the extra energy come from to permit attachment of a *new* one? The answer is simple: *there is no usable source of energy in this scenario.* All of the enzymes that polymerize DNA have the same general shape, as we discussed above, and the spatial arrangement of their catalytic domain simply will not bind to a DNA strand and a deoxynucleotide triphosphate in an orientation that would make this possible. The proofreading requirement prevents DNA polymerases from elongating a DNA strand at the 5′ end. If an organism existed that used the 3′-to-5′ polymerization strategy, each new nucleotide added could never be replaced if it was incorrect, and the error rate would likely be much higher than in organisms that can proofread. If an enzyme that could add nucleotides to the 5′ end ever did arise, it did not survive natural selection (perhaps because it could not proofread?), and now every organism on earth follows the same 5′-to-3′ rule.

Be sure this makes sense before moving on to the next section. This concept can be challenging to grasp at first, but it represents the culmination of a lot of the material we have discussed in this and the previous chapters. It is an excellent example of the three traits of proteins, in that it demonstrates that DNA polymerases can adopt a number of different stable shapes, but these shapes are not random; they have been subject to natural selection for billions of years, and there are currently no alternative methods for building DNA. This kind of specificity is a hallmark of proteins, and DNA polymerases are some of the most important proteins in any living organism. This also helps explain why some chemicals can increase the rate of mutation in cells: if they disrupt the shape of either the DNA or the polymerase enough, specificity (and hence accuracy) is lost.

DNA Replication Is Semidiscontinuous

The overall strategy cells use for replicating their DNA is rather complex, but it is also an elegant solution to the difficult problem of rapidly replicating nucleotide sequences with a very low error rate. Recall from Chapter 2 (*DNA Forms a Double Stranded Helix*) that the two strands in a double helix of DNA are oriented antiparallel to one another; that is, the 5′-to-3′ orientation of one of the two strands points in the opposite direction on the other strand. This has important implications for how cells copy their DNA. Because they must follow the 5′-to-3′ synthesis rule, DNA polymerases have to "walk" in opposite directions on the two strands, and they have to progress in only the 3′-to-5′ direction. (Don't forget that the two strands being copied are the "template" strands, so they remain intact throughout the entire process.)

Let's consider one "easy" way to possibly copy DNA while adhering to the 5′-to-3′ synthesis rule. Cells could theoretically pull the two hydrogen-bonded strands apart at each end, place a DNA polymerase on the 3′ end of each strand, and have them march toward the 5′ ends; at some point, they would pass by one another en route to the opposite ends of the DNA. Once they reached the end of their template strand, they would have synthesized an entirely new strand that complements their template (hence it is called the **complementary strand**), and the job would be done.

But unfortunately, that's not what happens, because of principle #3. If cells used this strategy, they would leave enormous regions of their chromosomes in a very vulnerable state: completely unwound, and single stranded. This form of DNA breaks very easily, and cells cannot afford to take such a risk; keeping DNA intact is priority one. This single-stranded DNA would only be "rescued" when the polymerase reading it made the complementary strand, allowing it to form a double helix again.

DNA Replication Begins at Sites on Chromosomes Called Origins of Replication

The solution to this problem is to begin replication at sites where DNA polymerases working on both strands cooperate to keep the amount of single-stranded DNA at an absolute minimum. The trick to accomplishing this is for both polymerases to move together along the DNA, *in the same direction,* and synthesize new DNA strands from both templates at the same time. At first glance, this seems impossible: How can the polymerases move in the *same* direction while copying DNA in *opposite* directions?

Recall from Chapter 2 that to be replicated, DNA must be in the form of euchromatin, and it must be unwound, down to the beads-on-a-string form. Replication begins at a specific site (i.e., nucleotide sequence) on DNA called an **origin of replication**. Organisms with relatively small genomes, such as the bacterium *E. coli,* have only one origin of replication on their chromosome (small circular DNAs in bacteria, called plasmids, each have their own single origin of replication as well). Eukaryotic cells, which typically contain much more DNA organized into separate chromosomes, have several origins on each chromosome. For simplicity, we will focus on what happens at a single origin of replication.

An origin of replication contains a distinctive DNA sequence called an **autonomously replicating sequence** (**ARS**). The first step in DNA replication is binding of the ARS by a protein complex known as the **origin recognition complex**, shown in **FIGURE 7-11**. In bacteria, this complex contains several copies of the same polypeptide subunit, but in eukaryotes the complex contains six different subunits.

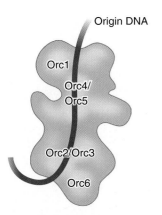

Origin DNA

Orc1

Orc4/
Orc5

Orc2/Orc3

Orc6

FIGURE 7-11 A model of the origin recognition complex in yeast, a simple eukaryote. Numerous proteins are contained in the complex.

During Replication, Specialized Proteins Unwind and Separate the Two Strands to Form a Replication Fork

Once the origin replication complex has bound to the origin of replication sequence, it changes shape, in accordance with the three traits of proteins. This shape change enables it to now bind more proteins, forming what is called a **prereplicative complex**. One of the proteins in this complex is called a **DNA helicase**, so named because it removes the "helix" trait of DNA; in simpler terms, it unwinds the double helix, as shown in **FIGURE 7-12**. Note that in eukaryotes, this unwinding can take place even though the double helix is wrapped around histones to form a nucleosome. Each prereplicative complex contains at least two helicase enzymes, and they unwind the DNA in opposite directions along the strand, giving rise to a small region of single-stranded DNA called a **replication bubble**. The area where the double-stranded DNA splits into two single strands is called the **replication fork**. Each replication bubble has two replication forks that progress in opposite directions along the double helix.

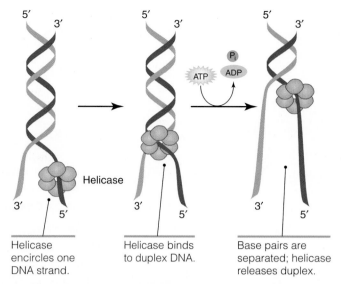

| Helicase encircles one DNA strand. | Helicase binds to duplex DNA. | Base pairs are separated; helicase releases duplex. |

FIGURE 7-12 A hexameric helicase moves along one strand of DNA. It probably changes conformation when it binds to the duplex, uses ATP hydrolysis to separate the strands, and then returns to the conformation it has when bound only to a single strand.

DNA Replication Requires an RNA Primer

A replication fork contains a number of proteins in addition to the DNA polymerases (one per strand) and helicase. To function properly, DNA polymerase cannot simply bind to a template strand and start synthesis right away—it must first bind to a double-stranded region on the DNA. But this presents a paradox: If the function of DNA polymerase is to build a double-stranded DNA molecule from a *single*-stranded template, how can it possibly require a *double* strand to get started? Where does this double strand come from? The answer, which likely reflects the ancient origins of this process, is that cells build a short, temporary "patch" of complementary RNA molecules to form a hybrid double strand. This short piece of RNA (approximately 30 nucleotides long) is the very first complementary strand made in the replication fork. It is called an **RNA primer**, shown in **FIGURE 7-13**, and the enzyme that synthesizes it is called **primase**. Once the primer has been attached to the template strand, DNA polymerase binds to the temporary double strand formed by the template and primer and begins attaching deoxyribonucleotides to the primer, in the required 5′-to-3′ direction. To do so, it moves along the template strand in the 3′-to-5′ direction.

Let's have a closer look at the replication fork in **FIGURE 7-14** to see how both template strands are read at the same time. On one side of the fork, a single primer is all that is needed to initiate replication, because as the replication fork moves forward, one of the DNA strands is unwound in the 3′-to-5′ direction. This makes it straightforward for DNA polymerase: all it has to do is bind a primer in a replication fork, start making DNA after the primer, and continue making a DNA strand as long as there is a template strand in front of it. This side of the parental DNA is called the **leading strand** for this reason. The product of a leading strand template is a single (usually very long) complementary strand. This side of the fork is relatively easy to build.

Things are quite different on the other side of the replication fork, however. To begin, the parental DNA template is

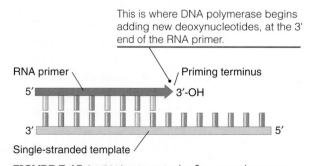

FIGURE 7-13 An RNA primer is the first complementary sequence synthesized in the complementary strand.

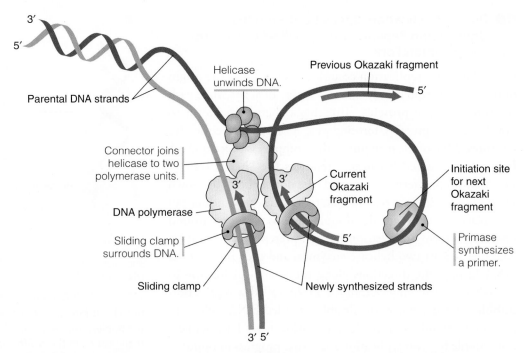

FIGURE 7-14 The helicase creating the replication fork is connected to two DNA polymerase catalytic subunits, each of which is held onto DNA by a sliding clamp. The polymerase that synthesizes the leading strand moves continuously. The polymerase that synthesizes the lagging strand dissociates at the end of an Okazaki fragment and then reassociates with a primer in the single-stranded template loop to synthesize the next fragment.

unwinding, but in the 5′-to-3′ direction, so DNA polymerase can't advance along it. The DNA polymerase on this side of the replication fork is actually coupled to the other DNA polymerase in the fork by a protein subunit committed to keeping the two polymerases attached to one another, so it is "trapped" near the beginning of the fork. Second, this unwound template is very susceptible to cleavage if it is not protected, so a protein, aptly named **single-strand binding protein** (SSBP), binds to it; several SSBPs bind to any single DNA strands created during replication. But SSBP has nothing to do with making new DNA, it simply protects the parental strands that are already there. How do we get this side started? First we need an RNA primer. So, primase synthesizes one, even though it appears to be facing in the "wrong direction."

What happens next is truly remarkable, and figuring it out was a tour de force for discoverers Reiji and Tsuneko Okazaki. They found that this side of the replication fork actually contains a short loop of DNA, and that by looping the DNA, a short segment of it can be oriented in the same direction as the leading strand. Now the primer can be lined up in the "right" direction, such that DNA polymerase recognizes it and begins synthesis. Eventually, however, the leading strand polymerase drags this polymerase forward in the fork, so it has to release the template and reattach closer to the fork. This interrupted DNA synthesis results in the formation of a short piece of DNA (~100 nucleotides) attached to a primer. The Okazakis next figured out that this loop–prime–synthesis process happens over and over again on this side of the replication fork, so that multiple DNA-RNA hybrid fragments are made, eventually forming complementary base pairs with every nucleotide in the template (called the **lagging strand**). This is why DNA synthesis is called *semidiscontinuous:* the complement to the leading strand is continuously elongated, but synthesis of the complement to the lagging strand is frequently interrupted (discontinuous). These discoveries were so important that the fragments were named **Okazaki fragments**.

◼◼◼ DNA Ligases Join Fragments of Single-Stranded DNA

Visualize what happens on the leading and lagging strands as helicase unwinds DNA and the replication forms on both sides of a replication bubble, using **FIGURE 7-15** as a guide. On circular (prokaryotic) DNA molecules, the two forks advance in opposite directions until they meet on the opposite side of the circle. On linear (eukaryotic) chromosomes, the forks advance in opposite directions until they either collide with another fork headed their direction or reach the end of the chromosome. The collision of two forks is not a problem; the proteins in each fork simply detach from the DNA, and can be recycled to build new replication forks elsewhere. When several origins of replication are present on a single chromosome, the replication bubbles grow until they merge, ultimately leading to two copies of the same chromosome. Because replication is semiconservative, both copies contain one intact DNA strand, as well as one strand made up of short (Okazaki) fragments and longer pieces synthesized on the leading strand of a replication fork. Each one of these fragments, either long or short, contains an RNA primer at its 5' end.

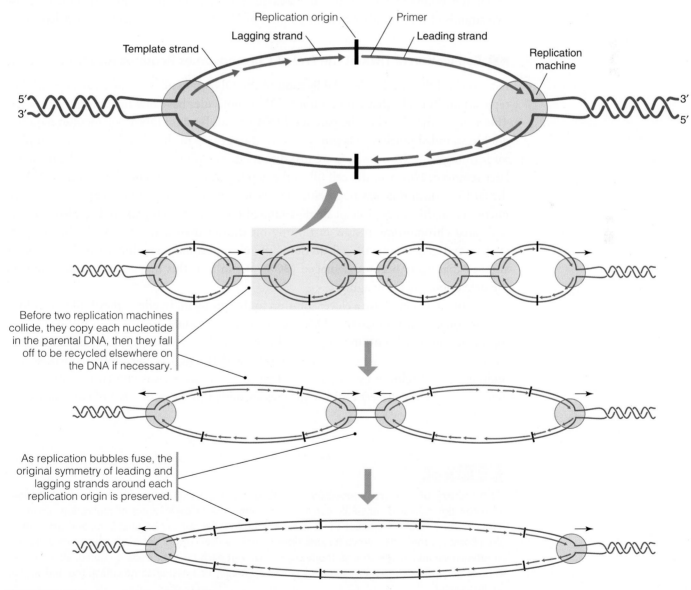

FIGURE 7-15 Fusion of replication bubbles.

Replication is not considered complete until all of the fragments on the newly synthesized strand are linked together to form a continuous sugar-phosphate backbone. A cell cannot simply link these fragments as they are, because each contains a small RNA primer. The primer has to be removed and replaced with DNA. The enzymes responsible for removing the RNA primers differ in prokaryotes and eukaryotes, but ultimately they achieve the same goal: they cut out the RNA portion of each fragment *without displacing the DNA portion.* This is critical, because it ensures that the proper base pairing between the old and new DNA strands is maintained.

Following removal of the primers, a form of DNA polymerase different from that found in a replication fork fills in the gap. At this point, all of the nucleotides in the newly synthesized strand are deoxyribonucleotides, and they are hydrogen bonded to their complementary nucleotides in the parental strand. All that remains to be done is the formation of covalent bonds between adjacent DNA fragments on the new strand; this is accomplished by a class of enzymes called **ligases**. This name is derived from the Latin word *ligare,* which means "to bind," and the verb *to ligate* is a popular way of describing the covalent attachment of one molecule to another. Ligases form the same kind of phosphodiester bonds that DNA polymerases do, but their substrates are different: they use two strands of DNA, rather than one strand of DNA and a single deoxyribonucleotide.

■■ Replication of DNA at the End of Chromosomes Requires Additional Steps

For circular DNA molecules, the ligation of the DNA fragments is the final step in DNA replication. But cells that contain linear DNA molecules have one final problem to solve. This arises at the 3′ end of the parental DNA strand. Because DNA polymerases require a double-stranded portion to begin synthesis, the very tip of the 3′ end serves as a template for forming an RNA primer, as we just discussed. The problem occurs when the primer is later removed: How can the cell fill in the gap left at this end? Since there is no DNA on the far (5′) side of this new fragment, there is no place for a DNA polymerase to bind. Left uncorrected, this short piece of single-stranded DNA in the template is degraded, and the replicated chromosome is now just a tiny bit shorter than it was before. If this happens over several generations of cell division, the chromosomes can shorten by such a large amount that genes become damaged (**BOX 7-3**); in fact, this may be one explanation for the deleterious effects of aging.

Fortunately, cells have a clever solution to this problem. In effect, they build extra-long chromosomes, with noncoding DNA sequences at the ends, so that even if the chromosomes are shortened, nothing of great value is lost. The enzyme complex that builds these chromosome extensions is called **telomerase**, and the noncoding DNA sequences they build are called **telomeres**. The name, derived from Greek, was chosen to indicate that many (*-mer,* as in *polymer*) DNA sequences are added to the end (*telo-*) of chromosomes.

BOX 7-3

The Library of Congress analogy, revisited. The Library of Congress analogy may help you visualize the danger of not fully copying a chromosome in each round of replication. Imagine photocopying all of a book but the last page, then lending out the original book and using only the photocopy thereafter. When it came time to make a new copy of the book, you would use your shortened version as the source. If one page is left out each time the book is copied, after several rounds of copying/shortening, the book would eventually be missing some critical text and would be less useful.

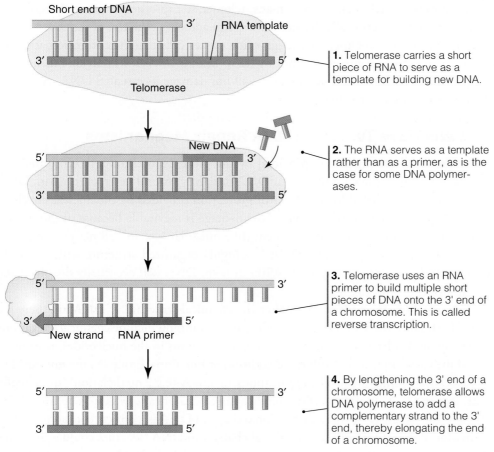

Short end of DNA

3'

RNA template

3'   5'

Telomerase

1. Telomerase carries a short piece of RNA to serve as a template for building new DNA.

New DNA

5'   3'

3'   5'

2. The RNA serves as a template rather than as a primer, as is the case for some DNA polymerases.

5'   3'

3'   5'

New strand   RNA primer

3. Telomerase uses an RNA primer to build multiple short pieces of DNA onto the 3' end of a chromosome. This is called reverse transcription.

5'   3'

3'   5'

4. By lengthening the 3' end of a chromosome, telomerase allows DNA polymerase to add a complementary strand to the 3' end, thereby elongating the end of a chromosome.

FIGURE 7-16 Four features of telomerase.

The mechanism telomerase uses to build telomeres is complicated and only partially understood, but four features deserve mentioning, as shown in **FIGURE 7-16**:

- First, rather than wait for another enzyme to build an RNA primer for it, telomerase actually carries its own piece of RNA; it binds to a piece of RNA so stably that this RNA is always available to form a double strand with whatever DNA the telomerase binds to.

- Second, the RNA is *not* the primer in this case. In effect, it switches roles: the 3' DNA is the primer, and the RNA is the template. This allows it to add more deoxyribonucleotides to the 3' end of the DNA. The resulting elongated 3' end can then be RNA primed and filled in by the usual primase/DNA polymerase mechanism, though it is not yet clear which polymerase is responsible. Once the DNA has been added up to the end of the RNA template, the telomerase slides forward and repeats this process. Telomeres typically contain thousands of base pairs.

- Third, telomerase builds DNA from an RNA template, and this is somewhat unusual. During transcription, RNA is synthesized off of a DNA template. Because this relationship is reversed in telomerase, this enzyme is said to have **reverse transcriptase** activity. Note that DNA is still synthesized in the 5'-to-3' direction.

- Fourth, the fact that the DNA is synthesized off a short RNA template means that telomeres are typically composed of short (usually 5–10 bases), highly repetitive DNA sequences. This high concentration of repeating sequences is distinct from the rest of the chromosome, and is actually targeted by telomere-binding proteins that perform a variety of functions, including protecting the ends of the telomere from digestion by exonucleases.

Not all cells express telomerase. In humans, telomerase expression is generally restricted to stem cells and the germ cells that generate sperm and eggs. The rest of the somatic cells do not usually express telomerase, though some cancer cells that arise from somatic cells can. The presence of telomeres does not by itself mean that a cell will divide indefinitely; it simply means that when it does divide it will not damage its essential DNA sequences. This too is evidence of principle #3.

■■ Cells Have Two Main DNA Repair Mechanisms

Despite the elaborate protection provided by the proteins in chromatin (and nucleoid in prokaryotes), plus the sequestering of DNA in tightly wound heterochromatin, DNA gets damaged. Even the additional layer of protection provided by the nuclear envelope fails to completely prevent this damage. The damage is inevitable because of the second law of thermodynamics: in a closed system (like our solar system), all matter progresses toward increasing entropy. Intact DNA is a highly organized structure with relatively low entropy, as we saw in Chapter 2. Without help, DNA will eventually dissolve back into the atoms composing it.

There are likely as many ways to alter the chemical structure of DNA are there are chemical reactions to assemble it. Some of the most common include chemical alterations induced by ultraviolet radiation, loss of some of the atoms in the nucleotides, rearrangement of the atoms in the nucleotides, or addition of functional groups to the nucleotides. Remember that some chemical modifications of DNA (e.g., methylation) are actually intentional and quite useful to cells; not all chemical modifications should be considered to be harmful. Because the three-dimensional shape of DNA is a critical component of its information storage function, unintentional changes to DNA structure could either alter or erase the information it stores, much like demagnetizing a hard drive in a computer. Cells are in a constant battle to preserve the structure of their DNA even as it is breaking down around them; in general, most cells win this struggle long enough to generate daughter cells, and it is this winning strategy that has established DNA as the genetic material in all known organisms. (For the purposes of this text, viruses—including those that use RNA to store genetic information—are not considered to be alive, though this is still a hotly debated issue.)

One of the keys to preserving DNA structure is to have readily available sources of useful energy. Recall from Chapter 1 (see *Cells Are Self-Replicating Structures That Are Capable of Responding to Changes in Their Environment*) that entire cells resist the drive toward chemical equilibrium (also a product of the second law of thermodynamics) by constantly investing energy in building chemical gradients and complex molecular structures. The same lesson applies to DNA. Rather than assemble DNA into even more complex chemical structures, cells spend some of this energy twisting DNA into highly organized, folded structures (recall the levels of DNA packing in Chapter 2). The remainder of the energy investment is spent directly replacing or repairing the damaged deoxynucleotide building blocks of DNA. The strategies for repairing DNA are numerous and sometimes quite specialized for certain types of cells, but most of them fall into two classes: excision and recombination.

■■ Excision Systems Remove One Strand of Damaged DNA and Replace It

Perhaps the most straightforward means of repairing DNA is to cut out (excise) the damaged portion and replace it with new, undamaged pieces of the same type. This is analogous to how many repairs are performed on everyday machines: as the parts wear out, new parts are swapped in to keep the machine running properly. The mechanisms that use this strategy in cells are called **excision repair systems**.

Excision repair takes place in four steps, shown in **FIGURE 7-17**. First, an endonuclease enzyme detects the damaged portion of the DNA (imagine the havoc this enzyme could wreak if it was not specific for damaged DNA; we will discuss how it finds these damaged regions in Chapter 13) and breaks the phosphoester bonds in the DNA backbone. Second, it excises the damaged portion, leaving a gap in the double-stranded DNA. Third, proper replacement nucleotides are inserted into the gap by DNA polymerases that specialize in repair, and the nucleotide sequence is determined by reading the complementary DNA strand, exactly like the process of DNA replication. Finally, a ligase reconnects the sugar backbone of the repaired strand. Nearly all of these activities require energy input, usually in the form of ATP.

Excision repair systems are especially useful for clearing up errors and damage immediately after DNA replication. Because mismatched (but otherwise structurally sound) base pairs frequently arise from replication errors that slip through the proofreading by DNA polymerase, **mismatch repair** is a well-known excision repair system. An important assumption made here is that the endonuclease enzyme recognizes which of the two strands is damaged or mismatched. If the system removes the wrong portion, the resulting change in DNA sequence (and structure) is often irreversible, and can easily cause mutations. In bacteria, if a region of the parental DNA is methylated, for a brief period following replication the parental strand, but not the new strand, is methylated, and this enables the enzymes to determine which strand to cut. In mammalian cells, the accuracy of the detection portion of step one is increased by enzymes that search for and remove damaged bases. The release of the base triggers the remainder of the steps.

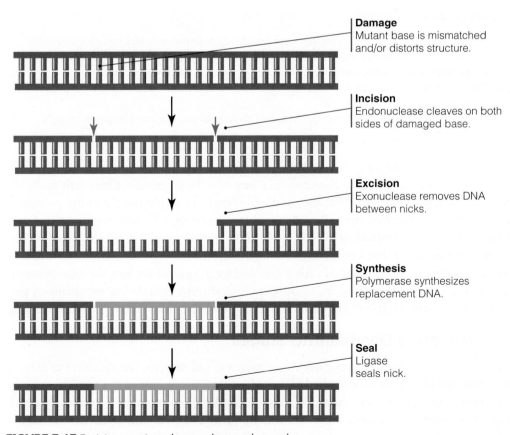

Damage
Mutant base is mismatched and/or distorts structure.

Incision
Endonuclease cleaves on both sides of damaged base.

Excision
Exonuclease removes DNA between nicks.

Synthesis
Polymerase synthesizes replacement DNA.

Seal
Ligase seals nick.

FIGURE 7-17 Excision repair replaces a damaged strand.

■■ Recombination Systems Repair Large-Scale DNA Damage

If a damaged or mismatched nucleotide is not corrected by excision repair, a host of backup systems are available for a last-ditch effort to fix the problem before a cell completes a new round of DNA replication during the next cell division. We will revisit these systems when we discuss the mechanisms controlling cell division in Chapter 13. These are called **recombination repair systems**, because they rely on genetic recombination to overcome complications caused by damaged DNA and restore proper replication of DNA. Genetic recombination takes place primarily during metaphase of mitosis and meiosis, when the arms of homologous chromosomes occasionally break and swap pieces with one another. We will briefly discuss recombination in the next section.

CONCEPT CHECK #2

Imagine that life is discovered on another planet, and that by some miraculous stroke of luck the organisms closely resemble eukaryotic life on earth, including the use of antiparallel double-stranded DNA. You are selected to study their genetic replication, and observe the following differences between their replication machinery and that of organisms on earth:
A. All alien cells express the exact same, single type of DNA polymerase.
B. Alien DNA contains more C–G base pairs than earthly organisms' DNA.
C. Alien chromosomes are never shorter after DNA replication.
D. Okazaki fragments are absent in alien cells.
Based on these observations, what differences would you expect to find between the mechanisms of alien DNA replication and that of earthly organisms?

▉ 7.4 ▉ Mitosis Separates Replicated Chromosomes

KEY CONCEPTS

▸ The function of mitosis is to safely separate replicated chromosomes into two daughter cells.
▸ Mitosis is divided into five phases, based largely on morphological changes in the location and arrangement of chromosomes.
▸ The microtubule cytoskeleton, including microtubule motor proteins, is essential for proper segregation of chromosomes.
▸ The actin cytoskeleton is required for the actual division of one cell into two daughter cells following mitosis.

Following completion of DNA replication, another major challenge for cells is to successfully sort and separate the chromosomes into two daughter cells. Once cells make the decision to replicate their DNA, they are committed to completing the entire process of cell division (often called the **cell cycle**), including mitosis. (We will examine the details of the mechanisms controlling the entire cell cycle, including mitosis, in Chapter 13.) The phases of mitosis are well known to most beginning cell biology students, so we will skip some of the basic descriptions of each phase and focus instead on how the chromosomes are attached to and moved on the mitotic spindle, thereby completing the theme of preserving DNA structural integrity throughout the most dynamic stages of DNA processing.

▉ ▉ Mitosis Is Divided into Stages

Like most major advances in the beginning stages of cell biology, the discovery of mitosis was reported by a microscopist. In 1879, Walther Flemming coined the term to describe the motion of what he described as "threads" (Greek, *mitos*) moving in an actively dividing cell, as shown in **FIGURE 7-18**. The activity he described is now divided into the five phases: prophase, prometaphase, metaphase, anaphase, and telophase. Keep in mind that

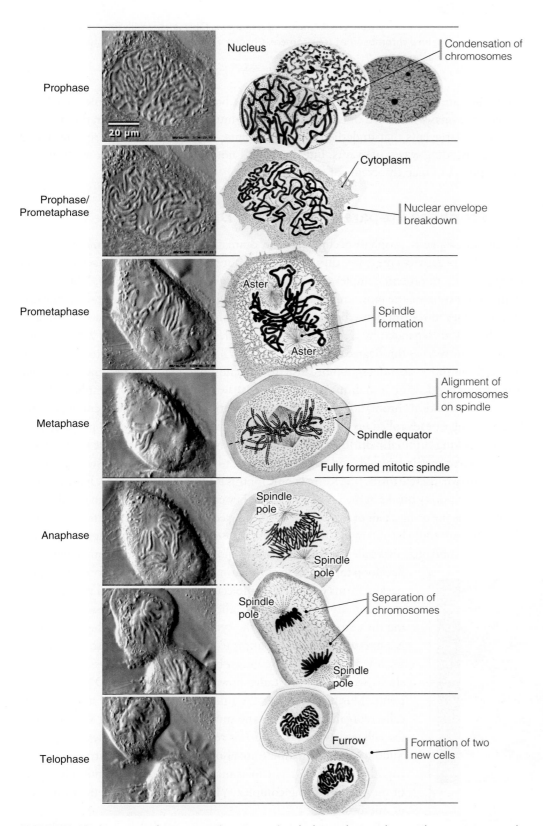

Prophase

Nucleus

Condensation of chromosomes

20 µm

Prophase/ Prometaphase

Cytoplasm

Nuclear envelope breakdown

Prometaphase

Aster

Spindle formation

Aster

Metaphase

Alignment of chromosomes on spindle

Spindle equator

Fully formed mitotic spindle

Anaphase

Spindle pole

Spindle pole

Spindle pole

Separation of chromosomes

Spindle pole

Telophase

Furrow

Formation of two new cells

FIGURE 7-18 The events of mitosis. In the top panel, only the nucleus is shown. The remaining panels show the entire cell. Photos © Conly L. Rieder, Wadsworth Center. Illustrations from W. Flemming, W. Archiv für Mikroskopische Anatomie. Berlin: J. Springer (1871).

the phases are not distinct events; the boundaries between phases are somewhat fuzzy, because the phases are based on visible movements humans can categorize, rather than on the molecular events that underlie these movements. A more realistic description would include cascading waves of events, but this would not be easy to classify, so we'll use the visible phases to map the events of mitosis.

Prophase Prepares the Cell for Division

As the name suggests, prophase occurs *before* any dramatic chromosome movement is visible. A great deal of activity takes place during prophase, but most of it involves the activation and reorganization of protein complexes too small to see with a light microscope. With respect to the chromosomes, the most important activity taking place is the folding of the replicated chromosomes into extra dense bodies known as **chromatids**. Each chromosome contains two chromatids separated by a constriction point called a **centromere** (**BOX 7-4**). When a chromosome reaches this degree of folding, virtually all gene expression is halted. Microscopists call this process condensation because the chromosomes gradually become visible, like water vapor condensing into drops, as shown in **FIGURE 7-19**. This a very complicated task, because the identical chromatids (called **sister chromatids**) are very prone to entanglement, like two long threads rolled up into a ball. Two proteins play a critical role in properly sorting and packing chromatin during condensation. One is called **topoisomerase II**. It is a part of the nuclear matrix scaffold, and uses ATP energy to sort the replicated sister chromatid DNA into two discrete forms (the formal name for this is **DNA decatenation**). It does this by forming temporary breaks in the DNA strands so they can be teased apart, then resealing the breaks. (Imagine using a pair of scissors to cut the tangled ball of threads at strategic points so the two threads could be untangled, then tying the loose ends together once they are free of the ball.) The second protein is called condensin. Condensin is a multi-subunit protein that stabilizes loop formation in chromatin, and therefore helps form the highly packed forms of DNA. Its molecular mechanism is still not clear.

Soon after condensation, the nuclear envelope dissolves, releasing the chromosomes into the cytosol. Keep in mind that the cytosol is a hazardous region for unprotected DNA, so the dissolution of the nuclear membrane represents a big step in the progression of mitosis, as well as demonstrates how well-protected the DNA is at this point. During prophase, the microtubule network is also dramatically reorganized to form the mitotic spindle. During the period of the cell cycle when the DNA is replicated, the centrioles lying in the middle of the centrosome are also replicated. Both centrioles are surrounded by the pericentriolar material that includes multiple copies of the γ-tubulin ring complex (γTuRC) that nucleates the formation of microtubules.

Formation of the spindle relies on the dynamic instability of microtubules and the force generation of microtubule-associated motor proteins. The cytosolic microtubule network dissolves, and new microtubules, called **astral microtubules**, grow from

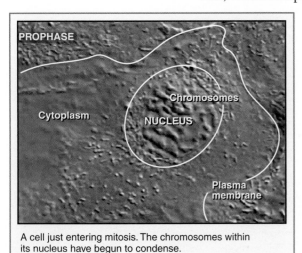

A cell just entering mitosis. The chromosomes within its nucleus have begun to condense.

FIGURE 7-19 The first frame of a video that follows the chromosomes through the intial stages of mitosis. Photo © Conly L. Rieder and Alexey Khodjakov, Wadsworth Center.

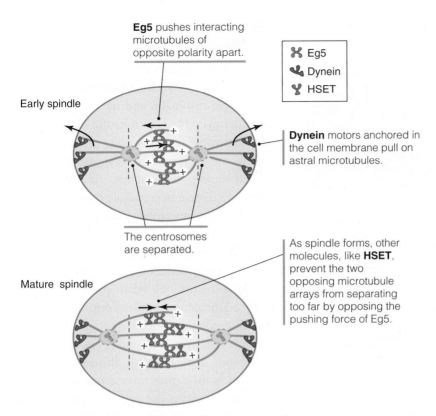

Eg5 pushes interacting microtubules of opposite polarity apart.

Eg5
Dynein
HSET

Early spindle

Dynein motors anchored in the cell membrane pull on astral microtubules.

The centrosomes are separated.

Mature spindle

As spindle forms, other molecules, like **HSET**, prevent the two opposing microtubule arrays from separating too far by opposing the pushing force of Eg5.

FIGURE 7-20 Microtubule motors help form the mitotic spindle.

each new centrosome. These microtubules come into contact with the plasma membrane and overlap one another. This distribution gives microtubule motors an opportunity to contribute to the formation of the mature spindle, as shown in **FIGURE 7-20**. Dynein motor proteins (see *Microtubule-Associated Proteins Regulate the Stability and Function of Microtubules* in Chapter 5) connected to the plasma membrane pull on the microtubules, drawing the two centrosomes apart. Simultaneously, **kinesin-related motor proteins** attach to overlapping regions of microtubules projecting from the two centrosomes: by "walking" toward the plus ends of both microtubules at the same time, the kinesins push the two microtubules apart. This complements the efforts of the dyneins, and spreads the centrosomes to opposite ends of the cell, giving rise to a full-size mitotic spindle.

Chromosomes Attach to the Mitotic Spindle during Prometaphase

The mitotic spindle contains three primary structural components. One is the array of astral microtubules that initiates formation of the spindle. Another is an array of overlapping microtubules called **polar microtubules**. These microtubules are attached to two types of kinesin-related motor proteins, called Eg5 and HSET, that "walk" in opposite directions, opposing one another and thereby stabilizing the entire spindle. The third component is a collection of microtubules that are directly connected to the chromosomes; these are called **kinetochore microtubules** because they attach to **kinetochores** on chromosomes. Kinetochores are compact protein complexes that attach to chromosomes on either side of the centromere at the onset of prometaphase and act as attachment sites for kinetochore microtubules.

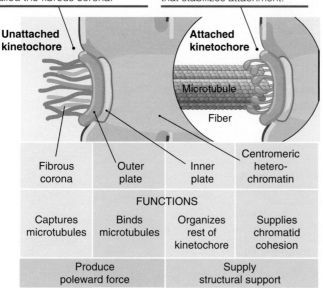

Unattached kinetochores contain two structures called inner and outer plates. The outer plates are connected to fibers called the fibrous corona.

When a kinetochore is attached to a microtubule, the fibrous corona forms a basket shape that stabilizes attachment.

Fibrous corona	Outer plate	Inner plate	Centromeric hetero-chromatin
FUNCTIONS			
Captures microtubules	Binds microtubules	Organizes rest of kinetochore	Supplies chromatid cohesion
Produce poleward force		Supply structural support	

FIGURE 7-21 The structure of kinetochores.

■■ Kinetochores Attach Chromosomes to the Mitotic Spindle

Kinetochores contain three characteristic structures, as shown in **FIGURE 7-21**. The inner plate lies between the centromere chromatin and the rest of the kinetochore; the outer plate binds to microtubules; and fibrous proteins extending from the outer plate, called the fibrous corona, are responsible for capturing unbound microtubules during prometaphase. The fibrous corona also contains a dynein microtubule motor.

During prometaphase, astral microtubules undergo extensive dynamic instability (see *The Growth and Shrinkage of Microtubules Is Called Dynamic Instability* in Chapter 5), thereby generating many rapidly growing and shrinking microtubules that "probe" the cytosol. Eventually, a kinetochore binds to one of these microtubules and immediately begins to migrate toward a spindle pole (centrosome), pulled along by the dynein in the fibrous corona. This is illustrated in **FIGURE 7-22**. While it is migrating toward one pole, an additional microtubule from the opposite pole attaches to the second kinetochore. Kinetochore microtubules can shrink and grow while remaining attached to the chromosome, as **FIGURE 7-23** shows. The dynein motors in both kinetochores pull in opposite directions, though typically not at the same time, and the forces eventually equilibrate when the two kinetochore microtubules are approximately the same length and the chromosome arrives at the midpoint between the two poles (also called the **spindle equator**). If microtubules from the same spindle pole attach to both kinetochores, there is a risk that a chromosome will be pulled toward the pole rather than toward the spindle equator, but fortunately this type of microtubule attachment is unstable and rapidly breaks apart to prevent this problem. The equatorial alignment takes place for each chromosome in a dividing cell, and the name used to describe this chromosome movement is **congression**.

■■ Arrival of the Chromosomes at the Spindle Equator Signals the Beginning of Metaphase

Another name for the spindle equator is the **metaphase plate**. Once all of the chromosomes arrive at the spindle equator, microscopists first thought they might be attached to a disk (or plate) because they aligned so well; we now know this is not the case, but the name persists. Metaphase is the period of mitosis (and meiosis) when most chromosome recombination takes place. Because recombination requires breaking and rejoining of chromosome fragments, this is another critical period for cells. Repair enzymes ensure that regardless of where the fragments end up, chromosomal integrity is maintained.

■■ Separation of Chromatids at the Metaphase Plate Occurs during Anaphase

The function of anaphase is to separate the replicated chromosomes from one another. After DNA replication is complete, the two copies of a chromosome are held together by the centromere. This explains why condensed chromosomes typically have an "X" shape: one

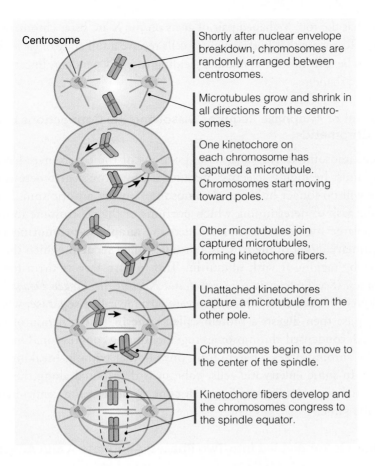

Centrosome

Shortly after nuclear envelope breakdown, chromosomes are randomly arranged between centrosomes.

Microtubules grow and shrink in all directions from the centrosomes.

One kinetochore on each chromosome has captured a microtubule. Chromosomes start moving toward poles.

Other microtubules join captured microtubules, forming kinetochore fibers.

Unattached kinetochores capture a microtubule from the other pole.

Chromosomes begin to move to the center of the spindle.

Kinetochore fibers develop and the chromosomes congress to the spindle equator.

FIGURE 7-22 Dynamic microtubules search for kinetochores throughout the cell by growing and shrinking in random directions from the centrosomes. Microtubules that encounter a kinetochore are captured and stabilized. This search-and-capture mechanism of spindle assembly allows a spindle to be formed regardless of the shape of the cell or the positions of the chromosomes at the start.

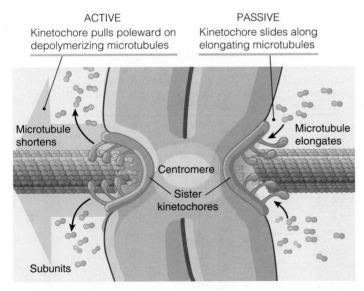

ACTIVE
Kinetochore pulls poleward on depolymerizing microtubules

PASSIVE
Kinetochore slides along elongating microtubules

Microtubule shortens

Microtubule elongates

Centromere

Sister kinetochores

Subunits

FIGURE 7-23 Once attached to the spindle, a kinetochore can exist in one of two activity states. One (left) allows it to move toward a pole on a shortening kinetochore fiber. The other (right) allows it to remain stationary or move away from its pole on an elongating fiber. The activities of the two sister kinetochores are somehow coordinated so that both kinetochores are rarely in the active state at the same time.

can imagine that the top, V-shaped pair of arms on the X are sister chromatids, as are the bottom arms. To ensure that both daughter cells receive a full copy of each chromosome, the X needs to be split down the middle, yielding two independent, linear chromosomes (resembling a | | shape).

■■ The Onset of Anaphase Requires Dissolving the Connections between Sister Chromatids

To split a chromosome during anaphase, portions of the centromere have to be cut. This cutting must be highly selective, because if the entire centromere is dissolved, a kinetochore will no longer hold the chromosomes on the mitotic spindle and mitosis will stop. The task of determining which portions of the centrosome to digest is performed by a large molecular complex called the **anaphase-promoting complex**, or **APC**. The primary target of the APC is a protein called **securin**, which the APC marks for digestion by tagging it with ubiquitin. This in turn targets them for destruction by proteosomes. (See *Proteins in the Cytosol and Nucleus Are Broken Down in the Proteasome* in Chapter 3). Securin inhibits a protease enzyme called **separase**; when securin is digested, separase then digests a protein called **cohesin**. The function of cohesin is to hold replicated, condensed chromosomes together during congression. Once the cohesin is cleaved, the chromosomes separate from one another without breaking, as shown in **FIGURE 7-24**. In some eukaryotic cells, cohesin is distributed along the entire length of the replicated chromosomes rather than concentrated at the centromere, but their function is the same.

■■ Anaphase Is Subdivided into Two Phases: Anaphase A and Anaphase B

Following digestion of cohesin, the next step of anaphase is to physically pull the two halves of a replicated chromosome to opposite spindle poles. To accomplish this, force must be generated at the kinetochore. Two types of force are generated during anaphase, arising from different protein complexes; although they both act at the same time, cell biologists have divided anaphase into two subphases, called **anaphase A** and **anaphase B**.

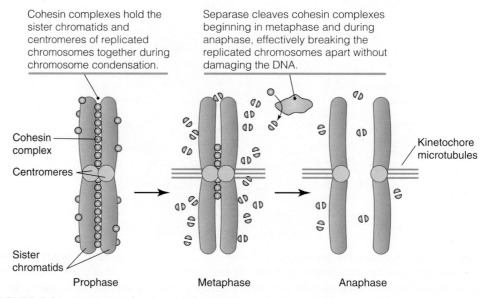

FIGURE 7-24 Separation of replicated chromosomes during anaphase.

During anaphase A, the kinetochores on sister chromatids are *pulled* apart by kinetochore dyneins and progressive shortening of kinetochore microtubules, the same forces that helped direct the chromosomes to the spindle equator during prometaphase. Because the cohesin proteins holding the centromere have been digested, application of these forces to both sides of the replicated chromosomes causes them to rip apart and begin migrating toward the spindle poles. During anaphase B, kinesin-like motor proteins *push* overlapping polar microtubules apart, while cytoplasmic dynein pulls on astral microtubules (as in prophase), further elongating the spindle. This is illustrated in **FIGURE 7-25**.

In addition, a third type of microtubule activity takes place in anaphase. Independent of anaphase A and B, bundles of overlapping microtubules called **stem bodies** form between the spindle poles. These microtubules are oriented with their minus ends pointed at the spindle poles, while their plus ends overlap near the spindle equator. They are not responsible for generating any chromosome motive force in anaphase, but do trigger the formation of a ring-shaped cortical actin–myosin filament network (called the **contractile ring**) that plays a crucial role after mitosis is complete and cytokinesis begins.

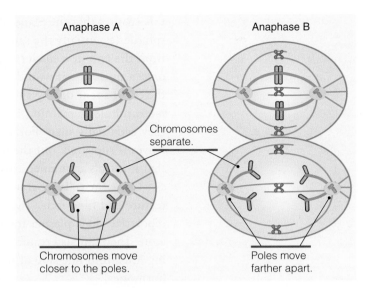

FIGURE 7-25 As the chromosomes move to the poles (anaphase A), the poles move farther apart (anaphase B), increasing the separation between the two groups of chromosomes. The poles are moved by pulling forces on the astral microtubules and motors in the center of the spindle that slide overlapping microtubules past one another. Together, anaphases A and B ensure that the two new nuclei are far enough apart so that the cell body can be reliably divided between them.

The Structural Rearrangements That Occur in Prophase Begin to Reverse during Telophase

Once the separated chromosomes have begun moving away from the spindle equator, telophase begins. The APC is responsible for triggering the onset telophase (it might therefore be thought of as the "anaphase AND telophase promoting complex," but this is not its official name). In effect, the purpose of telophase is to initiate reversal of the destructive activities that occurred in prophase: the nuclear membrane is reformed around each of the two sets of chromosomes, and the interphase cytoskeleton begins to reform. As the spindle poles move further apart, the spindle microtubules dissolve, the astral microtubules lengthen, and the overlapping stem bodies formed during anaphase condense to form a structure called the **midbody**. The midbody contains many additional proteins that control the timing and location of the final stage of cell division, cytokinesis.

Cytokinesis Completes Mitosis by Partitioning the Cytoplasm to Form Two New Daughter Cells

Having successfully accomplished the primary mission of mitosis (the safe segregation of replicated chromosomes into two new nuclei), the cell then squeezes itself in half to generate two new daughter cells. In keeping with the Greek origin of so many names in cell biology, this process is called cytokinesis (*cyto* = cell, *kinesis* = motion). As the contractile ring tightens (like purse strings), the cytosol is divided into two roughly equal portions. The contractile ring is induced to contract by proteins in the midbody, and this ensures that the plane of separation is perpendicular to the long axis of the mitotic spindle (essentially the same place where the spindle equator was), literally as far away from the two new nuclei as possible. Imagine the disaster if the contractile ring closed onto a nucleus—it would likely rupture the nucleus, killing the cell and wasting all of the effort spent during the cell cycle. Once the ring contracts down to a tiny circle, the microtubules remaining in the midbody are all that hold the two halves together, and they eventually snap. For a brief moment, the plasma membrane surrounding the two halves is ruptured, but the hydrophobic nature of the membrane phospholipids ensure that it reseals spontaneously, and the two cells are entirely separate.

Once the cells separate, they can reconstitute their interphase (nondividing) architecture and functions. This means they can re-form cell–cell and cell–ECM junctions, repolarize their plasma membrane, and rebuild their intermediate filament cytoskeleton. They can also resume whatever specialized functions they may have (e.g., pumping ions from one side of the plasma membrane to another, synthesizing and releasing hormones).

Fragmentation of Nonnuclear Organelles Ensures Their Equal Distribution in the Daughter Cells

During prophase, the entire endomembrane system dissolves: the endoplasmic reticulum, Golgi complex, endosomes, and so on, all fragment into tiny vesicles; only mitochondria and chloroplasts remain intact. During the remainder of mitosis, these vesicles randomly attach to the microtubules forming the spindle, awaiting the conclusion of mitosis. Once cytokinesis is underway, the vesicles reform, thereby reconstituting the endomembrane system. There are usually enough copies of mitochondria and chloroplasts to ensure that each daughter cell contains at least one, and they are capable of dividing to generate more if necessary.

CONCEPT CHECK #3

An often-overlooked property of mitosis is that most normal cell functions come to a halt during this period in the cell cycle. Complete the table below by indicating whether the listed proteins are likely to be active/functional during mitosis or not.

Protein(s)	Active during mitosis	Inactive during mitosis
Replication fork DNA polymerases		
Microtubule capping proteins		
Integrin complexes		
γTuRC		
Dyneins		
Hyladherins		
Glycosyltransferases		
Topoisomerase		
Cadherins		
Kinesins		
Nuclear pore complex proteins		
Proteosomes		
Primase		

7.5 Chapter Summary

Cells invest considerable energy in protecting the structural integrity of their DNA. Perhaps the most obvious form of protection is the nucleus in eukaryotic cells, which is a double-membrane-bound organelle devoted almost entirely to the replication, transcription, and storage of DNA. Even within the nucleus, subregions exist for carrying out a specific set of activities associated with DNA. Entrance to and exit from the nucleus is carefully controlled by nuclear pore complexes.

The most vulnerable time for DNA is when it is being replicated; this is because it loses the protection provided by the host of proteins that form the nuclear matrix. It is fully unwound and separated into two single strands, which serve as templates for semiconservative replication. The mechanisms for controlling DNA replication reinforce the importance of keeping DNA in a single-stranded form for as short a time as possible, even at the cost of replication efficiency. The replication fork, where DNA is actually copied, is a very complex (and somewhat convoluted) structure that nonetheless is capable of synthesizing DNA in the 5′-to-3′ direction on both antiparallel template strands at the same time. The machinery responsible for copying DNA this way likely contains some of the first proteins to appear in cells, and they are expressed in all organisms. The DNA polymerases in the replication fork possess a proofreading ability that helps keep the mutation rate in cells very low.

Mitosis is the name given to the elaborate restructuring of the cell interior that accompanies the separation of replicated chromosomes into two daughter cells. Nearly all organelles cease functioning during mitosis, and many are broken into small vesicles that coalesce to re-form organelles once mitosis is complete. The microtubule cytoskeleton plays a prominent role in shuttling the condensed chromosomes from the cytosol to the two nuclei that form at the conclusion of mitosis. The actin–myosin cytoskeleton constricts the plasma membrane at the end of mitosis, pinching the cytosol into two nearly equal halves, each of which contain a full complement of chromosomes and all of the necessary materials for building a fully functional cell.

CONCEPT CHECK ANSWERS #1

Let's start by remembering that prokaryotes and archaea were the first forms of life on earth, and they appeared billions of years before eukaryotes (and the nuclei that defined them). This means that at least some of the molecular machines could have evolved before nuclei even appeared, especially the replisome and RNA polymerase complexes, because DNA replication and transcription had to take place in these first cells. Whether these tasks were performed by the exact same complexes we now find in eukaryotes is not known, but it is highly unlikely; nonetheless, at least some form of DNA replication and RNA polymerase predated the nuclear matrix, simply because they were required for cells that had no nucleus.

The spliceosome is another matter altogether. Splicing of RNA molecules takes place only in eukaryotic cells, so we know this machinery must have evolved after nuclei appeared.

These arguments lead to the conclusion that primitive forms of at least some of the molecular machines in nuclei evolved before the nuclear matrix. Does this mean that the nuclear matrix arose *de novo* in eukaryotes? Not likely: remember, the proteins that compose the matrix bind to DNA. Chapter 5 ended with a discussion of the evolutionary precursors to the eukaryotic cytoskeleton; at least one type of nuclear matrix protein, the lamins, is thought to have evolved from DNA-binding proteins in prokaryotes. These observations suggest that the ancestors of the modern nuclear matrix genes were functioning in prokaryotes before nuclei appeared, and simply adapted once they were enclosed within the confines of an additional membrane (i.e., during endosymbiosis).

In conclusion, it appears that the nuclear matrix is an evolutionary adaptation that appeared in nuclei after some form of DNA replication and transcription machinery had already formed in prokaryotes. The spliceosome, however, must have evolved after the other structures, because it is unique to eukaryotes.

CONCEPT CHECK ANSWERS #2

Perhaps the most informative finding is that all alien DNA polymerases are the same. This implies that the alien DNA polymerase is very versatile, because earthly organisms express different specialized types of DNA polymerases. Some of the functions for DNA polymerases include leading and lagging strand synthesis in the replication fork, polymerization of the complementary strand formed by telomerase, and repair of damaged DNA; also, the absence of Okazaki fragments suggests that the alien polymerase possesses the ability to synthesize DNA in both the 5′ to 3′ and 3′-to-5′ direction. The fact that alien chromosomes never shorten also suggests either that all alien cells express telomerase, or that the polymerase does not require an RNA primer. The fact that the alien DNA has more C–G base pairs is of no consequence in this case.

CONCEPT CHECK ANSWERS #3

Protein(s)	Active during mitosis	Inactive during mitosis
Replication fork DNA polymerases		X
Microtubule capping proteins	X	
Integrin complexes		X
γTuRC	X	
Dyneins	X	
Hyladherins		X
Glycosyltransferases		X
Topoisomerase		X
Cadherins		X
Kinesins	X	
Nuclear pore complex proteins		X
Proteosomes	X	
Primase		X

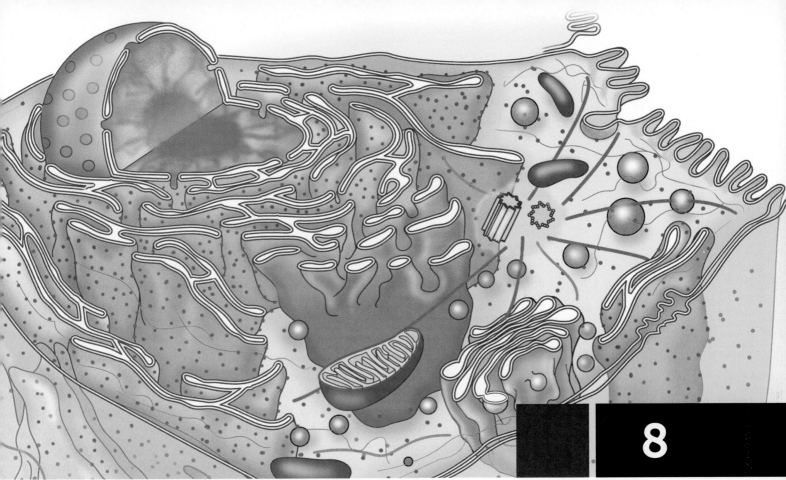

Protein Synthesis and Sorting

8.1 The Big Picture

In Chapter 7 we discussed the mechanisms cells use to replicate and segregate their DNA. The goal of this chapter and the next is to finish the "web strand" describing how DNA information is used. To do so, we will return briefly to the concept of DNA information transfer first discussed in Chapter 1, where we will now have a look at how this transfer occurs. The number of chemical reactions responsible for converting DNA information to cellular activity is tremendous, and defies memorization, so we will look for concepts that link these steps into logical units. To help us build these concepts, we will introduce cell biology principle #4: DNA encodes the function of RNAs and proteins.

Another way of expressing this principle is: polymers of DNA beget polymers of RNA, which beget polymers of amino acids (polypeptides). Historically, this is known as the **central dogma of molecular biology**, and it applies to every organism on earth (some also have the ability to convert RNA into DNA, and thus move information "backward," according to the dogma). Our focus, therefore, is on how these polymers are made. Fortunately, there is a temporal sequence for these events, so they naturally organize into a story-like format.

Some of the first modern biochemists and molecular biologists devoted their entire careers to developing and understanding this principle. Decades of work in this area by thousands of researchers have yielded thousands upon thousands of facts, and it is often tempting to line up these facts in the proper sequence, like a movie, in the hope that simply watching them play out from beginning to end will reveal the important themes that tie them together. In my experience, this does not work very well.

Instead, when I encounter a complex new subject, I apply a two-step strategy for navigating it. The first step is to consider the potential challenges or problems that this complex system must face. In this chapter, we'll focus on the issues cells confront as they convert cellular information from one physical form to another. In Chapter 7, we discovered that cells commit tremendous effort to ensuring that mistakes are kept to an absolute minimum when DNA is replicated. Given that DNA and RNA are also being read during the conversion process, we will propose three challenges cells may encounter: (1) Does converting DNA to RNA, or RNA to protein, subject cells to the same risk of making critical mistakes? (2) How can a cell ensure the right information is converted at the right time and place? and (3) How does a cell help ensure that the information is most efficiently passed from one form to another? The second step of the strategy is to propose possible solutions to these challenges. Note that we can do this regardless of how much background we bring to the subject: we are not focusing on accuracy. Instead, we're looking for a way to organize the facts as they accumulate. Technically, we are *generating hypotheses* to potentially help explain how the process works. These hypotheses do not need to be highly technical or sophisticated; some of the best scientific hypotheses are remarkably simple. Nor do our hypotheses need to be correct—learning that a possible explanation is incorrect can be a great help while learning a new subject. In cell biology, hypotheses are routinely disproven and replaced by others, as a matter of course.

Using the sample challenges introduced above, here are some examples of hypotheses to help explain how cells overcome them: (1) Cells may use a proofreading mechanism for ensuring accuracy during the conversion of DNA to RNA and RNA to protein. If true, we can look for parallels between this and the DNA replication proofreading we encountered in Chapter 7; (2) Just as DNA replication is performed by specialized replication factories at specific sites within the nucleus, so too might "conversion factories" be placed at the locations where the conversion products are needed most; and (3) The most efficient machines waste the least amount of energy, so we might encounter specialized molecular "machines" that convert stored energy into growing polymers with relatively little waste of energy.

This chapter picks up the thread where Chapter 7 left off: after a cell has successfully completed mitosis and cytokinesis, it's time for the daughter cells to start functioning, meaning the DNA has to be at least partially unwound, and genes expressed. The way cells accomplish this is distributed into three major sections in this chapter, each focusing on a discrete stage of the conversion process.

- The first section addresses the events that begin with the unwinding of the DNA and culminate with the production of functional RNA molecules; collectively, these events are called transcription. Here is where we begin addressing the three challenges discussed above, and where we can begin testing our accompanying hypotheses. Unlike in Chapter 7, we will emphasize important differences between prokaryotes and eukaryotes.

- The second section covers translation, the RNA-to-protein conversion events that take place right after transcription. The three challenges and hypotheses apply to this stage of conversion as well. I strongly recommend that, for students new to this subject, the first pass through this section be limited to looking for evidence of solutions to the problems transcription presents. With subsequent readings, we can start looking for evidence in support of (or rejecting) our hypotheses. The concept question that comes at the end of this section will help keep us focused.

- The third section of this chapter concerns where the products of translation, namely proteins, are sent to do their jobs. Again, cell biologists know a great deal about some parts of this topic, and the risk of getting lost in the minutia is ever-present. Applying the challenge-and-hypothesis approach can help keep things in perspective, by allowing us to focus on the *meaning* of the most useful facts, rather than on remembering as many of them as possible.

One technical note before we begin: the terms *polypeptide* and *protein* are often used interchangeably throughout this chapter, despite the fact that, technically, the words refer to different structures. Since every protein contains at least one polypeptide, even individual polypeptide subunits tend to be called proteins in this context. Our focus now is on how these structures find their primary destinations, not their polypeptide composition.

8.2 Transcription Converts the DNA Genetic Code into RNA

KEY CONCEPTS

▸ Transcription resembles DNA replication, in that DNA is separated into a "bubble" of single strands, and the single-stranded DNA serves as a template.

▸ Transcription differs from DNA replication, in that typically only one side of the transcription bubble is used as a template, and the bubble does not grow in size as transcription progresses.

▸ The steps of transcription are grouped into three stages, called initiation, elongation, and termination.

▸ Eukaryotic RNAs undergo posttranscriptional processing; mRNAs are the most studied forms of processed RNA.

▸ Following processing, RNAs are bound to several proteins and transported into the cytosol through the nuclear pore.

The term *transcription* refers to the processes whereby DNA, which is largely inactive in cells, is converted into very active RNA molecules (**BOX 8-1**). Fortunately, the mechanisms for converting a DNA sequence into an RNA molecule are quite similar to the DNA

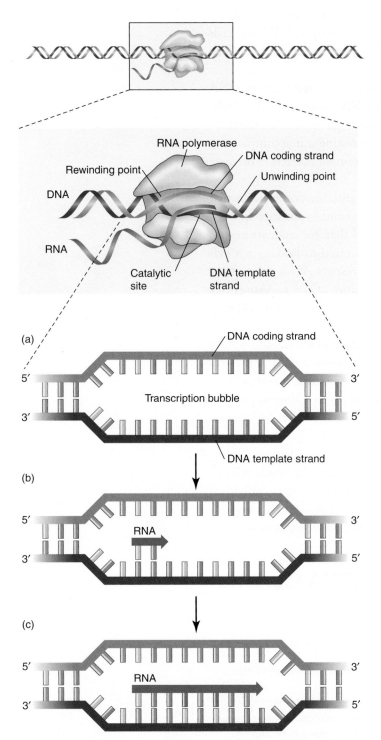

BOX 8-1 TIP

Transcription vs. translation. Students often have difficulty remembering the difference between the words transcription and translation, perhaps because they sound so similar. An easy way to remember the difference is that transcription appears before translation in a dictionary, so this means transcription happens before translation in a cell.

replication processes we covered in Chapter 7. The most important difference between transcription and DNA replication is that transcription is highly selective—only defined portions of the DNA are converted into RNA molecules.

RNA Polymerases Transcribe Genes in a "Bubble" of Single-Stranded DNA

The first clue that DNA replication and transcription are related is that transcription takes place in small regions of single-stranded DNA. A **transcription bubble** looks very similar to a DNA replication bubble, except that transcription bubbles do not expand in both directions. Rather, one of the single strands of DNA serves as a template for the synthesis of an RNA molecule, and once the complementary RNA strand is synthesized, the template is immediately rejoined to its complementary strand, as **FIGURE 8-1** shows. This means that the transcription bubble remains small—approximately 12–14 bases long—and it usually moves in one direction, following the length of the gene being transcribed. On rare occasions in prokaryotes, two genes encoded on different strands of DNA overlap, so a single stretch of DNA can actually have two replication bubbles cross it, reading different templates and moving in opposite directions.

Transcription Occurs in Three Stages

Mechanistically, transcription is much easier to follow than DNA replication. Because only a single strand of DNA in a transcription bubble is being read, it is always read in the 3′-to-5′ direction, and this permits the complementary RNA strand to be synthesized as a single strand in the 5′-to-3′ direction; there are no primers or Okazaki fragment-type RNA pieces created. What we called the "leading strand" during DNA replication serves as the template for generating RNA molecules. Transcription is typically divided into three stages, as illustrated in **FIGURE 8-2**. Before we go through them, let's introduce the players.

FIGURE 8-1 An overview of the transcription bubble. **(a)** DNA strands separate to form a transcription bubble. RNA is synthesized by complementary base pairing with one of the DNA strands. **(b)** Transcription takes place in a bubble, in which RNA is synthesized by base pairing with one strand of DNA in the transiently unwound region. As the bubble progresses, the DNA duplex re-forms behind it, displacing the RNA in the form of a single polynucleotide chain. **(c)** During transcription, the bubble is maintained within bacterial RNA polymerase, which unwinds and rewinds DNA and synthesizes RNA.

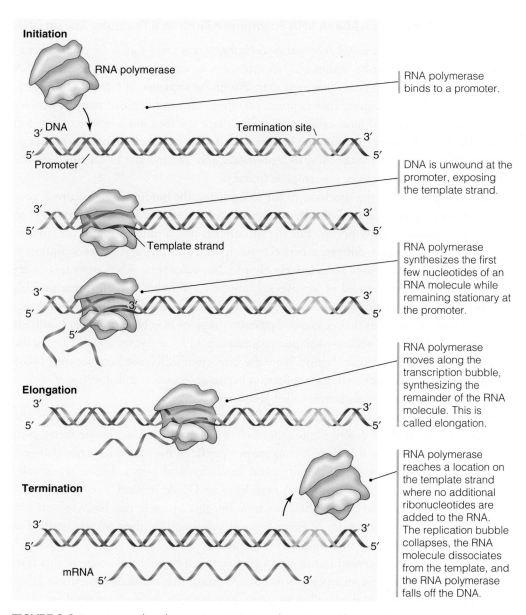

Initiation

RNA polymerase

RNA polymerase binds to a promoter.

DNA

3′
5′

Promoter

Termination site

3′
5′

DNA is unwound at the promoter, exposing the template strand.

3′
5′

3′
5′

Template strand

RNA polymerase synthesizes the first few nucleotides of an RNA molecule while remaining stationary at the promoter.

3′
5′

3′
5′

3′
5′

Elongation

3′
5′

3′
5′

RNA polymerase moves along the transcription bubble, synthesizing the remainder of the RNA molecule. This is called elongation.

Termination

3′
5′

3′
5′

RNA polymerase reaches a location on the template strand where no additional ribonucleotides are added to the RNA. The replication bubble collapses, the RNA molecule dissociates from the template, and the RNA polymerase falls off the DNA.

mRNA

5′

3′

FIGURE 8-2 Transcription has three stages: initiation, elongation, and termination.

▪▪ In Eukaryotes, Three Different RNA Polymerases Are Used to Transcribe Different Forms of RNA

Just as cells possess several different DNA polymerase enzymes that specialize in synthesizing DNA under different conditions, they also possess multiple enzymes for building RNA molecules. Eukaryotes express three different RNA polymerases, each specialized to generate a different type of RNA molecule. RNA polymerase I synthesizes a single large ribosomal RNA that is later cut into three of the rRNAs found in ribosomes. RNA polymerase II synthesizes messenger RNAs, small nuclear RNAs, and microRNAs. RNA polymerase III synthesizes transfer RNAs, one ribosomal RNA, and a few other small RNA molecules. These enzymes are often abbreviated as **pol I**, **pol II**, and **pol III**, and all three follow the same three stages during transcription. Since pol II is the polymerase that makes mRNAs (which are subsequently translated into polypeptides), it is by far the most studied of the three, and we will focus our discussion on pol II whenever possible.

■■ Transcription Begins After a RNA Polymerase Binds to a Promoter Site on DNA

Recall from Chapter 2 (see *DNA Information Is Packaged into Units Called Genes*) that genes are specific sequences that lie within a DNA molecule. In eukaryotes, most of each DNA molecule does *not* contain genes (e.g., less than 2% of the sequence of human DNA actually codes for RNA molecules). Transcription, therefore, does not begin at random sites on a chromosome; designated sites called **promoters** mark the location where transcription begins. The DNA forming the promoter is, of course, double stranded, so to begin transcription the RNA polymerase must bind to a promoter and unwind the DNA to create the transcription bubble and expose the template strand.

In prokaryotic cells, the mechanism for recognizing the binding site is relatively simple: the RNA polymerases bind directly to the promoter sequence of DNA, very close to where transcription actually begins. But the situation is much more complex for eukaryotic cells. One important difference between eukaryotic and prokaryotic transcription is that prokaryotic polymerases are relatively simple, but eukaryotic RNA polymerases are enormous molecules composed of, on average, about 12 different subunit proteins. This permits each subunit to specialize in a subset of the tasks necessary for transcription. A second important difference is that eukaryotic promoter regions may be spread over hundreds of base pairs in the DNA, while prokaryotic promoters tend to be shorter and lie close to the site where transcription actually begins. Note the important difference here between DNA replication and transcription: whereas replication begins at sequences called *origins of replication,* transcription begins at sequences called *promoters.*

Proteins that control transcription are called **transcription factors**. Historically, the name *factor* was used to describe molecules before much was known about them; even though we are now certain that these factors are polypeptides, the names have not changed. Transcription factors share very little structural similarity, and play a diverse set of roles during transcription. Many of them don't even bind to DNA; instead, they bind to the DNA-binding proteins and control their function. Imagine a protein that binds to a protein that binds to DNA—if this team of three controls transcription, they are all classified as transcription factors whether they bind DNA or not.

Describing all of the known transcription factors is well beyond this scope of this text, so instead, we will focus on an especially important group that serves as subunits of RNA polymerases. In eukaryotes, the smallest group of these proteins necessary to yield any transcriptional activity at all are called **basal transcription factors**, because they literally form the base upon which the regulatory structures are built. Collectively, they are called the **basal transcription complex**. For pol II, six of these factors, named transcription factors IIA, IIB, IID, IIE, IIF, and IIH (commonly abbreviated as TFIIA, TFIIB, TFIID, etc.), form the basal transcription complex (**BOX 8-2**). Some of these factors are, in turn, composed of several polypeptides, so the actual polypeptides in the basal transcription complex numbers nearly 30.

Stage 1 of transcription, also called **initiation**, begins when the basal transcription factors bind to the promoter nearest the site where transcription begins, as shown in Figure 8-2. The name of the site is the **transcription startpoint**, and it is often abbreviated

---

BOX 8-2 TIP

Here is another example of the apparent confusion between using the words *protein* and *subunit* that we discussed in Chapter 3. The naming scheme is not consistent with the convention used for most other proteins, but the names of these factors are fixed, so we simply have to get used to another exception to the rule.

in diagrams as an arrow pointing from left to right above a horizontal line representing the DNA strand, coupled to a vertical line that points to the location on the DNA where the startpoint is located. Note that the first few nucleotides that are transcribed into mRNAs, despite their proximity to the startpoint on the DNA, do not define the point where *translation* begins. This sequence simply marks where the first bases in the RNA are synthesized. By convention, the nucleotide where transcription begins is numbered "0" and the DNA nucleotides lying on the 5′ side of this point are called *upstream* and assigned negative numbers (e.g., −10 is 10 nucleotides 5′ of the transcription startpoint); likewise, those lying on the 3′ side are assigned positive numbers and called *downstream*.

The promoter is subdivided into distinct sequences, each with its own name. The portion of the promoter nearest the transcription startpoint is called the **core promoter**, because it is absolutely essential for transcription. (By definition, the core promoter is recognized as the entire DNA sequence upstream of the transcription startpoint necessary for stages 1–3 to take place.) Thus, *the basal transcription factors assemble on the core promoter,* as shown in **FIGURE 8-3**. The first of the factors to bind is TFIID, which recognizes nucleotide sequence 5′-TATAAA-3′ and is therefore called the **TATA box**. Three additional factors then bind, creating a "pocket" for pol II. After pol II binds, TFIIE and TFIIH attach to "cap" the polymerase and secure it in its proper location.

These last two factors play three important roles in regulating transcription:

- First, TFIIH binds specifically to the template strand, thereby ensuring that pol II reads the proper strand during transcription.
- Second, TFIIE binding partially unwinds the DNA double helix; TFIIH contains a helicase enzyme that uses ATP to complete the DNA unwinding necessary to expose the template strand as the transcription bubble progresses. The unwinding of the DNA is another property of initiation.
- Third, TFIIH contains a kinase enzyme (see Box 3-4 in Chapter 3) that adds phosphates to one of pol II's subunits, thereby inducing a shape change in pol II that simultaneously dissociates it from the basal transcription factors (including TFIIH) and activates it, as **FIGURE 8-4** shows.

Starting transcription is not as easy as it might seem at first. Even though the pol II enzyme is locked in place and activated, there is still a good chance that it will make mistakes. RNA polymerases do not have the proofreading abilities that DNA polymerases have, so rather than repair a mismatched nucleotide, they simply make several short pieces of RNA until they get it right, meaning until the newly formed RNA forms all of the proper complementary hydrogen bonds with the DNA strand being transcribed. The mismatched short strands are discarded, and once they make the proper short strand, the polymerases can move forward.

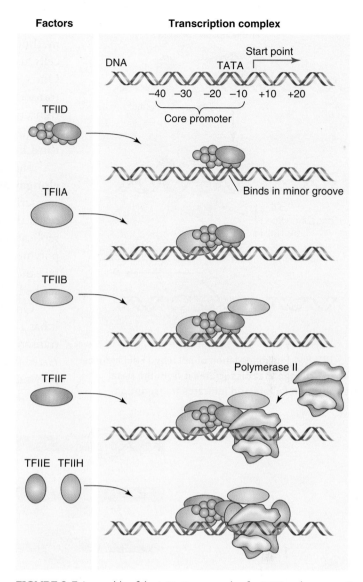

FIGURE 8-3 Assembly of the initiation complex for DNA polymerase II.

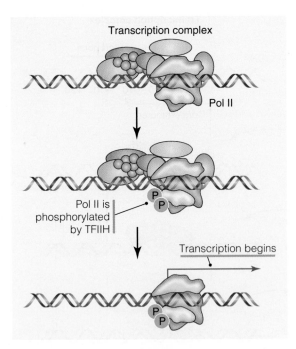

Transcription complex

Pol II

Pol II is phosphorylated by TFIIH

P P

Transcription begins

P P

FIGURE 8-4 Phosphorylation of pol II by TFIIH induces a shape change that dissociates it from the basal transcription factors and inititates transcription.

Before we move on, let's revisit the challenge-and-hypothesis method for understanding this material. Our first hypothesis, that cells use a proofreading mechanism during transcription, is false. Our second hypothesis, that cells carefully regulate when and where transcription occurs, is true: the promoter sequences control where transcription occurs, and the transcription complex controls when it begins. We have yet to encounter any discussion of efficiency (our third hypothesis), so we'll visit that issue later.

The act of moving the transcription complex forward, also called **elongation**, is stage 2 of transcription. During elongation, the RNA transcript is extended in the 5'-to-3' direction as the RNA polymerase reads the template DNA strand in the 3'-to-5' direction. The entire replication bubble is contained within the space occupied by the RNA polymerase, as **FIGURE 8-5** shows. Note that this strategy helps protect the single-stranded DNA from being damaged as it is being transcribed, illustrating principle #3 introduced in Chapter 7.

One important point to remember about the elongation stage is that it has a major impact on the structure of the entire DNA molecule being transcribed. Recall from Chapter 2 (see *DNA Forms a Double-Stranded Helix*) that double-stranded DNA naturally forms a coil (helix). As RNA polymerase moves along a DNA double helix, it unwinds the coil to form the transcription bubble, but this unwinding has an additional

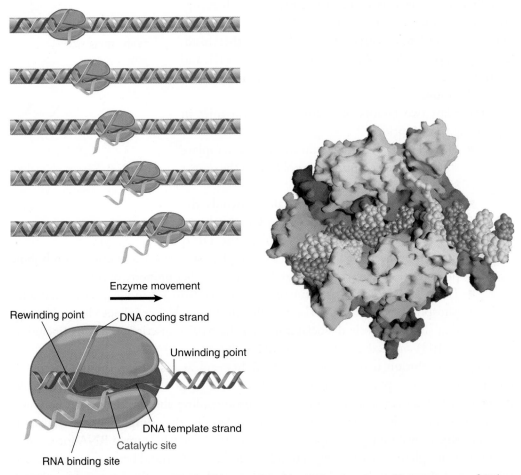

Enzyme movement

Rewinding point — DNA coding strand

Unwinding point

DNA template strand

Catalytic site

RNA binding site

FIGURE 8-5 The entire transcription bubble is enclosed by RNA polymerase. Photo courtesy of Seth Darst, Rockefeller University.

The rope analogy. An easy way to visualize why transcription can cause problems in DNA structure is to imagine holding onto a piece of rope at the end, by grasping two of the strands that form the rope. Note that the strands in the intact rope are twisted together to begin with. If you try to unwind the rope by pulling the two strands apart, like what happens to DNA during transcription, this will work for a little while, but it will also induce *extra* twisting of the existing rope as you pull; it won't take much pulling before you are left with two (or more) loose strands that form a nasty super-coiled knot at the point where the rope has uncoiled. You will also have two loose strands with no coiling whatsoever. Without additional help, transcription would eventually form these kinds of massive tangles in some regions of DNA, plus create floppy uncoiled regions next to them. That would, of course, be problematic because it would be nearly impossible to return such a mess to its original shape. There are only two ways to avoid this outcome: (1) have the entire DNA molecule spin on its axis as the strands are pulled apart, then have it spin back when transcription is done, or (2) cut out the section of DNA you want to unwind so it can be pulled apart without disrupting the remaining DNA, then retwist and reattach it when transcription is done. However, the first solution will not work for cells, because most DNA molecules are extremely long relative to the sequence being transcribed, so it's essentially impossible to twist the entire molecule every time a gene gets transcribed. Instead, cells use a variation of the second solution, by briefly cutting one or two of the strands and allowing the ends to twist as necessary to maintain the correct degree and orientation of coiling, then immediately resealing them.

effect: it induces even more coiling (called **positive supercoiling**) in the DNA just ahead of the replication bubble (**BOX 8-3**). In addition, after the replication bubble passes, the recently unwound DNA forms a double helix, but it has lost whatever positive coiling it had prior to the arrival of the bubble, and is now actually twisted in the opposite direction (also called **negative supercoiling**). Fortunately, cells can solve these problems with two simple enzymes: **gyrase** reverses the extra positive supercoiling ahead of the transcription bubble by inducing negative supercoiling, and **topoisomerase** relaxes the negative supercoiling behind the transcription bubble. Both enzymes work essentially the same way: they cut one of the two strands of a DNA helix, then twist it around the uncut strand in a positive (topoisomerase) or negative (gyrase) direction as necessary; some topoisomerases cut both strands of DNA then reseal them after the strands have unwound. A cartoon visualization of how a topoisomerase that cuts a single stand of DNA might work is shown in **FIGURE 8-6**.

Stage 3 in transcription is called termination because, at this point, the RNA polymerase stops synthesizing RNA. The result is the release of the RNA transcript and RNA polymerase from the DNA template, and reformation of double-stranded DNA. Prokaryotes and eukaryotes use many different mechanisms for inducing termination, and most of what is known about transcription termination comes from studies of bacteria, especially

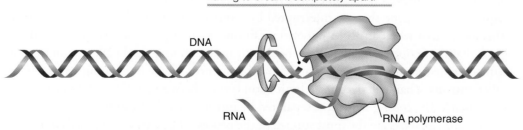

By breaking one strand and allowing it to spin around the other, the enzyme adds coiling to the molecule without having to break it completely apart.

DNA

RNA

RNA polymerase

FIGURE 8-6 Topoisomerase induces positive supercoiling of DNA.

the model organism *E. coli.* To avoid getting bogged down in species-specific details, we will highlight three features of termination that are especially important:

- First, in prokaryotes, termination is encoded by specific DNA sequences called **terminators**. These sequences lie within the transcribed region of a gene, suggesting that termination actually requires that these sequences to be transcribed. In many cases, the terminator portion of a gene encodes a stretch of RNA that can fold into a loop and actually form a semistable double-stranded RNA called a **hairpin loop**. Eukaryotic genes do not have terminator sequences.

- Second, some termination requires additional proteins to bind to RNA polymerase. These proteins may detect sequences in the transcribed gene and induce the polymerase to change shape and stop transcription. The best known of these proteins is the **rho protein** in bacteria.

- Third, terminators are not universally effective. In some cases, antiterminator proteins can bind to the terminator and suppress whatever mechanism is used to stop transcription; this results in a phenomenon called readthrough, and can cause more than one gene to be transcribed onto a single RNA molecule. A technical term for such an elogated RNA is **polycistronic RNA**, and these types of RNAs are especially important when several adjacent genes function together. A popular example is the **operons** in bacteria, which are clusters of genes that encode proteins involved in metabolic pathways; by controlling transcription of several genes devoted to a common purpose (e.g., using glucose or lactose as a preferred food source), the amount of regulation necessary to increase or decrease a metabolic pathway is reduced.

In eukaryotes, transcription typically continues well past the end of the gene coding sequence, and is only stopped after the RNA molecule is literally cut away from the RNA polymerase. After the transcript is cut, whatever RNA remains attached to the polymerase is degraded, and this eventually causes the polymerase to fall off the DNA. Here is a test of our third hypothesis: prokaryotes have a fairly sophisticated means of determining when and where transcription ends, but eukaryotes do not—compared to prokaryotes, their termination transcription is rather sloppy. So far, our hypothesis appears to be correct for prokaryotes and false for eukaryotes, with respect to transcription.

In Eukaryotes, Messenger RNAs Undergo Processing Prior to Leaving the Nucleus

One of the most fascinating differences between prokaryotic and eukaryotic transcription is that eukaryotes routinely modify the structure of their mRNA molecules before they are ever put to use. The general term for this activity is **RNA processing**, and it occurs only in eukaryotes. The easiest way to convert a DNA sequence into a polypeptide is to have a continuous coding sequence in a gene converted into a continuous sequence of mRNA, which is then translated into a continuous sequence of amino acids by a ribosome. This is what all prokaryotic organisms do. But eukaryotes take the hard route: the DNA coding sequences for mRNAs (and some tRNAs and rRNAs) are organized into several separate sequences in the gene that are interrupted by sequences that should never be translated. This means that when an RNA polymerase finishes transcription of a eukaryotic mRNA gene, that transcript is largely useless, because it is (1) too long, and (2) contaminated by sequences that do not encode any amino acid sequences. These noncoding sequences are called **introns**, a name derived from both Latin (*intra-,* between) and Greek (*-on,* an entity unto itself). The coding sequences separated by introns are called **exons**.

Because mRNAs are the most studied of the processed RNAs, we will focus our attention on them. The name given to the RNA that results from transcription of a eukaryotic gene encoding an mRNA is, unfortunately, not "mRNA"; instead, it is called a **primary transcript**, because it is the first product of transcription. To earn the name mRNA, three modifications, shown in **FIGURE 8-7**, need to be made to this molecule. Each modification has its own name, and these should be committed to memory because biologists from multiple disciplines use them routinely.

◼◼ The Spliceosome Controls RNA Splicing

Splicing is the term used to describe the complex mechanisms cells use to remove introns from the primary transcript and stitch the exons together into a continuous, coding RNA. Multiple mechanisms exist for splicing primary transcripts, but the most well understood is that performed by an enormous protein/RNA complex called the spliceosome. This structure recognizes signature (partially conserved) nucleotide sequences at the boundaries of introns and exons on mRNA, and facilitates the breaking and reforming of the sugar-phosphate backbone at these boundaries, thereby deleting the intron between two exons and linking the exons directly to one another, as **Figure 8-7** shows. Chemists have a technical term for this process that is also commonly used by cell biologists: **transesterification**. Remember that the sugar-phosphate backbone is held together by ester bonds between phosphate groups and the hydroxyl groups on the sugars; the word transesterification simply refers to the fact that an *ester* bond is broken in one place (intron–exon boundary) and *transferred* to another location (exon–exon boundary). The formal name for this newly formed bond is the **exon junctional complex**, or **EJC**.

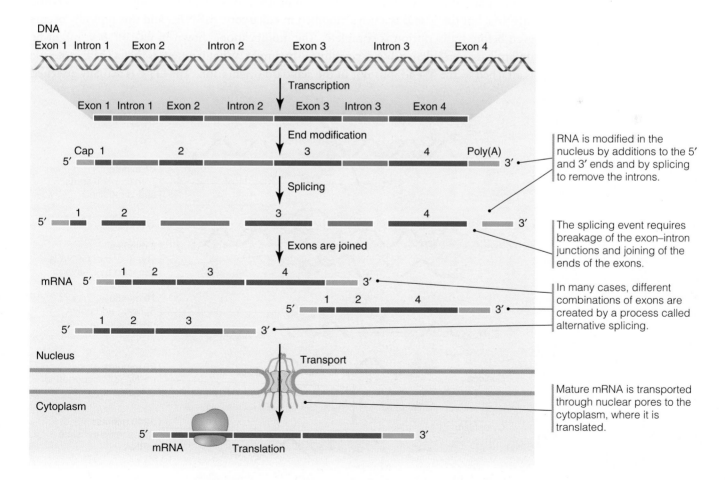

FIGURE 8-7 Eukaryotic mRNA is modified, processed, and transported.

Many eukaryotic genes contain more exons than are absolutely necessary to construct a functional mRNA. In these cases, the spliceosome has an extra job: it must not only find the intron–exon boundaries, it must also decide which of these to break and which to leave intact. By choosing to cut some and not others, the spliceosome can remove both introns and exons from the primary transcript, and thereby reduce a large set of exons into a smaller set that forms a "customized" transcript. By mixing and matching exons this way, a cell can generate several different mRNA sequences (and hence polypeptides) from a single gene, as Figure 8-7 shows. This can be useful, for example, when different cells require slightly different myosin proteins to properly contract and/or move—*Drosophila melanogaster* cells can generate at least 15 splice variants from a single myosin gene. This activity is called **alternative splicing**, and most eukaryotic genes encoding polypeptides can be alternatively spliced—over 90% of human genes generate at least two different mRNAs. While the exact mechanism of most alternative splicing is not well understood, it is clear that the decision to cleave a specific intron–exon boundary is affected by proteins that promote or delay the formation of the spliceosome at that boundary.

■■ The 5′ and 3′ Ends of Messenger RNAs Are Modified Prior to Export

Independent of the splicing taking place at the interior of a primary transcript, both the 5′ and 3′ ends undergo extensive modification prior to the export of the completed mRNA into the cytosol. This is illustrated in **FIGURE 8-8**.

The 5′ end is the first to be modified. Remember from our examination of nucleic acid structures in Chapter 2 that the 5′ end of a DNA or RNA strand always has a triphosphate group attached to it (see Figure 2-11). Technically, this is true for a short time in eukaryotic mRNAs, but the 5′ end is always modified in eukaryotic mRNAs, and this typically occurs even before transcription is complete. The modification, shown in **FIGURE 8-9**, might be unfamiliar at first glance, but is actually fairly simple. What happens is that a GTP molecule is attached to the 5′ end, but not in the standard orientation. The GTP is actually attached

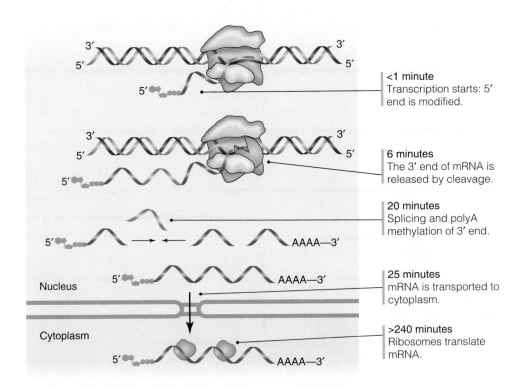

<1 minute
Transcription starts: 5′ end is modified.

6 minutes
The 3′ end of mRNA is released by cleavage.

20 minutes
Splicing and polyA methylation of 3′ end.

25 minutes
mRNA is transported to cytoplasm.

>240 minutes
Ribosomes translate mRNA.

FIGURE 8-8 5′ modification occurs before splicing and 3′ modification in the nucleus.

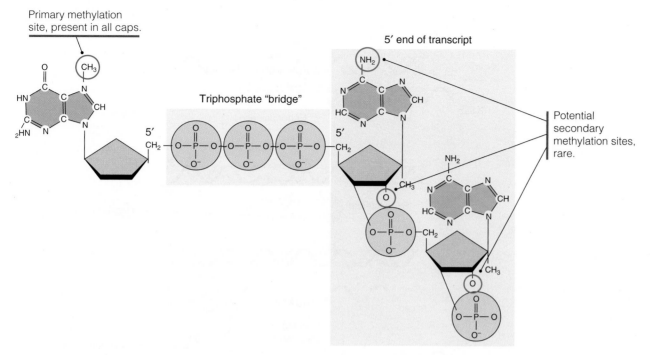

FIGURE 8-9 Eukaryotic mRNA has a methylated 5′ cap. The cap protects the 5′ end of mRNA from nucleases and may be methylated at several positions.

by its 5′ end, not its 3′ end; this means that two 5′ ends are linked directly to one another. Because both 5′ ends have three phosphates attached to them, there is the potential to have the two 5′ ends bridged by *six* phosphates, but that doesn't happen. Instead, the newly added GTP loses two of its phosphates, and the 5′ nucleotide on the mRNA loses one of its phosphates, resulting in a triphosphate bridge between the two sugars. It might help to circle the bases, sugars, and phosphates in Figure 8-9 to see how they are arranged.

Simply adding this GTP "in reverse" is not the end—once GTP is attached, it is then methylated (see *Nucleosome Structure Can Be Modified by Cells* in Chapter 2) on the 7 position of the guanine base, and this methylation is called a cap. All eukaryotic mRNAs have this 5′ methylguanosine cap, as it is formally called. Some mRNAs have even more methyl groups attached to nucleotides near the 5′ end, but these are fairly rare. The function of this cap is to disguise the 5′ end of the mRNA from enzymes that preferentially degrade "naked"(uncapped) 5′ ends. These enzymes are especially good at tearing up the 5′ end of virus mRNAs, and are therefore a cellular self-defense mechanism.

The 3′ end is modified in a completely different way, as illustrated in **FIGURE 8-10**. Recall from earlier in this chapter that eukaryotic primary transcripts are literally cut away from the RNA polymerase; this means that the exact location where the cut takes place can vary considerably between different transcripts. In most cases, however, this cleavage occurs far enough downstream of the coding sequence that there is a long stretch of noncoding RNA at the 3′ end. This 3′ end is subsequently cleaved in the 3′-to-5′ direction by an exonuclease contained in another enormous protein complex called the **polyadenyation complex**. In most eukaryotes, this digestion occurs until the nuclease reaches a specific sequence in the noncoding RNA, called a polyadenylation sequence (5′AAUAAA3′). Upon reaching this region, a different enzyme, called **poly(A) polymerase**, adds approximately 200 adenosines in a row to the polyadenylation sequence, giving rise to what is called a **poly(A) tail**. Think of the poly(A) tail as being similar to the tail one often sees attached to a kite: it isn't responsible for the overall behavior of the structure (i.e., the tail doesn't fly), but without it, the structure does not function properly—mRNAs without poly(A) tails do not function properly.

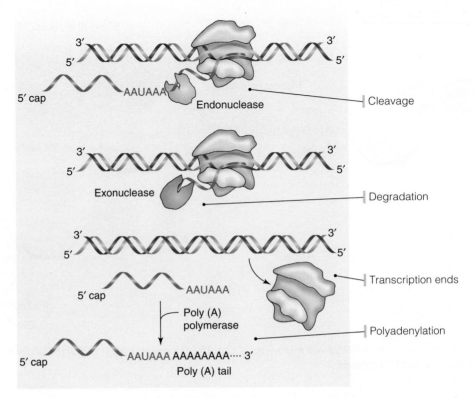

FIGURE 8-10 Exposure of the polyadenylation sequence by endonuclease and exonuclease cleavage triggers addition of the poly(A) tail by poly(A) polymerase.

Let's revisit our challenge-and-hypothesis strategy again. The information in this section demonstrates that transcription in prokaryotes and eukaryotes differs in some significant ways that impact our hypotheses. For example, prokaryotes and eukaryotes both address the issue of efficiency quite differently: whereas prokaryotes efficiently encode polypeptides with a single, continuous sequence of nucleotides, eukaryotes introduce great inefficiency by including noncoding sequences that must be removed before a functional RNA is produced. Thus, our third hypothesis is correct for prokaryotes, which use a single transcription complex to convert DNA sequences into RNA sequences. But our hypothesis is false for eukaryotes, which use different complexes to convert and then splice the RNA transcript. Including such a complex system also increases the likelihood that errors will occur, further disproving our first hypothesis.

■■ RNA Export Is Unidirectional and Mediated by Nuclear Transport Proteins

One of the most important functions of the poly(A) tail is to bind to a protein, appropriately called **poly(A)-binding protein** (PABP). Several copies of PABP can be bound to one mRNA, and the resulting complex actually helps cover the 5′ end as well, as **FIGURE 8-11** shows. Eukaryotic cells typically contain thousands of different mRNAs at any given time, and they all form this kind of structure. Because it contains so many different mRNAs, in addition to dozens of additional proteins, it is known by some scientists as a **heterogeneous nuclear ribonucleoprotein particle** (**hnRNP**). Others call it a **messenger ribonucleoprotein particle** (**mRNP**) once all the mRNA processing is completed. As accurate as these names are, they're a bit difficult to pronounce (at least 13 syllables), and even the acronyms don't help much, so, I usually just call it a "big ball of protein and mRNA," and abbreviate it as "the ball."

This ball has to form in order for the mRNA to be exported through the nuclear pore. Because some of the components of the ball include splicing proteins, splicing is an essential activity for export from the nucleus to take place. **FIGURE 8-12** shows many of the proteins that form the ball, arrayed on a stretched out mRNA for easier viewing. Notice that some of them fall out of the complex once it reaches the nuclear pore, while most pass into the cytosol along with the mRNA. One of the most important of these temporary proteins is **Tap**, which binds to both the nuclear pore fibers and to another protein in the ball. This is likely how the ball is targeted to nuclear pores. Once the ball reaches the pore, it is transported into the cytosol, where the proteins dissociate and are returned to the nucleus to form a new ball, and the mRNA is translated by a ribosome.

Let's finish this section with a question: Why do eukaryotes use such an error-prone, inefficient means of converting DNA information into RNA? The answer is surprisingly straightforward: this method allows eukaryotes to generate far more variations of RNAs from their genomes than prokaryotes can. The great increase in phenotypic variation that results from this outweighs the cost of inefficiency, because it better equips a population of organisms with the potential

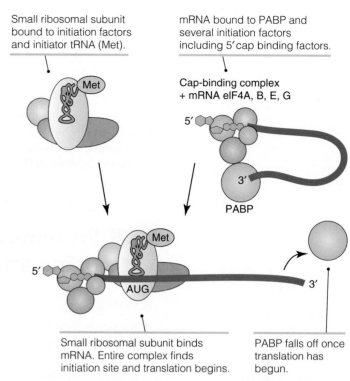

Small ribosomal subunit bound to initiation factors and initiator tRNA (Met).

mRNA bound to PABP and several initiation factors including 5′ cap binding factors.

Cap-binding complex + mRNA eIF4A, B, E, G

Small ribosomal subunit binds mRNA. Entire complex finds initiation site and translation begins.

PABP falls off once translation has begun.

FIGURE 8-11 Initiation of translation in eukaryotes. PABP binds to the poly(A) tail of mRNA and forms a complex with other proteins. This complex binds to another associated with the small ribosomal subunit, and together they scan for the initiation site on the mRNA.

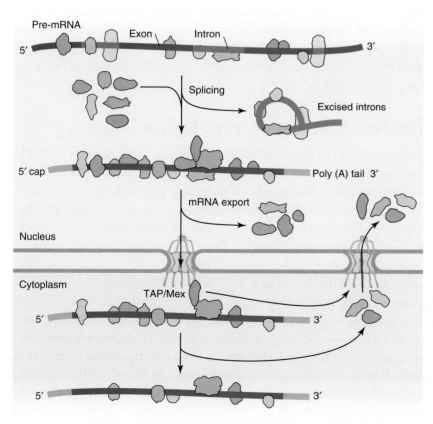

FIGURE 8-12 Some portions of the hnRNP are recycled into the nucleus after passing through the nuclear pore. Others remain associated with the mRNA and impact translation.

of finding a way to successfully compete during natural selection. In fact, some scientists believe this is an important cause of speciation (the appearance of new species) in eukaryotes, though the issue is still hotly debated among evolutionary biologists.

CONCEPT CHECK #1

Why is it important for RNA processing to take place in the nucleus, rather than in the cytosol? Processing in the nucleus requires that the RNA processing proteins be synthesized in the cytosol and imported into the nucleus—wouldn't it be more efficient to just leave these proteins in the cytosol and bring the RNAs to them? Use cell biology principle #3 to help answer this question.

8.3 Proteins Are Synthesized by Ribosomes Using an mRNA Template

KEY CONCEPTS

▸ Translation is the term used to describe the conversion of mRNA information into polypeptides.
▸ Translation requires cooperation between ribosomal RNAs, transfer RNAs, messenger RNAs, and numerous proteins.
▸ Translation is performed by one of the largest molecular complexes in cells: the ribosome.
▸ The steps of translation are grouped into three stages: initiation, elongation, and termination. These are very different from the identically named stages of transcription.

Having reached the cytosol, an mRNA now has the opportunity to perform its function. Notice that the DNA that encodes mRNAs includes the necessary information for ensuring it binds to the right proteins in the nucleus and transporting it into the cytosol, consistent with principle #4. mRNAs have only one job: they instruct ribosomes about how to build polypeptides. The conversion of mRNA information into a polypeptide sequence is called translation.

Translation Occurs in Three Stages

With the exception of the few proteins synthesized in mitochondria and chloroplasts, all polypeptide synthesis in eukaryotic cells begins in the cytosol; the focus here will be on cytoplasmic translation. The number of individual chemical reactions, protein–protein binding events, and protein–RNA interactions necessary for translation is far too large for us to discuss them all here. Fortunately, these events are organized by convention into three stages, with the exact same names as those used for transcription, and we will highlight the essentials necessary to illustrate principle #4.

Before we describe the three stages, let's invoke the challenge-and-hypothesis method for navigating the complexity we'll encounter. For those new to this subject, keeping track of the number of molecular players involved and their multisyllabic names is not as important as understanding the overall strategy and why it works. So, we can ask ourselves: (1) Does proofreading take place during translation?; (2) Does translation take place at defined locations in the cell, and is translation performed at specific times?; and (3) Does translation include any method for maximizing thermodynamic efficiency? If we keep our eyes on these questions, a lot of the details fade a bit. Once the answers to these questions become clear, we can take another pass through this section to pick up the details necessary to understand the molecular mechanisms.

First, let's identify the key players involved. We already know the structure of mRNA and understand how it is delivered to the cytosol. To complete translation, three more structures are required.

The largest structure involved is the ribosome. Recall that ribosomes are composed of two subunits that in eukaryotes are synthesized in the nucleus, specifically in nucleoli (further evidence supporting our second hypothesis). The size and exact composition of ribosome subunits varies between prokaryotes, mitochondria, chloroplasts, and eukaryotic cytosol, but they all play the same roles in translation. The two subunits are named according to the techniques used to isolate them. Both subunits are much larger than most cellular structures, as **FIGURE 8-13** shows, and this makes it relatively easy to isolate them by a technique known as **gel filtration**, or **size exclusion chromatography**. Based on the time it takes for the subunits to migrate through a viscous substance (a gel), one can estimate their size, which is expressed in units called the **Stokes' radius**, abbreviated simply as S. Prokaryotic subunits are typically 70S in size, composed of 50S and 30S subunits (the Stokes' radii are not additive, for reasons not worth discussing here). Eukaryotic ribosomes are 80S in size, containing 60S and 40S subunits. Both prokaryotic and eukaryotic ribosomes contain several ribosomal RNA molecules and a multitude (50–80) of proteins. The "S" nomenclature has become the standard for describing ribosomes and their constituents, so it is important to understand what these numbers mean.

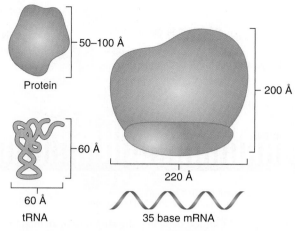

FIGURE 8-13 The relative sizes of components of the cellular translation machinery.

The smallest major players in translation are the transfer RNAs, or tRNAs, as Figure 8-13 shows. The function of tRNA is to deliver amino acids to the ribosomes. Why can't the amino acids enter the ribosomes by themselves? There are two important reasons. First, by coupling the amino acid to a tRNA molecule, the ribosome can easily align the ribosome with a specific region of the mRNA. This occurs by base pairing of the three-letter **codon** sequences in the mRNA with their corresponding three-letter anticodon sequence in the tRNAs. By aligning the codons and anticodons, the ribosome ensures that the code in the mRNA is reflected by the sequence of amino acids in the polypeptide. The second reason amino acids must be coupled to tRNAs during translation is that the amount of energy required to form a new peptide bond between the existing polypeptide bond (or the first amino acid, when translation begins) and the newly arrived amino acid is too high for the ribosome to generate it itself. Building a peptide bond takes energy, and this energy comes from the bond that holds the tRNA and amino acid together. For this reason, tRNAs that are coupled to amino acids are said to be "charged," like a battery, and are assigned the name **aminoacyl tRNAs**. The enzyme that charges tRNAs is called **aminoacyl tRNA synthetase**, and it uses ATP energy to couple the amino acid to the hydroxyl group (–OH) at 3′ end of the tRNA. There are as many different synthetases as there are different amino acids, so each specializes in attaching one amino acid to one specific tRNA.

The last group is called translation factors. These proteins are further divided into two classes, called **initiation factors** and **elongation factors**. The name of most of these proteins begins with "IF" or "EF," respectively, to indicate which class a factor belongs to. As is often the case, most of the information regarding translation comes from studies of prokaryotes, especially bacteria; occasionally, the eukaryotic factors contain an "e" at the beginning of the name to indicate it is found in eukaryotes. We saw above that poly(A)-binding protein binds to processed mRNA in eukaryotes, and is transported into the cytosol along with the mRNA. This protein also binds to these factors, and thus forms a critical link between RNA processing and translation.

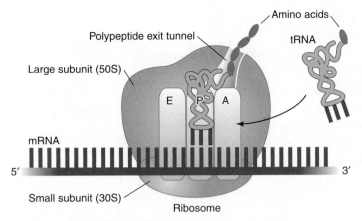

Amino acids

tRNA

Polypeptide exit tunnel

Large subunit (50S)

E P A

mRNA

5′   3′

Small subunit (30S)

Ribosome

FIGURE 8-14 Cartoon depicting the structure of an intact ribosome coupled to an mRNA. Notice that the E, P, and A site can contain tRNAs, but the tRNA in the E site has released its amino acid.

■■ The Ribosome Has Three tRNA-Binding Sites

A map of an intact ribosome attached to an mRNA is shown in **FIGURE 8-14**. Notice that the large and small subunits converge on the mRNA, with the smaller subunit primarily responsible for holding onto it during translation. The large subunit contains four important sites where the polypeptide is built from individual amino acids. The first is called the **A site**, and this is where *aminoacyl* tRNAs first attach to the ribosome. The second site is called the **P site**, where the newly arrived amino acid is removed from its charged tRNA and added to the growing amino acid chain via formation of the *peptide* bond we discussed in Chapter 3 (see *A Peptide Bond Joins Two Amino Acids Together*). The third site is called the **E site**, and this is where the spent tRNA is *ejected* from the ribosome. Finally, the fourth site is called the **polypeptide exit tunnel**, which is where the growing polypeptide is guided to the surface of the ribosome and ultimately released. We will explore what happens at the A, P, and E sites in later sections.

■■ Initiation Requires Base Pairing between mRNA and rRNA

The first stage of translation is called **initiation**. The purpose of initiation is to bring all of the elements necessary for translation together into a single giant cluster. Initiation begins when the smaller of the two ribosomal subunits attaches to an mRNA molecule. This interaction can, of course, occur between the subunit and any portion of the mRNA since the collision is random, but translation begins at a very specific point. It is therefore essential that a ribosome find the right location on the mRNA before beginning polypeptide synthesis.

The first goal of initiation is for the ribosomal subunit to find this landmark, called the **ribosomal binding site**. How bacteria and eukaryotes accomplish this differs considerably. Bacteria include a specific nucleotide sequence in their mRNA, called a **Shine-Delgarno sequence**, that forms complementary base pairs with a portion of the rRNA contained in the small subunit. This sequence lies upstream of the translation start sequence, also called the **initiation site**. This site is always the sequence 5′-AUG-3′, which in the triplet code of genes encodes a methionine amino acid. Once the proper base pairing is achieved, the mRNA and small subunit are properly aligned, the first tRNA (called the **initiator tRNA**) binds to the AUG, and the large ribosomal subunit clamps down on the small subunit, forming an intact ribosome.

The eukaryotic strategy for assembling a ribosome is far more complicated. First, the base pairing method of alignment is not used, because eukaryotic cells do not contain any sequences like the Shine-Delgarno in their genes. Instead, they capitalize on the methylguanosine cap that is added to the 5′ end of mRNA. The small subunit in eukaryotes binds to a number of initiation proteins (eIFs), and some of these recognize and bind specifically to the cap. Once the cap is identified, the small ribosomal subunit is thought to crawl toward the 3′ end of the mRNA until it reaches the initiation site. AUG is a fairly common sequence in mRNA, so the small ribosomal subunit does not necessarily stop at the first AUG it encounters; what makes the AUG at the initiation site special is that it is surrounded on either side by a specific sequence of nucleotides. As in prokaryotes, once the initiation site is found, the initiator tRNA binds to it, and the large ribosomal subunit encloses the tRNA. Eukaryotic cells also have at least a dozen different translation factors attached to the complex at this time, and many of these will be used later. The formation of this large complex marks the end of the initiation stage of translation.

At this point, we can see that both prokaryotes and eukaryotes expend considerable effort to ensure that translation begins at exactly the right location, lending support to our first two hypotheses. An error such as beginning translation even one nucleotide upstream or downsteam of the correct starting point could be catastrophic, because the resulting polypeptide would likely contain a very different sequence of amino acids, translated from an incorrect set of triplet codons. Simply put, there is no room for error during initiation.

In both prokaryotic and eukaryotic initiation, the initiator tRNA is located in the P site of the ribosome. This leaves the A and E sites empty, and these will be used in conjunction with the P site in the next stage. Let's now focus on what happens at each of these three sites when the polypeptide is growing.

■■ Elongation of a Polypeptide Occurs When an Amino Acid Is Added to the Carboxy Terminus of the Polypeptide in the A Site

Initiation describes the events responsible for correctly localizing the first codon, and thus the first amino acid, in the translation of an mRNA. The elongation stage is characterized by the addition of all of the other amino acids to the polypeptide; FIGURE 8-15 outlines the eight major steps of elongation in bacteria (because the bacterial system is more thoroughly understood, we will use it as a model for how all elongation takes place). Notice that the steps are organized as a loop—this means that all of the events illustrated in the figure happen each time an amino acid is added to a polypeptide, and every time the loop is completed the polypeptide grows longer by one amino acid. Polypeptide synthesis always takes place from the amino terminus to the carboxy terminus, so the first amino acid defines the amino end and the last one added defines the carboxy end of a polypeptide. Notice that each round requires three inputs: the charged tRNA (attached to an amino acid) coupled to a protein complex called **EF-Tu-GTP**, and a translation factor complex called **EF-G-GTP**. Next, notice the three different products for each round: a "spent" tRNA, EF-Tu-GDP, and EF-G-GDP. This tells us that the only input that is not subsequently released is the amino acid attached to the tRNA; this makes sense, because that amino acid is added to the polypeptide. Also, notice that the translation complexes entered the cycle with a GTP molecule and exited with a GDP molecule. This suggests that the cleavage of GTP to GDP provides the energy and/or protein shape changes necessary to complete the loop.

Let's work our way through the loop, beginning with the top left image, and moving clockwise. This image reflects what elongation looks like after a complete trip through the loop, so we will return to it when we have completed the entire loop to explain some of the details. The first step is addition of a new charged tRNA to the A site. Notice that the A site encloses the triplet codon immediately downstream of the initiation site. This is how the triplet code of genetics is translated into polypeptides: each group of three nucleotides downstream of the initiation site codes for a single tRNA, and thus a single amino acid, until the ribosome reaches a specific stopping point. The charged tRNA bearing the appropriate anticodon that corresponds to the triplet codon is inserted into the A site. Note too that the charged tRNA does not come alone—it is bound to a protein called elongation factor tu (or EF-Tu; Tu is an acronym for thermally unstable), which is also bound to a GTP molecule. EF-Tu is required for proper insertion of the tRNA into the A site, but the exact purpose of the GTP is not known.

Cleavage of the GTP to yield GDP that is still attached to EF-Tu is the second step in elongation. This cleavage is performed by the large ribosomal subunit, and it triggers the release of the charged tRNA from the elongation factor, which is step three. The cleavage only occurs when the tRNA anticodon forms a perfect match with the mRNA codon; if the match is not perfect (the wrong tRNA is in the site), the tRNA remains attached to

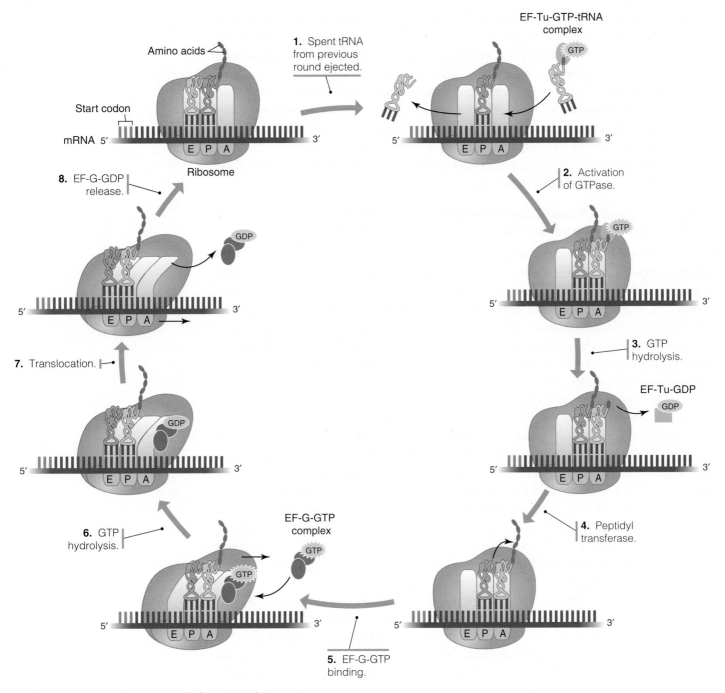

FIGURE 8-15 The elongation cycle during translation.

the EF-Tu and is eventually discarded. This is how the ribosome ensures that the correct amino acid is matched to each codon, which is clear evidence that proofreading takes place during translation, thereby supporting our first hypothesis.

Step four is defined by the formation of a peptide bond between the amino acids attached to the charged tRNAs in the A and P sites. This is not a random event: once the A site is properly occupied by a charged tRNA, the P site in the ribosome undergoes a dramatic shape change, such that it stretches its tRNA out like a rubber band and orients the carboxyl group of the P site amino acid with the amino group of the A site amino acid. This makes the formation of the peptide bond easier (such efficiency supports our third hypothesis), and ensures that the orientation of the two reactants is always the same. This is very important to understand, because it explains how the triplet code in the mRNA, which reads from 5′ to 3′, is reflected in the sequence of amino acids in a polypeptide, which is built from amino terminus to carboxy terminus. Notice too that because the carboxyl end of the amino acid in the P site attaches to its tRNA, the addition of the amino acid from the A site does not disturb this bond; the newly added amino acid in the A site has an amino acid attached to its amino group and a tRNA attached to its carboxy group. This means that there are now two amino acids attached to the tRNA in the A site. The enzyme that actually forms the peptide bond, called a **peptidyl transferase**, is a part of the large ribosomal subunit. It is important to understand these four steps in the loop before reading any further.

Once the amino acid is released from the tRNA in the A site, the ribosome has to remove it to make room for more charged tRNAs. To do so, it needs to move the spent tRNA to the E site. Two events help make this possible. First, another translation factor called EF-G, which also binds to GTP, inserts itself into the A site, effectively shoving the tRNA in the A site into a highly unstable shape that sits between the A and P sites. In response, the tRNA that previously occupied the P site is also shoved over a bit, helped by its snap back from the stretched shape it had when it was donating its amino acid to the A site. This is the fifth step, and it lasts for only a very short time. Almost immediately afterward, step six takes place, wherein the GTP attached to EF-G is cleaved, and in step seven the energy released by this reaction literally forces the small subunit of the ribosome to slide forward by one codon. Once the small subunit catches up with the big subunit, the EF-G-GDP is released, and this is the final, eighth step. When this sliding of the ribosome was first observed, scientists called it translocation.

An easy way to visualize this loop is to form a fist (the large subunit) with the right hand, and lay it on the palm of the left hand (the small subunit), which is stretched out flat. Align the hands so the thumbs are able to line up next to each other. During elongation, the large subunit moves first: slide the fist forward, out of the left palm and onto the left fingers; this is the effect of EF-G-GTP binding. Next, slide the palm forward to again touch the fist; this is the effect of GTP hydrolysis. If the hands were sandwiched around a smooth thin object, such as a hose or pipe (the mRNA), these motions could simulate the "crawl" along the mRNA.

Now let's have another look at the top left image in Figure 8-15. Notice that because of the crawling motion of the two ribosomal subunits, the two tRNAs are in different locations on the ribosome than when we started, but in fact they have not really moved at all: they are both still attached to their codons on the mRNA, and the mRNA never moved at any step in the loop—it was the ribosome that moved. Because of the translocation, they have slipped to the left by one space: the tRNA that was in the P site is now in the E site, and the tRNA in the A site is now in the P site. It is critical to understand why the tRNA in the E site has *no* amino acid attached to it, while the one in the P site has several amino acids linked together. This top left drawing assumes that the loop has completed eight cycles, so there are nine amino acids attached to the tRNA in the P site. By the time we get to the drawing after step 7, there are now 10 amino acids on that tRNA. It is this string of amino acids that fills the polypeptide exit tunnel: depending on the organism, this tunnel is about 30 to 50 amino acids long, so no polypeptide actually sprouts from the surface of the ribosome until at least 30 rounds of elongation have taken place. Finally, it is worth noting that as we begin another pass through the steps, the tRNA in the E site is ejected. This doesn't have to happen in strict conjunction with the A site entry of a new charged tRNA, but for simplicity we have included both in the same step. Recall that some polypeptides are thousands of amino acids long, so ribosomes are able to repeat this loop with tremendous precision and almost no errors.

■■ Termination Occurs When the Bond Holding the Polypeptide to tRNA Is Hydrolyzed

Eventually, the process of adding amino acids has to stop, but not because the ribosome crawls all the way to the 3′ end of an mRNA. Remember from our description of transcription that the coding sequence is only a part of the mRNA molecule. Both prokaryotes and eukaryotes have untranslated regions on either side of the coding sequence. To stop translation, cells use specific sequences that lie immediately downstream of the codon encoding the carboxy terminus of the polypeptide. These sequences use the same three letter code (UAA, UAG, UGA) that the mRNA and tRNA molecules use, and are therefore called **stop codons**. The name is a bit misleading though, because they aren't codons at all; they don't bind to any tRNA molecules. Instead, they attract proteins called **release factors**. Not surprisingly, these factors resemble the elongation factors EF-Tu and EF-G we just discussed; this means they can easily fit into the A site of a ribosome.

Release factors stop translation by a very simple, and very clever, mechanism. To fully appreciate it, let's briefly review how peptide bonds are formed. In Chapter 3 (see *A Peptide Bond Joins Two Amino Acids Together*) we characterized the formation of a peptide bond as a dehydration reaction, because one of the products is a water molecule. By the same token, cleavage of polypeptides is often called hydrolysis, because a water molecule is introduced back into the polypeptide, and this breaks a peptide bond. In short, making and breaking peptide bonds is as simple as subtracting or adding water.

Now let's have another look at Figure 8-15. Notice that when a new amino acid is added to a growing polypeptide, the reaction proceeds in the same direction: the carboxy terminus of the existing polypeptide in the P site is attached to the amino terminus of the new amino acid in the A site. Release factors, therefore, work as follows: one type closely resembles EF-Tu, so it slips into the A site and binds to the stop codon (since no tRNA recognizes a stop codon, it is not competing with tRNA for access). Essentially, it fakes out the peptidyl transferase enzyme in the P site: when the transferase cuts the polypeptide off of the tRNA in the P site, it turns and looks for another amino acid in the A site, but instead finds a release factor in its place. And here is the beauty of the mechanism: when the peptidase "hands over" the polypeptide to the A site, the release factor offers *water* as the acceptor. Not another amino acid, just a simple water molecule. Not knowing the difference, the peptidase attaches the carboxy terminus to the water molecule (thereby generating a carboxylic acid on the polypeptide and a hydroxyl ion), then releases the polypeptide. Now, because there is no tRNA to hold the polypeptide in place, it can float away, leaving the ribosome empty handed.

To finish the whole story off, another protein factor called **ribosome release factor** then occupies the A site. The ribosome of course cannot "know" that anything is amiss, so it proceeds as if elongation were still in progress. After the EF-G-GTP molecule enters and induces the large subunit to slide over (step five), ribosome release factor uses the energy from the GTP cleavage step (step six) to separate the large and small ribosome subunits. The ribosome subunits separate but are otherwise undamaged, and the tRNAs and mRNAs are likewise released unharmed; this means they can all be recycled for another round of translation, perhaps even on the same mRNA.

CONCEPT CHECK #2

What evidence can you find suggesting that eukaryotes inherited their translation machinery from prokaryotes? Which portions of the machinery do you think evolved first? What changes did eukaryotes introduce to further fine-tune this translation machinery to fit their extra needs?

CHAPTER 8   Protein Synthesis and Sorting

8.4 At Least Five Different Mechanisms Are Required for Proper Targeting of Proteins in a Eukaryotic Cell

KEY CONCEPTS

▸ Virtually all protein synthesis is centralized in the cytosol for eukaryotic cells, and many of these proteins are targeted to specific cellular locations by signal sequences.

▸ Proteins that enter and leave the nucleus are maintained in a functional shape at all times.

▸ Proteins enter the peroxisome in a functional, folded state, but this transport is unidirectional. Peroxisomal proteins appear to originate from several sources, including the cytosol.

▸ Proteins enter the endoplasmic reticulum (ER) cotranslationally, and are folded into their final shape as they enter the ER lumen. They also undergo extensive posttranslational modification.

▸ Distinct hydrophobic sequences in transmembrane polypeptides are responsible for stabilizing them in membranes.

▸ Proteins enter mitochondria and chloroplasts through very similar posttranslational mechanisms, suggesting they share a common (prokaryotic) origin. Chaperone proteins in the cytosol and interior of these organelles help maintain these proteins in an unfolded and folded state, respectively.

▸ Some mRNAs can be localized to specific regions of the cytosol, thereby controlling where the resulting proteins are concentrated. The actin and microtubule cytoskeletal networks assist in this.

Having described the construction of a polypeptide from a gene, we now turn our attention to the fate of this polypeptide. Let's apply the challenge-hypothesis strategy for keeping the big picture of this subject in focus. According to our principle, the DNA encodes the *function* of these polypeptides, and this means that DNA must endow proteins with a specific set of skills. Inherent in this principle is the notion that these tasks cannot be performed just anywhere. Just as a carpenter and a pilot cannot likely perform their functions in the same location, the same applies to proteins. This is our first challenge: if nearly all proteins are synthesized in the same cellular location, how do cells control where specific proteins function?

Many different hypotheses could be proposed to explain how they solve this. I'll propose two hypotheses that we can check as we move deeper into the details. First, it seems possible that proteins only adopt functional shapes in the regions where they work. By this reasoning, we hypothesize that the *chemical environment* of each "job" location is optimally suited to its constituent proteins. Alternatively, it may be possible that proteins adopt functional shapes almost anywhere in a cell, but they are targeted to specific cellular locations in such a way that they cannot easily leave them. If this is true, we hypothesize that proteins contain instructions for properly targeting the correct location. Such a strategy would enable us to classify proteins according to their structure, rather than by the chemical environment where they function.

We have already seen evidence supporting both hypotheses in Chapter 6, when we saw that extracellular matrix proteins do not become functional until they reach their final destination, the extracellular space (supporting the first hypothesis), and they are targeted first to the endoplasmic reticulum before reaching the extracellular space (supporting the second hypothesis). Look for additional evidence supporting or rejecting these hypotheses in the sections below. When proteins pass through more than one location, we'll refer to the first target as the primary destination, and any subsequent location as a secondary destination. In this section we will focus on how cells reach their primary destinations, and devote the entirety of the next chapter to the mechanisms used to sort them between different secondary destinations.

The "cell as a busy city" analogy, revisited. Let's modify our "cell as a busy city" analogy to illustrate the concept of protein targeting. Imagine that our city now is in the future, and all work is performed by robots; proteins in a cell are represented by these robots. If the city is functioning like a cell, these robots are built in one large outdoor location in the city, say a large park (our equivalent to the cytosol). Once the robots are made, some remain in the park, and others begin marching through the streets and yards, doing whatever they are programmed to do (e.g., clean streets, paint buildings, mow the grass, build more robots). These "outdoor workers" are the cytosolic proteins: they work in the same cellular space where they were made. Many other robots march off (or are shuttled) to buildings, and they enter the building to begin their work. These "indoor workers" represent the proteins that are built in the cytosol and then transported into organelles, where they assist is the specialized functions of the organelles.

Before we start, let's compile a list of some themes that we have seen repeatedly in Chapters 1–7 that will help us understand the examples below.

- The most important theme is that the sequence of bases in the coding region of a gene is directly reflected by the sequence of amino acids in a protein. This conversion from DNA to protein brings with it a powerful tool: differences in amino acid *sequence* result in great variation in protein *shape*. Two genes may look physically quite similar, but could encode two proteins with radically different shapes.
- Second, controlling protein *shape* is the mechanism cells use for controlling protein *location*.
- Third, controlling protein *shape* is the mechanism cells use for controlling protein *function*. This means the *shape, location,* and *function* of a protein are directly linked. They are all encoded by the gene that gave rise to the protein, illustrating how DNA information is converted into specialized cellular machines.

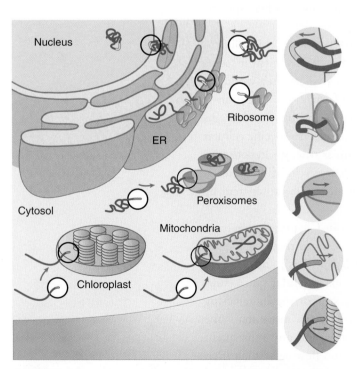

FIGURE 8-16 An overview of protein targeting in eukaryotic cells. Note that signal sequences lead the insertion into most target organelles.

Signal Sequences Code for Proper Targeting of Proteins

Let's begin by discussing what **protein targeting** means (**BOX 8-4**). So far we have established that essentially all mRNA in eukaryotic cells is made in the nucleus and translated into proteins in the cytosol. In fact, the cytosol is the *default destination* for all proteins synthesized there. But every chapter in this book mentions proteins that perform their jobs in organelles and/or the extracellular space. The term *targeting* refers to the mechanisms cells use to move proteins from the cytosol to these destinations.

The concept of protein targeting is illustrated in **FIGURE 8-16**. Proteins that have no specific destination remain in the cytosol. If they have a primary destination, each destination has its own distinguishing sorting mechanism. In many cases, the mechanisms use simple sequences of amino acids that are part of the primary structure of these proteins; the structures these amino acids form act as signals to indicate where proteins should go after they are synthesized, and so are called **signal sequences**. Proteins without signal sequences stay in the cytosol after they are synthesized (**BOX 8-5**).

The ticket analogy. One way to understand how signal sequences work is to imagine that proteins are like people, and that signal sequences are like concert or movie tickets. To get into a particular venue, a person needs a ticket for the event; to get into an organelle, a protein needs a signal sequence for that organelle. At the door, tickets are usually ripped in half so they can't be reused; at the membrane of most organelles, the signal sequences are cut off the proteins so they can't be reused. To be more accurate, we have to push the credibility of the analogy and say that people are *born* with tickets built right into their structure, and they only get to go to the one venue that is imprinted on their built-in ticket, which is then cut off. Those who don't have tickets stay at home (in the cytosol).

As you read on in this section, you will encounter two exceptions to this. One is nuclear transport, where many proteins get to enter and exit whenever they like. The analogy for this situation is the backstage pass, which usually permits repeated entry and exit from a carefully guarded door, and is never taken away. Another exception is uncleaved signals in the ER, such as signal anchor sequences. The analogy for this is bringing a passport onto an international cruise ship: you will never part with it as long as you are onboard, and proteins with uncleaved signal sequences never part with these sequences as long as they are in their target location(s).

How these signals are detected and what happens as a result varies according to the type of signal they contain and their target organelle. An easy way to organize protein sorting mechanisms is by primary target destination. We will examine five ways that proteins find their primary destinations.

The Nuclear Import/Export System Regulates Traffic of Macromolecules through Nuclear Pores

One unique property of nuclear transport is that it is bidirectional (meaning traffic between the cytosol and the nucleoplasm occurs in both directions), and many proteins routinely cycle into and out of the nucleus. We have already encountered a good example of this behavior in this chapter: many of the proteins that form the hnRNP (a.k.a. mRNP, or "big ball") accompany mRNA on the journey from the nucleoplasm to the cytosol, then return to the nucleoplasm to accompany another hnRNP later on. To function properly, these proteins must be able to shuttle back and forth.

Proteins Are Transported into and out of the Nucleus in Their Properly Folded, Functional State

Because many nuclear proteins must perform a function on *both* sides of the nuclear envelope, they must maintain their stable shape on both sides as well. This means that, unlike all other mechanisms that target proteins to organelles, nuclear transport (in either direction) does not permanently disrupt the tertiary (or even quaternary) structure of the proteins passing through the nuclear membrane. This is especially astonishing when one considers how big some of the protein complexes are: have another look at the ribosomal subunits in Figure 8-13, and consider how large the hole has to be to permit something that big to pass, unharmed, through an intact, functional (double) membrane. To help visualize this, imagine if an average (5–10 nm diameter) protein were the size of an average-sized adult, then the large and small ribosomal subunits would be approximately 2–4 times that size, or nearly 8 meters (approximately 26 ft) in diameter in human terms. That would require a *big* hole, which in the membrane would be 10 times the size of a typical phospholipid. A corollary to this is that small molecules (anything smaller than 9 nm diameter, including small proteins) can easily diffuse through the pore complex.

Nuclear Localization Sequences (NLS) and Nuclear Export Signals (NES) Are Amino Acid Sequences Recognized by NLS and NES Receptors, Respectively

The signal sequences that permit entry to and exit from the nucleus are called **nuclear localization sequences** (**NLS**) and **nuclear export sequences** (**NES**), respectively. The sequences are very small, typically about 10 amino acids in length. Proteins that make only the one-way journey into the nucleus (e.g., histone proteins in chromatin) have an NLS, but no NES. These signal sequences permit transport through the nuclear pore because they bind to receptor proteins (called **importins** and **exportins**); these receptors act as chaperones to ensure the transported proteins make the journey through the nuclear pore complex without clogging the pore or being damaged in transit. Amazingly, if biochemical or molecular biology methods are used to attach just the NLS to proteins (or even to inanimate objects like gold particles), these hybrids are transported into the nucleus (assuming that they are no larger than 40 nm in diameter).

Importins and exportins accompany their cargoes through the nuclear pore complex, and then return to their original location. They are called soluble proteins, indicating they are not bound to any membranes or large molecular complexes. (The term "soluble" actually refers to the fact that these proteins can be extracted from cells with an aqueous buffer—that is, they are *soluble* in such a buffer—but the word is now used routinely to describe the cellular location of a protein, along with its actual solubility.) The fact that they are soluble means they can diffuse easily in the cytosol, so their transport into and out of the nucleus is very rapid.

The Direction of Nuclear Transport Is Controlled by the G-Protein Ran

Five key points are necessary to demonstrate how nuclear transport illustrates principle #4:

- First, the signal that permits transport through the nuclear pore, in either direction, is built into the transported protein, in the form of an NLS/NES.
- Second, the importins and exportins that bind to the NLS and NES are themselves proteins, and thus their structure and function are reflected by their primary sequence, again encoded by DNA. These proteins can simultaneously bind to the transported cargo protein and to proteins in the inner or outer rims of the nuclear pore complex, called **nucleoporins**.
- Third, the importin and exportin proteins are only responsible for delivering their cargo to the rim of the nuclear pore complex, a process called **docking**. A different set of interactions is responsible for actually moving the cargo–importin/exportin complex through the channel of the pore. The mechanisms responsible for this are still not known.
- Fourth, transport of proteins into the nucleus requires energy, specifically GTP. Initially it was thought that the nucleoporins used the energy released by GTP to physically move the proteins through the nuclear pore complex, but that is not the case.
- Fifth, to explain the GTP dependency, we have to introduce a new player, a protein called **Ran**, which controls the transport of these proteins through the nuclear pore. Ran acts like a waterwheel, moving in and out of the nucleus in a cycle powered by GTP.

Use **FIGURE 8-17** as a guide as we describe how Ran functions.

Let's focus on import first. After importin accompanies its cargo into the nucleus, its next job is to release the cargo and return to the cytosol for another round of importing. Ran binds to either GTP or GDP, like the elongation factors we discussed earlier. In the nucleus, Ran-GTP helps this dissociation happen by binding to importin, causing it to change shape and release its cargo molecule.

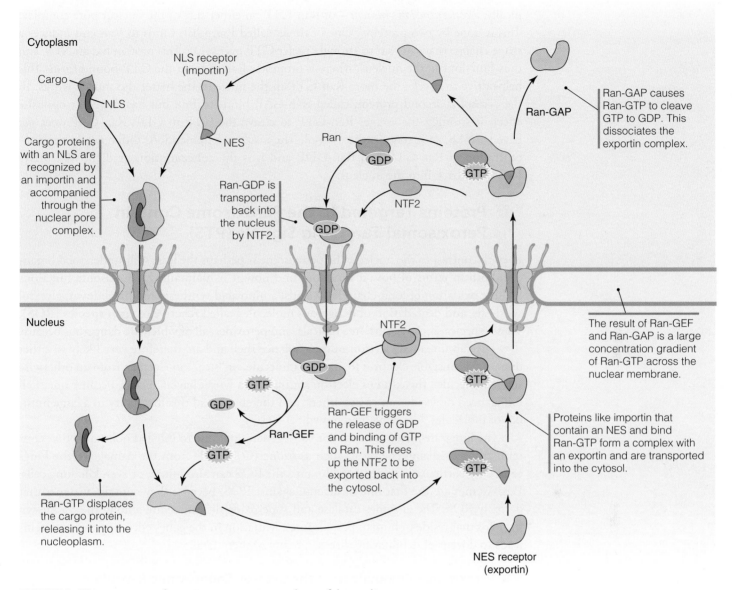

Cytoplasm

Cargo

NLS

NLS receptor
(importin)

NES

Cargo proteins
with an NLS are
recognized by
an importin and
accompanied
through the
nuclear pore
complex.

Ran-GDP is
transported
back into
the nucleus
by NTF2.

Ran

GDP

GDP

NTF2

GTP

GTP

Ran-GAP

Ran-GAP causes
Ran-GTP to cleave
GTP to GDP. This
dissociates the
exportin complex.

Nucleus

GTP

GDP

GDP

NTF2

GTP

The result of Ran-GEF
and Ran-GAP is a large
concentration gradient
of Ran-GTP across the
nuclear membrane.

Ran-GEF

Ran-GEF triggers
the release of GDP
and binding of GTP
to Ran. This frees
up the NTF2 to be
exported back into
the cytosol.

GTP

GTP

Proteins like importin that
contain an NES and bind
Ran-GTP form a complex with
an exportin and are transported
into the cytosol.

Ran-GTP displaces
the cargo protein,
releasing it into the
nucleoplasm.

NES receptor
(exportin)

FIGURE 8-17 An overview of protein transport into and out of the nucleus.

Now let's look at export. Any protein in the nucleus that has an NES (including importin) is escorted out by binding to exportin, entirely analogous to the way importin helps bring proteins into the nucleus. But there is an extra requirement: this export also requires Ran-GTP to bind to the exported cargo. Whereas cargo coming into the nucleus is bound to importin only, cargo leaving the nucleus is bound to exportin *and* Ran-GTP.

What happens to Ran next explains the waterwheel analogy. Once it reaches the cytosol, Ran-GTP cleaves its GTP to GDP. This causes a shape change that releases the importin and exportin, and the released importin can now find a new partner with an NLS for another trip into the nucleus. Ran-GDP has completed its task. To trigger more transport, the Ran has to release its GDP and bind a fresh GTP, but this does not take place in the cytosol; it occurs in the nucleus. Ran-GDP binds to an importin called NTF2, and this carries Ran-GDP back into the nucleus, where it is able to swap the old GDP for a GTP. The Ran cycle is now complete: Ran-GTP promotes export of proteins, then it converts to GDP in the cytosol and is carried back into the nucleus where it picks up a fresh GTP molecule and repeats the cycle.

The key to using the Ran-GTP shuttle system is to keep an imbalance—or, more technically, a *concentration gradient*—of Ran-GTP on either side of the nuclear pore complex. This is done by two proteins. One of these, called Ran-GEF, binds to Ran and induces a shape change that causes it to strongly prefer GTP over GDP. This protein has an NLS, and thus functions in the nucleus, where it promotes Ran to be in the GTP-bound form. This helps drive the cycle; the more Ran-GTP in the nucleus, the better the shuttle works. In the cytosol, a second protein, called Ran-GAP, binds to Ran but has the exact opposite effect: it strongly encourages Ran-GTP to cleave the GTP to GDP. Ran-GAP does *not* have an NLS, so it stays in the cytosol. The result is a tremendous difference in the concentration of Ran-GTP and Ran-GDP, and it is the concentration gradient that drives Ran-GDP back into the nucleus.

Proteins Targeted to the Peroxisome Contain Peroxisomal Targeting Signals (PTS)

In stark contrast to the nucleus, the peroxisome is perhaps the least well-understood organelle, both in terms of how it functions and how it is maintained. Peroxisome functions include oxidation of long-chain fatty acids, amino acid synthesis and breakdown, glycerol synthesis, and degradation of dangerous molecules called **reactive oxygen species** (**ROS**). These "species" include ions, free radicals, and peroxides, all of which are dangerous because they have an unpaired electron in their valence shell and are therefore very likely to either donate this unstable electron to another molecule, or "steal" an electron from an otherwise stable molecule. This kind of electron shuffling can wreak havoc in cells, because one such unbalanced molecule can create others, and thereby spread this instability to a large number of molecules, like panic in a crowd.

The lungs are a common site for ROS generation, due to inhaled pollutants that contain strong oxidizing chemicals such as ozone (O_3). Aside from the damage to the lungs caused by particulate matter in pollutants, the ROS can also injure, or even kill, lung cells. Peroxisomes are the first line of defense against ROS, because they contain enzymes that detoxify ROS. The enzymes catalase and superoxide dismutase are among the 85 known proteins that reside in human peroxisomes. Mutation in the genes encoding these proteins leads to dozens of debilitating diseases, many of them fatal.

Peroxisomes Originate from the Cytosol, Endoplasmic Reticulum, and Possibly Mitochondria

Part of the reason peroxisomes are rather mysterious is that we don't yet know where they come from, or how they are made. They contain no DNA or ribosomes, so every protein in them must originate elsewhere in the cell. The most obvious place to look is the cytosol, and indeed many peroxisomal proteins are synthesized there and delivered directly to peroxisomes. However, a growing body of evidence suggests that peroxisomes are assembled from and maintained by several sources, and this does not fit easily into our current understanding of organelle behavior. Nonetheless, proteins that are commonly found in peroxisomes are also located in the ER, suggesting that the two structures are somehow linked, though no vesicular traffic between them is known. When cells are manipulated in the laboratory to deprive them of all peroxisomes, they grow new ones that bud from the ER membrane. Existing peroxisomes can also grow and divide like mitochondria, and indeed some mitochondria contain peroxisomal proteins. Whether this represents one-way traffic from the peroxisome to the mitochondria, or two-way traffic between them, remains to be determined. Like both the ER and mitochondria, peroxisomes are transported through the cell by kinesin and dynein motor proteins bound to microtubules (see *Microtubule-Based Motor Proteins Transport Organelles and Vesicles* in Chapter 5).

Proteins Are Transported into the Peroxisomal Matrix in Their Properly Folded, Functional State

The best understood mechanism for peroxisomal protein transport is that used to deliver cytosolic proteins. A generalized model of peroxisomal transport is shown in **FIGURE 8-18**. Most known peroxisomal proteins are synthesized by "free" ribosomes (not attached to any membranes) in the cytosol. These proteins are transported in their native (fully folded) shapes. Even multimeric complexes are transported into peroxisomes this way, as is the case for nuclear transport, but there is no equivalent to the nuclear pore complex in peroxisomes. So even through the mechanisms appear at first glance to be quite similar, they must be radically different.

Aside from the approximately 50 metabolic enzymes in peroxisomes, the remaining proteins are essential for peroxisomal biogenesis, as demonstrated by experimental mutations (primarily in yeast cells). These essential proteins are called **peroxins**, and in the tradition of yeast genetics, they are abbreviated as Pex, followed by a number and sometimes a letter. In the mid-1980s, scientists discovered two peroxisome signal sequences, called **peroxisomal targeting signals** (**PTS**). PTS1 is found at the C terminus of most Pex proteins, and PTS2 is found at the N terminus of the rest.

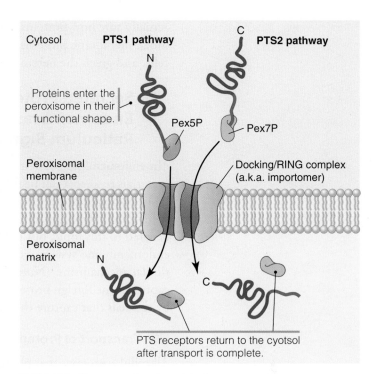

FIGURE 8-18 A generalized model of peroxisomal protein import.

Consistent with nuclear transport, these signal sequences on cargo proteins (a generic term for proteins and other structures that are transported; the same term is used to describe the material carried by microtubule motor proteins) are bound to PTS receptors in the cytosol. The PTS1 and PTS2 receptors are called Pex5p and Pex7p, respectively. Once the receptors bind the signal sequences, they bind to a **docking complex** on the cytoplasmic surface of the peroxisome composed of other, membrane-bound Pex proteins that vary according to species. These proteins, in turn, are bound to yet another group of Pex membrane proteins called a **RING complex**. Together, the docking complex and RING complex import the cargo into the interior of the peroxisome, which is called the matrix (the same term is used to describe the interior of mitochondria). The docking/RING combination is named **importomer**.

PTS Receptors Return to the Cytoplasm After Delivering Their Cargo

To be useful, the PTS receptors have to be able to release their cargo once it has bound to the docking complex and return to the cytosol to find another cargo molecule. Careful study of the timing of peroxisomal import reveals that most of these receptors release their cargo before the cargo is imported, but some remain attached to the cargo even during the importation process. Biochemical studies show that the receptors can bind stably to RING proteins (at least in some cell types), but why this happens is not well understood. Regardless, even if the receptor is carried into the peroxisomal matrix, it is eventually transported back out into the cytosol. Note that this bears some resemblance to nuclear transport, in that receptors are transported into and out of the target organelle.

Import of Peroxisomal Membrane Proteins Is Not Well Understood

The discovery of the docking and RING proteins is a relatively recent event, and unfortunately these proteins do not behave consistently in all the organisms that have been studied—even different species of yeast do not behave alike. The result is that no one is actually sure how proteins enter the peroxisomal matrix. The same is true for the peroxisomal membrane proteins. It is possible that there is no single mechanism shared by all cells, and given the heterogeneous origin of peroxisomes, this would not be surprising.

Secreted Proteins and Proteins Targeted to the Endomembrane System Contain an Endoplasmic Reticulum Signal Sequence

The endomembrane system is a network of organelles and vesicles that constantly exchange materials to keep a cell healthy and functional. The major constituents of the endomembrane system are the ER, the Golgi complex, endosomes, lysosomes, and the plasma membrane. How they function and interact is so complex that the entire next chapter is devoted to this topic, but before we get there, let's address how cellular proteins enter the endomembrane system. Fortunately, all of these proteins enter at the same site, via very similar mechanisms. (Note that part of the function of the endomembrane system is to capture some foreign proteins from the external environment, and since they are not made by the cells that capture them, we can ignore how they are made.)

Transport of Proteins into the ER Membrane or ER Lumen Is Cotranslational

All cellular proteins that populate the endomembrane system enter via the ER. The mechanisms used to insert these proteins differ in three significant ways from the mechanisms we have discussed for nuclear and peroxisomal transport:

- First, transport into the ER is **cotranslational**. This means that the protein is inserted into the ER *as it is being synthesized* by a ribosome. This requirement explains why the rough ER is dotted with ribosomes. The two processes are tightly linked: if a ribosome stops synthesizing an endomembrane system protein, the protein no longer enters the ER, and if the ribosome is detached from the ER but allowed to continue synthesis of the protein, the resulting polypeptide will also not be imported. Some primitive, single-celled eukaryotic organisms are able to insert some proteins into their ER posttranslationally, but this is so rare that we will ignore it.
- Second, transport into the ER requires that a protein be in its unfolded state. Native (fully folded) proteins do not enter the ER; each subunit of a multmeric protein must be imported separately. One way to help ensure that a protein or subunit is unfolded as it enters the ER is to insert it as it is being synthesized. The protein then folds as it enters the inner compartment of the ER, called the **lumen**.
- Third, transport into the ER is unidirectional. Once a protein enters the ER, it never comes back into the cytosol unless it fails to fold properly.

Transport of Proteins into the ER Is a Multistep Process

Proteins destined for any portion of the endomembrane system express at least one signal sequence, called the **ER signal sequence**. ER signal sequences are typically located at or near the amino terminus of a polypeptide. To understand how the ER signal sequence targets proteins into the endomembrane system, let's follow an imaginary polypeptide through the steps of ER insertion, as illustrated in **FIGURE 8-19**.

Step 1: A Free Ribosome Synthesizes the ER Signal Sequence

Once a gene for an endomembrane system protein is transcribed and the processed mRNA is transported into the cytosol, it binds to a ribosome through the initiation stage of translation we described earlier in this chapter. The elongation phase then begins, and a growing polypeptide eventually emerges from the polypeptide exit tunnel. At some early point in elongation, the ER signal sequence emerges as a part of the growing polypeptide.

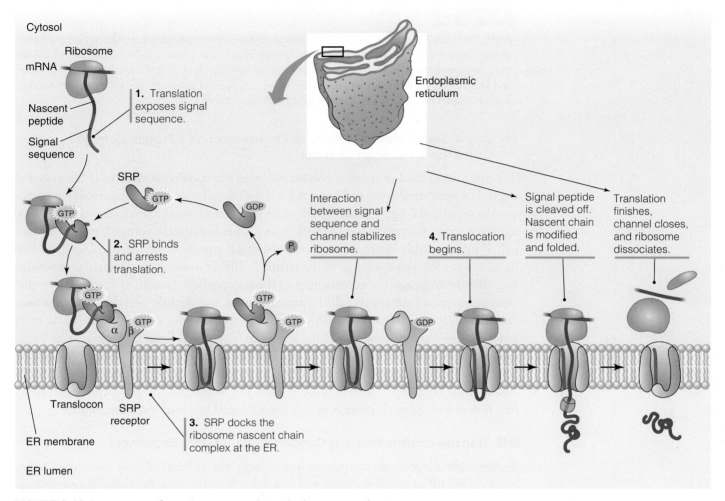

FIGURE 8-19 An overview of protein import in the endoplasmic reticulum.

▪▪▪ Step 2: The Signal Recognition Particle Binds to the ER Signal Sequence

Almost immediately after the ER signal sequence emerges, its receptor binds to it. The name of this receptor is **signal recognition particle** (**SRP**). Unlike the other receptors for signal sequences we have encountered thus far, this receptor is not a simple protein: it is composed of six polypeptide subunits and a form of RNA called small cytosolic RNA (scRNA). Once SRP binds to the ER signal sequence, it interacts with the ribosomal subunits and slows translation dramatically, most likely by interfering with binding of translational elongation factors. This is the first time we have encountered a mechanism that slows translation.

▪▪▪ Step 3: Binding of SRP and Its Receptor Allows Proteins to Dock at the ER Membrane

SRP undergoes a shape change when it binds to the ribosome; this form of SRP is able to bind to a membrane protein in the ER called the **SRP receptor**. The SRP receptor is composed of two polypeptide subunits (called α and β), and its function is to transfer the ribosome and growing polypeptide to the structure that permits the polypeptide to enter the ER. This structure is called the **translocon**, and it is composed of multiple proteins that form a channel through the ER membrane. Immediately after the SRP receptor transfers the SRP–ribosome bundle, a large complex forms, consisting of the mRNA, its initiation and elongation factors, the ribosomal subunits, SRP, SRP receptor, and all of the translocon-associated proteins. At this point the SRP and SRP receptor leave the complex by undergoing shape changes induced by the cleavage of GTP molecules that they carry with them (**BOX 8-6**).

Step 4: Translocation of Proteins Occurs through a Protein Complex Called the Translocon Channel

The mechanism used to insert a polypeptide into the translocon channel is illustrated as a series of numbered steps in Figure 8-19. First, the ribosome interacts weakly with the translocon, and the signal sequence lies outside the translocon channel. Second, the signal sequence enters the translocon channel, where it binds a protein component of the channel. Third, the stable interaction between the signal sequence and the translocon channel induces a conformation change in the channel, further opening the channel and permitting the translocation of the remainder of the polypeptide. Fourth, as the newly synthesized polypeptide (sometimes called a **nascent chain**, to reflect the fact that it has just been "born") enters the lumen of the ER, it begins to fold up into its mature shape. At some point during translocation, the signal sequence is removed from the polypeptide by an enzyme attached to the translocon called **signal peptidase**. The final product is a folded, soluble polypeptide. According to the three traits of proteins (see *Proteins Are Polymers of Amino Acids That Possess Three Important Traits* in Chapter 3), this polypeptide can either function as a monomeric protein or serve as a subunit in a multimeric protein.

Transmembrane Proteins Contain Signal Anchor Sequences

Transmembrane proteins are present throughout the endomembrane system, and they enter via the ER in much the same way as soluble proteins, with the following exceptions.

Integration of Transmembrane Proteins Requires Specific Amino Acid Sequences

All transmembrane proteins in the endomembrane system contain at least one **transmembrane domain**. Typically, transmembrane domains are clusters of approximately 30 predominantly hydrophobic amino acids in the primary sequence of transmembrane protein. When these sequences reach the translocon channel, they attach themselves to the wall of the translocon, where they remain until the translocon releases them into the membrane.

Many of these proteins contain a second sequence, called a **signal anchor sequence**. A signal anchor sequence resembles a traditional ER signal sequence in that it is recognized by SRP, and binds to the interior of the translocon channel. Two major differences between a signal anchor sequence and a traditional ER signal sequence are (1) signal anchor sequences are not found near the N terminus of the polypeptide, and (2) they are not cleaved by signal peptidase. In effect, a signal anchor sequence is both a transmembrane domain and a signal sequence.

Transmembrane proteins are classified into types, based on their topological orientation in membranes. One of the most commonly used classification schemes was developed by the same Dr. Jonathan Singer (of "fluid–mosaic model" fame) we mentioned in Chapter 4 (see *The Fluid–Mosaic Model Explains How Phospholipids and Proteins Interact within a Cellular Membrane*). According to his system, illustrated in **FIGURE 8-20**, **type I transmembrane proteins** have a single transmembrane domain, and are oriented so that their amino terminus is in the lumen of the ER and their carboxy terminus is in the cytosol. **Type II transmembrane proteins** also have a single transmembrane domain, but are oriented in the opposite direction. **Type III** and **type IV transmembrane proteins** both span the plasma membrane multiple times, but can be distinguished on the basis of how the transmembrane regions interact in the membrane. If they form a central aqueous channel, they are classified as type IV; if not, they are type III.

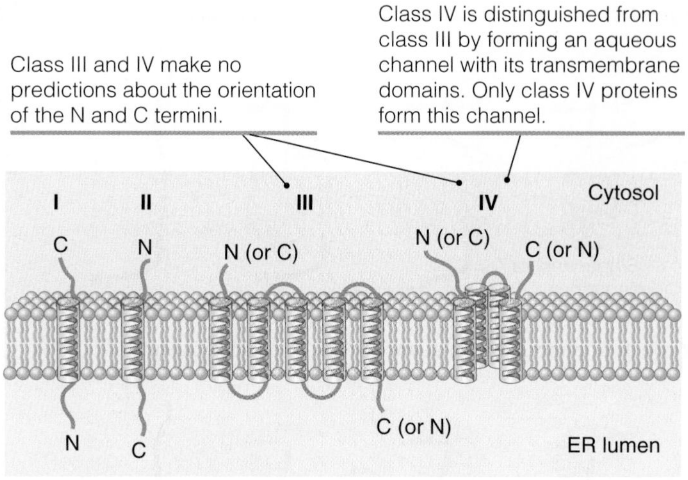

FIGURE 8-21 illustrates how a protein that contains both a traditional ER signal sequence and a single transmembrane domain can result in a type I orientation. Because the signal sequence is cleaved by signal peptidase, the amino terminus of the polypeptide must be in the ER lumen. The presence of a transmembrane domain in the middle of the polypeptide forces the portion of the polypeptide synthesized after it to remain in the cytosol. Notice that the same translocon binds to both the signal sequence and the transmembrane domain. Once a polypeptide stops moving through a translocon, it dissociates into two pieces that separate somewhat like a clam shell, releasing the transmembrane domain into the phospholipid bilayer.

FIGURE 8-22 illustrates how a polypeptide with a signal anchor sequence and no N-terminal signal sequence can adopt a type I or type II orientation. For reasons that are still not completely understood, signal anchor sequences are able to insert themselves into the translocation channel in either direction. If it is oriented with its amino end pointed toward the ER lumen, whatever portion of the polypeptide that was synthesized before it can slip by through the translocation channel and end up in the lumen, giving rise to a type I orientation. If this signal anchor sequence inserts itself into the translocation channel in the opposite orientation, the portion of the polypeptide synthesized before it remains in the cytosol and the portion synthesized after it slips by in the translocation channel and ends up in the lumen, resulting in a type II orientation.

Unfortunately we don't know how the orientation of type III and type IV transmembrane proteins is determined. **FIGURE 8-23** shows one model as to how these proteins may be organized. For this to be true, transmembrane domains must have the ability to orient into membranes in either direction, like signal anchor sequences.

Class III and IV make no predictions about the orientation of the N and C termini.

Class IV is distinguished from class III by forming an aqueous channel with its transmembrane domains. Only class IV proteins form this channel.

FIGURE 8-20 The four classes of transmembrane proteins, according to the Singer classification system.

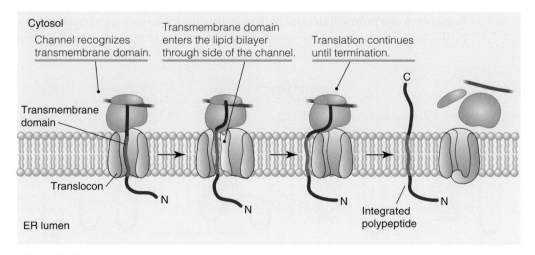

FIGURE 8-21 Integrating a Type I transmembrane protein with a signal sequence in a single transmembrane domain.

Type I N terminus of signal anchor sequence translocates

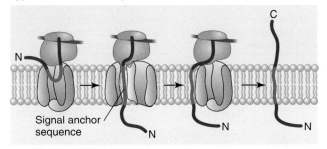

Signal anchor
sequence

Type II C terminus of signal anchor sequence translocates

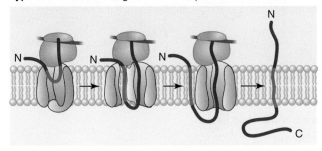

FIGURE 8-22 Integrating membrane proteins with a signal anchor sequence. Depending on the orientation of the signal anchor sequence, a Type I or Type II orientation results.

■■■ As Proteins Enter the ER Lumen, They May Be Posttranslationally Modified

Once a polypeptide is fully synthesized and inserted into the ER lumen, it can undergo at least four posttranslational modifications. These are:

- Most proteins targeted to the ER by an N-terminal signal sequence have this sequence cleaved off by the signal peptidase attached to the lumenal side of the translocon.
- A protein complex called the **N-oligosaccharyl transferase (OT) enzyme complex** is also attached to the translocon channel. Its function is to attach a branched cluster of sugars called the **core oligosaccharide** to some polypeptides. This core oligosaccharide is synthesized on a lipid "anchor" called **dolichol phosphate** by enzymes on both sides of the ER membrane (the dolichol phosphate is flipped from the cytosolic leaflet to the lumenal leaflet by a flippase, see *The Smooth Endoplasmic Reticulum and Golgi Apparatus Build Most Eukaryotic Cellular Membrane Components* in Chapter 4), then transferred en mass to the incoming polypeptide, as shown in **FIGURE 8-24**. The OT complex recognizes a characteristic sequence of three amino acids in translocating polypeptides called a **sequon**. Three different seqons are recognized by the OT complex, and all three contain the amino acid asparagine (abbreviated as N in the single-letter code for amino acids). Not every polypeptide has these sequons, but those that do are attached to the core oligosaccharide via the asparagine, hence the process is called N-linked glycosylation. The core oligosaccharide undergoes a great deal of processing as the polypeptide passes from the ER to the Golgi complex as part of the endomembrane system. We will discuss this processing in greater detail in Chapter 9.

- A phospholipid called glycophosphatidylinositol (GPI) is synthesized on a phosphatidylinositol "anchor" by enzymes on both sides of the ER membrane, very similar to the way the core oligosaccharide is made. The GPI is added by a group of proteins called a **transamidase complex**, which recognize a specific hydrophobic sequence of amino acids. In soluble proteins, this sequence tends to be at the C terminus of a polypeptide, and in transmembrane proteins it appears somewhere in the lumenal portion. The transamidase complex cuts the polypeptide at this

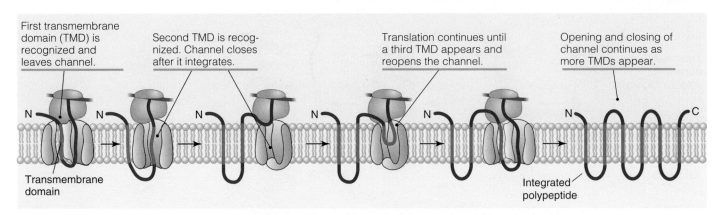

First transmembrane domain (TMD) is recognized and leaves channel.

Second TMD is recognized. Channel closes after it integrates.

Translation continues until a third TMD appears and reopens the channel.

Opening and closing of channel continues as more TMDs appear.

Transmembrane domain

Integrated polypeptide

FIGURE 8-23 A model for the integration of multi-spanning membrane proteins.

CHAPTER 8 Protein Synthesis and Sorting

sequence, generating a new C terminus called an omega (ω) site, and attaches the GPI to this site. This results in a polypeptide that does not interact directly with the ER membrane, but is anchored to it by the GPI molecule, as shown in **FIGURE 8-25**.

- Disulfide bonds are created in polypeptides by a family of enzymes called protein disulfide isomerases (PDIs). These bonds are formed between the side chains of cysteine amino acids, either within the same polypeptide or between two different polypeptides. These bonds have a significant impact on the folding of polypeptides and the shape of proteins, so their correct placement is crucial. PDIs can detect incorrect disulfide bonds and rearrange them until the proteins fold properly, as shown in **FIGURE 8-26**.

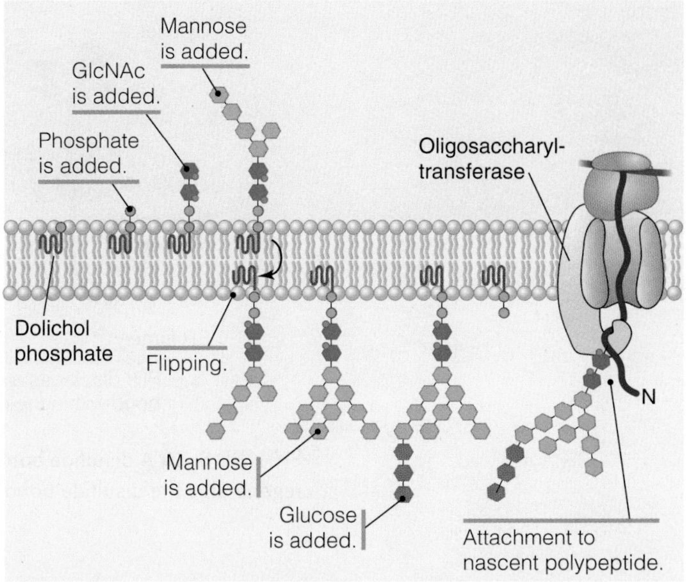

■■ Chaperone Proteins Assist in the Proper Folding of ER Proteins

PDIs are just one family of proteins that assist protein folding in the ER. One of the most abundant groups is the heat shock proteins, so named because they protect proteins from denaturing when they are heated. The heat shock proteins that specifically participate in the proper folding of

FIGURE 8-24 N-linked glycosylation of polypeptides in the ER. In a series of steps, an elaborate carbohydrate structure is synthesized on the phospholipid dolichol phosphate. Synthesis begins on the cytoplasmic face of the membrane of the ER but is completed on its lumenal face. The entire oligosaccharide structure that results is transferred onto translocating proteins by the enzyme OST. Modification takes place on asparagine residues that appear in a particular sequence.

nascent polypeptides are called chaperone proteins, or chaperonins. One of the best characterized chaperonins in the ER lumen is a protein called BiP. BiP binds to hydrophobic regions in nascent polypeptides as they enter the ER lumen, then uses ATP hydrolysis to detach and

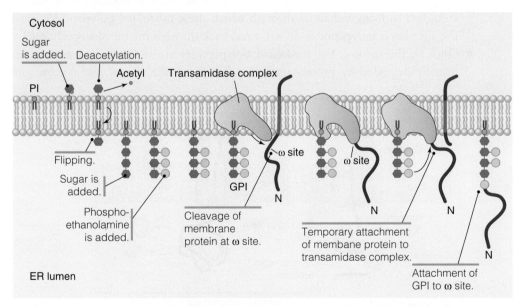

FIGURE 8-25 GPI synthesis and modification of proteins. GPI a synthesized in a series of steps. The first several steps occur on the cytoplasmic face of the ER membrane, and later ones on its lumenal face. When complete, GPI is covalently attached to a polypeptide by a transaminase complex. The complex first cleaves the polypeptide at the omega site, then attaches the GPI to the carboxy terminus of the cleaved polypeptide.

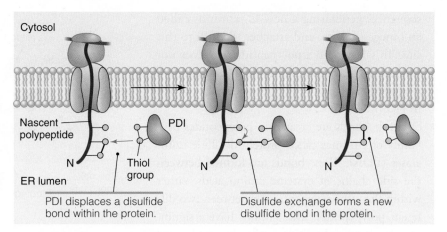

Cytosol

Nascent
polypeptide

PDI

N

Thiol
group

ER lumen

PDI displaces a disulfide
bond within the protein.

Disulfide exchange forms a new
disulfide bond in the protein.

N

N

FIGURE 8-26 A disulfide bond in PDI is used to form one in the nascent polypeptide. An enzyme regenerates the disulfide bond in PDI afterward so that it can act multiple times.

reattach to the polypeptides as they fold. This means that protein import into the ER is ATP-dependent. As the mature protein reaches its properly folded state, the hydrophobic patches are buried within the interior, and BiP no longer binds. This is illustrated in **FIGURE 8-27**.

■■■ Terminally Misfolded Proteins in the ER Are Degraded in the Cytosol

Despite the best efforts of chaperone proteins, some polypeptides that enter the ER do not fold up properly. Because they are trapped in the ER, there is a risk that these useless polypeptides may clog the tightly orchestrated translocation, posttranslational modification, and folding mechanisms taking place there. For a very long time, the fate of these misfolded polypeptides was a mystery. Very recently, some light has been shed on this subject, but at this point we are still left with more questions than answers. We do know the following:

- Misfolded polypeptides in the ER are "reverse translocated" back into the cytosol. Little is known about how this takes place, but one protein in yeast called Sec61p is known to form a channel through which these misfolded polypeptides leave the ER. These polypeptides do *not* travel back through the translocon channel.
- Once in the cytosol, the misfolded polypeptides are ubiquitinated and subsequently degraded by proteosomes (see *Proteins in the Cytosol and Nucleus Are*

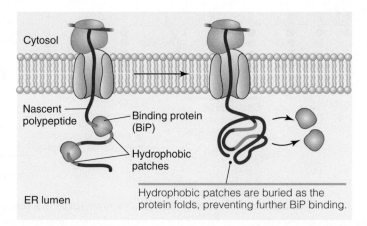

Cytosol

Nascent
polypeptide

Binding protein
(BiP)

Hydrophobic
patches

ER lumen

Hydrophobic patches are buried as the
protein folds, preventing further BiP binding.

FIGURE 8-27 BiP binds to exposed hydrophobic patches in recently translocated proteins. After a protein has folded, its hydrophobic patches are buried within its structure and are no longer accessible to BiP.

Broken Down in the Proteasome in Chapter 3). The process of identifying, reverse translocating, and destroying these polypeptides is called ER-assisted degradation, commonly referred to by the acronym ERAD.

- A communications system exists between the ER and the nucleus that triggers the expression of heat shock protein genes when misfolded proteins begin to accumulate in the ER lumen. This system is named the **unfolded protein response**.

Similar questions arise about the fate of misfolded proteins in other organelles as well. Currently, we understand even less about how these proteins are degraded. It is likely that this area of research will receive considerable attention in the near future.

Cytosolic Proteins Targeted to Mitochondria or Chloroplasts Contain an N-Terminal Signal Sequence

Targeting of cytosolic proteins to mitochondria and chloroplasts is complicated by the fact that both of these organelles are enclosed by two membranes; as such, proteins may be targeted to the outer membrane, inner membrane, intermembrane space, or interior of these organelles. Chloroplasts have an additional membrane-bound compartment called the thylakoid, giving proteins two additional possible destinations. Due to this increased complexity, both organelles contain several different translocating complexes specially tuned for each destination.

Transport of Proteins into Mitochondria and Chloroplasts Is Posttranslational and Energy Dependent

The overall mechanisms controlling import of cytosolic proteins into mitochondria and chloroplasts share similarities with the mechanisms used by peroxisomes and the ER. The essential features of these mechanisms are:

- Cytosolic proteins targeted to these organelles contain short signal sequences at or near their amino terminus. These sequences are named either **mitochondrial signal sequences**, or **transit peptides** for chloroplasts.
- The signal sequences are recognized by receptors. However, unlike the receptors for transport into nuclei, peroxisomes, and the ER, these receptors are *membrane proteins* located on the outer membrane of the organelles.
- Transport into these organelles occurs posttranslationally, and the polypeptides being transported are *not* in their mature, functional shapes. In fact, multiple copies of a chaperone protein called heat shock protein 70 (Hsp70) bind the polypeptides to keep them in their unfolded state. The interior of both organelles contains specialized forms of Hsp70, called mitochondrial or chloroplast Hsp70, that help the polypeptides fold into a stable shape.
- Like transport into mitochondria, transport into these organelles requires ATP energy. Mitochondria also have a chemical gradient of protons across their inner membrane, and this gradient (a form of potential energy) helps drive import into the mitochondrial interior. An additional form of energy is required for import into mitochondria, but it is not yet understood.

The steps required for transport of a polypeptide through both membranes of a mitochondrion are:

- A polypeptide bearing a signal sequence is completely translated in the cytosol. As it is being translated, it is bound by cytosolic Hsp70 to prevent it from folding.
- **FIGURE 8-28** illustrates the remaining steps. The signal sequences recognized by a membrane protein called a TOM (<u>t</u>ranslocase of the <u>o</u>uter <u>m</u>embrane). Several different TOM proteins form a complex that both recognizes and translocates the polypeptide through the outer membrane.

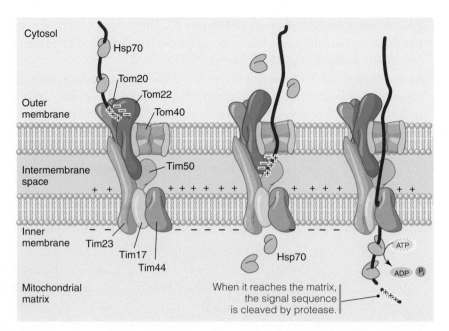

FIGURE 8-28 A protein to be translocated into the mitochondrial matrix is recognized at the outer membrane nand then passed directly between channels in the inner and outer membranes. When it reaches the matrix, the signal sequence is removed. Transport is driven both by ATP hydrolysis by chaperones in the matrix and and by the membrane electrical potential across the inner membrane. The Tim23 protein is integrated into both membranes and links the channels in them. Adapted from M. J. Baker, et al., Trends Cell Bio. 17 (2007): 456–464.

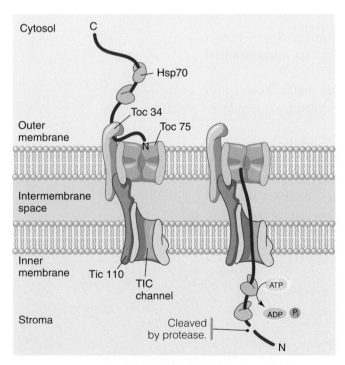

FIGURE 8-29 Chloroplast proteins must cross two membranes to enter the stroma. Proteins to be imported are kept unfolded in the cytoplasm by chaperones. The proteins move through connected channels in the outer and inner membranes. When they have entered the stroma, their transit peptides are removed. A source of energy is required for import but has not yet been identified. Adapted from P. Jarvis, New Phytol. 179 (2008): 247–565.

- Some of the TOM proteins bind to similar proteins that form a complex on the inner membrane. These proteins are called TIMs. This ensures that the translocating polypeptide is passed directly from one complex to the other.
- The polypeptide passes through a translocating channel formed by the TIMs, and enters the interior of the mitochondrion. This space is called the **mitochondrial matrix**. We will spend a great deal of time discussing what happens in this space in Chapter 10.
- Once inside the mitochondrial matrix, mitochondrial Hsp70 begins assisting the folding of the polypeptide into a stable shape. At some point, the mitochondrial signaling sequence is cleaved by a protease.

FIGURE 8-29 summarizes the mechanisms for transport into the interior of chloroplasts, which is called the **chloroplast stroma**. Notice how similar this diagram is to Figure 8-28. Instead of having TIM and TOM protein complexes, chloroplasts have TOC and TIC complexes that function in much the same way. Chloroplast Hsp70s help fold the proteins once they reach the stroma.

Proteins targeted to the thylakoids in the stroma must have an additional signal sequence, and this is usually detected once the transit sequence is cleaved by a protease in the stroma. At least three different mechanisms are used to translocate these proteins into the thylakoids. (Interestingly, one uses a SRP-based method for cotranslational import of proteins encoded by the chloroplast DNA, and an SRP-like particle in mitochondria and prokaryotes reveals the common evolutionary ancestry of these eukaryotic organelles and prokaryotes.)

▪▪ The Cytoskeleton Immobilizes and Transports mRNAs

For many years, scientists have observed that not all mRNAs located in the cytosol "float" through the cytosol. The generally accepted view of translation that we covered in this chapter does not take into account that mRNAs preferentially localize to specific areas in the cytosol, but some do. In fact, they associate with both the actin and microtubule cytoskeleton, and are actually transported on actin filaments and microtubules by motor proteins. This kind of movement occurs in animal, plant, and yeast cells and is thought to be widespread in eukaryotes.

One excellent example is the mRNA encoding an isoform of actin called β-actin. A region of the 3′ untranslated portion of this mRNA is called a **zipcode sequence** because it binds to a protein that links it to the branching actin network that forms at the leading edge of a migrating cell (see *Migrating Cells Produce Three Characteristic Forms of Actin Filaments: Filopodia, Lamellopodia, and Contractile Filaments* in Chapter 5). Depending on the cell type, the β-actin mRNA is dragged to distinct cellular locations by myosin motors on actin filaments, or kinesin and dynein motors on microtubules. It is based on the concept that transcription and translation factors can remain attached to mRNA for extended periods of time, even when transcription is complete and translation is not currently occurring. By binding to a different set of these factors, different mRNAs can create their own unique address that motor proteins will deliver them to. This is a rapidly growing area of research, and we expect to understand the mechanisms governing these behaviors much better in the near future.

CONCEPT CHECK #3

Explain in a few sentences the relationship between a signal sequence and protein's primary destination, without using an analogy. How does the shape of protein impact its mechanism of targeting? List evidence supporting the "environmental conditions" and "protein shape" hypotheses presented at the beginning of this section.

8.5 Chapter Summary

This chapter continues the story we began in Chapter 7, concerning how DNA information is handled in cells. Whereas Chapter 7 dealt with the behavior of DNA during cell replication, this chapter concerns how that information is transcribed and translated into RNA molecules and polypeptides, respectively. The story of DNA information transfer will conclude in Chapter 9, where we discuss the sorting and processing mechanisms cells use to distribute some of their proteins into the endomembrane system.

The first section of this chapter concerns transcription, the process whereby DNA information is converted into RNA information. Because this follows immediately after our discussion of DNA replication in Chapter 7, one convenient way to learn the steps of transcription is to compare it to the steps of DNA replication. Single-stranded DNA again serves as a template, but the product is not double-stranded DNA—it is a single-stranded RNA molecule. One very distinctive feature of transcription is that in eukaryotes, the RNA molecules generated from the DNA templates undergo considerable rearrangement and modification prior to their use. These modifications take place inside the nucleus, and the most common finished products (rRNA, tRNA, mRNA) are then exported into the cytosol to participate in the next step of information transfer: translation.

The second section of this chapter concerns translation. Like transcription, the steps of translation are organized into three stages, called initiation, elongation, and termination. The mechanisms prokaryotic and eukaryotic cells use to complete these stages are quite different, but they follow a common theme reflected in the structure and function of specific sites on ribosomes. Translation is a cyclical process, whereby long strings of amino acids are joined together one amino acid at a time.

The final section of this chapter concerns the first stages of protein targeting. A central theme of protein targeting is that proteins synthesized in the cytosol and destined for specific organelles contain at least one sequence of amino acids, called a signal sequence. Signal sequences are recognized by receptor proteins specific for each organelle. The exact mechanisms used to deliver targeted proteins differs considerably from organelle to organelle, ranging from cotranslational unidirectional import in the endoplasmic reticulum to posttranslational bidirectional transport in the nucleus. Chapter 9 picks up this story after proteins are inserted into the endoplasmic reticulum, and follows these proteins as they are sorted and transported to their final destinations.

CONCEPT CHECK ANSWERS #1

Energetically, yes, it would be more efficient to have the nucleus generate primary transcripts, and then have them processed in the cytosol. But doing so introduces a major problem: the proteins in the cytosol assume that the RNAs they encounter are already processed, so they do not lose any time putting them to work. If a cell had a mixture of primary transcripts, partially processed transcripts, and fully processed transcripts in the cytosol, then the cytosolic proteins that use these RNAs would be forced to distinguish between the forms that are useful and those that are still being processed. This would be complicated. The point is that the nuclear membrane and nuclear pore complex constitute a functional boundary, as well as a physical one. Any RNAs that appear in the cytosol are ready to go, and those that are not are kept concealed so as to not confuse the cytosolic machinery that converts RNA information into proteins.

CONCEPT CHECK ANSWERS #2

Evidence supporting a prokaryotic origin for eukaryotic translation:
▸ Use of a single-stranded RNA template that is read in the 5′- to-3′direction
▸ AUG initiation codon
▸ Use of large and small ribosomal subunits
▸ Same genetic code
▸ Same chemical reaction joining amino acids together (formation of peptide bond)
▸ Common termination mechanism
Evidence that eukaryotes have modified prokaryotic translation:
▸ Requirement for processing (5′ and 3′ modifications) of mRNA
▸ Assembly of nuclear export proteins with processed mRNA
▸ Different types of elongation factors
▸ Different mechanisms of initiation

CONCEPT CHECK ANSWERS #3

Sorting by generating a chemically distinct environment (hypothesis one) is an attractive way to explain protein sorting, but none of the sorting mechanisms we have covered in this section rely on this method. (We will see in the next chapter that this method is used for sorting to secondary locations, but not to primary destinations.) Instead, transport to primary destinations is mediated by signal sequences. A signal sequence is a sequence of amino acids in a protein that adopts a specific shape. This shape is recognized by a receptor protein, which must have a complementary shape that recognizes and binds the signal sequence. Binding of the signal sequence to the receptor changes the shape of the receptor so that it can be recognized by the next protein in the targeting mechanism, and this recognition leads to additional rounds of binding and changes in protein shape. Because the signal sequences for each target organelle are different, the specificity of protein targeting is ensured by the specificity of protein shape changes. As such, all sorting mechanisms that use signal sequences support our second hypothesis in this section.

CHAPTER 8   Protein Synthesis and Sorting

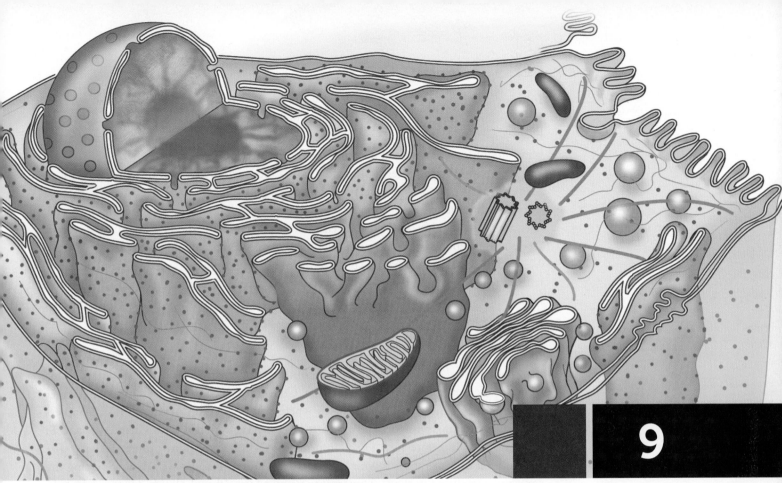

The Endomembrane System
and Membrane Trafficking

■9.1■ The Big Picture

In this chapter, we will finish the story we began in Chapter 6 concerning how cells use DNA information; the principles we identified in Chapters 7 and 8 apply equally as well here. What is new is the concept of vesicle-mediated transport of proteins and membrane phospholipids, which we will need to explain how many proteins reach their final (secondary) destinations after entering the ER. This is more complex than the signal sequence–mediated transport of cytosolic proteins, so we will spend the majority of this chapter discussing how vesicles are formed, moved, targeted, and fused with their destination compartments. As a guide, we will use cell biology principle #5: the endomembrane system serves as the cellular import/export machinery for most macromolecules.

Because most of this chapter concerns the movement of membranes and membrane proteins, it relies heavily on the material covered in Chapter 4. Keep the figures in Chapter 4 in mind when we discuss membrane dynamics, for example, curvature of the membrane at the trans-Golgi network. Also, remember that a given protein may make several different trips, in different vesicles, on its journey to its final destination.

This chapter is divided into seven major sections, yet it is shorter in overall length than many of the chapters before it. Both of these facts matter, for two reasons. First, this chapter is densely packed with information. Collecting seven major ideas into one chapter requires a lot of integration, and this is where comprehension of the previous chapters becomes increasingly important. The first level of integration is with Chapters 7 and 8; together, they illustrate how DNA information is converted into protein activity. The second level of integration is with Chapters 1, 3, and 4, because we are now putting the concepts introduced in those chapters into practical use. Phospholipids, for example, help determine the shape and binding partners of a membrane, and we will see an example of this in sections 9.3 and 9.6. The second reason for the large number of sections is that they divide a very long journey through the network into smaller discrete parts.

Let's briefly discuss how these seven sections are organized:

- The first section defines the endomembrane system and introduces the general concepts of vesicle-mediated traffic that we will need to explain how materials are transported between the organelles in that system. Special emphasis is placed on the principles that govern how vesicles are made, targeted, and fused with their destination compartments. This is a central theme throughout the entire chapter, so it is important to be comfortable with this material before moving on to later sections.

- The second section resumes the narrative in Chapter 8, beginning in the endoplasmic reticulum and ending with the plasma membrane. This section follows a temporal sequence of events much like those in Chapter 8, and again it may be tempting to memorize this sequence. But the same caveat applies from Chapter 8: don't try to learn every detail at the outset. The number of facts is far more challenging to remember than the concepts that link them, so begin this section by asking what one can expect to remember after reading through it the *first* time. This section was written with the expectation that it will be read more than once, and that comprehension will grow in stages with each rereading. While the individual steps may be a bit overwhelming at first, the reason for their existence should not change much no matter how many facts one can assimilate, so start with the *purpose*, then add the steps of the *mechanism* as they become clear.

- The third section is focused exclusively on one organelle, the Golgi apparatus. This is by far the most complicated organelle discussed in this chapter, so it is covered in its own section, in between the sections concerning export and import of materials. This also emphasizes that the Golgi apparatus is not subservient to any given vesicle pathway: in fact, it plays a central role in directing nearly all vesicle traffic. In my classes, I call it the "traffic cop of the cell." Because it plays such an essential role, it is important to understand *what* functions it performs, *how* it performs them, and *why* the Golgi apparatus continues to surprise biologists. Keeping track of what it does can be made easier by partitioning its activities into those associated with export of materials and those with import. The material covering sorting in the trans-Golgi network is especially detailed.

- The fourth section finishes off the export pathway, after the trip through the Golgi apparatus has been completed. We will focus entirely on the path from the Golgi apparatus to the plasma membrane here, but will then have to return to the Golgi apparatus in the next section to help explain the import pathway as well.

- The fifth and sixth sections concern the mechanisms of importing proteins and other materials from the plasma membrane and extracellular space, and then sorting them to their appropriate destinations. In many ways, import resembles the export pathway running in reverse: all materials originate from one location (the plasma membrane) and get sorted to final destinations. The key here is that the primary sorting organelle is *not* the Golgi apparatus; instead, it is a structure called the endosome. Having described how an organelle sorts in section 9.3, we now apply those lessons to a new organelle here. Notice that this section is much shorter than most of those before it, because we are now building on concepts introduced in earlier sections.

- The seventh, final section is again devoted to a single organelle, the lysosome. This is the final destination for some of the proteins sorted by the endosome, and thus represents the end of their journey. It relies heavily on all of the sections before it, and thereby serves as a good test of one's comprehension of all that comes before this section.

My advice for this chapter is to be patient, take it slowly, read one section at a time, and fully expect to reread portions of the material. Look for links to the previous chapters: the events taking place in this chapter rest on the foundation they provide. Stop and ask how a given fact in this chapter is related to the material in Chapter 2, or Chapter 5, for example. The *really* big picture emerges when these connections appear.

9.2 The Endomembrane System Is a Network of Organelles in Eukaryotic Cells

KEY CONCEPTS

▸ The endomembrane system is a set of interconnected organelles that readily exchange materials.
▸ The primary functions of the endomembrane system are to control the export (exocytosis) and import (endocytosis) of materials to/from the extracellular space.
▸ Membrane-bound compartments called vesicles shuttle between organelles in the endomembrane system and are responsible for carrying material from one organelle to another.
▸ The creation, transport, targeting, and fusion of vesicles occurs in nine steps.

Before we begin tracing the journey made by proteins traveling through the endomembrane system, let's take a brief overview of the entire network, and examine how vesicles permit molecular exchange between the organelles.

The Endomembrane System Controls Molecular Transport into and out of a Cell

As we discussed in Chapter 1 (see *The Plasma Membrane Is a Semipermeable Barrier between a Cell and the External Environment*) and Chapter 4 (see *Phospholipid Bilayers Are Semipermeable Barriers*), the phospholipid bilayers that constitute the bulk of cellular membranes are semipermeable barriers that help separate a cell from its external environment. Due to their chemical makeup, they are impermeable to most types of molecules found in cells. Yet cells clearly require a wide variety of molecules, including those that are not permeable, to remain alive and perform their functions. How do the nonpermeable molecules pass into and out of cells and their compartments? As we will see in Chapter 10, small molecules such as ions, monosaccharides, and amino acids can be transported through phospholipid bilayers by transport proteins embedded in membranes, and the resulting gradients of these molecules across a membrane are common forms of stored potential energy. In this chapter, we will examine how cells transport the macromolecules that are too large to be transported by membrane proteins alone.

This kind of molecular transport is much more complicated than that for small molecules, because large pieces of membrane must be moved from one location to another. Several words have been created to describe this kind of transport. The import and export of these large materials is called endocytosis and exocytosis, respectively (the names have spawned verbs to describe the actions involved: *to endocytose, to exocytose*). The progressive movement of materials and membranes during import and export are called the **endocytic** and **exocytic pathways**, as **FIGURE 9-1** shows. Activating either pathway can

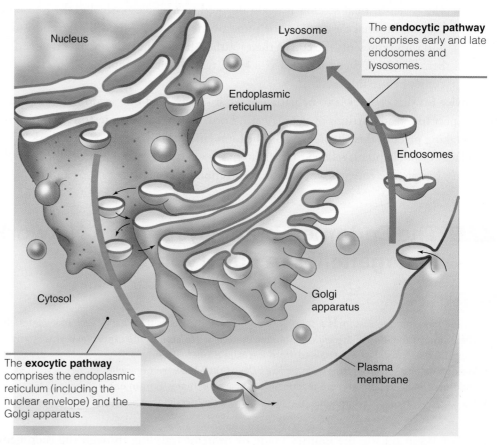

FIGURE 9-1 The endocytic and exocytic pathways.

lead to a significant rearrangement of a cell's interior, so the pathways are carefully controlled. Several organelles cooperate to form the pathways, including the endoplasmic reticulum, Golgi apparatus, endosomes, and lysosomes. These organelles are collectively called the endomembrane system, to emphasize that while each organelle is specialized to perform a subset of functions, they are only truly functional when they cooperate as a system.

Vesicles Shuttle Material between Organelles in the Endomembrane System

One of the most distinctive features of the endomembrane system is the method it uses to move material between compartments. This method is based on the formation, by pinching off a segment of an existing membrane, of small, membrane-bound compartments called vesicles. Due to the hydrophobic nature of the lipid bilayer, the detached membrane segment spontaneously forms a vesicle, somewhat like the way a soap bubble spontaneously forms when one blows on a bubble-making toy. These vesicles are usually shortlived, and so do not typically have any organelle-like functions. Instead, their job is to transport material from one membrane-bound compartment (often called the **donor compartment**) to another (the **acceptor compartment**). Some organelles are composed of several distinct compartments, and vesicles ensure that each compartment "sees" the transported materials in the correct order.

The movement of material in vesicles is typically called **vesicle-mediated transport**, as illustrated in **FIGURE 9-2**. Look closely at the topology of the vesicles and their contents, and notice that once vesicles reach their target destination, they fuse with the target membrane and literally become a part of the acceptor compartment—the contents of the vesicle are actually dumped into the compartment. Both endocytosis and exocytosis require the formation of several different vesicle types that cooperate to carry the transported material (often called cargo) for segments of the overall journey, somewhat like members of a relay team. Each vesicle type always carries its cargo in one direction; no vesicles perform import and export at the same time. To maintain a balance between endocytosis and exocytosis, most organelles in the endomembrane system both generate and accept vesicles travelling in opposite directions. In the exocytic pathway, the forward movement of vesicles (i.e., toward the plasma membrane) is called **anterograde transport**, and the reverse movement is called **retrograde transport**. This two-way traffic is called **vesicle shuttling**.

Exocytosis Is Vesicular Transport of Molecules to the Plasma Membrane and Extracellular Space

Exocytosis is a one-way journey from the endoplasmic reticulum through the Golgi apparatus to the plasma membrane. Upon fusion of the final vesicle types (often called **exocytic vesicles**) with the plasma membrane, membrane-bound proteins become part of the plasma membrane, while soluble proteins are released into the extracellular space. The release of soluble proteins is called **secretion**, the vesicles holding them are called **secretory vesicles**, and the released proteins are called **secretory proteins**.

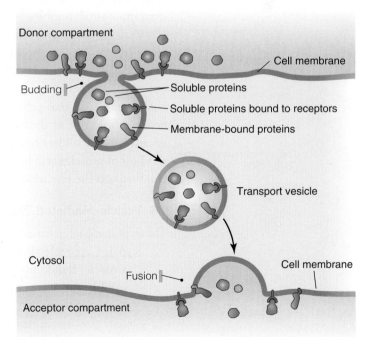

FIGURE 9-2 In vesicle-mediated transport, a membrane-bound vesicle buds from one compartment and fuses with another.

A shorthand description of exocytosis is that it is a method for *adding* materials to the plasma membrane and the extracellular space. A good example of exocytosis, taken from Chapter 6, is the secretion of ECM proteins such as collagen and fibronectin, and fusion of their corresponding receptor proteins (integrins) with the plasma membrane (see Figures 6-4 and 6-6). Cells use exocytosis to modify their surface and the environment that surrounds them: as our body grows during childhood, for example, our cells secrete additional materials (e.g., bone and cartilage proteins) to support the structures they are forming.

■■ Endocytosis Is Vesicular Import of Molecules from the Plasma Membrane and Extracellular Space

Endocytosis is the complement to exocytosis, in that it begins at the plasma membrane and proceeds inward. It is also a way of *subtracting* materials from the cell surface and extracellular space. However, it is not simply exocytosis in reverse, because the endocytic vesicles that form at the plasma membrane do not travel to the same organelles that compose the exocytic pathway. Most of the material imported via endocytosis is quite different from the material that cells secrete via exocytosis, and much of this imported material is degraded or even recycled once it is internalized.

■■ A Host of Proteins Control Vesicular Traffic in the Endomembrane System

Formation, targeting, and fusion of vesicles is not simple. One easy way to visualize the formation of vesicles is to imagine a layer of soap bubbles on the top surface of a glass of water. Imagine that the larger bubbles are organelles, and the smaller bubbles are vesicles: What is the likelihood that a large bubble will split into two smaller bubbles? In my experience, this has never happened spontaneously; the only reliable way to split big bubbles into smaller ones is to vigorously stir the entire mixture. I have tried to pinch off a portion of a large bubble with a spoon, and while I can usually deform the bubble surface somewhat, it invariably slips away from the spoon, pops, or reverts to its original shape once I give up.

Fusing bubbles is even harder. No matter hard I push, the bubbles in my glass won't fuse. Stirring the glass make matters even worse, of course, because now I have created even more small bubbles.

Cells face the same difficulties. They obviously can't stir their entire contents to generate or fuse vesicles, nor would they want to; instead, they have to employ a host of proteins that are specialized to carefully separate or join pieces of membrane from an organelle without causing either the pieces or the organelle to "pop." This is one of the most structurally challenging tasks that cells face, yet every eukaryotic cell does this routinely. Figuring out how has been one of the most difficult challenges cell biologists have faced in the past few decades. As the subject has received considerable attention, a great deal is now known about this host of proteins and how they interact. For the purposes of illustrating our principle, we will focus on the essentials and leave the rest to other, more advanced texts.

■■ Vesicle-Mediated Transport Occurs in Nine Steps.

The steps involved in formation, transport, and fusion of vesicles are illustrated in **FIGURE 9-3**.

> Step 1: Cargo selection. It is important that only the correct molecules are contained in a vesicle. Randomly plucking pieces of membrane off a donor, or enclosing whatever soluble contents of the donor compartment that lie near the membrane, would be a terribly inefficient way to transport molecules between organelles. As such, cells use sorting mechanisms to select only the right molecules for transport. The best understood mechanism is based on recognition of signal sequences in the cytosolic tails of transmembrane receptors for soluble proteins, similar to

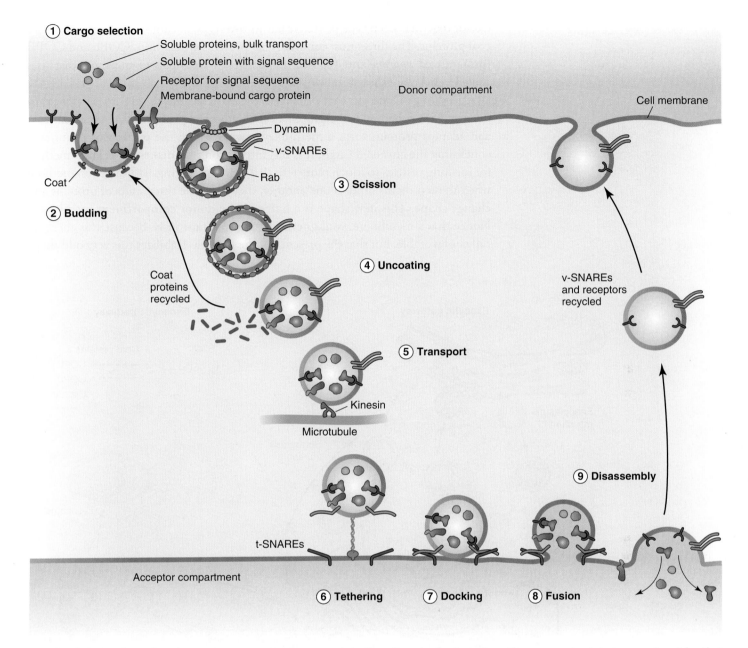

① Cargo selection

- Soluble proteins, bulk transport
- Soluble protein with signal sequence
- Receptor for signal sequence
- Membrane-bound cargo protein

Donor compartment

Cell membrane

Coat

Dynamin
v-SNAREs
Rab

③ Scission

② Budding

Coat proteins recycled

④ Uncoating

v-SNAREs and receptors recycled

⑤ Transport

Kinesin
Microtubule

⑨ Disassembly

t-SNAREs

Acceptor compartment

⑥ Tethering   **⑦ Docking**   **⑧ Fusion**

FIGURE 9-3 Vesicle-mediated transport occurs in nine steps that allow the selective inclusion of cargo proteins into transport vesicles that target and fuse with specific target membranes.

those we discussed in Chapter 8 (see *Signal Sequences Code for Proper Targeting of Proteins*). One such anterograde signal is simply two leucine amino acids in succession: NH_3^+ . . . –Leu–Leu– . . . COO^- (abbreviated LL). Other transmembrane proteins that are transported in vesicles contain these signal sequences, too, even if they don't bind to cargo.

An essential part of cargo selection is the binding of **coat proteins** in the cytosol to the sorting signals. The coat proteins cause the receptors and other proteins with signal sequences to cluster into a patch on the surface of the donor compartment. Recall from Chapter 4 (see *The Fluid–Mosaic Model Explains How Phospholipids and Proteins Interact within a Cellular Membrane*) that membrane proteins can diffuse throughout the plane of a phospholipid bilayer, and this allows them to cluster together. Each type of vesicle has its own type of cargo, cargo receptors, and coat proteins. The three most common coat proteins are clathrin, COPI, and COPII (pronounced *kop-one* and *kop-two*, respectively). Examples of transport that uses coated vesicles are shown in **FIGURE 9-4**.

Step 2: Budding. Once the coat proteins have clustered the cargo into a patch, they interact with other cytosolic proteins called **adaptor proteins**. Together, the coat and adaptor proteins form a mesh-like network lining the portion of membrane containing the clustered cargo. It is this interaction that makes the cellular method for forming vesicles so much more effective than my stirring-with-a-spoon method: when these proteins bind to one another, they obey the three traits of proteins and change shape. This new shape is a bulge in the donor compartment membrane. Notice that a membrane would never do this spontaneously, because it is energetically unfavorable, but that the presence of the proteins stabilizes this very odd shape

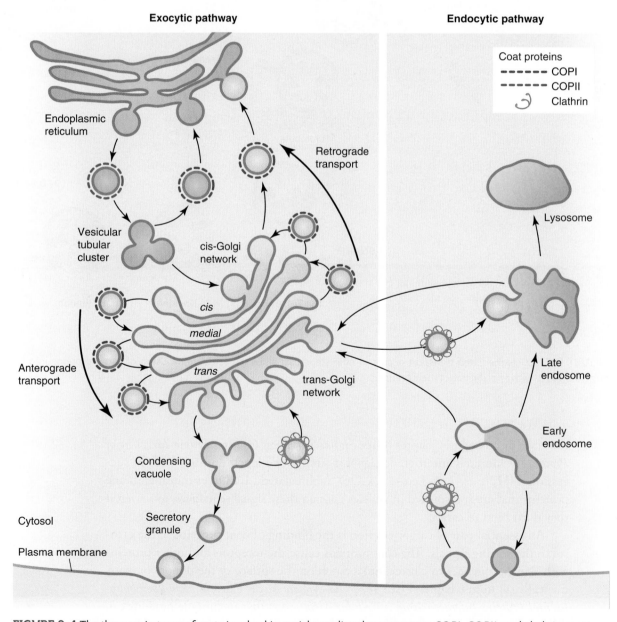

FIGURE 9-4 The three main types of coats involved in vesicle-mediated transport are COPI, COPII, and clathrin coats.

in the membrane. As more adaptor and coat proteins are added to the complex, the bulge grows until it resembles a bud projecting off the donor compartment.

Step 3: Scission. To convert the bud into a vesicle, it has to be cut off of the donor compartment membrane. A specialized set of scission proteins surrounds the "neck" of the bud and tightens until the neck narrows enough to break, somewhat like how an apple falls from a tree when its stem breaks. This releases the bud and causes it to spontaneously form a vesicle. Many biologists confuse the term *scission* with *fission*, so this step is sometimes mistakenly referred to as fission.

Step 4: Uncoating. Remember that the function of the coat proteins is to assist in the deformation of the donor membrane. Now that the vesicle is separate from the donor compartment, these coat proteins are no longer needed. They are removed from the vesicle and recycled for forming more vesicles. Without uncoating, the remaining steps would not occur.

Step 5: Transport. Vesicles can attach to motor proteins that interact with microtubules and actin filaments (see *Microtubule-Based Motor Proteins Transport Organelles and Vesicles* and *Actin-Binding Motor Proteins Exert Force on Actin Filaments to Induce Cell Movement* in Chapter 5), and these motors assist in the transport of the vesicles to their destinations. In most cells, microtubule-based transport is more common than actin based. Transport via these cytoskeletal networks is not absolutely required, but significantly increases the efficiency of vesicle delivery.

Step 6: Tethering. Attachment of a vesicle to its target acceptor-compartment membrane occurs in two steps. In the first, small GTP-binding proteins on the vesicle, called Rab proteins, and tether proteins on an acceptor membrane form a temporary complex that holds the vesicle close to the acceptor membrane, allowing it to "search" or "sample" the acceptor compartment for the correct target proteins. If the vesicle fails to find them, the Rab-tether complex dissociates and the vesicle is free to move on to find another acceptor.

Step 7: Docking. The "searching" referred to in step 6 involves the interaction of two types of proteins, called SNAREs. Those on vesicles are called v-SNAREs, and those on target (acceptor) compartments are called t-SNAREs. Each v-SNARE binds to its complementary t-SNARE, so if the respective t- and v-SNAREs on a tethered vesicle and its target are complementary, they bind to one another, and this stabilizes the temporary associations formed by Rabs and tether proteins.

Step 8: Fusion. The interactions of several different pairs of v-SNAREs/t-SNAREs lead to the fusion of the vesicle membrane with the acceptor compartment.

Step 9: Disassembly. Once fusion has occurred, the SNARE pairs have to be uncoupled to permit them to participate in another round of vesicle fusion. Proteins called NSF and SNAP separate the pairs, and the v-SNAREs are recycled back to the donor compartment in vesicles moving in the opposite direction.

When we move on to the details of exocytosis and endocytosis in the final three sections of this chapter, be sure to refer back to these steps as necessary to keep track of the order of events. The details for different types of vesicles vary, but the overall scheme is always the same.

CONCEPT CHECK #1

Let's say vesicles are "cargo ships" carrying material between organelles. While they may look very similar under a microscope, they must be specialized enough to discriminate one organelle from another. How many distinguishing properties of vesicles can you imagine? Once you compile a list, estimate how many different vesicle types an average cell could make.

9.3 Exocytosis Begins in the Endoplasmic Reticulum

KEY CONCEPTS

▸ Newly synthesized endomembrane proteins are modified in the ER.
▸ Signaling sequences in the newly made proteins signal their secondary destinations. Some have retention signals that keep them in the ER.
▸ Proteins that leave the ER enter the cis-Golgi network via COPII-coated vesicles.
▸ Retrograde vesicle transport returns ER-resident proteins from the Golgi apparatus to the ER.

The function of the exocytic pathway is to correctly sort and distribute newly made proteins in the endomembrane system. All protein synthesis in the endomembrane system takes place on the surface of the rough endoplasmic reticulum, according to the steps outlined in Chapter 8 (see *Secreted Proteins and Proteins Targeted to the Endomembrane System Contain an Endoplasmic Reticulum Signal Sequence*). We will now continue on with this story, beginning with a very small amount of overlap with Chapter 8 to get our bearings.

Newly Synthesized Proteins Begin Posttranslational Modification as ER-Resident Proteins Help Them Fold Properly

Recall that several proteins that are permanent residents of the ER can begin acting on newly synthesized proteins as soon as they pass through the translocon. One is signal peptidase, which cleaves the amino-terminal signal sequencer from most proteins that possess one (see Figure 8-21). A second is N-oligosaccharide transferase, which adds a core oligosaccharide to the side chain of asparagine amino acids if they are present in an N-linked glycosylation sequon (see Figure 8-26). Others include transamidase, disulfide isomerase, and chaperonins like BiP. Keep in mind that these proteins do not interact with *all* newly synthesized polypeptides: a transmembrane protein that contained a signal anchor sequence would not be recognized by signal peptidase, for example.

Let's assume that all of the folding and processing of the proteins in the ER is complete, and continue the journey to the plasma membrane. Keep in mind that a good deal of the experimental evidence for these steps has been obtained in yeast cells, and we do not yet know how applicable the specifics are to other eukaryotes.

COPII-Coated Vesicles Shuttle Proteins from the ER to the Golgi Apparatus

The next stop for these proteins is the Golgi apparatus. The Golgi is a complex organelle, composed of a single compartment that is folded into sacs that appear to be stacked on top of one another. Each sac (sometimes called a cistern; plural, cisternae) is considered a separate subcompartment of the Golgi, and the chemical reactions that take place in each sac are different, performed by a unique complement of enzymes. The Golgi cisternae are arranged in a specific order, so that proteins that enter one side of the Golgi apparatus move progressively through adjacent cisternae until they emerge from the opposite side. The side facing the nucleus is called the **cis face**, and the opposite side is called the **trans face**.

Proteins enter the Golgi apparatus from the ER via the cis face. The very first cisterna on the cis side is elaborately branched, is much larger than most of the other cisternae, and is called the cis-Golgi network (CGN) or the ER Golgi intermediate compartment (ERGIC). To reach the CGN, the proteins in the ER must enter a vesicle targeted to it.

The formation and delivery of the vesicle follows the nine steps outlined in the previous section and illustrated in Figure 9-3.

In some cells, vesicles form at specific locations on the ER membrane called **ER export sites**. In others, the sites of vesicle formation are called **transitional ER**. Step 1, cargo selection, takes place through a variety of mechanisms; some transmembrane receptors bind to soluble cargo molecules, other transmembrane proteins are themselves cargo, and still others are concentrated by "packaging proteins" that are not transported themselves—these proteins contain **ER export signals**. Some soluble proteins are simply trapped in the vesicle, even though they are not selected at all. Carrying this nonspecific cargo is sometimes called **bulk transport**, and it occurs rather frequently in ER-to-Golgi transport. Regardless of what they bind to or how they are concentrated, the cargo molecules are enclosed in a coat of COPII coat proteins, as shown in **FIGURE 9-5**. Notice that a v-SNARE protein is included in the selected cargo, though the mechanism for how this takes place is not yet known.

Step 2, budding, occurs when some COPII proteins interact with adaptor proteins (some researchers do not make a distinction between coat and adaptor proteins, and call the entire complex of five proteins the **COPII complex**). Step 3, scission of the ER-to-Golgi vesicle, is not well understood, so we will not discuss it further.

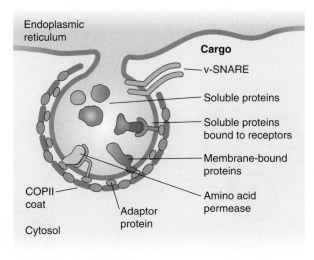

FIGURE 9-5 Cargo proteins are incorporated into nascent COPII vesicles by different mechanisms.

The next 3 steps center around the activity of 1 protein in the COPII complex, called Sec23p in yeast (or Sec23 in other cell types). Step 4, uncoating, occurs when Sec23 cleaves GTP to GDP, resulting in a shape change that releases at least some of the COPII proteins from the adaptors. Step 5, transport, occurs when Sec23 binds to a protein called **dynactin**, which can bind to dynein and kinesin motor proteins. This enables the vesicles to be transported in either direction along microtubules. Step 6, tethering, occurs when Sec23 interacts with at least 2 tethering proteins on the CGN. This, in turn, permits them to bind to a Rab protein, effectively tethering the vesicle. Step 7, docking, takes place when at least 4 proteins interact, including (in yeast) the v-SNARE Bet1p. Step 8, fusion, occurs through specific interactions between 2 pairs of v-/t-SNAREs, 1 of which includes Bet1p in yeast. A recent study found that the Golgi apparatus in these yeast cells contains 7 different v-SNARES and 21 different t-SNAREs, underscoring just how selective these interactions are. Presumably, the remaining v- and t-SNAREs mediate docking and fusion of vesicles that travel between Golgi cisternae or from the Golgi to other organelles. Finally, step 9, disassembly, is mediated by the proteins NSF and SNAP.

Resident ER Proteins Are Retrieved from the Golgi Apparatus

Not all proteins that arrive in the CGN are supposed to be there. Often, proteins that function only in the ER get carried to the CGN by bulk transport, and other proteins, such as v-SNAREs that are required to find the CGN, need to get back to the ER to be useful again. They do so by being included in vesicles undergoing Golgi-to-ER retrograde transport. This means, of course, that all nine steps of vesicle transport must be taken to properly deliver the cargo back to the ER. Rather than retrace the steps, I'll just highlight some interesting points that distinguish these nine steps from those we just covered.

- First, to be included in retrograde transport, recycled proteins must contain what is known as an **ER retention signal**. Three such amino acid sequences are known. One is Lys–Asp–Glu–Leu (abbreviated KDEL according to conventional

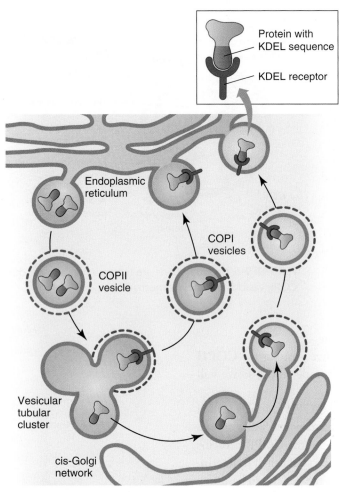

FIGURE 9-6 ER-resident proteins are occasionally carried to the CGN and must be retrieved. Some of these proteins can have a KDEL amino acid sequence that is recognized by the KDEL receptor, which returns them to the ER in COPI-coated vesicles.

single-letter names for amino acids), and the remaining two are dilysine (Lys–Lys–X–X or Lys–X–Lys; abbreviated KKXX and KXK) and diarginine (Arg–Arg or Arg–X–Arg; RR and RXR), where X varies from one cell type to another. These are found near the carboxy terminus of soluble ER proteins (e.g., BiP) and type I transmembrane proteins (see Figure 8-22) that function only in the ER. A transmembrane receptor for proteins bearing the KDEL sequence is located in the CGN. The path taken by proteins bearing the KDEL sequence is shown in **FIGURE 9-6**.

- Second, at least in yeast cells, a protein that functions as a t-SNARE during ER-to-CGN transport appears to serve as a v-SNARE for retrograde transport, permitting it to target both compartments in different roles.

- Third, these receptors interact with proteins in the COPI (*not* COPII) coat complex, which triggers the budding (step 2). So far, no KKXX or RR receptor is known, but there is some evidence that proteins in the COPI complex may bind directly to KKXX- and RR-containing proteins.

- Fourth, the vesicles are transported back to the ER on microtubules, powered by kinesin motor proteins.

CONCEPT CHECK #2

Imagine a new gene was genetically engineered to encode a protein containing an ER retention sequence but no ER signal sequence. If this gene was added to some yeast cells and successfully transcribed and translated, where would this protein be found in the cells?

9.4 The Golgi Apparatus Modifies and Sorts Proteins in the Exocytic Pathway

KEY CONCEPTS

▸ The Golgi apparatus is organized into discrete compartments called cisternae. The cisternae are stacked on top of one another, and are classified as cis, medial, or trans according to their relative location within the overall Golgi structure.
▸ Golgi-resident proteins are primarily responsible for modifying proteins undergoing exocytosis. They are retained in the Golgi apparatus by transmembrane Golgi retention sequences.
▸ The extreme ends of the Golgi apparatus are elaborated into long, tubular structures called the cis-Golgi network and trans-Golgi network.
▸ Both Golgi networks sort proteins into vesicles targeted to different locations. The trans-Golgi network is especially effective at sorting a large number of proteins into many distinct vesicle types.

The heart of the endomembrane system is the Golgi apparatus, because nearly all vesicular trafficking pathways intersect with it. I firmly believe that the Golgi is the most complicated and most fascinating organelle in cells, for the simple reason that the combined power of modern molecular biology, genetics, biochemistry, and microscopy has not yet solved the most obvious question: How does the Golgi work? The cornucopia of modern techniques has uncovered a great deal about *what* functions the Golgi performs, and even the proteins that are involved, but we are still not sure how the proteins perform these functions (**BOX 9-1**). Rather than be discouraged by this, I find the situation very exciting, because finding the answers will require some very unconventional thinking that may very well change the way we see cells in general.

The Golgi Apparatus Is Subdivided into Cis, Medial, and Trans Cisternae

Let's briefly discuss some of the functions the Golgi plays with respect to exocytosis. To do so, we'll need to take a closer look at its structure. Depending on the cell type and organism, the size of the Golgi apparatus can vary tremendously, from as few as three cisternae to well over a dozen; in general, the more exocytosis a cell does, the more elaborate the Golgi structure is. Cells that secrete proteins almost constantly, such as those in the exocrine pancreas, are very popular with Golgi specialists for this reason. Yeast geneticists still prefer their organisms for dissecting the functional role of specific genes and proteins, and so far the two camps have helped spur each other forward without significantly confusing the overall picture. Thus, we can now paint a composite image of the Golgi compiled from unicellular and multicellular organisms, as shown in Figure 9-4.

BOX 9-1

The car engine analogy. To help you understand the current state of our knowledge of how the Golgi functions, let's compare the Golgi apparatus to a car engine. If you were tasked with learning how an engine worked, what steps would you take? You might start by simply looking at the shape of the engine and the parts you can easily see; this is equivalent to the early electron microscopy studies (1950s–1970s) that outlined the overall shape of the Golgi. Based solely on its shape, you might have trouble reaching many conclusions about an engine: there are several parts with distinctive shapes (wires, hoses, etc.), so it is likely a complicated device, but that's about as far as you could confidently go. This was the same conclusion the microscopists reached, for the same reason. Some of the parts (membranes, vesicles) were recognizable because they could be found elsewhere in a cell, but the particular arrangement of even the most familiar parts didn't offer many other clues.

The next step might be to take the engine apart, examine the pieces, and then try to infer how they work together. Biochemists took the lead here, and in the 1960s to 1980s were able to identify numerous proteins, glycoproteins, and lipids in the Golgi, but it was not easy to see how they all fit together to perform a specific function. Reducing a complex machine into simple parts is often not enough to reveal the bigger picture. (I once had a roommate who completely disassembled a car's transmission to see how it worked, and he had the same challenge.)

Knowing the parts, however, is a major step forward, because now one can experiment by tinkering with the parts (usually, one at a time) and testing to see what effect this has on the engine. Pulling a wire on a spark plug will result in lowered engine efficiency, so we can conclude that the wire, the spark plug, or both contribute to the efficiency of the machine. Pulling all the plugs stops the engine dead. Scientists can do the equivalent test by mutating genes that encode the proteins in the Golgi. Yeast geneticists could go a step further by inducing random mutations and then selecting mutant yeast strains that have obvious problems in their Golgi apparatuses. Although this is somewhat like swinging a hammer blindly and then testing which kinds of resulting damage affect the car engine, the power of yeast genetics allows scientists to swing a lot of hammers with very little effort. Once they have several different mutant strains, they can cross them to see if they have additive or complementary effects.

We are currently at the stage where we know most of the molecular parts in the Golgi "engine" and we have observed the effects of mutating many of them. As a result, we can say with confidence that certain proteins play a role in the function of the Golgi apparatus, because when we mutate them, the Golgi doesn't work properly. But that still doesn't tell us *how* the proteins work, any more than a hammer swing tells us how a spark plug works. This explains why we can mention many proteins that operate in the Golgi without explaining how they function.

Vesicular traffic within the Golgi apparatus occurs in both anterograde and retrograde directions, following the nine steps discussed earlier. These vesicles are coated by COPI; hence, ER-to-Golgi and intra-Golgi anterograde transport are mediated by different coat proteins. The cisternae in the Golgi are named according to their relative location and protein composition. The trans-Golgi network (TGN) resembles the CGN in appearance, but sits at the opposite end of the Golgi. Between the CGN and TGN, the cisternae are classified as cis, medial, or trans, according to the enzymes each cistern contains. There is some evidence that, while a specific enzyme may exist in more than one cistern, each cisterna contains a unique mix of enzymes, such that every one is specialized to perform a specific subset of functions. A good example is O-linked glycosylation, wherein a group of sugars are attached to some serine amino acid side chains. This process only takes place in the medial-Golgi cisternae. Another example is the phosphorylation of a mannose sugar on some glycoproteins, which we will discuss later in this chapter.

■■ Posttranslational Modification of Proteins in the Golgi Is a Stepwise Process

The primary function of the Golgi cisternae immediately following the CGN is post-translational modification of proteins undergoing exocytosis. These cisternae contain dozens of different enzymes that add and subtract sugars from the core oligosaccharides attached to specific arginines by N-oligosaccharyl transferase in the ER (see *As Proteins Enter the ER Lumen, They May Be Post-translationally Modified* in Chapter 8). The steps involved are complex, vary considerably from cell to cell, and are not well understood. They most likely require a careful spatial distribution of enzymes so that sequential steps in the processing can take place in the correct order, while still allowing the vesicular trafficking to proceed smoothly. This is an excellent example of our knowing *what* the Golgi does, but not yet clearly understanding *how*.

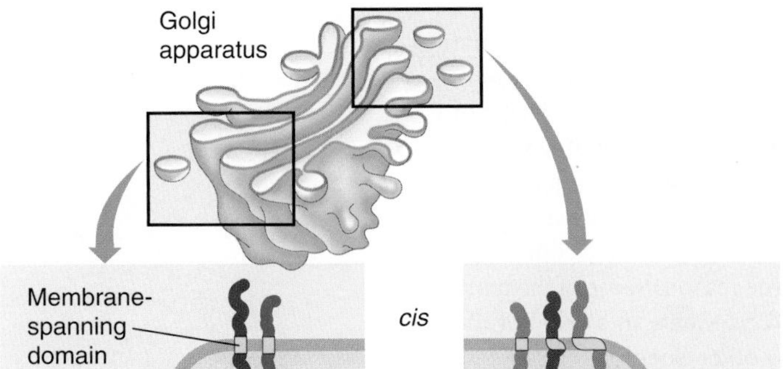

■■ The Transmembrane Region of Resident Golgi Proteins Serves as a Retention Sequence

Just as ER-resident proteins contain retention signal sequences, proteins that reside in the Golgi also contain **Golgi retention sequences**. For transmembrane proteins, the sequence includes the signal anchor sequence that initially targets it to the ER. This sequence, plus the amino acids that flank it on either side of the membranes, are responsible for targeting the protein to the ER, determining the topological orientation of the protein (type I, type II, etc.), and retaining the protein in specific compartments of the Golgi.

Two models have been proposed to explain how these sequences retain proteins, as shown in **FIGURE 9-7**. One, called the **kin recognition model**, is based on the aggregation of proteins that recognize each other's membrane-spanning domains. As these proteins are carried through the Golgi apparatus by anterograde vesicles, they encounter other proteins with similar transmembrane domains and bind to them. Eventually, by a mechanism

Kin recognition model
Proteins recognize each other via their membrane-spanning domains, preventing forward movement due to aggregation.

Bilayer thickness model
Proteins move along secretory pathway until length of the membrane-spanning domain matches thickness of the bilayer.

FIGURE 9-7 Two models for retention of transmembrane proteins in the Golgi.

still not fully understood, this complex no longer enters vesicles and is therefore trapped in a specific cistern. A second model capitalizes on the observation that the amount of cholesterol in Golgi membranes increases from cis to trans; the concentration of cholesterol contributes to the overall thickness of a membrane (see *The Fluid–Mosaic Model Explains How Phospholipids and Proteins Interact within a Cellular Membrane* in Chapter 4), so there is a gradient of the membrane thickness in the Golgi. According to the **bilayer thickness model**, the length of the transmembrane domains in Golgi proteins is tailored to match the thickness of the region in the Golgi where these proteins are retained. To put it more simply, proteins with relatively short transmembrane domains localize preferentially to the cis side of the Golgi, those with intermediate-length transmembrane domains localize to the medial cisternae, and those with the longest domains are concentrated in the trans-cisternae.

Note that neither of these models requires there to be a receptor for the Golgi retention sequences. In almost every instance we encounter in this book, proteins that contain some sort of targeting sequence bind to a corresponding receptor in the target, yet this is likely not the case for Golgi retention. This is an excellent example of how creative thinking is helping to uncover how the Golgi works.

The Trans-Golgi Network Sorts Proteins Exiting the Golgi Apparatus

The TGN is even more complex than the CGN, in both structure and function. It was traditionally thought of as the last cisterna in the exocytic pathway, but we now know that it is composed of more than one cisterna, and is very dynamic: its structure in individual cells can change considerably, according to the kind of vesicle traffic it carries at any given time. These cisternae are characterized by their elongated, tube-like extensions that project out into the cytosol and are often found lying very close to the ER.

This proximity to the ER is important, and raises an important point about the differences between how we learn about the Golgi and how it actually functions. By convention, it is easiest to learn about the endomembrane system by examining the sequence of compartments that a trafficked protein visits as it passes through the system. If such a journey is diagrammed for exocytosis, it looks much like that shown in Figure 9-1. This is easy to visualize, because it implies that the physical movement of the proteins begins in the deep interior of a cell, then radiates outward, in a fairly straight path, until the protein reaches the plasma membrane / extracellular space. According to this logic, the ER and TGN should be far apart, separated by the CGN and the rest of the Golgi cisternae. But this is not the case for most cells. Recall from Chapter 5 (see *Microtubule-Based Motor Proteins Transport Organelles and Vesicles*) that both the ER and the Golgi are attached to microtubules, primarily by kinesin and dynein motor proteins, respectively. This has important implications for where these organelles are positioned in cells: the kinesin motors tend to stretch the ER out into a web by pulling it toward the plus ends of the microtubules, while the dynein motors tend to condense the Golgi on top of the centrosome. Therefore, a protein could be synthesized on a portion of the ER held near the cell periphery by kinesin motors, then enter a vesicle that moves into the interior to find the CGN via a dynein motor. Once it completes the journey through the Golgi cisternae, it enters a vesicle that departs the TGN and moves back out to the periphery to fuse with the plasma membrane. In short, vesicle traffic during exocytosis is *not* linear, even though it is tempting to visualize it as such. Keep in mind that the vesicles carrying proteins in the exocytic pathway actually follow a complex, sometimes convoluted path en route to the plasma membrane.

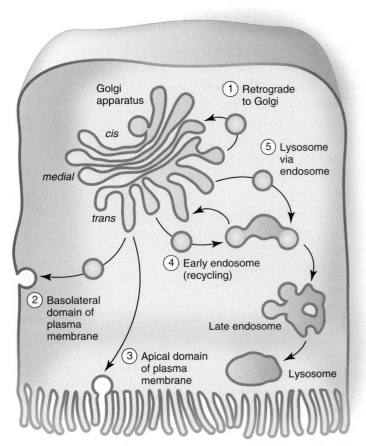

FIGURE 9-8 Proteins leaving the TGN can travel to at least five different destination compartments.

When a protein leaves the Golgi, it can be targeted to as many as five different destination compartments, as shown in **FIGURE 9-8**. We have already discussed three of these acceptors in this and earlier chapters: the first and second are represented by specific subdomains of the plasma membrane, for example, the apical and basolateral domains of epithelial cells (see *Tight Junctions Form Selectively Permeable Barriers between Cells* in Chapter 6). Retrograde transport to the Golgi is a third destination. We will discuss the fourth and fifth compartments when we cover endocytosis later in this chapter.

The great variety of destinations for vesicles leaving the TGN means that the TGN is especially skilled at sorting proteins into appropriate vesicles. How this sorting occurs confounded cell biologists for decades, because traditional sorting mechanisms alone simply could not explain why the TGN functions so efficiently. According to the standard protein targeting models (such as those discussed in Chapter 8 see *Signal Sequences Code for Proper Targeting of Proteins*), proteins destined for each location would contain a sorting sequence, and would have to be recognized by at least one receptor protein in the TGN. If this were true, the next logical question to ask is, "How did the seven (or more) receptor proteins get sorted to the TGN in the first place?" This would require yet another signal for these receptors, and then additional receptors for their signals, and so on *ad infinitum*. Any model that requires an infinite number of receptors for still more receptors for cargo proteins would quickly collapse under its own weight. Fortunately, we now know why this is not the case.

The TGN uses several different mechanisms for sorting proteins (also known as cargo selection, which is step 1 in vesicle trafficking) without falling into the infinite receptor trap. For example:

- Posttranslational modifications (for review, see *Cells Chemically Modify Proteins to Control Their Shape and Function* in Chapter 3) can act as sorting signals. These include N- and O-linked glycosylation, phosphorylation, and ubiquitination.
- Protein aggregation, somewhat like the kin recognition model discussed above, also helps sort proteins. Both kin recognition and clustering by nontransported cargo recognition proteins, such as those we discussed for vesicle formation in the ER, contribute to aggregation-based sorting.
- The traditional signal-receptor method we just dismissed actually does apply to some proteins passing through the TGN. We will address why this makes sense later in the chapter when we discuss endocytosis.

An entirely new kind of sorting mechanism has also been found in the TGN that is based on the formation of membrane microdomains such as lipid rafts. The basis of this **lipid raft hypothesis** is that the TGN assembles patches of membrane enriched in cholesterol and other lipids, and that some proteins have an especially high affinity for these rafts. For example, a protein known as ARF acts as an adaptor to cluster cargo proteins that synthesize a specific phospholipid called phosphatidylinositol 4-phosphate (PIP_4). The PIP_4, in turn, binds to lipid-binding proteins to change the local phospholipid composition of the TGN membrane, thereby promoting the formation of a lipid raft. The raft then attracts the GPI-bound proteins we discussed in Chapters 4 (see *Membrane Proteins Associate with Membranes in Three Different Ways*) and Chapter 8 (see *As Proteins Enter the ER Lumen, They May Be Posttranslationally Modified*) and cytosolic proteins that specifically target it to the plasma membrane. The key to this method of protein sorting is that the TGN is found very near the ER, such that lipid transfer proteins carry a set of lipids from the ER (where they are synthesized; see *Most Membrane Assembly Begins in the SER and Is Completed in the Target Organelle* in Chapter 4) directly to the TGN. This mechanism is best understood in polarized epithelial cells, which selectively target lipid rafts to the apical domain of their plasma membrane.

Collectively, these mechanisms compose only the first part of step 1, cargo selection. Attachment of the coat proteins occurs next, and this is somewhat more complicated for the TGN than for other compartments, because vesicles leaving the TGN are attached to different coats, based on their destination. Those targeted to the Golgi via retrograde transport bind to COPI, but the rest bind to clathrin. Adaptor proteins play an important role in selecting the type of coat protein for each cargo cluster. The clathrin adaptor proteins that form exocytic vesicles are not yet known.

■■■ Budding of Vesicles at the TGN Likely Occurs by Several Different Mechanisms

Having selected cargo for transport, the next step in exocytosis is step 2, budding. Budding in the TGN is somewhat unusual, in that the buds are often long, tube-shaped structures rather than the pea-shaped buds formed by the ER and other Golgi cisternae. This likely occurs because the shape of the buds is actually an important part of the cargo sorting process, so that different shapes of buds hold different types of cargo proteins.

At least three different mechanisms for budding have been proposed:

- One involves the recruitment of proteins that actively bind phospholipids and thereby induce curvature in the TGN membrane. How these curvature-inducing proteins are recruited to patches of TGN membrane is not yet known.
- A second, and quite novel, mechanism is modification of the phospholipids. An enzyme called phospholipase A_2 cuts one of the two acyl chains off of phospholipids in the cytosolic face of the TGN membrane, leaving a wedge-shaped "half phospholipid" (technically called a lysolipid) in the membrane with its acyl end pointed inward, toward the exoplasmic face of the membrane. When enough of these wedges accumulate in the cytosolic face, they begin to pack together and, simply because of their shapes, they force the membrane to curve.
- The third possibility is that **phospholipid asymmetry** could cause the bending. This asymmetry could arise from phospholipid flippases (see Figure 4-10) or lipid transfer proteins preferentially adding phospholipids to the cytosolic face of the TGN. How these proteins could target specific sites is not yet known.

■■■ Scission of Vesicles Budding from the TGN Often Requires Microtubule Motor–Based Forces

Inducing curvature in the TGN membrane is not sufficient to explain how the extended tubules at the edge of the TGN form. Just as microtubule-based motors are responsible for determining the location of the ER and Golgi apparatus in the cytosol, they also provide the traction force necessary to elongate the curved patches of TGN membrane. Several models have been proposed to explain how these motors preferentially associate with the

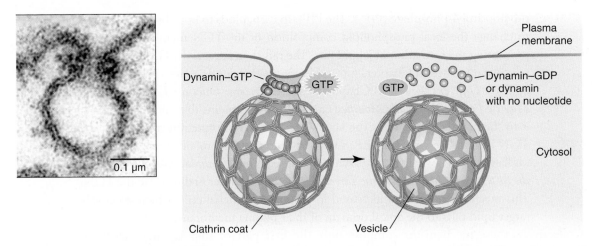

FIGURE 9-9 Dynamin uses GTP to regulate scission of a vesicle from a donor compartment. Photo courtesy of Pietro de Camilli, Yale University School of Medicine.

TGN buds. The simplest is that the cytosolic tails of cargo receptors and transmembrane cargo proteins bind directly to the motor proteins, based on evidence that dynein motors do so in some experimental systems. Second, the motors may bind to adaptor proteins that in turn bind to specific types of cargo. Regardless, it is important that these regions of the TGN interact with microtubule motors.

The actual scission events are best understood for clathrin-coated vesicles targeted to the plasma membrane. These vesicles vary in shape and size, because the elongated tubes formed by the microtubule motors can be cleaved at any point, yielding small round vesicles or more elongated tubes. (This is how the elongated fibripositor that carries collagen to the plasma membranes [see Figure 6-4] is formed.) The protein that actually cleaves the membrane is called **dynamin**. I particularly like dynamin, because one can see how it works by simply looking at it, as **FIGURE 9-9** shows. Dynamin forms a loop around the neck of a budding clathrin-coated vesicle, and then uses GTP cleavage to squeeze tight, choke off, and eventually snap the membrane between the bud and the rest of the compartment. Additional proteins are likely required, because mutation or inhibition of their activities stops exocytosis at the TGN, but exactly how they participate is still unknown.

We will cover the remaining steps of post-TGN vesicle transport during exocytosis in the next section.

CONCEPT CHECK #3

Recall that the Golgi cisternae are connected to the microtubule network by motor proteins. The drug nocodazole can bind to tubulin subunits and prevent them from polymerizing, and this eventually leads to a complete loss of microtubules in cells. What effect do you think nocodazole treatment would have on the sorting of cargo in the TGN and targeting of vesicles leaving the TGN?

▇ 9.5 ▇ Exocytosis Ends at the Plasma Membrane

KEY CONCEPTS

▸ Cells regulate the last stage of exocytosis (fusion) for most exocytic vesicles, to control when and how much material is released into the extracellular space and to control the delivery of membrane-associated proteins to the plasma membrane.

▸ Controlled secretion is also called regulated secretion, and is under the control of signaling pathways.

The culmination of exocytosis is the arrival of a vesicle at the plasma membrane, followed by fusion of the vesicle. This releases the soluble contents of the vesicle (e.g., collagen) into the extracellular space, and inserts membrane proteins (e.g., collagen receptors) into the plasma membrane. How these vesicles arrive at the plasma membrane is not completely understood, but it is likely that steps 4 (uncoating) and 5 (transport) are closely linked. While the best known example of clathrin uncoating comes from studies of endocytosis rather than exocytosis, it is quite likely that the mechanisms are similar. A protein called hsc70 uses ATP energy to shatter the clathrin coat of endocytic vesicles, and frees it to fuse with a target membrane. Once uncoated, vesicles can be transported to the plasma membrane by microtubule-based motors or, in some instances, myosin-like proteins that move along actin filaments.

Cells Use Two Mechanisms for Controlling the Final Steps of Exocytosis

Steps 6–9 for post-TGN vesicles vary considerably. The main reason for this is that not all secretion is the same. In some cases, it is beneficial for cells to fuse their exocytic vesicles almost immediately after they arrive at the plasma membrane. Liver cells must secrete albumin proteins into the blood almost constantly, so requiring the vesicles carrying albumin to run a gauntlet of regulatory steps would not be of any benefit. On the other hand, secretion of neurotransmitters must be carefully timed to control the activity of the nervous system and the muscles and glands it controls.

Albumin and neurotransmitter secretion represent the opposite ends of a spectrum of secretory behaviors that cells exhibit. Secretion that is not controlled is called **constitutive secretion**. The word *constitutive* isn't used much in everyday language, but it is quite apt for this process; derived from the same linguistic source that gave us the word *constituent*, which means "an essential part of the whole," biologists like the term because it indicates what came first, and usually implies that it is uncontrolled. When applied to secretion, it simply means it is the foundation for all other types of secretion. In evolutionary terms, there is little doubt that it was the first type of secretion to appear.

On the other end of the spectrum lies **regulated secretion**. The type and degree of regulation varies widely between the types of molecules and the cells that secrete them. It is generally better for cells to control the final steps of exocytosis, so even those that permit some constitutive secretion put limits on most secretory vesicles. Cells would have a difficult time controlling their external environment if they couldn't control what and when they add to it. Cells that specialize in regulated secretion often store the materials they secrete in a concentrated form in customized secretory vesicles, sometimes called **zymogen granules**. The intermediate state between a dilute exocytic vesicle and a condensed secretory vesicle is sometimes called a **condensing vacuole**. **FIGURE 9-10** shows examples of these structures in a pancreas cell.

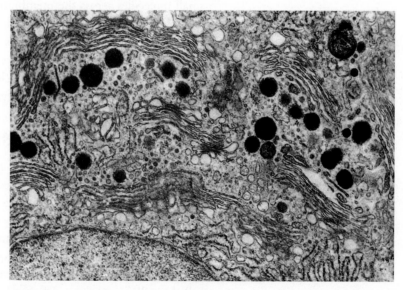

FIGURE 9-10 Examples of the golgi cisternae, rough ER (RER), and various stages of secretory granule formation. © Don W. Fawcett/Photo Researchers, Inc.

Regulated Exocytosis Is Controlled by Cellular Signaling Proteins

Recall that GTP-binding proteins called Rabs play a major role in tethering vesicles. Because they use GTP binding and cleavage as a regulatory mechanism, they are relatively easy to control by groups of proteins that influence their GTP/GDP binding state. As a result, cells can regulate their secretion by controlling the shape and activity of the proteins required to complete exocytosis. These regulatory proteins are, in turn, under control of other regulatory proteins. We will focus exclusively on this regulatory system, made up of networks of interacting signaling proteins, in Chapter 11.

▮ 9.6 ▮ Endocytosis Begins at the Plasma Membrane

> **KEY CONCEPTS**
>
> ▸ The onset of endocytosis is most often indicated by the clustering of cargo receptors on the plasma membrane, accompanied by the assembly of a clathrin coat on the cytosolic face of the cluster. In micrographs, this structure resembles a pit in the membrane, so it is often called a coated pit.
> ▸ Coated pits complete the nine steps of vesicle transport and deliver the vesicle to an organelle called the endosome.

While exocytosis adds materials to the extracellular space and plasma membrane, endocytosis removes them. Before we address how this occurs, let's briefly discuss why cells do this. The import/export analogy that began this chapter provides a clue. A city can use importation to acquire items that it needs but cannot make itself, and cells can use endocytosis for a very similar reason.

Two of the most important functions of endocytosis do not easily translate into analogy. First, cells in multicellular organisms use endocytosis to destroy dangerous objects (including foreign cells and viruses) in the extracellular space. In doing so, they keep their surroundings in good order. Second, endocytosis provides a method for remodeling the cell surface. Remember that the plasma membrane is the interface between a cell and the environment, so it is typically crowded with a myriad of proteins (and in eukaryotes, proteoglycans) responsible for sensing environmental cues, attaching to other cells and objects, transporting small molecules across the membrane, etc. As conditions in the environment change, it is important for a cell to adapt by readjusting the types and amounts of proteins on its surface.

With these goals in mind, let's follow the path of endocytosis to understand how it takes place. As with exocytosis, a central feature of endocytosis is the selective formation, transport, and fusion of vesicles. The same nine-step mechanism used for forming and targeting vesicles during exocytosis also applies to endocytosis.

▮▮ Clathrin Stabilizes the Formation of Endocytic Vesicles

Most of what we know about the nine steps of vesicle transport comes from studies of endocytosis. Clathrin was discovered in egg cells (oocytes) by microscopists in the 1960s, as shown in **FIGURE 9-11**. As cargo receptors, driven by adaptor proteins in the cytosol, cluster on the surface, the clathrin begins to condense on the cytosolic surface of the plasma membrane. This causes the membrane to deform a bit, forming what is called a **coated pit**. Eventually, the coated pit grows into a complete coated vesicle, pinched off the membrane by dynamin. The coat is then released by the hsc70 protein when it interacts with a protein called **auxillin**, and the vesicle is transported into the cell interior. In some cases, this occurs on actin filaments; in others, the driving force for transport is not known. The journey of an endocytic vesicle ends when it fuses with the next compartment in the endocytic pathway, the endosome, as summarized in **FIGURE 9-12**. The tethering, docking, fusion and disassembly of the transport machinery proceeds as shown in Figure 9-4.

9.7 The Endosome Sorts Proteins in the Endocytic Pathway

KEY CONCEPTS

▶ The endosome is formed by the fusion of endocytic vesicles with specific vesicles that bud from the TGN.

▶ The endosome sorts materials arriving from the plasma membrane; cargo receptors are recycled to the plasma membrane, while cargo remains in the endosome.

▶ The lumen of the endosome is slightly acidic relative to the extracellular space, and this acidity is key to the sorting mechanism.

▶ This sorting mechanism is very different from the sorting mechanisms used in the Golgi apparatus.

Not all proteins captured in endocytic vesicles remain in the cell interior. Some are receptors for molecules in the extracellular space, so they can be recycled by moving them back to the plasma membrane, similar to how coat proteins can be reused. This recycling requires a sorting mechanism to separate the receptors from their cargo, but it is quite different from those in the CGN and TGN.

The Endosome Is Subdivided into Early and Late Compartments

One obvious difference between sorting in endocytosis and exocytosis is that endocytic sorting requires multiple compartments to complete. Collectively, these sorting compartments are called endosomes. The function of endosomes is to properly separate the imported cargo from their receptors, and then sort the receptors into a vesicle that returns to the cell surface while leaving the cargo behind. The separation and sorting take place in different compartments, which are called early and late endosomes.

Early endosomes are derived from vesicles that originate on the plasma membrane. These clathrin-coated vesicles contain Rab proteins, adaptors, and SNAREs, and follow the usual 9 steps to form, target, and fuse with either specific vesicles from the TGN (to create new early endosomes), or with preexisting early endosomes. It is important to note that even though the overall steps are the same as for other types of

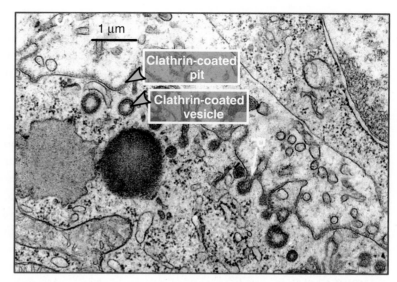

FIGURE 9-11 Transmission electron micrograph of clathrin-coated pits and vesicles at the oocyte surface. Photo © Roth and Porter, 1964. Originally published in The Journal of Cell Biology, 20: 313-332. Used with permission of Rockefeller University Press.

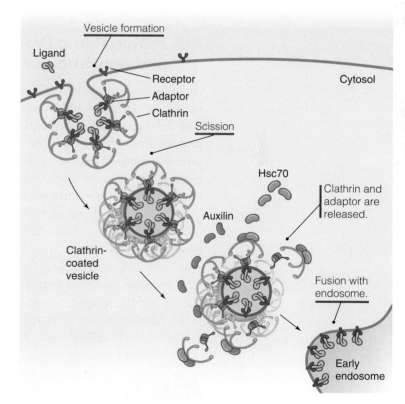

FIGURE 9-12 The role of clathrin in endocytosis. Endocytic vesicles follow the typical nine steps, with clathrin providing the coat. Auxillin interacts with hsc70 to trigger uncoating.

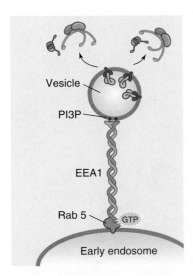

FIGURE 9-13 Clathrin-coated vesicles are uncoated and tethered to early endosomes via EEA1, which binds to phosphatidylinositol-3-phosphate and Rab5.

Vesicle

PI3P

EEA1

Rab 5 — GTP

Early endosome

vesicle transport and targeting, each vesicle type contains its own specific complement of these proteins, and the unique combination of proteins in each vesicle allows it to fuse selectively with the correct target. For example, over 30 different proteins are known to participate in the formation of clathrin-coated vesicles at the plasma membrane. In the case of vesicle fusion with early endosomes, the specificity is provided by a tethering protein called early endosomal antigen-1 (EEA-1), which binds to a t-SNARE called syntaxin, a Rab protein called Rab5, and a specific phospholipid called phosphatidylinositol-3 phosphate (PIP$_3$) on early endosomes, as shown in **FIGURE 9-13**. This specific combination of interacting molecules helps ensure that the correct vesicles fuse with vesicles from the TGN or the early endosomes. The entire 9-step process takes less than 1 minute in many cells.

Once the vesicles fuse together, the cargo and their receptors dissociate. This happens because the pH of the early endosome is slightly lower (6.4–6.8) than the pH of extracellular space (7.0–7.4). The increased acidity changes the charges on the receptor and cargo molecules, so they change shape. (To review the impact of pH on protein shape, see *Proteins Change Shape in Response to Changes in Their Environment* in Chapter 3.) These new, acid-induced shapes do not bind to one another, so the molecules separate. In the case of soluble cargo molecules, the cargo releases from the membrane receptor and diffuses into the interior of the early endosome. The receptors are then clustered together and returned to the cell surface in yet another type of vesicle; about 75% of recycled receptors return in these vesicles. Very little is known about this return-journey pathway, in part because it happens so quickly. Some evidence suggests it requires clathrin coats and/or transport by microtubule motors, but no general consensus has emerged yet.

Proton Pump Proteins Play a Central Role in the Sorting and Activation of Endosomal Contents

An important question to ask about the sorting mechanisms in endocytosis is why the pH in vesicles and endosomes is lower than in the extracellular space. The simplest answer is that the vesicles from the TGN also contain transmembrane proteins that actively pump protons from the cytosol into the vesicle interior (lumen). The more of these proton pump proteins an endosome has, the more acidic it becomes over time. We will discuss the way these pumps work when we discuss transport of molecules across membranes in Chapter 10, and how they get to the early endosome in the next section of this chapter.

The activity of the pump proteins in the endosomal membranes helps define the difference between early and late endosomes. As the early endosome continues to pump protons into its lumen, the pH continues to drop. Cell biologists have decided that when the pH drops below 6.0, the early endosome is now called a **late endosome**. It is generally assumed that by the time an endosome has matured to the late stage, the bulk of the receptor–cargo dissociation and sorting has already taken place. In most cells, this maturation takes only a few minutes. In others, it takes over an hour, resulting in an intermediate stage that is sometimes called a **recycling endosome**.

9.8 Endocytosis Ends at the Lysosome

KEY CONCEPTS

▸ Complete degradation of endocytosed materials takes place in an organelle called the lysosome.
▸ The lysosome is likely generated from the endosome in several ways, and requires fusing a vesicle from the TGN that contains essential proton pump proteins and digestive enzymes with the endosome.

- The digestive enzymes and proton pumps are carried to the endosome by receptors that recognize a mannose-6 phosphate tag on proteins in the TGN. The mannose-6 phosphate receptors are sorted and returned to the TGN, while the hydrolases and pumps remain in the endosome. The pumps are responsible for triggering the drop in pH in the endosomes.
- Once the plasma membrane cargo receptors and mannose-6 phosphate receptors have been removed from the endosome, it becomes progressively more acidic; when it reaches pH 5.0, shape changes in the hydrolases are triggered, which activate the endosomes and allow them to begin digesting cargo.
- Once digested, the cargo building blocks (sugars, nucleosides, amino acids, etc.) are transported into the cytosol for reuse.

If a protein does not leave the endosome before it becomes a late endosome, it likely will never leave, and will instead be degraded. Recall that the reason a cell sorts its recycled receptors is to prevent them from falling into this late stage, because from this point on, there is no going back. While the precise mechanisms for generating lysosomes from endosomes are still not completely understood, it is clear that lysosomes originate from late endosomes. One popular explanation of lysosome generation is called the maturation model. According to this model, the late endosome continues to acidify, and when it reaches a pH near 5.0, it "matures" and is called a lysosome. (Other models suggest that endosomes and lysosomes are separate compartments linked by vesicle trafficking.) The name lysosome was created to emphasize that this "body" (Greek, *soma*) is responsible for lysing (Greek, loosening) molecules. In plain English, a lysosome *digests* molecules. The maturation from endocytic vesicle to lysosome is illustrated in **FIGURE 9-14**.

The chemistry underlying this digestion is remarkably simple. This is an excellent time to review how sugars, nucleic acids, proteins, and phospholipids are synthesized by cells (Chapters 1–4). Polysaccharides, nucleic acids, and proteins are polymers of building blocks held together by covalent bonds. These covalent bonds are generated by joining a –OH group with either another –OH group (polysaccharides, nucleic acids) or an –NH$_2$ group (proteins). Phospholipids are formed by joining two fatty acids to a phosphoglycerol molecule using the same type of reaction, then adding a head group to the phosphate by the same type of reaction yet again. All of these reactions are called dehydration reactions because they create a covalent bond with a water molecule as a byproduct. Therefore, breaking all of these bonds is quite straightforward: a group of proteins adds water back to these chemical bonds, breaking the polymers back into their building blocks. Because the addition of water breaks these bonds, the process is called **hydrolysis**. The enzymes that perform hydrolysis are therefore called hydrolytic enzymes, or hydrolases.

So far, our journey along the endocytic pathway has not encountered any hydrolytic enzymes, but we know they are present and active in lysosomes. To explain where they come from, we need to back up a bit and return to the Golgi apparatus. This is where the exocytic and endocytic pathways overlap, so we need to proceed carefully at this point to avoid confusion. Have another look at Figure 9-14 as we go through this.

Endogenous Proteins Destined for the Lysosome Are Tagged and Sorted by the Golgi Apparatus

The last stage of endocytosis can be confusing, because it results from the intersection of two membrane-trafficking pathways. The first pathway begins at the plasma membrane, then moves to the early and late endosomes, and ends in the lysosome. This pathway explains where the material to be digested comes from, but it does not explain where the proteins

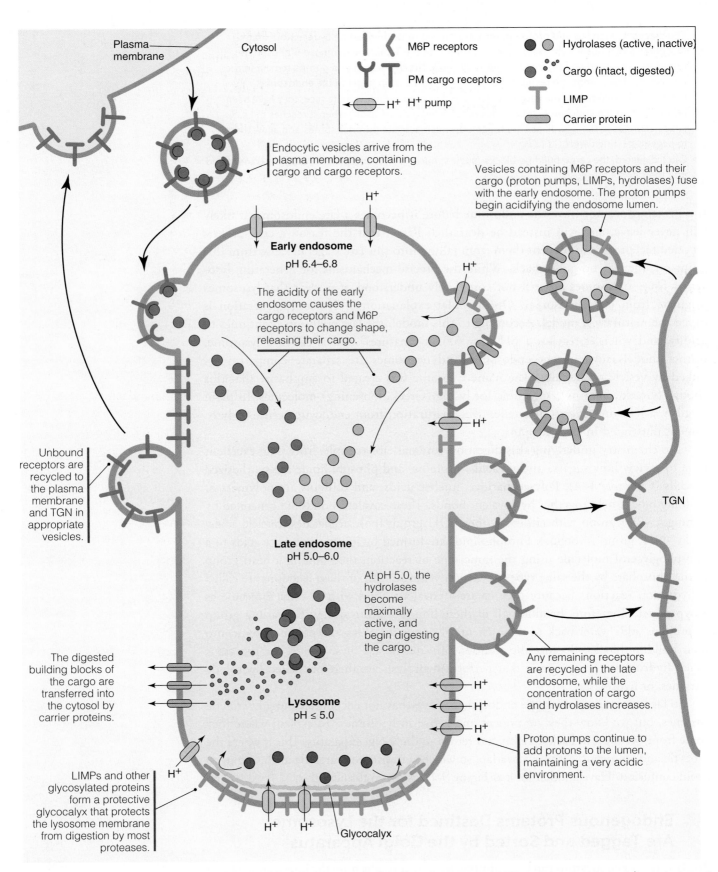

Legend:
- ⟨ M6P receptors
- Y T PM cargo receptors
- ← H⁺ H⁺ pump
- Hydrolases (active, inactive)
- Cargo (intact, digested)
- T LIMP
- Carrier protein

Plasma membrane

Cytosol

Endocytic vesicles arrive from the plasma membrane, containing cargo and cargo receptors.

Vesicles containing M6P receptors and their cargo (proton pumps, LIMPs, hydrolases) fuse with the early endosome. The proton pumps begin acidifying the endosome lumen.

H⁺

Early endosome
pH 6.4–6.8

The acidity of the early endosome causes the cargo receptors and M6P receptors to change shape, releasing their cargo.

Unbound receptors are recycled to the plasma membrane and TGN in appropriate vesicles.

Late endosome
pH 5.0–6.0

At pH 5.0, the hydrolases become maximally active, and begin digesting the cargo.

TGN

Any remaining receptors are recycled in the late endosome, while the concentration of cargo and hydrolases increases.

The digested building blocks of the cargo are transferred into the cytosol by carrier proteins.

Lysosome
pH ≤ 5.0

Proton pumps continue to add protons to the lumen, maintaining a very acidic environment.

LIMPs and other glycosylated proteins form a protective glycocalyx that protects the lysosome membrane from digestion by most proteases.

H⁺   H⁺   H⁺

Glycocalyx

FIGURE 9-14 Maturation of endosomes to form lysosomes. Sorting of cell surface receptors and M6P receptors requires acidic environment provided by proton pumps.

that actually do the digesting come from. These hydrolytic enzymes do not come from the extracellular space, so they are not endocytosed; instead, they are made by the cell that captures the endocytosed material. This means that these proteins are encoded by mRNA in the cytosol, translated by ribosomes on the rough ER, and complete the journey from the ER to the TGN, alongside those proteins that are destined for the plasma membrane. They follow all the steps we discussed in Chapter 8, and follow the same steps of exocytosis described in this chapter, until they reach the TGN.

We have already described several ways that the TGN accurately sorts proteins into the proper destinations. Now we have to add two more destinations to Figure 9-9: the endosome and the lysosome. The TGN can target proteins to the endosome by the same nine-step vesicle transport process we have used throughout this chapter, but we need to go through this slowly, because we will be treading some of the same territory that we covered in exocytosis. The bottom line is this: proteins that travel the ER–CGN–Golgi–TGN route can contribute to exocytosis *or* endocytosis. Be careful not to fall into the trap of thinking that "ER–CGN–Golgi–TGN" is shorthand for "exocytosis." Traditionally (and in this chapter as well), that series of compartments is first used to describe exocytosis, but it is not used exclusively for that purpose. We don't normally call the TGN-to-endosome/lysosome transport endocytosis, because nothing being transported in this pathway has been "imported," but this transport is nevertheless absolutely necessary for endocytosis to function properly. Step 1, cargo selection, is mediated by a cargo receptor called the mannose-6 phosphate receptor. This transmembrane protein binds to proteins in the TGN that contain a very specific sorting signal. This is completely unlike any of the signals we have encountered so far, because it isn't an amino acid sequence; instead, it is a modified sugar, mannose-6 phosphate (typically abbreviated M6P). This sugar is made by enzymes in the Golgi apparatus that modify the high-mannose form of the oligosaccharide on N-glycosylated proteins, as shown in **FIGURE 9-15** (**BOX 9-2**).

The M6P receptor does have an amino acid sorting sequence (dileucine), and this is recognized by the adaptor protein GGA (this name represents the acronym of a very long term we need not discuss). When GGA binds to M6P receptor, it recruits clathrin and forms a coated pit. Steps 2 and 3, budding and scission, proceed normally to form a coated vesicle.

If we skip ahead to the point where the vesicle has successfully fused with the endocytic vesicles or endosome, we can now better explain why endosomes are acidic. Look again at Figure 9-14. The proton pump proteins we previously discussed have M6P tags, so they are carried by the M6P receptors to the endosome. The longer they reside there, the more acidic the endosome becomes. But acid alone does not degrade the endocytosed cargo. Hydrolytic enzymes also contain the M6P tag, so they too are sorted at the TGN into endosome-bound vesicles, sometimes the same ones that carry the proton pumps. The acidic environment inside the endosome causes the M6P receptors to release their M6P-containing cargo, and they are sorted back to the TGN. Notice that this is the only vesicle we have described so far that reaches the TGN by retrograde transport.

If we think carefully about the mechanism for adding these enzymes to endosomes/lysosomes, a problem arises. If the proton pumps and hydrolytic enzymes are made in the ER, then shipped all the way through the Golgi before reaching the endosomes, why aren't the ER and Golgi apparatus acidic, and why aren't the hydrolytic enzymes active in these compartments? Of course, if that were the case, the entire endomembrane system would shut down, because the cell would literally digest it. Let's rephrase the question: how does the cell prevent these proteins from destroying the endomembrane system? The answer, illustrated in Figure 9-14, is very clever: the proton pumps are not very active in the ER and Golgi apparatus because they are not in an active shape. Remember that the three traits of proteins requires every protein to have at least one inactive and one active shape, and these pumps are

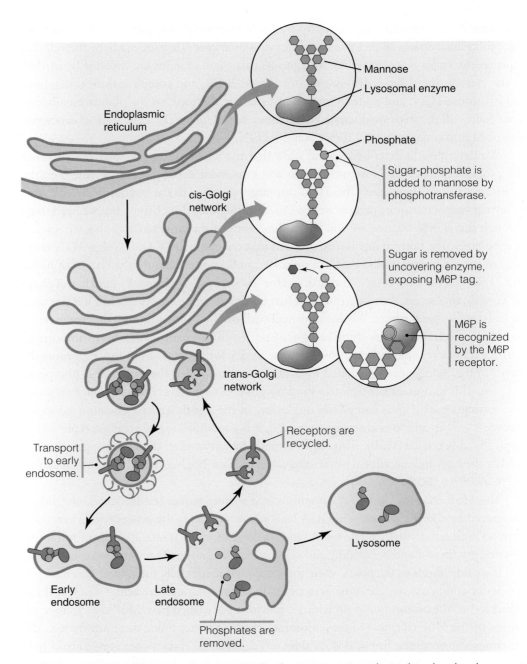

FIGURE 9-15 Assembly of the mannose 6-phoshate sorting signal. A phosphorylated sugar is attached to the high-mannose form of N-linked oligosaccharide, then the sugar is removed in the TGN, revealing a mannose 6-phosphate.

BOX 9-2 TIP

Here is an excellent example of why N-linked glycosylation matters, and it gives us another opportunity to review what we have covered. Here's a test I like to give my students: I throw the most jargon-laden, complex sentence I can think of at them and see if they can actually translate it back into plain English. In this case, we can cover the entire content of this chapter in one sentence. I view it like a series of dominoes tipping into one another: "Without N-glycosylation in the ER, the Golgi would not be able to create the M6P, and without M6P, there would be no cellular contribution to the endosomes by the TGN, so the pH of the endosomes and lysosomes would not change, and there would be no dissociation of the cargo and cargo receptors; everything endocytosed would end up in the endosome, but wouldn't be digested because there would be no hydrolytic enzymes present." If this intentionally long, awkward sentence makes sense to you, you're ready to move on.

in an inactive (or very low activity) shape when they are initially made. They still get their M6P tag as they move through the Golgi, but this does not fully activate them. (The TGN is more acidic than the rest of the Golgi and the ER, so the pumps evidently do have some activity by the time they reach it.) Here is the clever part: they adopt a fully active conformation only when they reach an acidic compartment (like the endosome); in other words, the pumps create an environment that activates the pumps. The hydrolytic enzymes are likewise controlled by the same mechanism: in the ER and Golgi apparatus, they are in an inactive conformation, but they switch to an active shape in the acidic environment created by the pumps. This also helps explain why these pumps and hydrolytic enzymes do not linger in the ER or Golgi: once they are made, they get processed, sorted, and shipped immediately to the endosome.

Another important point illustrated in Figure 9-15 is that not all acidic environments are the same. The acidity in the endosome is sufficient to separate the M6P receptors from their cargo, and to activate the proton pumps, but it is *not* strong enough to activate the hydrolytic enzymes. Remember that the late endosome only becomes a true lysosome when the pH reaches 5.0, which is the amount of acid necessary to finally activate the enzymes, and digestion commences. As added protection, some hydrolases initially contain extra amino acids that must be cleaved off before the enzyme becomes fully active, similar to the way extra amino acids prevent collagen from assembling in the ER and Golgi (see Figure 6-4). Polypeptides containing these extra amino acids are often described as being in a "pro-" form (derived from Latin, to indicate it is in an early, or unprocessed form). Cleavage of these amino acids in the hydrolases (i.e., conversion from the pro-form to the mature form) can only happen in the lysosome, so the cell is further protected from renegade enzymes.

So, pH is used three ways to control endocytosis:

- The acidic environment in endosomes helps sort cargo from receptors.
- The relatively neutral pH of the ER and Golgi apparatus keeps the hydrolytic enzymes from digesting these organelles.
- The enzymes' requirement for a strong acid environment protects the endomembrane system from digesting itself.

Once these hydrolases become active, they digest whatever they find in the lysosomes. This is illustrated in **FIGURE 9-16**. Proteases digest proteins, including themselves. The steps involved in digesting complex oligosaccharides on glycoproteins are about as complex as those involved in building the oligosaccharides in the ER and Golgi apparatus, and are mediated by enzymes called **glycosidases**. Digestion of nucleic acids is performed by nucleases. The primary nuclease for digesting DNA is an enzyme called **DNAse II**; this breaks DNA down into nucleosides (i.e., a nucleotide lacking a phosphate group). The phosphates are removed from nucleotides by **phosphatases**.

Digestion of phospholipids is a bit trickier, because the lysosome itself is enclosed by a phospholipid bilayer. If **phospholipases** were allowed unrestricted access to the lysosomal phospholipids, they would likely eat a hole in the membrane, depleting the accumulated acid and releasing

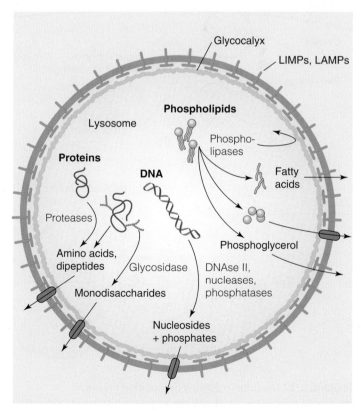

FIGURE 9-16 Digestion of cargo molecules in the lysosome.

the digestive enzymes into the cytosol where they could do great damage. To solve this problem, lysosomes contain **lysosomal integral membrane proteins** (**LIMPs**) and **lysosomal associated membrane proteins** (**LAMPs**) that are heavily glycosylated. The extensive oligosaccharides on these proteins form a "sugar coat" (technically called a **glycocalyx**) around the lumenal side of the lysosome membrane, as shown in Figure 9-17. This coat prevents the lipases (and proteases) from accessing the membrane.

To break down cellular membranes targeted for digestion in the lysosome, the membranes must first form a structure called a **multivesicular body** (**MVB**). The trigger for MVB formation is ubiquitination of the cytoplasmic tail of transmembrane proteins in a vesicle or an endosome (see *Proteins in the Cytosol and Nucleus Are Broken Down in the Proteasome* in Chapter 3 for another use of ubiquitin during protein digestion). A large complex of proteins detects the ubiquitin tags and removes them, while also inducing a pocket to form in the membrane. This pocket eventually pinches off to form the vesicle-within-the-vesicle that defines the MVB. These structures are then enveloped by lysosomes, and digestion begins. Lipases activated by the low pH have free access to the vesicle membranes, and they readily digest the phospholipids. An example of how a vesicle containing a protein, in this case one called epidermal growth factor receptor, and its transmembrane receptor is digested is shown in **FIGURE 9-17**.

Failure to fully digest materials in lysosomes has very serious consequences. In humans, over three dozen diseases, some of them fatal, have been linked to lysosomal defects. Similarly, uptake of indigestible materials (e.g., cigarette tar) can damage even healthy lysosomes, leading to a host of related medical problems.

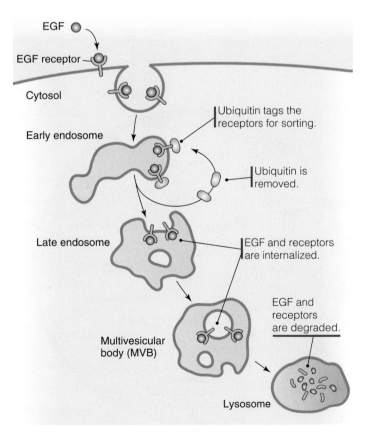

FIGURE 9-17 Ubiquitin-mediated degradation occurs in multivesicular bodies, which are late endosomal compartments of the endocytic pathway.

Digested Material Is Transported into the Cytsol

Once the cargo materials have been digested by the hydrolases, the digestion products are transported into the cytosol by transmembrane carrier proteins as illustrated in Figure 9-17. (We will discuss carrier proteins in greater detail in Chapter 10.) Hundreds of integral membrane proteins have been detected in lysosomes, but the functions of most are unknown. At least 10 different lysosomal transporters have been identified for amino acids and small peptides in humans, but we still cannot account for the transport of all 20 amino acids. Similarly, 3 different sugar carrier proteins have been found in human lysosomes that can collectively transport all 11 different monosaccharides that these cells use. The primary transporter of nucleosides in lysosomes is called an **equilibrative nucleoside transporter** (**ENT**). The products of phospholipid digestion (fatty acids, phosphoglycerol, head groups) either diffuse directly into the cytosol or leave the lysosome via transport proteins.

Lysosomes Can Also Degrade Some Resident Organelles

Over time, organelles simply wear out, or they become damaged and need to be replaced. In cases where cells are deprived of nutrients, they can literally eat their own organelles until either the nutrient supply is restored or the cells die. This kind of organelle digestion is called **autophagy** (pronounced ah-TOFF-ah-jee). The mechanism of autophagy varies from one condition to another, but all types share a few features. First, organelles that are digested are enclosed by a new membrane that forms in the cytosol. The partially assembled membrane is called a preautophagosome. Once it encloses the organelle to be digested (e.g., a mitochondrion), it fuses with a lysosome to form a **phagosome**, also called an autophagosome (from Greek, *phagos,* meaning to eat). The glycocalyx on the lysosomal membrane proteins protect them from digestion. Once phagosomes complete digestion, they usually disappear.

Have another look at the entire network of membrane trafficking to organize all the details we have covered. Try drawing a copy of this diagram, and then add the details in Figure 9-14. Note that almost all membranes can generate and accept vesicles, so two-way traffic is the norm, and it keeps the system balanced. The ER is the origin of all cellular proteins and membranes that contribute to the system, and lysosomes represent the end.

CONCEPT CHECK #4

To help you gather all of the information from Chapters 7–9 together, consider the following journey. Some lysosomal hydrolases (and their M6P receptors) are accidentally carried to the plasma membrane by bulk transport. Follow the journey of one of these wayward hydrolases as it is transcribed, translated, transported, recaptured, further transported, and ultimately sorted into a lysosome. Describe all of the events that take place along its journey. (Assume it has an N-terminal signal sequence.)

▮9.9▮ Chapter Summary

The topics covered in this chapter complete a three-chapter sequence that follows the path of DNA information as it is converted into RNA, and then translated into a polypeptide that eventually becomes a protein. In this chapter, our focus is on how these proteins reach their final destinations in the endomembrane system. We begin with the primary destination for these proteins, the ER. Two pathways govern how these proteins reach their destination: exocytosis (export) and endocytosis (import).

The first pathway we follow is exocytosis. Following modification in the ER, the proteins are carried by vesicles to the Golgi apparatus, where they enter the CGN. This journey is divided into nine steps, which apply to every vesicle that travels in the endomembrane system. Following sorting to return stray ER-resident proteins to the ER, the exocytosed proteins pass through each cisterna of the Golgi apparatus via vesicle transport. When they arrive in the TGN, they are sorted into vesicles targeted to at least seven different locations. The mechanisms of this sorting are only partially understood, but include some very atypical methods for targeting proteins. Once the proteins leave the TGN, they arrive at the plasma membrane via either constitutive or regulated secretion. Their final destination is either the extracellular space (if they are soluble proteins) or the plasma membrane (if they are membrane proteins).

The second pathway is endocytosis, which begins at the plasma membrane and ends at the lysosome. Following the formation of a clathrin-coated vesicle at the plasma membrane, endocytic vesicles form endosomes. Endosomes are sorting organelles capable of separating cargo from cargo receptors; the receptors are recycled to the plasma membrane, and the cargo remains in the endosome. The cargo is digested by hydrolytic enzymes synthesized at the ER and carried through the Golgi to the TGN, where they are sorted into endosome-targeted vesicles by virtue of their bearing a mannose-6 phosphate sugar. When these proteins arrive in the endosome, they uncouple from their M6P receptors and begin digesting the cargo when the lumen of the endosome reaches a pH of 5.0. This acidification is provided by proton pump proteins, which are likewise delivered from the TGN by M6P receptors. According to one popular model, after both the cargo receptors and M6P receptors have been removed from the endosome membrane, the endosome "matures" to form a lysosome. The lysosome digests the cargo into the molecular building blocks we described in Chapter 1–4, and these are carried across the lysosomal membrane by membrane proteins specializing in this kind of transport. Once in the cytosol, these building blocks can be used to build new macromolecules in the cell.

CONCEPT CHECK ANSWERS #1

Distinguishing properties of vesicles can include:
▶ Size/shape: Are the vesicles large or small? Are they spherical or distorted?
▶ Membrane components: What kind of phospholipids does the vesicles contain? How much cholesterol is in their membrane? What classes of membrane proteins (integral, peripheral, type I, II, III, IV, etc.) do they contain?
▶ Contents: What molecules are in the vesicle lumen? What is the pH of the vesicle lumen?
▶ Adaptor and coat proteins: Is the vesicle coated? If so, what type of coat protein encloses it? What type of adaptor proteins are linking to coat to the vesicle membrane?
▶ Motor proteins: What kind of motor proteins is attached to the vesicle? Is it transported on actin, microtubules, both, or neither?
▶ Targeting and tethering proteins: What kind of Rab, SNARE, and tethering proteins is on the vesicle surface?
Based on these variables, cells must be able to make thousands of different vesicle types.

CONCEPT CHECK ANSWERS #2

Targeting sequences are hierarchical, and the ER signaling sequence is more important than the ER retention sequence. If a protein contained no ER signal sequence, it would never enter the ER when it was being made. Since it was never in the ER, it would never get transported to the Golgi, so the ER retention sequence would never be recognized in the Golgi. It would be found in the cytosol.

CONCEPT CHECK ANSWERS #3

Treatment with nocodazole would effectively eliminate any contribution microtubules (and their motors) make to vesicle trafficking. By carefully checking the nine steps of vesicle trafficking, we see that microtubules are not essential for any of them, despite the fact that microtubules play an important role in maintaining the overall structure of the ER and Golgi apparatus. So, adding nocodazole should not, by itself, impact either the sorting of cargo or the targeting of the resulting vesicles. If it had any effect, it would be on the transport of the vesicles (step 5), by slowing down the rate of transport of any vesicles that could be carried on microtubules.

CONCEPT CHECK ANSWERS #4

▶ The hydrolase gene is transcribed in the nucleus, after the promoter region is bound by the appropriate transcription factors and RNA polymerase is allowed to unfold, copy, and refold the DNA. Transcription proceeds through initiation, elongation, and termination.
▶ The resulting primary transcript undergoes three forms of posttranscriptional processing: addition of a 5′ methyl guanosine cap, splicing of introns, and polyadenylation at the 3′ end.
▶ The fully processed mRNA is bound by several transcription and translation factor proteins and the entire molecular complex is transported into the cytosol through the nuclear pore complex.
▶ The mRNA is recognized by the initiation tRNA, the small ribosomal subunit, and eventually the large ribosomal subunit. Translation passes through initiation to elongation phase.
▶ The signal sequence of the hydrolase is bound by SRP, and the ribosome is then bound to the surface of the ER, where elongation and termination of translation are permitted to take place.

- As the N terminus of the hydrolase passes into the lumen of the ER, the signal sequence is cleaved by signal peptidase, and the hydrolase is N-glycosylated. BiP helps the completely synthesized protein fold into a stable (but inactive) shape.
- The core oligosaccharide is trimmed down a high-mannose form in the ER.
- The partially processed hydrolase is carried by bulk flow into the CGN via COPII-coated vesicles undergoing anterograde transport. The nine steps of vesicle transport are completed each time a vesicle makes a journey between compartments.
- In the medial-Golgi, the high-mannose sugar is modified by the addition of a phosphorylated sugar.
- In the TGN, the modified sugar is removed, leaving behind a phosphorylated mannose (M6P). This is the signal to be carried to the endosome.
- During sorting in the TGN, the hydrolase is not properly recognized by the M6P receptor, and instead is carried by bulk transport to the plasma membrane, where it is released into the extracellular space upon fusion of the exocytic vesicle.
- At some point while it is floating in the extracellular space, the hydrolase is bound by a wayward M6P receptor that was also mistargeted to the plasma membrane.
- The M6P receptor is endocytosed via a clathrin-coated pit, and it carries the hydrolase with it.
- The clathrin-coated vesicle uncoats and fuses with an endosome because it carries the appropriate SNARE and binds the appropriate tethering protein.
- In the endosome, the acidity causes the M6P receptor and the hydolase to dissociate. The M6P receptor is properly sorted to return to the TGN, while the hydrolase remains in the endosome.
- Upon further acidification of the endosome, the hydrolase becomes activated, and the endosome becomes a lysosome. Our wayward hydrolase has finally come home.

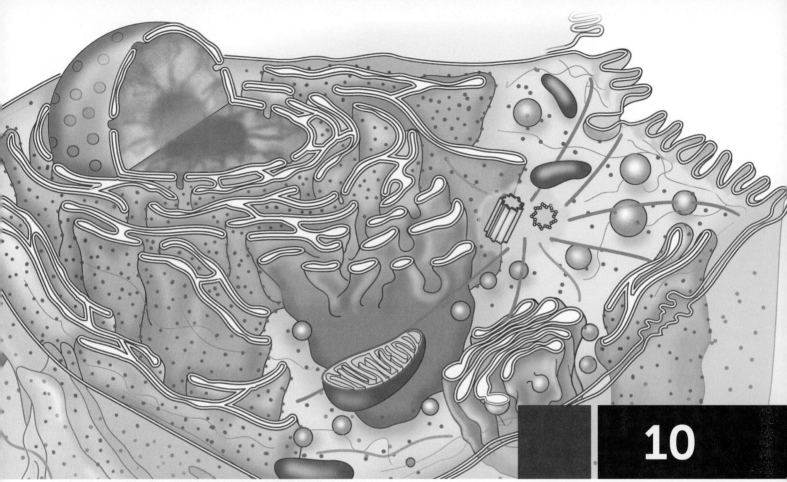

Cellular Metabolism and Energy Storage

■10.1■ The Big Picture

This chapter introduces a third strand devoted to explaining how an average cell functions. Like the strand linking Chapters 7 through 9, it represents a journey, with a beginning and an end: this story concerns how energy cycles into, through, and out of living organisms. This cyclical journey began several billion years ago, is repeated trillions of times a day, and forms the chemical basis for the existence of life. It's a long journey, filled with a multitude of chemical structures and reactions, but the rationale for this complexity is straightforward: to remain alive, cells must capture, store, and use energy.

Our task here is to understand how this happens by examining what and where events take place, and why these events help cells. To avoid getting buried in the details, we'll limit the amount of *how* information and keep our focus on the *what, why,* and *where* questions. For example, rather than spend a lot of time explaining how sugars are made, we'll look at the strategic question of why sugar is a good form of energy storage, and cover just enough of the mechanistic details to keep the story moving. I like to view this "energy cycle" as an epic narrative, with electrons as the protagonists.

To help keep things in perspective, let's introduce cell biology principle #6: chemical bonds and ion gradients are cellular fuel. What this means is that electrons, which are conduits for all energy in cells, sometimes use this energy to build chemical bonds. In other cases, this energy builds ion gradients, which are then used to build new chemical bonds by energizing electrons again. So most of this journey involves moving energy from one physical form to another, onto and off of electrons, and using the energy to do work along the way.

This chapter is divided into six major sections. Before we begin the journey, the first two sections introduce some important concepts that we'll need to explain what is happening along the way.

- The first section covers the basics of energy flow and introduces the first two laws of thermodynamics, which define the rules for energy flow.
- The second section introduces specialized proteins that play a central role in energy transduction, called the membrane transport proteins.

Consider the first and second sections background reading before we embark on the electron flow journey. The journey is presented in four phases, and each is covered in a separate section.

- Phase I, discussed in the third section, focuses on the events taking place in the chloroplast, where electrons are ripped away from water and begin the long voyage through cell energy transducers. This is where the detailed chemistry first appears, but be careful to avoid being sidetracked by memorizing structures until the path that the electrons take through each step of photosynthesis is clear. This section is the longest in the chapter, and it devotes a lot of attention to themes that appear in later sections. Take plenty of time to understand this section. I expect most students to read through it several times if it is a new subject area for them.
- Phase II, discussed in the fourth section, relies heavily on the material presented in the second section and builds an integrated network of membrane transport proteins for moving the food molecules generated by Phase I into and out of cells. By convention, we will focus exclusively on the transport of glucose. This provides an opportunity to test comprehension of the second section.
- Phase III, discussed in the fifth section, is concerned with the first 10 enzymatic reactions in the metabolism of the glucose discussed in Phase II. It introduces an important concept in energy transduction: the mechanisms used to both build

If at anytime the details begin to confuse you, ask yourself these questions: (1) Where are the electrons in question?; (2) Do they have a lot of energy, or a little energy?; (3) If they started out with high energy and now have low energy, where did this energy go?; and (4) If they started out with low energy and now have high energy, where did this energy come from? The answers to these questions are actually quite limited: the electrons are usually on (a) a food molecule, (b) an energy carrier molecule, (c) an ATP or GTP molecule, (d) a component of an electron transport chain, or (e) water. Water and CO_2 always have the lowest energy electrons. When the energy is not in the electrons, it's either in an ion gradient or, in one case, in photons of sunlight. Like the previous chapters, I expect most students to reread this a few times to fully understand it. Don't try to absorb every detail the first time you go through it. Be patient.

An important feature of this chapter is that it provides you an opportunity to apply much of what we have covered in previous chapters to the subject at hand. There's a lot of material in the first nine chapters in this book, and we want to make sure it actually matters, so in this and the next four chapters I'm going to devote a lot of effort toward tying the subjects together. Remember, cells are not divided into chapters, only books are; in reality, nearly every event discussed in this book is going on at the same time in a single cell. For example, mitochondria import proteins (Chapter 8) at the same time that they are generate ATP (this chapter); in fact, that's a major reason they import proteins in the first place. I find this kind of integration is one of the most difficult concepts for my students to grasp, because they tend to group subjects according to when they study them. If you are falling into this habit, try following the links in this chapter to break it.

This chapter also gives you an opportunity to apply the basics of chemistry and physics that you will need to become a practicing life scientist. For example, we'll see how redox potential and the laws of thermodynamics directly impact cell biology.

food molecules and take them apart are very similar in organization, but run in opposite directions on the energy spectrum.

- Phase IV continues this mirror-image theme by comparing the cyclical enzymatic reactions in mitochondria to those that take place in chloroplasts during photosynthesis. I've divided the events in Phase IV into four stages; pay attention to the difference between phases and stages here. If the events in Phase I make sense, the similarity between Phases I and IV should be apparent. If the description of Phase IV sounds unfamiliar, read about Phase I again (**BOX 10-1**).

Lastly, we will encounter discoveries by 8 scientists that led to 5 Nobel Prizes in the past 60 years. Considering that the work of only about 100 Nobel Prize winners directly impacts biology, it is amazing that 8 of these individuals will serve as guides in this one story.

10.2 Cells Store Energy in Many Forms

KEY CONCEPTS

- Energy exists in three forms: kinetic, potential, and heat.
- The laws of thermodynamics define the rules for energy transfer.
- Cells remain alive by converting environmental energy sources into cell-accessible energy forms.
- High-energy electrons and ion gradients are the most common forms of cellular energy storage.
- The amount of energy in an ion gradient is expressed as an electrical potential.

At a very fundamental level, the universe (including cells) is composed of matter and energy. Most of this text is concerned with the structure and function of the matter (molecules) in cells, but in this chapter we turn our attention to energy. The laws of physics classify the most common forms of energy into three types: **kinetic energy**, **potential energy**, and **heat energy**. Despite the fact that all forms of energy are invisible, we can still "see" them through the matter they affect. Cellular structures store energy as potential energy, use kinetic energy to perform *work* (moving a mass over a distance), and generate heat energy when potential energy is converted into kinetic energy, and viceversa. The acquisition, storage, and controlled usage of energy are fundamental traits of all living organisms. All molecules contain energy, but cells are fairly selective for the molecules that store and use it.

▮▮ The Laws of Thermodynamics Define the Rules for Energy Transfer

The laws of thermodynamics, first developed by physicists in the 19th century to help explain energy transfer (especially heat energy) during work performed by machines, form the basis for cellular energetics as well. The first two of these laws help explain why the strategies cells have developed for handling energy are so universal. Let's quickly review them before we embark on our examination of cellular energy transfer and work.

The first law is commonly expressed as "Energy can neither be created nor destroyed. It can only change forms." When energy changes forms, at least some of it is always converted into heat. Of the three forms of energy, heat energy is least useful to most cells, so they have developed elaborate strategies for generating and converting potential and kinetic energy while minimizing the amount of heat generated. A good rule of thumb to apply here is that *most cells do not want to generate heat*. The little they do generate is useful to them, but any excess could be harmful, as we will see. We'll focus our attention on cellular strategies to efficiently interconvert potential and kinetic energy.

The second law is frequently expressed as "Energy systems have a tendency to increase their entropy." For this statement to make sense, we need to understand what entropy is. The formal definition of the entropy of a system is the measure of the unavailability of its thermal energy for conversion into mechanical work. Less formally, we can consider entropy to be energy that cells can't use. The entropy of an energy system is a function of the heat it contains: its entropy is proportional to the percentage of its energy is in the form of heat. We can interpret this law to mean that cellular structures (or "energy systems") naturally tend to convert their kinetic and potential energy into heat, most of which is useless. Together, these two laws of thermodynamics demonstrate why cells, like all other forms of matter, decay over time.

The trick to staying alive is to delay this decay as long as possible. All living organisms do this by capturing the useful energy emitted by other matter as it decays, and using this captured energy to (temporarily) reverse their own increase in entropy. For example, photosynthetic cells capture light energy emitted by the sun as the sun's entropy increases, and heterotrophic cells (e.g., those found in animals) digest complex organic matter to hasten its decay and capture the useful energy that is released. The most important concept to remember here is that cells combat their inevitable increase in entropy by continually consuming new energy (i.e., food, sunlight) to repair or replace their structures that decay. In short, a cell keeps its entropy low by capitalizing on the increase in entropy of its energy sources. If a cell loses its ability to maintain this linkage to the decay of other matter, it dies (**BOX 10-2**).

▮▮ Fats and Polysaccharides Are Examples of Long-Term Energy Storage in Cells

According to the second law of thermodynamics, low-entropy molecules store the most-useful energy. Cells favor storing energy in molecules that are relatively small and highly ordered. Recall from Chapter 1 (see *Lipids Are Carbon-Rich Polymers That Are Insoluble in Water* and *Sugars Are Simple Carbohydrates*) that fats and oligosaccharides are polymers composed of several small repeating structures. Fats contain long-chain fatty acids, which typically contain over a dozen alkane or alkene groups (Table 1-1) joined together in a linear, highly organized fashion. Fats spontaneously organize into hydrophobic lipid droplets, and this further decreases their entropy. Polysaccharides and oligosaccharides typically contain multiple repeats of a single disaccharide, and the disaccharides are themselves composed of numerous alkane and alkene groups. Because fats have less entropy than poly/oligosaccharides, they carry more useful energy than sugars. Note that this does *not* mean that large molecules have any more or less energy than small molecules.

Nucleic acids and proteins have much higher entropy than lipids and sugars. However, some cells (e.g., skeletal muscle) are absolutely packed with protein, so during starvation, cells will resort to digesting some of their own proteins to stay alive. Because it is an irreplaceable source of information (cell biology principle #3), DNA in living cells is not a source of stored energy, even during starvation.

High-Energy Electrons and Ion Gradients Are Examples of Short-Term Potential Energy in Cells

While fats and oligosaccharides are effective long-term storehouses of potential energy, accessing their stored energy is relatively difficult and time consuming. Typically, several chemical reactions are required to convert these molecules into some form of kinetic energy, so, cells keep a separate pool of readily accessible potential energy that needs few if any chemical reactions to be useful. This pool exists in two forms: (1) high-energy electrons in special molecules called **energy carriers**, and (2) ion gradients across biological membranes.

Energy carriers are so-named because they carry one or two electrons in covalent bonds that are notably difficult to form, and hence release a large amount of energy when they are broken. The free energy (a thermodynamic unit) of a single carbon–carbon (C–C) bond is approximately 80 kilocalories (kcal) per mole. When a C–C bond is broken in cells, some of the released energy can be captured by energy carriers. The free energy of the high-energy covalent bonds in electron carriers ranges from 30 to 50 kcal/mol, so these molecules store a considerable amount of the C–C bond energy. When these high-energy bonds are later broken, the released energy can be used to create ATP from ADP and P_i (inorganic phosphate), which requires approximately 7 kcal/mol; several ATP can often be generated from one broken C–C bond.

To distinguish the high and low energy forms of these carriers, we sometimes borrow some technical terms from chemistry: those storing this potential energy are often called **reduced energy carriers** to indicate that they are holding high-energy electrons, while those that have released the electrons are called **oxidized energy carriers** (BOX 10-3). The names of most energy carriers are rather long, so we use their acronyms instead; they are typically named as pairs to represent the reduced and oxidized forms, respectively. Examples of redox pairs of energy carriers include NADPH/NADP$^+$ + H$^+$, NADH/NAD$^+$ + H$^+$, and FADH$_2$/FAD + 2H$^+$. We will encounter each of these carriers later in this chapter (by convention, most of the time we ignore the protons and just call the oxidized forms NADP$^+$, NAD$^+$, and FAD).

Ion gradients are the most rapidly accessible form of potential energy. They are formed by simply creating a concentration imbalance of an ion across a membrane. Recall from Chapter 1 (see *Cells Are Self-Replicating Structures That Are Capable of Responding to Changes in Their Environment*) that to remain alive, a cell must maintain disequilibrium with its environment, and most of this disequilibrium is in the form of concentration gradients across the plasma membrane. For example, in the human body, the concentration of sodium ions (Na$^+$) is much higher in the extracellular fluid cells than inside cells (142 mM vs. 10 mM). The concentration of potassium ions (K$^+$) is conversely higher inside our cells than in the extracellular fluid (148 mM vs. 5mM). Inside cells, mitochondria and chloroplasts form proton (H$^+$) gradients across their membranes to store potential energy. These gradients are formed by transmembrane proteins that use energy in electron carriers to pump the ions from one side of a membrane to another. We'll have a closer look at these pump proteins later in this chapter (**BOX 10-4**).

Not all chemical gradients are useful energy storage in cells. For example, there is a gradient of DNA across the nuclear envelope, but cells cannot make use of this concentration difference to store or generate energy because DNA does pass in and out of the nucleus (look through Chapters 2, 6, and 7 to find evidence for this statement). Similar concentration gradients for other macromolecules exist between the cytosol and interior of organelles (e.g., sugars, proteins) and these gradients, too, are typically of little to no use as a storehouse of potential energy in most cells. Keep the focus on *ions* when considering how cells store potential energy with gradients.

Cells Couple Energetically Favorable and Unfavorable Reactions

To obey the second law of thermodynamics, cells must constantly consume potential and kinetic energy to decrease their entropy. Because this is moving against the general trend of the universe, the chemical reactions required to accomplish it are not spontaneous—to take place, these reactions require an input of energy. Cells achieve this by *coupling* a thermodynamically unfavorable reaction to one that is favorable. For example, the kinesin and dynein motors that organize the distribution of the ER and Golgi apparatus in the cytosol (see *Microtubule-Based Motor Proteins Transport Organelles and Vesicles* in Chapter 5 and *The Trans-Golgi Network Sorts Proteins Exiting the Golgi Apparatus* in Chapter 9) consume some of the energy released by the energetically favorable hydrolysis of ATP; this ensures that some of the energy released by the favorable reaction is immediately applied to stimulate the unfavorable one. This ability to couple reactions is one of the unique properties of living organisms. Nearly all of this coupling is performed by proteins, though in some cases, ion gradients can supply this energy directly, for example by coupling the energetically favorable flow of ions through a membrane (down a concentration gradient) to an unfavorable reaction, such as the addition of a third phosphate group to an ADP molecule, yielding ATP.

Nucleotide Triphosphates Store Energy for Immediate Use

ATP, an energy carrier molecule, is perhaps the most familiar form of cellular energy storage. While ATP is used as an energy source for a large number of chemical reactions and cellular work, it is not the only nucleotide triphosphate that serves this purpose. Dissociation of the SRP/SRP receptor complex from the ER membrane, and translocation of the small ribosomal subunit during translation (see Figure 8-20), are examples of cellular work coupled to GTP hydrolysis. All four nucleotide triphosphates (often collectively abbreviated as NTPs, where N represents any of the four bases) supply the energy necessary for adding themselves to growing strands of DNA and RNA (see *DNA Polymerases Add New Deoxyribonucleotides to the 3′ End of a DNA Strand* and *Transcription Occurs in Three Stages* in Chapter 7). Enzymes in the cytosol readily interconvert NTPs by simply moving phosphates from one nucleoside to another. Note that deoxyribonucleotide triphosphates (dNTPs) are *not* used as general energy stores; they are committed exclusively to the synthesis of DNA.

We will encounter additional energy carriers later in this chapter when we discuss the pathways of energy flow through cells.

The Amount of Potential Energy Stored in an Ion Gradient Can Be Expressed as an Electrical Potential

Harnessing the energy stored in separated electrical charges was one of the most significant technological advances of the 18th century. Since that time, the vast and complex fields of physics, chemistry, and electrical engineering have evolved to develop ever more efficient and specialized ways to capture, store, and harness this energy. Along the way, these scientists developed new jargon to help them explain the fundamental rules of electricity. Once biologists discovered that cells make regular use of electrical power to keep themselves alive and functioning, they began borrowing some of this jargon to describe cell behavior.

To accurately describe the mechanisms cells use to accomplish these tasks in the rest of this chapter, we need to define a few essential terms. The term *electricity* refers simply to the nature and behavior of electrons, those negatively charged particles that orbit around the nucleus in every atom; in biology, we expand on this a bit to mean the behavior of all charged atoms, and the molecules they form. Thus, a protein has an **electrical charge** associated with it, which is simply the sum of its positive and negatively charged atoms. The term *electrical potential* refers to a quantity of electrical energy stored as potential energy, and it is expressed in units called **volts**. The difference in electrical potential between two points in space is called **voltage**. Cell biologists commonly use voltage as a means of defining the electrical potential across a membrane; because biological membranes are selectively permeable to some charged molecules, this voltage becomes a useful source of energy, independent of the chemical gradient that may accompany it. In cases where both an electrical and chemical imbalance is present in the same atoms or molecules, the two terms are combined and called an **electrochemical gradient**, which possesses an electrochemical potential. Membranes themselves are considered to have no electrical charge, and hence no electrical energy, and the fact that a membrane separates electrical charges makes it an **insulator**. An insulator that keeps electrical charges separate until their energy is needed is called a **battery**, and most cellular membranes may be referred to as batteries for the same reason.

In conventional physics, when electrons flow from a negatively charged point in space to a positively charged point, the movement of these electrons is called an **electrical current**. In cells, it is more common to observe entire atoms moving from negative to positive spaces (and vice versa), rather than just the electrons, but we often use the term "current" to describe this movement anyway because it is measured in the same units, called **Amperes** or **Amps**. In cells, the magnitude of the electrical flow is small enough that the most common unit is the **milliAmp** (1/1000th of an Amp).

10.3 Gradients across Cellular Membranes Are Essential for Energy Storage and Conversion

KEY CONCEPTS

- ▸ Membrane transport proteins are responsible for moving ions through the phospholipid bilayer of cellular membranes.
- ▸ Membrane transport proteins are organized into three groups: channels, carriers, and pumps.
- ▸ All channels dissipate gradients, all pumps build gradients, and most carriers only dissipate gradients. Some carriers can build gradients as well, using indirect active transport.

Now let's discuss how gradients are built, maintained, and used by cells. The principle behind gradient formation is fairly straightforward: if one separates two populations of a molecule with a barrier that is permeable to the molecule (and keeping all other environmental conditions the same), the second law of thermodynamics requires that the molecules strive to achieve an equal concentration on either side of the barrier (equilibrium). They do so by passing back and forth between the barrier until there is no concentration difference between the two groups.

Recall from Chapter 4 (see *Phospholipid Bilayers Are Semipermeable Barriers*) that a phospholipid bilayer is impermeable to ions, as illustrated in **FIGURE 10-1**. Without some way to make this bilayer permeable to ions, differences in ion concentration on the two sides would be energetically meaningless. Cells can customize their membranes by adding **membrane transport proteins** that are highly selective for specific molecules and that permit these molecules to pass through the lipid bilayer. In other words, one membrane can be made permeable to chloride ions (Cl^-) by adding a special chloride-binding protein to it that permits the chloride ions to pass through the bilayer; another membrane can be kept virtually impermeable to Cl^- by simply not adding this protein. We discussed how proteins are inserted into biological membranes in detail in Chapter 8 (see *Signal Sequences Code for Proper Targeting of Proteins* and *Transmembrane Proteins Contain Signal Anchor Sequences*). Using this simple approach, cells can generate membranes with a multitude of different, highly selective molecular permeabilities. This customization is an important trait of all multicellular organisms.

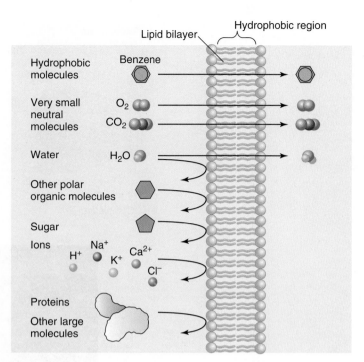

FIGURE 10-1 Permeation of lipid bilayers by biologically important molecules.

Protein Channels, Carriers, and Pumps Regulate the Transport of Most Small Molecules across Membranes

Not all membrane transport proteins are the same. Some only transport molecules in one direction across the membrane, while others permit movement in either direction. Some only permit molecules to travel *down* a concentration gradient, while others only move molecules *up* a gradient. Some carry the molecules across the membrane one at a time, and others permit a steady stream of movement. To keep track of the different types of these proteins, scientists have grouped them into three classes, based on their structural features.

Protein Channels Dissipate Gradients

The channels are a group of proteins that form a tube-like shape, called a **pore**, in a membrane. All cellular membranes likely contain at least one channel protein (over 140 have been identified so far), and channels are some of most critical proteins in all living organisms. Their significance was underscored by the awarding of the 2003 Nobel Prize in Chemistry to Peter Agre and Roderick MacKinnon, who were instrumental in defining how channels function.

Some channels are composed of a single polypeptide, while others contain multiple polypeptide subunits. But unlike most everyday tubes, these channels are highly selective. Molecules that fit into the confines of the pore must also fit through the narrowest portion, called the **selectivity filter**, to pass through a membrane. A common structural feature in channel proteins is the presence of at least one α-helix in the transmembrane domain of each polypeptide.

In cases where a channel pore is composed of several α helices, such as the Cl^- channel shown in **FIGURE 10-2**, it is fairly easy to visualize how they can combine to create a hydrophilic environment in the hydrophobic interior of the phospholipid bilayer. Because the R groups project outward in an α helix (Chapter 3), the surface of an α helix can be customized to form a channel, as illustrated in Figure 10-2. Most of the R groups can be hydrophobic to stabilize the helix in the membrane, but by inserting periodic hydrophilic amino acids into the helix, one "stripe" of each helix can be hydrophilic. When the hydrophilic stripes of several helices converge in a membrane, they naturally form an interior channel away from the lipid portion of the membrane. Most channels are selective for ionic forms of single atoms, such as Na^+, K^+, Ca^{+2}, Cl^-, H^+, an so on, so each ion channel protein permits the passage of a single type of ion. Larger ionic molecules, such as PO_4^{-2}, SO_4^{-2}, and NH_4^+, do not pass through ion channel proteins.

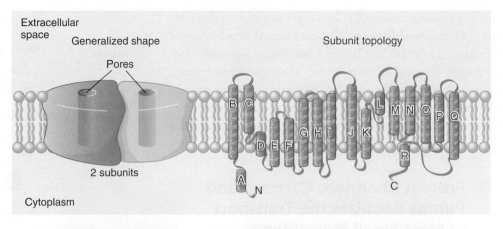

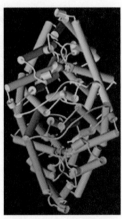

Extracellular view

FIGURE 10-2 Different views of a Cl⁻ transporter. Note how several transmembrane alpha helices combine to form the pore, including the selectivity filter. Bottom structures adapted from R. Dutzler, et al., Nature 415 (2002): 276-277.

One especially interesting group of channels is the **aquaporins**. These proteins selectively permit water molecules to pass through a membrane while preventing the passage of solutes (such as ions) dissolved in the water. Nearly all aquaporins are expressed exclusively on the plasma membrane, thereby regulating the water content of the entire cell. In humans, many of the known aquaporin proteins are expressed in the kidney, and they assist in the resorption of water; other aquaporins are found in the skin, blood vessels, cornea, and multiple cell types in nervous tissues. Several different diseases are known to result from mutations in these aquaporins, though their exact role in disease progression is not clear.

Because they function like tubes, channel proteins only facilitate transport driven by an existing concentration gradient. This is called **facilitated diffusion**, to distinguish it from the **simple diffusion** of molecules that easily pass through phospholipid bilayers, such as gasses (O_2, CO, CO_2, etc.). This means that channels permit ions to pass through in either direction, depending on the orientation of the gradient. Channel proteins *do not* form the gradient; to be truly effective, channels must be functionally coupled to some mechanism for generating the gradient that they dissipate.

An important property of all channel proteins is that they open and close by very precise mechanisms. This makes sense because cells have to carefully regulate the type and magnitude of chemical gradients they store to remain alive. To open and close, the proteins must change shape, so they are sensitive to many of the posttranslational modifications used to control protein structure and function (see *Cells Chemically Modify Proteins*

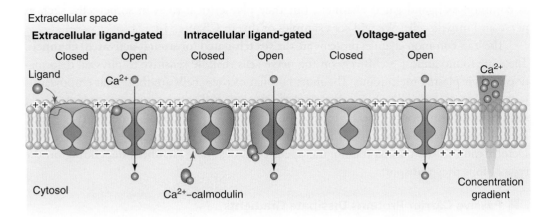

FIGURE 10-3 Three methods for controlling the opening and closing of channels. Ca^{+2} channels are used as examples.

to Control Their Shape and Function in Chapter 3). For example, some channels, known as ligand-gated channels, are regulated by binding to a signaling molecule (a.k.a. ligand), so the signal acts as an on/off switch for controlling whether an ion gradient dissipates. This is illustrated in **FIGURE 10-3**. Over 70 different types of these channels are known, making them the most common types of channels in cells. They are essential to many signaling pathways, as we will see in Chapter 11.

Another means of controlling the shape of channels is to alter the electrical state of the membranes they penetrate. Local changes in voltage cause these **voltage-gated channels** to open or close, as shown in **FIGURE 10-4**. When they are open, they permit K$^+$, Ca^{+2}, or Na$^+$ ions to pass through them. The resulting movement of these charged atoms alters the voltage of the nearby environment as well, so in some cases the channels can form a positive feedback loop wherein ions flowing through one open channel will alter the voltage enough to open a second voltage-gated channel nearby, and so on. Voltage-gated channels are not

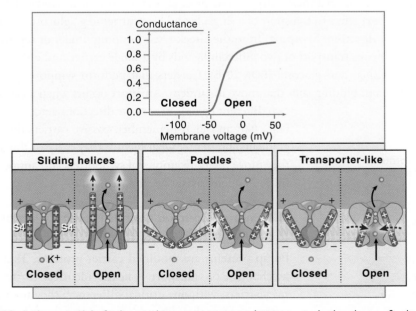

FIGURE 10-4 Three models for how voltage across a membrane controls the shape of voltage-gated channels. Different types of K$^+$ channels are shown as examples.

as common as ligand-gated channels, but they play critical roles in some cells, such as nerve and muscle cells. We will see examples of this in Chapter 14.

The least common channel proteins are the **stretch-gated** (or **stretch-activated**) **channels**. These are found almost exclusively on the nerve cells that are sensitive to physical deformation of their plasma membrane. The most familiar of these cells are those that empower us with the sense of touch, while others include the hair cells of the inner ear, which detect sound via deformation of special microtubule-containing cilia, and some pain receptors. Like voltage-gated channels, each is selectively permeable to a single ion type (e.g., K^+). Deformation of the plasma membrane opens or closes these channels, depending on the type of ions they transport.

■■■ Passive Carrier Proteins Dissipate Gradients

Unlike channel proteins, carrier proteins do not form pores. They have a completely different structure, and use a much more complicated method of transporting molecules across membranes. It is important to note that carrier proteins are classified into two groups: those that dissipate gradients and those that build gradients. The carriers that dissipate gradients are called *passive* carriers, and they only participate in facilitated diffusion.

Carrier proteins contain very selective binding sites for the molecules they transport, somewhat analogous to the active sites on enzymes. In fact, the molecules they bind to are sometimes called ligands because of this selective binding, but unlike enzymes, these proteins do not have active sites, and they do not modify the molecules they bind to. Instead, they simply carry them from one side of a membrane to the other, as shown in **FIGURE 10-5**. When they are inactive, they adopt an open conformation that permits their ligands to access the binding sites. Once the binding sites are occupied, the carrier proteins obey the three traits of proteins and immediately change shape. In this case, the shape change is rather dramatic, in that the binding sites that faced one side of the membrane when unoccupied now face the other direction, as seen in **FIGURE 10-6**. The ligands then detach and diffuse away, and the carrier resumes its open configuration again. Because passive carriers can transport their ligands in either direction across a membrane (depending on the orientation of the ligand concentration gradient), they can assume their open configuration on either side of a membrane (**BOX 10-5**).

Scientists use a special vocabulary to more precisely define how carriers move their ligands. The term ***uniport*** refers to transport of a single ligand by a carrier (e.g., glucose), and this can occur in either direction. **Symport** (from the Greek *sym-*, meaning similar or together) refers to the simultaneous transport of two different ligands by a single carrier, and only in the *same* direction (e.g., Na^+ and glucose) (**BOX 10-6**). Carriers that perform symport must have at least two different binding sites that move in tandem. **Antiport** occurs when a carrier moves two different ligands in *opposite* directions across a membrane (e.g., Na^+ and H^+). Remember, passive carriers do not build gradients, no matter what ligands they carry or the directions they move them. Examples of symporters and antiporters that move Na^+ ions are shown in **FIGURE 10-7**.

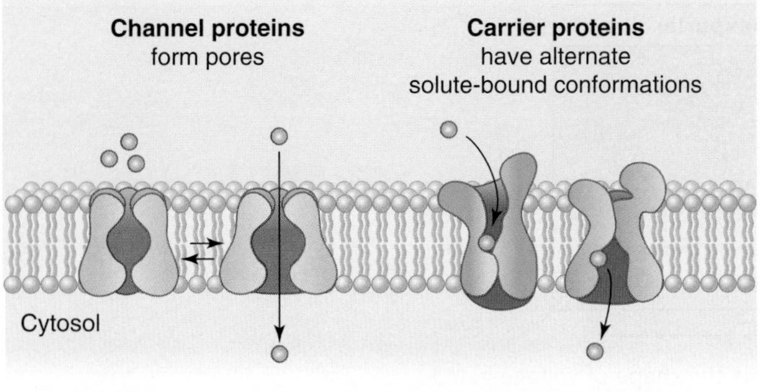

■■ Energy-Coupled Carrier Proteins (Pumps) Build Gradients

Pump proteins are modified carrier proteins. Their general mechanism of transport is nearly identical to that of passive carriers, with one critical difference: they use metabolic energy to force their ligands across a membrane, and thereby *build* concentration gradients. The source of this metabolic

FIGURE 10-5 A comparison of channel and carrier proteins.

Extracellular space

Channel proteins form pores

Carrier proteins have alternate solute-bound conformations

Cytosol

energy is almost always ATP; therefore, pump proteins *are* enzymes, and they must have at least two binding sites: one active site for the ATP cleavage, and at least one for the ligand being transported. One of the most common pump proteins in mammals has six binding sites: one for ATP, three for Na^+, and two for K^+. This **sodium–potassium pump** (or **Na^+/K^+ ATPase**) uses the ATP energy to transport three Na^+ ions from the cytosol into the extracellular space, while it simultaneously pumps two K^+ from the extracellular space into the cytsosol, as shown in Figure 10-7. It is therefore classified as an antiporter. Every cell in mammals has between 800,000 and 30 million of these pumps in its plasma membrane, and these pumps consume a significant amount of the ATP a cell generates. The heat lost during the conversion of the potential energy in ATP into the kinetic energy of pumping is the primary source for body heat in warm-blooded animals. Another important class of pump proteins is the H^+ pumps that acidify the TGN, endosomes, and lysosomes (Chapter 9).

FIGURE 10-6 An example of a conformation change in a carrier protein. This protein carries the sugar lactose. A model of the structural arrangement of the protein in a membrane is shown at right. Right structure from Protein Data Bank 1PV6. J. Abramson, et al., Science 301 (2003): 610-615.

Two more terms in the membrane transport lexicon are used to distinguish different methods for building gradients. The easier of the two is **direct active transport**, often referred to as active transport. This refers to the direct building of one or more gradients by a pump protein that consumes ATP energy as part of the transport mechanism. The activity of the sodium–potassium pump is an excellent example of this kind of transport. It's as simple as it appears: one protein uses ATP to pump a molecule *and* creates at least one gradient. But the second term, **indirect active transport**, often gives students trouble the first time they hear of it. Before we describe how it happens, let's list some important differences between the two mechanisms:

- *One* pump protein is required for direct active transport, but at least *two* different proteins are needed to perform indirect active transport.

BOX 10-5

The folded hands analogy. To visualize how a carrier protein works, press your two hands together at the base to form a "V" shape that exposes your palms. This structure represents a carrier protein that is open and oriented away from you. Then, use your fingers to pick up a small, solid object that fits into your palms. This is the closed configuration. Then, press the tips of your fingers together and spread the base of each hand . . . the object should fall out of your palms. This is the open configuration in the opposite orientation. By switching between the base-touch and fingertips-touch configurations, you can begin to see how the binding site (your palms) can move an object from one side of a membrane to another.

BOX 10-6

More name games. Be careful about this naming system: the "sym-" prefix indicates only that the ligands are carried in the same *direction*, and does not imply that the same *ligands* are being carried. If two identical or different ligand molecules are carried in the same direction, this is called uniport. Likewise, antiport refers only to the fact that two different ligands are being carried in opposite directions. Carrying two identical ligands in opposite directions at the same time would be utterly pointless, and this does not happen.

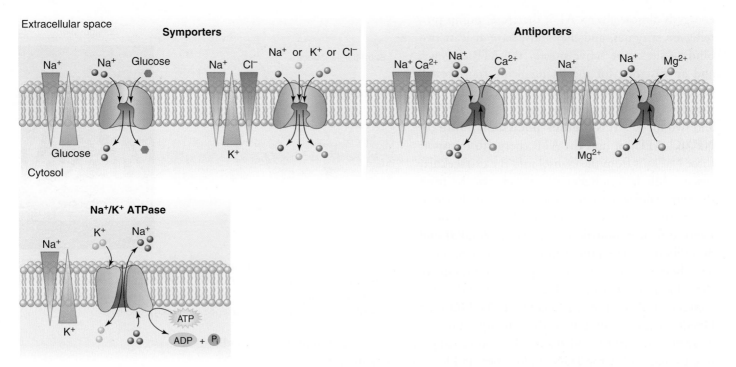

FIGURE 10-7 Some examples of Na⁺-dependent transporters. The energy of the plasma membrane Na⁺ gradient maintained by the Na⁺/K⁺ ATPase is used by these and many other transport systems.

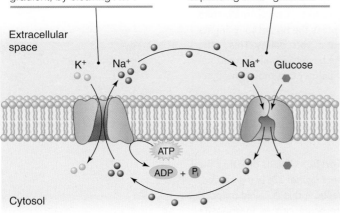

This **Na⁺/K⁺ antiporter pump** uses direct active transport to create both a Na⁺ and a K⁺ gradient, by cleaving ATP.

This **Na⁺/glucose symporter** uses indirect active transport to build a glucose gradient by expending a Na⁺ gradient.

FIGURE 10-8 The relationship between direct and indirect active transport.

■ In nearly all direct active transport, only *one* gradient is made, namely the one that is needed. In indirect active transport, at least *two* gradients are made. The first, a pump, builds a gradient by using ATP energy, and the second, a carrier, simultaneously dissipates the first gradient while building a second gradient.

The second difference is where the greatest potential for confusion lies, so let's proceed slowly here. A good way to sort this out is to start from the end and work backward, using **FIGURE 10-8** as a guide. Let's say a cell wants to build a gradient of glucose, by transporting it from the extracellular space into the cytosol. It can't rely on diffusion, of course, because it's seeking more glucose inside the cell than outside; it's going to have to *force* the glucose in. Let's put a carrier protein on the plasma membrane that can carry glucose from outside the cell to inside. Now we have to ask, what is going to supply the energy necessary to force the glucose in?

If we were using *active* direct transport here, the answer would be easy: we'd just have our carrier bind to ATP and cleave it during transport. But in this example we're not using direct active transport, so that energy has to come from somewhere else. Remember that ATP is useful as an energy source because it stores potential energy, but it is not the only form of potential energy cells use. Another easy place to find potential energy is in chemical gradients across membranes, so our glucose carrier can possibly use the potential energy of one chemical gradient to build another. This is the key to indirect active transport: *one gradient supplies the energy necessary to build another.* To make things

340   CHAPTER 10   Cellular Metabolism and Energy Storage

The Na⁺/K⁺ ATPase is an active transporter, since it uses ATP energy, but do we call the Na⁺/glucose symporter an active transporter or a passive transporter? Recall from above that passive carriers *never* build gradients. Interpreted literally, this rule means the Na⁺/glucose symporter cannot be passive. Instead, it uses potential energy (Na⁺ gradient) to make a glucose gradient, and therefore qualifies as an active transporter. But cell biologists don't ever call this kind of gradient building "active"; they reserve the term *active* solely for carriers that directly use ATP. Instead, they either call this two-carrier system *indirect active*, as we do here, or *coupled* transport, and so the protein that actually builds the desired gradient is called an indirect transporter or a coupled transporter. This may sound like a trivial distinction, but it matters to those who study it.

easy, let's use the Na⁺ gradient that exists across the plasma membrane of every cell in our body: there is more sodium outside the cell than inside, so if we let it in through our carrier, that flow can power the simultaneous flow of glucose. The carrier protein that permits this is therefore called the Na⁺/glucose symporter. Finally, we have to explain where the Na⁺ gradient came from, and the obvious answer is the Na⁺/K⁺ ATPase. Note that this carrier *does* use ATP.

Now that we have worked backward through the mechanism, let's describe it in a forward direction to make sure it all fits together; again, use Figure 10-8 as a guide. First, we start indirect active transport by performing direct active transport. The Na⁺/K⁺ ATPase uses ATP to create two forms of potential energy: a Na⁺ gradient and a K⁺ gradient. We will use the energy in the Na⁺ gradient to drive the second carrier protein as it builds yet another gradient (glucose), so the cleavage of ATP by one carrier leads, *indirectly*, to the generation of a glucose gradient by a second carrier. That's why we call it indirect active transport (**BOX 10-7**).

◼◼ Ion Transport Is Also Used in Cellular Signaling Pathways

In addition to powering the transport of glucose and other molecules, ion gradients can also serve as a means of communication in and between cells. In some cases, the transport of a given ion serves as a signal because the ion binds to a protein and changes its shape. This is most commonly done with Ca^{+2} ions, which are typically kept at a very low concentration in the cytosol. Upon receiving the appropriate signal, a **calcium channel** will often open, allowing Ca^{+2} ions to stream into the cytosol from the extracellular space (and in some cells, from the endoplasmic reticulum). These ions bind to calcium-binding proteins, altering their shape and function. In this way, Ca^{+2} ions can serve as an "on" switch for some proteins. Calcium pump proteins then push the calcium back into the extracellular space (and ER), thereby shutting the signaling proteins off. Other channels that control the flow of Na⁺, K⁺, and Cl⁻ ions also contribute to propagating signals, most often by altering the voltage across a small portion of a membrane. We will discuss the details of this signaling in Chapter 11.

CONCEPT CHECK #1

As a warm-up for examining how cells build up energy stores and use them to perform the work necessary to stay alive, try drawing a model cell with the following proteins on its surface:
▸ One antiporter pump protein that pumps molecule A into the cytosol and molecule B into the extracellular space.
▸ One symporter that transports molecule B and molecule C into the cytosol.
▸ One channel that is permeable to molecule A.
Follow the paths each molecule follows to answer this question: What is the net impact of all three proteins working *together*?

10.4 Storage of Light Energy Occurs in the Chloroplast

KEY CONCEPTS

▶ Chloroplasts capture kinetic energy in photons of sunlight and convert it into an ion gradient and high-energy electrons, which are stored on the electron carrier NADPH.

▶ The machinery that converts sunlight into these energy forms is a cluster of proteins in the thylakoid membrane inside chloroplasts. Collectively, they are known as the thylakoid electron transport chain.

▶ The ion gradient energy is converted into ATP by an enzyme called ATP synthase.

▶ The energy in ATP and NADPH is used to convert atmospheric CO_2 into a carbon-containing macromolecule called glyceraldehydes-3 phosphate via set of chemical reactions called the Calvin cycle.

Having identified the key players and methods for storing and using metabolic energy in the previous section, let's turn our attention to the processes involved in capturing this energy from the environment. To help keep things simple, I've divided the journey into four numbered phases, and limited our version of the journey to eukaryotic organisms, as including the multitude of ways that single-celled organisms complete this journey is far beyond the scope of this chapter.

In this section, we'll examine phase I (energy capture), namely how a carbon atom enters a cell from the extracellular environment and becomes trapped in a macromolecule. In the remaining sections of this chapter, we'll follow this macromolecule through the complex series of chemical reactions it undergoes as it is converted into the building blocks of cells we discussed in Chapters 1–4, and is then broken down into simple molecules that are released into the environment, completing the journey. As we go through this, we'll follow a temporal sequence of events just as we did in Chapter 9.

Chloroplasts Have Three Membrane-Bound Compartments

The star character of phase I is the chloroplast. Like the mitochondrion, this organelle is thought to be derived from a photosynthetic bacterium that was engulfed by another cell and established a symbiotic relationship with it. Over the course of evolution in autotrophic eukaryotes, the chloroplast shed most of its self-sustaining functions (and the corresponding DNA) to instead specialize in the conversion of CO_2 and H_2O into organic macromolecules. The overall structure of a typical chloroplast is shown in **FIGURE 10-9**. It is characterized by the presence of three different membranes: an inner and outer membrane that enclose the entire organelle, and the membrane that forms the innermost compartment, called the **thylakoid**. The inner compartment of the thylakoid is called the **thylakoid lumen**, and the thylakoid resides in a space called the chloroplast stroma.

Chloroplasts Convert Sunlight into the First Forms of Cellular Energy

Energy capture requires two sets of chemical reactions, which take place in different chloroplast compartments. The first set, sometimes referred to as the **light reactions** because they require sunlight to proceed, occur in the thylakoid membrane; a more descriptive name for these is **energy transduction reactions**. The second set take place in the stroma, and are called the **carbon assimilation reactions**, which are often referred to as **dark reactions** to indicate that they do not need sunlight. (Often the name is misinterpreted to suggest that these reactions require darkness to proceed, but this is not true. The dark reactions function

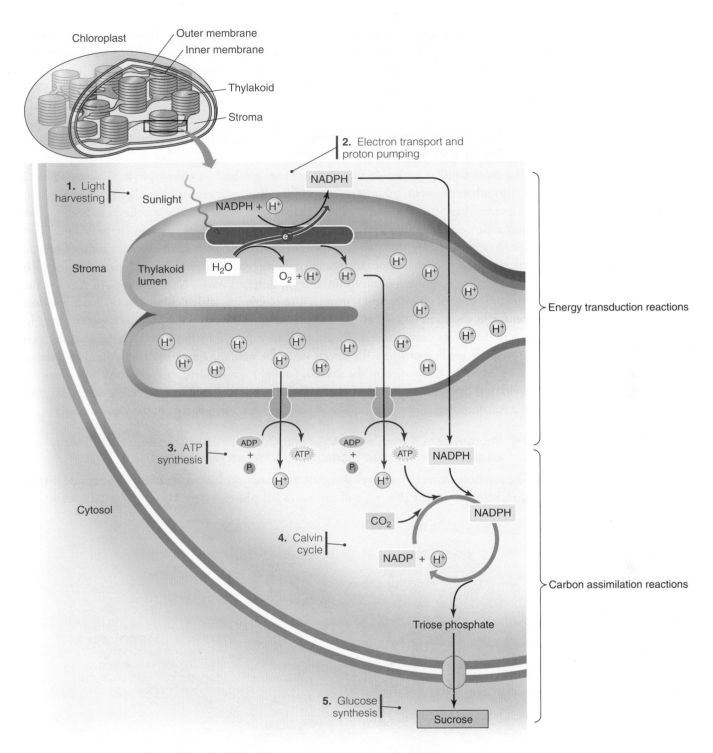

FIGURE 10-9 The overall structure of a chloroplast.

perfectly well in sunlight.) Also, there is no direct coupling of the two sets of reactions: if the dark reactions stop, the light reactions can continue, and viceversa.

▰▰ The Energy Transduction (Light) Reactions Convert Sunlight Energy into Stored Potential Energy

Let's begin this journey at the very start, even before carbon enters the cell. To couple carbon atoms together and construct cellular building blocks, energy must be available to help form the bonds holding these carbons together. Chloroplasts build these energy stores by converting sunlight into both forms of cellular potential energy, by first building an ion gradient across the thylakoid membrane by moving protons into the thylakoid lumen, and then converting the energy in this gradient into reducing electrons on NADPH and the terminal phosphate on ATP.

■■■ Chlorophyll and Other Pigments in the Thylakoid Membrane Capture Sunlight to Excite Their Electrons

The most familiar molecule involved in photosynthesis is **chlorophyll**. Chlorophyll is a hydrocarbon **pigment**, meaning it is especially effective at capturing sunlight energy. Sunlight is a somewhat mysterious form of energy for many of us, because it behaves as both a wave and a particle. For the purposes of this chapter, let's consider it a particle moving at light speed (approximately three million meters per second), so it is an excellent form of kinetic energy. Most photosynthetic eukaryotes contain several different chlorophyll molecules, each specializing in capturing photons traveling in different waves. Most chlorophyll captures the kinetic energy of photons and transmits it to highly specialized chlorophyll called a **reaction center**, as shown in **FIGURE 10-10**. To visualize how this happens, imagine a spider sitting in the middle of its web: as insects fly into and are captured by the strands of the web, the web vibrates. Regardless of where the insects strike the web, the spider in the middle always senses it, based on the orientation of the strands. Sunlight is equivalent to a bombardment of insects (photons) hitting strands (chlorophyll) in this web (called a **light-harvesting complex**), and the reaction center sits in the middle, where the vibrational energy in the web is strongest.

■■ Photosystem II Is a Large Molecular Complex That Transfers High Energy Electrons from Excited Pigments to a Proton Pump

The reaction center chlorophyll is tightly associated with a cluster of 25 polypeptides in the thylakoid membrane that are collectively called **photosystem II**, shown in **FIGURE 10-11**.

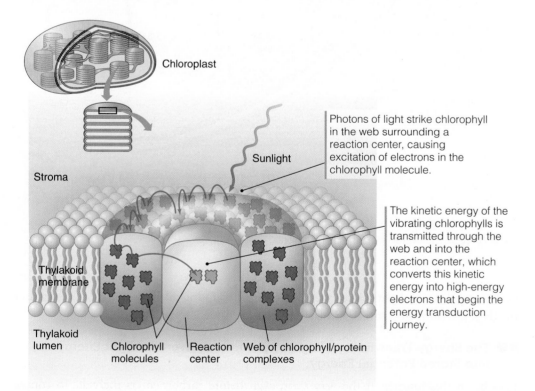

FIGURE 10-10 How chlorophyll transmits sunlight energy to a reaction center.

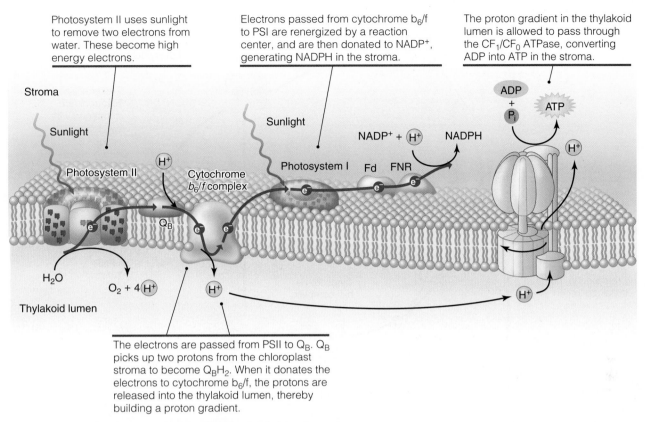

Photosystem II uses sunlight to remove two electrons from water. These become high energy electrons.

Electrons passed from cytochrome b_6/f to PSI are renergized by a reaction center, and are then donated to $NADP^+$, generating NADPH in the stroma.

The proton gradient in the thylakoid lumen is allowed to pass through the CF_1/CF_0 ATPase, converting ADP into ATP in the stroma.

Stroma

Sunlight

Sunlight

ADP + P_i

ATP

$NADP^+ + H^+$ NADPH

Photosystem I   Fd   FNR

H^+

Photosystem II   Cytochrome b_6/f complex

Q_B

H_2O

$O_2 + 4 H^+$

H^+

H^+

H^+

Thylakoid lumen

The electrons are passed from PSII to Q_B. Q_B picks up two protons from the chloroplast stroma to become Q_BH_2. When it donates the electrons to cytochrome b_6/f, the protons are released into the thylakoid lumen, thereby building a proton gradient.

FIGURE 10-11 The electron transport chain in the thylakoid membrane.

The overall function of this complex is quite simple: it uses the kinetic energy of the chlorophyll reaction center to literally pull electrons off of water and attach them to the first of a long string of proteins that pass the electrons from one to another in a regimented order, bucket brigade–style. This is how the bulk of atmospheric oxygen is generated on earth, and it is here that kinetic energy from the environment is converted to cellular potential energy. The chemical reactions taking place in photosystem II can be summarized by a simple equation:

$$H_2O + photosystem\ II\ (oxidized) \rightarrow O_2 + 2H^+ + photosystem\ II^*\ (reduced)$$

The electrons are passed from one polypeptide to another in photosystem II for a very simple reason, illustrated in **FIGURE 10-12**. To visualize this, let's turn these molecules into characters and give them "personalities" to help understand their preferred redox state. Recall that every redox pair of molecules has a **redox potential**, which is a measure of how much they "prefer" to be in the oxidized or reduced state. By convention, the *higher* the redox potential, the more a given pair prefers to be in the *reduced* state; put another way, redox potential is a measure of how much they "like" additional electrons. Redox potentials range from +800 to −800 millivolts (mV) for most cellular redox pairs. $H_2O/O_2 + 2H^+$ has a very high redox potential (+800 mV), meaning that water is very reluctant to give up its electrons, so removing these electrons requires a lot of energy, and that's what the kinetic energy in the reaction center supplies. Imagine a polypeptide in photosystem II and a water molecule in a tug-of-war over these electrons, and when the polypeptide gets a big energy boost from the excited chlorophylls in the reaction center, it wins. It takes this extra energy to tear the electrons away from the oxygen in the water molecule, and this energy actually resides in the electrons. This is why we refer to these electrons as high-energy electrons.

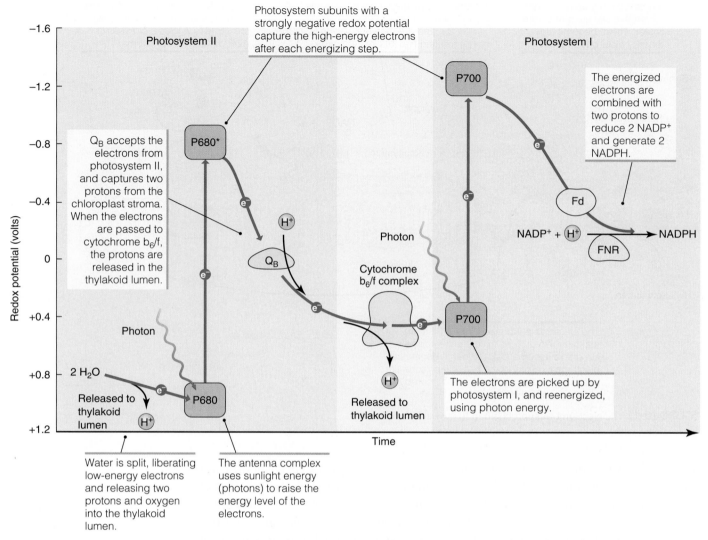

FIGURE 10-12 The redox potential of reactants indictates the energy level of the electrons moving through the thlakoid electron transport chain.

The key to understanding the ordered transfer of these electrons from one polypeptide to another is that the first polypeptide to be reduced, called P680, actually has a very negative redox potential (−800 mV). In other words, it actually "hates" the electrons it just picked up. This only makes sense if one considers how much energy it receives from the reaction center to receive them, and further underscores how critical chlorophyll is. Once it receives these electrons, chlorophyll immediately "looks around" for something else to hand them off to. The rest of the polypeptides involved in this electron transport have a negative redox potential, too, but not as low as that for P680, so they hate electrons too, just not as much. The one that hates electrons the most, but not as bad as P680, gets the electrons next, who then passes them on to the one that hates it the most after it, and so on. In short, one can line up all of the polypeptides that carry these electrons according to their redox potential, and the electrons are passed from the most negative to the second most negative, and so on, as shown in Figure 10-12.

The technical description for this series of reactions is more direct: electrons stripped from water are donated to a series of polypeptides, each of which has a progressively higher redox potential. This ensures that the transport is strictly linear. The official name for a series of polypeptides arranged in this fashion is *electron transport chain (ETC)*. The ETC in the thylakoid membrane is composed of about 50 polypeptides.

This begs the question: Why not just give the electrons to the polypeptide that actually wants it most? Why all the intermediate steps? The answer is that this process isn't concerned with getting the electrons onto a friendly molecule; its purpose is extracting as much energy as possible from the electrons while they are still in the high-energy state. According to the second law of thermodynamics, if this change in potential energy happens in a single step, whatever energy that cannot be captured by a protein is lost as heat. In technical terms, the change in free energy resulting from a single-step reaction moving the electrons from water to a final electron acceptor is so large most of this energy cannot be captured by proteins and is lost as heat.

In this context, heat is useless, so cells use the opposite strategy: hand the electrons from one polypeptide to another that has a *very similar* and slightly higher redox potential. This allows each polypeptide to capture a small amount of the energy stored in these electrons, with minimal heat loss. By the time they complete the entire energy transfer cycle in cells, the electrons return to their base energy state and the molecule they came from: water. We'll see them rejoin water at the end of the chapter.

The proteins/polypeptides in the ETC react differently to undergoing reduction and oxidation. All of them undergo a shape change, but the effect of the change differs in every different polypeptide. A summary of the major changes is given in Figure 10-11. One of the most important shape changes is that associated with a membrane protein called plastoquinol (abbreviated Q_B). When Q_B receives the high-energy electrons from photosystem II, it becomes reduced and picks up two protons from the chloroplast stroma, becoming Q_BH_2. This causes a shape change that permits it to donate the electrons to a structure called the cytochrome b6/f complex. When it donates the electrons, it reverts to its oxidized state and releases the two protons, but due to the shape change it undergoes while reduced, the release of electrons occurs when these protons face the thylakoid lumen. The net result is that, as a consequence of passing electrons along the ETC in the thylakoid membrane, protons are moved from the stroma to the lumen. A simpler way of describing this activity is that protons are forced from one side of the membrane to another, and Q_B is called a proton pump for this reason. (Some researchers include the cytochrome b6/f complex as part of the pump machinery, because without its oxidation of Q_B no protonswould be moved. For our purposes, this doesn't matter.) The most important consequence of this pumping is that some of the energy in the electrons is converted into a proton gradient. We'll see how the potential energy in this gradient is used by the dark reactions in later sections

■■■ The CF1/CF0 ATP Synthase Complex Uses the Proton Gradient to Make ATP

The potential energy in the proton gradient is transferred to an ATP molecule by a transmembrane protein complex called the CF_1/CF_0 ATP synthase, shown in Figure 10-11. This complex behaves like a proton pump that runs in reverse: rather than consume ATP to build a proton gradient, it allows protons to run down a gradient, and uses this energy to attach a third phosphate group to ADP. Thus, the protons make a complete round trip: they are pumped from the stroma into the thylakoid lumen by Q_B, and then allowed to flow back into the stroma. Remember that the thylakoid membrane is fluid, and transmembrane proteins can diffuse within the plane of the phospholipid bilayer. Figure 10-11 shows that, when rendered in three dimensions, the spatial arrangement of the thylakoid membrane proteins is quite random. Thus, the proton gradient is converted to ATP even as the electrons that generated it continue their journey along the ETC.

■■ Photosystem I Captures Light Energy to Reenergize Electrons after Proton Pumping

Generating ATP from a proton gradient is only one of the two jobs carried out by the ETC; the second job is reducing the energy carrier $NADP^+$. To do this, high-energy electrons are needed. Most of the energy added to the electrons stolen from water by photosystem II is used to pump protons, so they are not ready to be added to $NADP^+$ at this point. What is needed is another boost of energy, and this is provided by a second light-absorbing chlorophyll-protein complex called photosystem I, shown in Figure 10-11. (The numbers I and II reflect the order in which they were discovered, not the order of how they function. Now that we know the order in which they operate, it would be handy if the numbers of the two photosystems were swapped, but that hasn't happened.) For our purposes, we can think of photosystem I as being essentially identical to photosystem II, with three exceptions:

1. Photosystem I receives electrons from a protein called plastocyanin, rather than from water. Plastocyanin is an electron carrier that accepts the electrons from cytochrome b6/f complex, thereby continuing the ETC.
2. The electron acceptor in photosystem I is called P700 rather than P680.
3. The electrons excited by photosystem I are not used to pump protons.

■■ High-Energy Electrons Are Stored by the Electron Carrier Molecule NADPH

High-energy electrons leaving photosystem I are passed to two more proteins before ending their journey; the first is called ferredoxin, and the second is called ferredoxin-$NADP^+$ reductase. Based on the name alone, it is fairly easy to understand what ferredoxin-$NADP^+$ reductase does: it oxidizes ferredoxin and reduces $NADP^+$, thereby generating NADPH. This reduction takes place on the stromal side of the thylakoid membrane.

Before we move on, let's summarize what has happened so far. Figure 10-11 demonstrates the physical path travelled by electrons in the ETC: make sure to note that the same electrons that leave water at photosystem II end up in NADPH. A second way of summarizing the events in the ETC is shown in Figure 10-12, which shows how the *energy* of the electron pair taken from water (expressed as electrical potential of the redox pairs they encounter) changes as it travels along the ETC. Notice the two massive boosts provided by the photosystems, and the gradual drop that follows each boost. The characteristic zigzag pattern that we see on such a plot is often called the **Z-scheme** for this reason.

■■ The Carbon Assimilation (Dark) Reactions Convert Stored Potential Energy into Macromolecules

The products of the energy transduction reactions (ATP, NADPH) are relatively short-lived intermediates in our energy transfer journey. Immediately after they are generated in the chloroplast stroma, they are put to use by yet another series of proteins to undergo a set of reactions that are collectively called the **Calvin cycle**, named in honor of Dr. Melvin Calvin, who figured out the complex chemistry behind these reactions in the 1950s, and was awarded the Nobel Prize in Chemistry in 1961 for this outstanding achievement. Thanks to his efforts, we can now summarize the complexity of this cycle as a simple four-step process, as shown in **FIGURE 10-13**. It is difficult to overestimate the importance of this cycle, because all life depends on it. For the purposes of this chapter, the most important points to notice about this cycle are as follows:

- The simple gas CO_2, which easily diffuses across all cell membranes, is captured by the enzyme ribulose bisphosphate carboxylase/oxygenase and, along with water, is added to a five-carbon phosphorylated keto sugar (see *Sugars Are Simple*

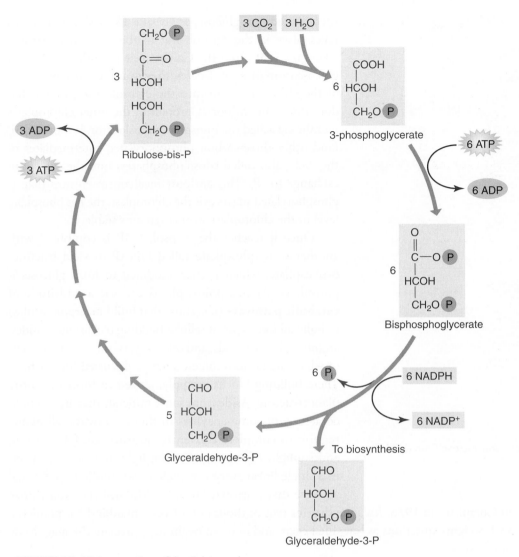

FIGURE 10-13 An overview of the Calvin cycle.

Carbohydrates in Chapter 1) called ribulose-1,5-bisphosphate. Biologists have fortunately developed a shorthand name for this enzyme: **rubisco**. Because it is present in the leaves of all plants, it is likely the most abundant enzyme on earth. This enzyme also is the link between inorganic carbon-containing molecules and organic molecules.

- Rubisco generates several copies of a three-carbon phosphorylated aldo sugar (the stoichiometry is not important for us), which are in turn converted into glyceraldehyde-3 phosphate (G3P). G3P is the source of all the other organic (biological) cellular building blocks we discussed in Chapters 1–4.

- Both ATP and NADPH are oxidized at different steps in the cycle. This energy is what drives the entire process.

Glucose Is One of the First Carbohydrates Synthesized from G3P

The final products of the Calvin cycle are G3P, $NADP^+$, ADP, and P_i. We have already described how the $NADP^+$, ADP, and P_i are recycled to make more NADPH and ATP. The next steps in our energy journey focus on the fate of G3P, the most common being that it

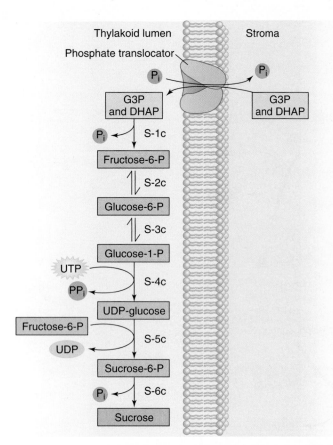

Thylakoid lumen   Stroma

Phosphate translocator

P_i P_i

G3P   G3P
and DHAP   and DHAP

P_i ← S-1c

Fructose-6-P

S-2c

Glucose-6-P

S-3c

Glucose-1-P

UTP

PP$_i$ ←   S-4c

UDP-glucose

Fructose-6-P

UDP ←   S-5c

Sucrose-6-P

P_i ← S-6c

Sucrose

FIGURE 10-14 The synthesis of glucose and sucrose from G3P in the cytosol.

is converted into glucose, as **FIGURE 10-14** shows. But this takes place in the cytosol, not in the chloroplast stroma where the G3P from the Calvin cycle is made, so to build glucose from these molecules, they must be transported out of the chloroplast and into the cytosol. The protein that does this is an antiporter protein in the inner chloroplasts membrane called the phosphate translocator. It moves G3P (and other three-carbon phosphorylated intermediates of the cycle, also called triose phosphates) into the cytosol in exchange for P_i. This antiport mechanism ensures that as phosphorylated sugars exit the chloroplast, the net phosphate level in the chloroplast stroma remains stable.

Once it reaches the cytosol, G3P is combined with another triose phosphate called DHAP to yield fructose-6-phosphate, which is then modified to form glucose-6-phosphate. At this point, plants can use a multitude of **catabolic pathways** (i.e., those that build macromolecules) to build all four types of cellular building blocks: nucleotides, amino acids, fatty acids, and other sugars, the enzymatic steps of which are far too numerous to be discussed further here. These building blocks are polymerized to form the entire plant structure. Aside from some minerals that are required cofactors for some enzymes in these pathways, all plants require to complete this entire journey are CO_2, water, and sunlight. Tying together the light and dark reactions in a single linear story was such an astounding intellectual advance that it earned Dr. Peter Mitchell the Nobel Prize in Chemistry in 1978. Imagine that for tens of thousands of years mankind has pondered such weighty questions as how life began, and one can begin to appreciate the magnitude of his work.

10.5 Cells Use a Combination of Channel, Carrier, and Pump Proteins to Transport Small Molecules across Membranes

KEY CONCEPTS

▶ The majority of the macromolecules made by cells can serve as food energy for other cells. To access this energy, the chemical bonds holding these macromolecules must be broken.

▶ In animals, macromolecules are broken into cellular building blocks (via digestion) in the extracellular space.

▶ Cellular building blocks (e.g., glucose) are transported across the plasma membrane by an integrated system of channels, carriers, and pumps.

The next step in our energy journey is the transfer of macromolecules from one cell to the next; we'll call this phase II (macromolecule transport). This takes place between all cells, but we will limit the story to what happens in multicellular heterotrophs such as ourselves. More simply, let's find out what happens when we eat the plant that just completed photosynthesis in the last section.

For most heterotrophic animals, the initial input of energy is in the form of the complex macromolecules that the plant makes during its lifetime. Remember that the second law of thermodynamics explains why these macromolecules are good energy stores, *if* they can be broken down. Plants synthesize lots of digestible molecules, but humans can't break down the $\beta1-4$ bonds in cellulose (see *Oligosaccharides and Polysaccharides: The Storage, Structural, and Signaling Components of Cells* in Chapter 1), so the complex structures made from cellulose are of no caloric value to us. They still serve a useful dietary function (as fiber), but we will dismiss them here. We'll focus instead on the remaining macromolecules that we can digest. By convention, virtually every cell biology textbook that covers this topic picks sugars (specifically glucose) as the representative food molecule, but any of the others (fats, proteins, nucleic acids) would serve just as well. We'll stick with the convention.

Most sugars we eat are more complex than glucose, which is a monosaccharide. Sucrose, what we commonly refer to as table sugar, is a disaccharide, and starches are larger complexes of disaccharides. Based on what we've discussed in previous sections of this chapter, we know that to get the sugar into our cells, we have to break it down into monosaccharides, because there are no disaccharide (or larger) carrier proteins in the plasma membranes of our cells. The steps involved in digesting everyday food into monosaccharides, amino acids, nucleotides, and fatty acids are too complex to be covered here, so we'll summarize them by simply stating that this digestion takes place outside of our cells, in our gastrointestinal tract.

We pick up the story in our small intestine, where digestion is completed. To "eat," a cell that lines our small intestine has to capture the food molecules as they flow by, as shown in **FIGURE 10-15**. Because food spends a limited amount of time in the digestive tract, intestinal cells don't wait for these molecules to flow down their concentration gradients across the plasma membrane, so they *force* them across, or *pump* them. By using this term, we can now access the information discussed earlier in this and other chapters to help explain how we absorb and store glucose. Armed with this information, let's build a scheme for transporting glucose from the lumen of the small intestine across the apical plasma membrane of gut epithelial cells, and then send it across the basolateral domain and into the bloodstream. Finally, we'll deliver it to the liver, where it will be stored as glycogen. If necessary, review the section *Protein Channels, Carriers, and Pumps Regulate the Transport of Most Small Molecules across Membranes* earlier in this chapter to recognize the three types of transport proteins we have at our disposal, plus the concepts of direct and indirect active transport. It may also be helpful to review the organization of an epithelial cell plasma membrane (see Figure 6-25) and the sorting of membrane proteins by the TGN (see *The Trans-Golgi Network Sorts Proteins Exiting the Golgi Apparatus* in Chapter 9).

■■ The Na⁺/K⁺ ATPase Maintains the "Resting Potential" across the Plasma Membrane

Let's start by establishing the beginning state of our cells. The key to pumping any molecule across a membrane is to use a carrier or pump protein that can access the potential energy necessary to build a gradient. The most common gradients across the plasma membrane of our cells are the Na^+ and K^+ gradients established by the Na^+/K^+ ATPase pump. The gradients this pump creates contribute to the overall imbalance of charge across the plasma membrane caused by every impermeable molecule in every one of our cells. Negative and positive charges on amino acids, proteins, fatty acids, nucleotides, and other ionized atoms (Ca^{+2}, H^+, etc.) all contribute to this membrane potential. For humans, the average of this so-called resting potential is -60 to -70 mV. This is our beginning state.

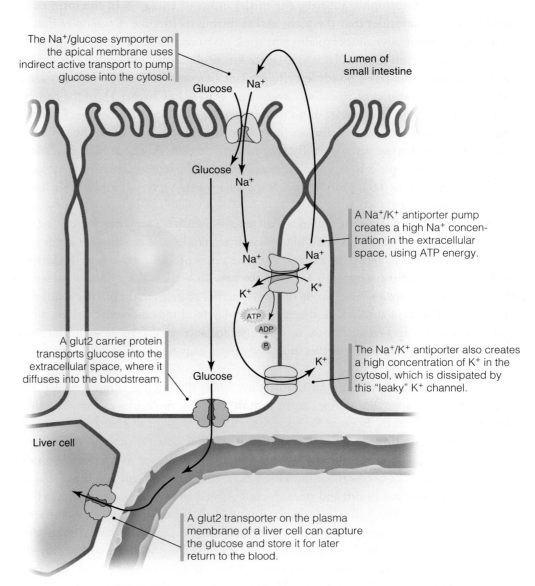

The Na⁺/glucose symporter on the apical membrane uses indirect active transport to pump glucose into the cytosol.

Lumen of small intestine

Glucose   Na⁺

Glucose
Na⁺

A Na⁺/K⁺ antiporter pump creates a high Na⁺ concentration in the extracellular space, using ATP energy.

Na⁺   Na⁺

K⁺   K⁺

ATP

ADP + Pᵢ

A glut2 carrier protein transports glucose into the extracellular space, where it diffuses into the bloodstream.

Glucose

K⁺

The Na⁺/K⁺ antiporter also creates a high concentration of K⁺ in the cytosol, which is dissipated by this "leaky" K⁺ channel.

Liver cell

A glut2 transporter on the plasma membrane of a liver cell can capture the glucose and store it for later return to the blood.

FIGURE 10-15 How glucose is transported from the intestinal lumen into the blood stream.

In the Vertebrate Gut, a Leaky K⁺ Channel, a Na⁺/Glucose Symporter, and a Passive Glucose Carrier Work Together to Move Glucose from the Gut Lumen to the Bloodstream

Our strategy for transporting glucose out of the small intestine, into the bloodstream, and finally into the liver requires four different membrane proteins. Follow along with Figure 10-15 as we introduce each one into our scheme. The first protein we will add is the Na⁺/K⁺ ATPase pump, to establish the Na⁺ and K⁺ gradients. (We'll use both Na⁺and K⁺ here.) This is a classic transmembrane protein in the endomembrane system, and is targeted by the TGN to vesicles that fuse with the basolateral domain of the intestinal epithelial cell. Once inserted, it begins cleaving ATP to maintain the two gradients.

The second protein we'll use is called a leaky K⁺ channel. This, too, is targeted to the basolateral domain. It is called "leaky" because it does not close completely under resting conditions, so the K⁺ that was pumped into the cytosol by the Na⁺/K⁺ ATPase simply leaks back out into the extracellular space. This may seem like a waste, but it is actually a very useful addition: by keeping the K^+ concentration in the cytosol low, it stimulates the Na^+/K^+ ATPase to work constantly to compensate. As a result of this K^+ "short circuit," the Na^+ gradient for these cells is especially steep.

Next, we'll add the Na^+/glucose symporter we discussed earlier in this chapter (see Figure 10-8). This protein capitalizes on the steep Na^+ gradient to very quickly force glucose from the intestinal lumen into the cytosol via indirect active transport. The faster this works, the more efficient our intestines are at absorbing the sugars we just digested.

After adding these three proteins, we can explain how glucose is pumped into the epithelial cell, but now we need to get it out into the bloodstream. Because it is so concentrated in the epithelial cell, a simple carrier protein in the basolateral domain of the epithelial cell (the fourth protein in our scheme) will allow the glucose to flow out of the cell, without reentering the intestinal lumen, where it gets picked up by the bloodstream. This **glucose transporter** is part of a family of related carriers called **gluts** (pronounced *gloots*), illustrated in **FIGURE 10-16**. Each glut has a different affinity for glucose, and thereby needs a different sized concentration gradient to operate. The glut in the epithelial cells (called glut2) has a low affinity for glucose, so it takes a large concentration difference to make it work. That large difference is there, courtesy of the Na^+/glucose symporter.

Once the glucose enters the bloodstream, it circulates throughout the body and is transported into cells by different gluts. Some cell types (e.g., neurons in the brain) consume glucose at a tremendous rate, so they soak up just about all of the glucose that flows by them, while other cells (e.g., bone cells) have a very low metabolism and can be more picky about when they want to import some glucose. Eventually, though, the glucose we absorb in the gut from a single meal is far more than our cells need during the period when the absorption takes place. This can create quite a problem: if our cells don't absorb the glucose, the blood glucose levels can rise to dangerous levels, and can over time cause a myriad of problems including blindness, kidney failure, heart disease, and nerve damage. Our bodies need a temporary storehouse for this glucose.

Liver (and skeletal muscle) cells play the role of short-term warehouse. As glucose streams by liver cells in the blood, a glut2 is activated when glucose levels rise. The glucose is carried across the plasma membrane into the cytosol of the liver cells where it is converted into a short-term storage molecule called **glycogen**. A few hours after we eat and our blood sugar levels begin to fall, the liver breaks down the glycogen, and the freed glucose is returned to the bloodstream. This is how our liver buffers our blood glucose levels between meals. The sensing of blood glucose levels and regulation of liver transport is under the control of the hormones insulin and glucagon, which are secreted by the endocrine pancreas via mechanisms we do not have room to discuss here.

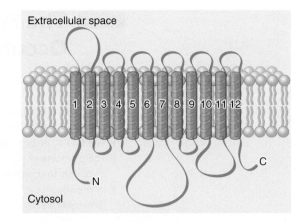

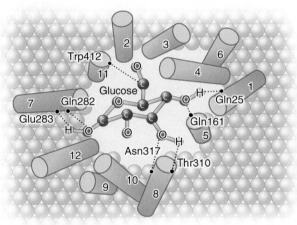

FIGURE 10-16 Two views of the glut1 transporter. Notice how a portion of several alpha helices interacts with glucose to form the binding site.

CONCEPT CHECK #2

How would the transport of glucose in our scheme be affected if the TGN failed to properly sort the glut2 protein in intestinal epithelial cells? What impact would this failure have on the rest of the cells in the body?

10.6 The First Phase of Glucose Metabolism Occurs in the Cytosol

> **KEY CONCEPTS**
>
> ▸ The steps taken to extract energy from glucose are very similar to the steps chloroplasts use to build glucose from G3P, only in reverse order.
> ▸ The first 10 enzymatic steps in the digestion of glucose are called glycolysis.
> ▸ The products of glycolysis include the molecule pyruvate, which must be metabolized to keep glycolysis from stalling.
> ▸ In the absence of molecular oxygen (O_2), pyruvate is metabolized by a process called fermentation. Two different methods of fermentation have evolved in different organisms.

Phase III of our journey (anaerobic metabolism) begins in the cytosol of one of our cells. Let's assume that this cell expresses the proper glut transporter and thus is able to maintain a stable concentration of glucose in the cytosol; now our goal is to understand how that glucose is converted into the potential energy that the cell needs to survive. Before we dive into the details, consider what we expect to see. We saw that the chloroplast undergoes a complex set of chemical reactions to build potential energy stores that were then used to make glucose. Now we are asking the opposite question: How do cells use glucose to make potential energy stores? Our expectation is that we should see a strategy that looks something like photosynthesis running in reverse.

How might this strategy work? Let's start at the end of photosynthesis and work backward. We saw that several chemical reactions were needed to convert triose phosphates into glucose, and this took place in the cytosol (Figure 10-14). Run in reverse, we'd expect to see that several steps are used to convert glucose into triose phosphates, and this too would take place in the cytosol. And that's exactly what happens. Because we are breaking glucose up by hydrolyzing the bonds holding it together, this process is called **glycolysis**.

Why a Stepwise Method of Metabolizing Glucose Is Necessary

Glycolysis is a 10-step process. There is a very good reason for taking glucose apart in so many steps, which is grounded in the same thermodynamic principles that explained the multiple steps in the thylakoid ETC: extracting energy in multiple small steps is more effective than doing it in one big step. By taking small steps, the heat lost during the conversion of one form of potential energy to another can be spread out over time and space. If glucose was oxidized in a single step, no single protein could ever trap all of the useful energy released, so it would instead be lost as heat. Glucose has a lot of energy: if I drop a match onto a pile of pure glucose, it will burst into flames. To be blunt: *if cells chose the short route of metabolizing glucose, they could burst into flames.* Even after three-billion-plus years of evolution by natural selection, nature has yet come up with a better solution to that problem. So, slow and steady is the name of the game, and every cell in every living organism does it the exact same way.

The 10 Chemical Reactions in Glycolysis Convert a Glucose Molecule into Two Three-Carbon Compounds, Two NADH Molecules, and Two ATP Molecules

In keeping with our focus on the concept of energy transport, we'll avoid looking at the actual structures of the intermediates and the enzymes that generate them. Instead, let's look at the overall reaction, as illustrated in **FIGURE 10-17**. First, we see that glucose, a 6-carbon sugar, is broken into 2 3-carbon sugars called **pyruvate**. This results in a small liberation of energy: for every glucose that undergoes glycolysis, the net output is just 2 ATP molecules and 2 NADH electron carriers (notice that NADH is *not* NAD*P*H, which is used in photosynthesis and other reactions), and no CO_2 is generated. Back when life first began and glycolysis was possibly the primary, if not only, means of extracting energy from glucose, this was a mighty accomplishment; it obviously was enough to propel the evolution of life, since every living organism now uses it. As we shall soon see, this 10-step method has, over time, been extended to over 30 steps, and the complete pathway yields 10-fold more useful energy than glycolysis alone.

Glycolysis Is Subdivided into Three Stages

Biochemists traditionally divide the 10 steps of glycolysis into the 3 stages shown in Figure 10-17. The first stage, which contains the first 5 reactions, is somewhat surprising, in that ATP is actually spent, rather than created, to generate phosphorylated sugar intermediates. The products of stage 1 are 2 G3P and 2 ADP molecules. Notice that from this point on, every reaction that takes place occurs *twice* for every glucose molecule, because 1 glucose has been split into 2 G3Ps. It is likely no accident that G3P appears as an intermediate step in the digestion *and* creation of glucose, given that the earliest photosynthetic organisms arose from those that underwent glycolysis. In fact, many of the intermediate molecules in both photosynthesis and glycolysis are identical, or nearly so.

Stage 2 of glycolysis comprises the sixth and seventh reactions, which represent the first energy harvesting steps. For each of the two G3P molecules produced during stage 1, one NADH and one ATP are generated. The ATP-generating step eliminates the energy deficit generated by stage 1. The final three steps compose stage 3, and they yield one additional ATP for each of the two reactants. The creation of pyruvate marks the end of glycoslysis.

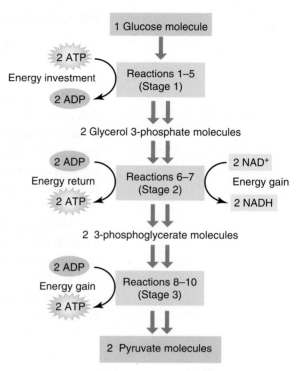

FIGURE 10-17 The three stages of glycolysis. Note that the total energy extracted is 2 ATP and 2 NADH.

Pyruvate Is Not an Endpoint in Glucose Metabolism

It is likely that even in the earliest cells capable of metabolizing glucose, pyruvate was not the final product. In the cells of most modern organisms, stopping at pyruvate would be fatal, the reason being based on simple stoichiometry: every enzymatic reaction during glycolysis is reversible, depending upon the concentration of the reactants. Glucose is more abundant than its first glycolytic derivative (glucose-1-phosphate) in the cytosol, so the enzyme that catalyzes the conversion of one to the other favors the reaction from glucose to glucose-1-phosphate. Likewise, the balance between glucose-1-phosphate and its derivative favors its conversion into the derivative, and so on. Follow this reasoning to the end of glycolysis: if the last step yields pyruvate, and nothing consumes the pyruvate, it will eventually build up to such a concentration that the enzyme that makes it will stop. That pyruvate precursor would then build up until it as no longer made, and so on; eventually, all 10 reactions of glycolysis would stop, no energy would be generated, and the cell would starve to death. It is therefore mandatory that something be done with the pyruvate. Current organisms have 2 solutions for lowering their pyruvate concentration.

■■ In the Absence of Molecular Oxygen, Pyruvate Undergoes Fermentation

The first solution to appear during evolution is now called **fermentation**, and is illustrated in **FIGURE 10-18**. This arose in organisms well before molecular oxygen accumulated in the atmosphere, and so is often referred to as **anaerobic metabolism**. During fermentation, the NADH molecules generated by glycolysis are used to convert pyruvate into one of two different molecules, depending on the organism. The first strategy yields CO_2 and ethanol as end products, and both of these molecules are membrane permeable: they can simply diffuse out of the cell, and thus keep the cytosol clear of glycolysis metabolites. Yeasts, the modern day descendants of the organisms that developed this strategy, are routinely used to convert the sugars in some beverages into the ethanol that makes these beverages alcoholic. If the released CO_2 is properly captured as well, bubbling beverages result.

The second fermentation strategy consumes the NADH to convert pyruvate into **lactic acid**, or lactate. This also consumes the NADH generated by glycolysis, so the net yield is still two ATP per glucose molecule. Lactate fermentation likely arose later, because lactic acid is not membrane permeable. To make this strategy work, the cells that generate lactic acid must also express a **lactic acid transporter protein** on their plasma membrane to carry it from the cytosol to the extracellular environment. Bacteria that use this method are

Aerobic conditions
In the presence of oxygen, many organisms convert pyruvate to an activated form of acetate known as acetyl-CoA. In this reaction, pyruvate is both oxidized (with NAD^+ being reduced to NADH) and decarboxylated (liberation of a carbon atom as CO_2). Acetyl-CoA then becomes the substrate for aerobic respiration, where NADH is oxidized back to NAD^+ by molecular oxygen.

Anaerobic conditions
When oxygen is absent, pyruvate is reduced so that NADH can be oxidized to NAD^+, the form of this coenzyme required in reaction Gly-6 of glycolysis. Common products of pyruvate reduction are lactate (in most animal cells and many bacteria) or ethanol and CO_2 (in many plant cells and in yeasts and other microorganisms).

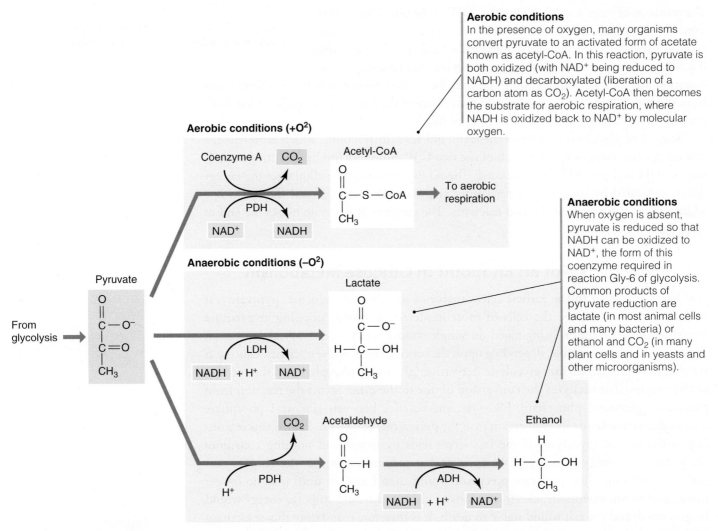

FIGURE 10-18 Three fates of pyruvate. Notice that the top reactions only take place in the presence of oxygen, while both fermentations take place in the absence of oxygen.

commonly used to ferment milk to make cheese, and human cells use this method when they do not receive enough oxygen from the bloodstream. The importance of oxygen in glucose metabolism will be discussed in the next section.

10.7 Aerobic Respiration Results in the Complete Oxidation of Glucose

KEY CONCEPTS

- ▸ The appearance of molecular oxygen in the atmosphere allowed some organisms to harness the strong electronegativity of oxygen atoms to extract 18-fold more ATP energy than glycolysis alone.
- ▸ Aerobic respiration takes place in mitochondria, and occurs in four stages.
- ▸ Stage one converts pyruvate into acetyl-CoA, the substrate of a metabolic cycle called the Krebs cycle (stage two).
- ▸ The Krebs cycle resembles the Calvin cycle run in reverse.
- ▸ During stage three, the high-energy electrons removed from pyruvate and acetyl-CoA are passed through an electron transport chain in the inner mitochondrial membrane, similar to that used by thylakoids. At the end of the electron transport chain, the electrons are returned to an oxygen atom to rebuild the water molecule that was oxidized at the beginning of photosynthesis, thereby completing the cyclic journey of electrons.
- ▸ In stage four, the proton gradient formed by the electron transport chain is converted into ATP by and ATP synthase.

Phase IV of our journey (aerobic metabolism) likely evolved later than the first three phases. Geological evidence demonstrates that about 2.5 billion years ago, enough molecular oxygen (7% O_2) accumulated in the atmosphere to induce permanent changes in the earth's crust that we can now detect; this suggests that oxygen accumulation began earlier, before it left any detectable evidence in the geological record. The fact that it increased at all is evidence that anaerobic organisms (those that can or must live in the absence of appreciable amounts of O_2) predated the appearance of aerobes (organisms that can or must live in the presence of O_2). Most of this oxygen likely arose from early photosynthetic organisms, which split water into protons and O_2 using sunlight as the energy source, much like modern photosynthetic organisms. The appearance of this oxygen was a watershed moment for life on earth, because it provided a new opportunity for extracting energy from biomolecules. This new oxygen-based method worked so well it is now essential for nearly all heterotrophic organisms, including humans. In other words, most (but not all) life on earth depends on it to survive.

The chemical basis for this method is the high electronegativity of oxygen atoms, which we discussed in Chapter 1 (see *Water Is the Most Common Compound Found in Cells*). Oxygen is by far the most electronegative atom commonly found in cells (only fluorine has a higher electronegativity value), so by extension, many molecules that contain oxygen atoms (e.g., water) share this trait, as reflected by their redox potentials. This means the H_2O/O_2 redox pair is an excellent candidate for receiving electrons at the *end* of an ETC. The ancestors of all modern aerobic organisms developed such an ETC, and understanding how it works is our goal in this section.

Aerobic Respiration Occurs in Four Stages

I've broken the journey from pyruvate through the final ETC into four stages, to help keep the sequence of events in perspective. In each stage, pay special attention to what molecules enter, what molecules emerge, and where the events take place; also, keep in mind the ultimate goal: extracting ATP from food while completing the electron journey.

■■ Stage 1: Pyruvate Is Transported into the Mitochondrial Matrix and Is Converted into Acetyl-CoA

The first stage of aerobic metabolism is converting the energy in pyruvate into reduced electron carriers. Notice that this is the *third* conversion of pyruvate we have discussed, after ethanol and lactate fermentation; one important difference between this conversion and the first two is that it is done in the mitochondrial matrix, as shown in **FIGURE 10-19**.

Consider for a moment how life evolved on earth, and this makes perfect sense. Before prokaryotic organisms developed the symbiotic relationship that gave rise to eukaryotic organisms (which occurred approximately 2.5 billion years ago), they were all unicellular. Based on the likely time when atmospheric oxygen first began to accumulate, the organisms that developed this third conversion method most likely were still unicellular. If one of these clever cells is engulfed by another cell that hasn't developed the strategy, the benefit of symbiosis becomes obvious. The host cell continues to gather biomolecules as food, while the engulfed cell specializes in the aerobic metabolism of said food: the host benefits from the extra energy yield, and the engulfed cell benefits by not having to find its own food anymore. Plus, by living inside the host, this cell evades most of the environmental dangers unicellular organisms have to face. Eventually, the engulfed cell specializes to the point where it becomes a mitochondrion. Evidence for this scenario is provided by modern aerobic bacteria, which resemble mitochondria in many ways. The metabolic proteins found in the plasma membrane and cytosol of these bacteria closely resemble those in the inner membrane and matrix of mitochondria.

Regardless of how this cooperative arrangement arose, the functional link between the host and mitochondrion is the transport of pyruvate into the mitochondrion interior (matrix). We, of course, do not know how this occurred in these first organisms, but in modern mitochondria, the transport takes place via a pyruvate transporter protein in the inner mitochondrial membrane. The pyruvate crosses the outer mitochondrial membrane through a porin, and then enters the mitochondrial matrix by following the concentration gradient formed by the generation of pyruvate (via glycolysis) in the cytosol and its consumption inside the mitochondria. This transporter is structurally related to carrier proteins for other small metabolites such as amino acids.

Once inside the matrix, pyruvate is modified by an enormous protein aggregate called the **pyruvate dehydrogenase complex**. (We'll skip the chemical reactions and focus on the reactants and products on this complex.) Pyruvate, a three-carbon compound, enters the complex along with an NAD^+ molecule and a molecule called coenzyme A, commonly abbreviated **CoA**. Like all enzyme cofactors, CoA is an essential ingredient for some chemical reactions, but is not modified by the enzymes that bind to it. What emerges from the pyruvate dehydrogenase complex are NADH, CO_2, and a molecule called acetyl-coA. An acetyl group contains two carbons (see Table 1-1). So, the three carbons in pyruvate are converted into one CO_2 and an acetyl-CoA, and this cleavage of pyruvate has already paid off, by forming an NADH.

The fact that NADH is created by this complex illustrates an important point. Let's change the way we describe what is happening to our food molecules: instead of saying we are digesting or cleaving them, let's now say we are *oxidizing* them. This term is sometimes hard to get used to, because it is hard to visualize an object being oxidized. We'll use the verb *oxidize* often for the duration of this section, but if at any time it doesn't make sense, substitute the word "digest" when we are talking about food molecules. The reason I want to switch words at this point is to emphasize that the most important part about digesting food is not the actual destruction of food molecules, but instead the extraction of high-energy electrons from them. When we take electrons from a molecule, we oxidize it. In biology, any carbon atom is in its most oxidized form when it is bound to *only oxygen atoms* (hence the term oxidize), and whenever we reduce a molecule, such as NAD^+, those electrons have to come from somewhere, and in this context it's usually a food molecule.

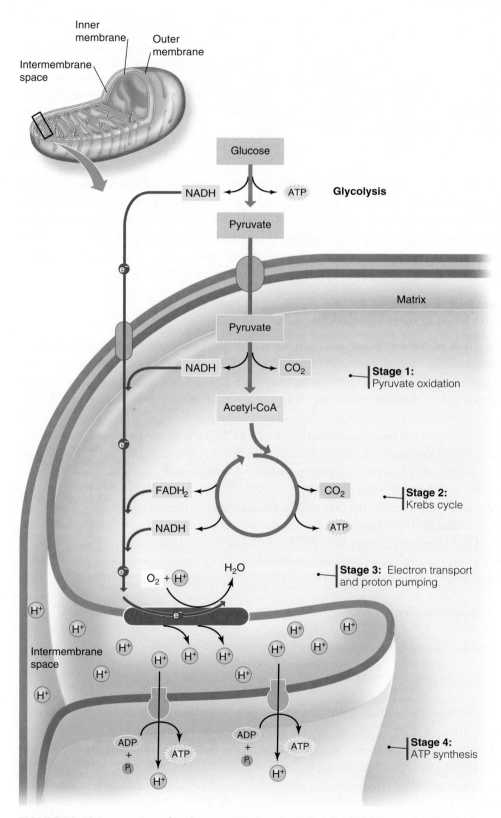

FIGURE 10-19 An overview of oxidative respiration. Note that the four stages take place in the mitochondrion.

So, let's rephrase what the pyruvate dehydrogenase complex does: it oxidizes pyruvate to yield a reduced energy carrier (NADH), and the carbon that donates those electrons is released in its most oxidized state, CO_2 (**BOX 10-8**). This leaves two more carbon atoms (in the form of an acetyl group) to be oxidized later, attached to a CoA molecule (for reasons that we have yet to address). This is how professional biochemists describe what happens in stage 1, and to be conversant with them, we need to learn that jargon.

■■ Stage 2: Acetyl-CoA Is Fully Oxidized to Yield CO_2, GTP, and High-Energy Electron Carriers

The function of the second stage of aerobic metabolism is to complete the oxidation of our food molecule, which enters this stage as acetyl-CoA. Before we get to question of *how* this stage is completed, let's make sure we understand *what* is happening and *why*. According to our rule, complete oxidation means converting the carbons in acetyl-CoA into CO_2, so we expect to see two CO_2 molecules emerge at the end. The extracted electrons end up in reduced energy carriers (NADH and $FADH_2$). As an added bonus, we even get a nucleotide triphosphate out of the process (GTP), yet another form of stored potential energy. This covers the *what*; the *why* is simple: we want to pull electrons off of our food so we can send them through an ETC.

How does this happen? An easy way to visualize it is to compare it to the Calvin cycle in photosynthesis. The function of the Calvin cycle is to oxidize NADPH to reduce CO_2 and combine it with the energy from ATP to make carbon-based macromolecules. In many ways, it sounds like the exact opposite of what we want to accomplish here, so let's run with that idea. If we look at the Calvin cycle and imagine it running in reverse, it looks like this:

$$\text{Triose phosphate} + NADP^+ + ADP + P_i \rightarrow CO_2 + H_2O + NADPH + ATP$$

Now let's look at what happens in stage 2 of oxidative metabolism:

$$\text{Acetyl-CoA} + NAD^+ + GDP + P_i \rightarrow CO_2 + NADH + GTP + CoA$$

They're almost identical (notice that CoA emerges unchanged). So, to explain the complete oxidation of acetyl CoA, we need a set of reactions that resembles a reversed Calvin cycle. And it turns out that one exists, called the **Krebs cycle** (after Sir Hans Adolph Krebs). **FIGURE 10-20** summarizes the events taking place in the Krebs cycle. Like the Calvin cycle, it's a series of complex chemical reactions that took extraordinary effort to unravel, and it earned Sir Krebs the Nobel Prize in Medicine in 1953. Other names for this cycle include the tricarboxylic acid (TCA) cycle and the citric acid cycle, referring to intermediate compounds in the cycle. In deference to the remarkable insights of Sir Krebs, I'll call it the Krebs cycle from here on. Just as the Calvin cycle takes place in the chloroplast stroma, the Krebs cycle takes place in the mitochondrial matrix. The parallels between these two cycles are not accidental: both are very likely derived from a common primitive cell that developed the ability to metabolize carbon compounds, and later mutations in this pathway survived natural selection to yield two slightly different pathways adapted to specifically reducing and oxidizing carbon compounds, respectively. A single photosynthetic cell can contain both: one captures sunlight to reduce CO_2 and assemble it into macromolecules, and the other oxidizes the macromolecules made by the first. Modern autotrophs (e.g., plants) are common examples.

Let's keep track of the yield of the glycolysis-aerobic metabolism pathway so far: for every glucose, we get two pyruvates, two NADH, and two ATP from glycolysis. Each pyruvate yields one NADH from stage 1, and three more NADH in the Krebs cycle, plus an $FADH_2$ and a GTP. Per glucose, we generate 10 NADH, 2 $FADH_2$, 2 ATP and 2 GTP so far, as shown in **TABLE 10-1**. Because GTP and ATP are functionally equivalent in terms of energy storage, we group them together as 4 ATP in the table.

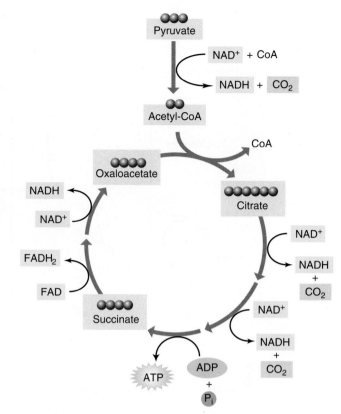

FIGURE 10-20 An overview of the Krebs cycle.

▣▣ Stage 3: The Electron Transport Chain Uses High-Energy Electrons from NADH and $FADH_2$ to Build a Proton Gradient across the Inner Mitochondrial Membrane

Just as the Krebs cycle has a cousin in chloroplasts, the ETC in chloroplasts has a close relative in mitochondria. And, in keeping with the theme of the second stage, the mitochondrial ETC in the third stage of aerobic metabolism is somewhat similar to the thylakoid ETC. Some of the important features of the mitochondrial ETC are:

- The input to this ETC is electrons from energy carriers, not electrons from water. They already have high energy, and thus do not need sunlight to excite them.
- The electrons traveling along the mitochondrial ETC are used to pump protons, just as in the thylakoid. This forms a gradient across the inner mitochondrial membrane, with the protons most concentrated in the intermembrane space.

TABLE 10-1. The net yield of products from glycolysis and the first two stages of oxidative respiration.

Stage of Glucose Metabolism	Stored Energy Products per Glucose Molecule
Glycolysis, Stage 1	−2 ATP (ATP consumed)
Glycolysis, Stage 2	2 ATP 2 NADH
Glycolysis, Stage 3	2 ATP
Aerobic Respiration, Stage 1 (Conversion of pyruvate to Acetyl CoA)	2 NADH
Aerobic Respiration, Stage 2 (Krebs Cycle)	6 NADH 2 FADH2 2 GTP
Sum	2 ATP + 2 GTP ≈ 4 ATP 10 NADH 2 FADH2

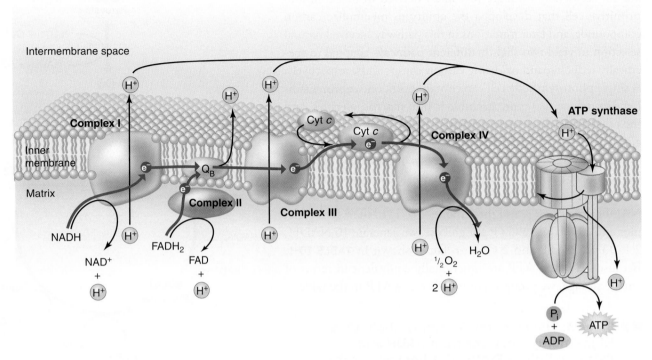

FIGURE 10-21 The mitochondrial electron transport chain.

- Once the electrons pass through proteins in the mitochondrial ETC, they are accepted by O_2 to make H_2O. This ETC is diagrammed in **FIGURE 10-21**. Notice how similar it appears to the thylakoid ETC in Figure 10-11.
- Large protein complexes are interspersed with small electron carriers, which diffuse between them to deliver the electrons from one complex to the next. The complexes are named in a similar fashion, with Roman numbers, which fortunately are in a more logical order: electrons from NADH enter at respiratory complex I, those from $FADH_2$ enter at complex II, and both proceed through complexes III and IV before ending their journey in a water molecule in the matrix.
- Just as in the thylakoid ETC, the sequence of proteins visited by the electrons is determined by the redox potential of these proteins, as shown in **FIGURE 10-22**. Compare Figure 10-22 with Figure 10-12 (thylakoid redox potential graph), and notice how similar they are. The second law of thermodynamics is a powerful tool for understanding both ETCs.
- As electrons pass through the protein complexes, protons are pumped across the inner mitochondrial membrane. Just as Q_B pumps protons in the thylakoid, so does coenzyme Q (CoQ) here. It is notable that three different complexes (I, II, and IV) pump protons in the mitochondrial ETC, while protons are only pumped once in the thylakoid. This too makes sense, because all of the energy in the mitochondrial ETC is devoted to generating this proton gradient, whereas only the first half of the thylakoid ETC builds a proton gradient (the second half reduces NADPH).
- The mitochondrial ETC is at least as complex as that in the thylakoid. Collectively, the four respiratory complexes contain over 70 polypeptides.

There is one feature in the mitochondrial ETC that is especially important. Not all of the electron carriers inject their electrons at the same point. NADH unloads its electrons into respiratory complex I, but $FADH_2$ unloads its electrons at respiratory complex II; as

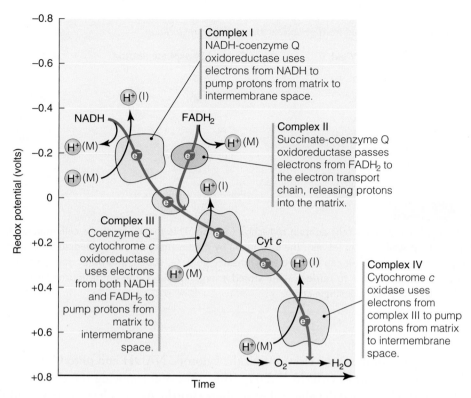

FIGURE 10-22 The redox potential of reactants indicates the energy level of the electrons moving through the mitochondrial electron transport chain.

a result, the electrons from NADH travel through all four respiratory complexes, but those from $FADH_2$ pass through only three of these complexes. This means that the electrons from $FADH_2$ cause fewer protons to be pumped across the inner mitochondrial membrane, resulting in a lower yield of ATP.

■■ Stage 4: The F1/Fo ATP synthase Uses the Proton Gradient to Make ATP

That final stage in our journey is very simple: the proton gradient generated by the ETC is used to synthesize ATP. Specifically, the protons are allowed to flow back into the mitochondrial matrix through an ATP synthase, which harnesses the energy in the proton flow to phosphorylated ADP to make ATP. One curious note about this final step: the ATP synthase actually *spins around* as it makes ATP, as shown in **FIGURE 10-23**. Thus, for one brief moment, the potential energy in the proton gradient is converted into *kinetic* energy that is then rapidly converted back into potential energy (ATP). This is one of the very few examples where an analogy like a waterwheel is actually very close to biologic reality.

Let's finish this off with a final accounting of the ATP yield from aerobic metabolism. Due to the difference in the amount of proton pumping their electrons can support, the average yield from NADH and $FADH_2$ is three and two ATP, respectively. When we multiply the NADH and $FADH_2$ in Table 10-1 by these numbers, the result is 38 ATP per glucose. However, in many cells yield is actually closer to 36 ATP, as shown in **TABLE 10-2**. The reason this number is lower is that the NADH generated during glycolysis cannot enter the mitochondrial matrix to contribute its electrons to the ETC, because the inner mitochondrial membrane is impermeable to NADH. Cells

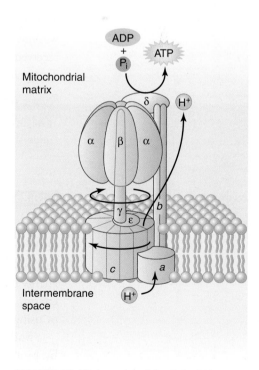

FIGURE 10-23 A model of the F₁Fo ATPase. Note that the central rod actually rotates as protons flow through the protein.

TABLE 10-2 Net ATP yield from glycolysis and oxidative respiration.			
	Initial Yield	**After Glycerol Phosphate Shuttle**	**ATP Yield**
NADH	10	8	$\times 3 = 24$
FADH2	2	4	$\times 2 = 8$
ATP	4	4	4
			36

> **BOX 10-9 TIP**
>
> Don't forget that all plant cells contain mitochondria, too. This means for those cells that contain chloroplasts as well (e.g., in leaves), the G3P made by the Calvin cycle can immediately be used in glycolysis if necessary. Or, it can be used to synthesize glucose, starch, and other macromolecules that can be digested later to retrieve their stored energy. In short, the entire journey we've just travelled can take place in a single cell.

must use enzymes to move the electrons off the cytosolic NADH and onto G3P, which is permeable, and then from G3P to $FADH_2$. This strategy for moving the electrons from NADH to $FADH_2$ is called the **glycerol phosphate shuttle**. As a result, two of the NADHs in Table 10-1 are converted into $FADH_2$, thus lowering the yield to 36 ATP per glucose (**BOX 10-9**).

Just as Dr. Mitchell was able to stitch together the details of the light and dark reactions to generate a clear comprehensive picture of photosynthesis, he was also able to link the Krebs cycle with the mitochondrial ETC to produce a comprehensible explanation for how oxygen consumption led to ATP synthesis. He called this coupling **oxidative respiration**. The final puzzle, how the ATP synthases work, was solved a few years later. And this effort, too, was rewarded with a Nobel Prize in Chemistry in 1997 for three scientists: Dr. Jens C. Skou received the largest portion of the credit, along with Drs. Paul D. Boyer and John E. Walker. This marks the fifth Nobel Prize we have encountered on our journey in this chapter.

CONCEPT CHECK #3

Explain why we have to breathe oxygen. First, use everyday words that your family and friends would understand; then, generate the most technical answer you can, citing specific molecules.

10.8 Chapter Summary

The principles of physics and chemistry provide the rules for energy transfer in cells. The laws of thermodynamics make it imperative that cells couple energetically unfavorable reactions, such as building cellular molecules, to favorable reactions, such as the conversion of potential energy to heat energy when ATP is cleaved to ADP + P_i. Cells use electrons to store potential energy, and pass energy from one molecule to another either by forming new covalent bonds or using an intermediate form of potential energy, the ion gradient.

Both direct transfer of energy and harnessing energy in an ion gradient are performed by proteins. One set of essential proteins for managing ion gradients are the membrane transport proteins. While channels permit molecules to flow through a pore in a membrane, carriers and pumps physically translocate each molecule they transport. Pumps use ATP energy to build gradients, while carriers and channels dissipate gradients. During indirect active transport, a carrier can capitalize on a molecular gradient formed by a pump to build a second molecular gradient by dissipating the first.

The journey of energy flow in living organisms begins with the splitting of water atoms, and this reaction is powered by sunlight. It takes place in the thylakoid membrane of chloroplasts. The electrons removed from water by this reaction then embark on an elaborate system of electron flow from one molecule to the next, passing through the thylakoid electron chain and Calvin cycle to create the high-energy compound G3P. G3P is then converted into a host of macromolecules, including sugars, amino acids, and fatty acids. These form the building blocks for plant cell structures.

The electron flow journey passes into heterotrophic cells (e.g., in animals) when they eat the macromolecules built in plants. The high-energy electrons initially taken from water in plants now begin the long route back to water in animals. This journey resembles that taken in plants, only run in reverse. The counterparts of G3P catabolism, the Calvin cycle, and the thylakoid electron transport chain in chloroplasts are glycolysis, the Krebs cycle, and the mitochondrial electron transport chain in mitochondria. At the end of the mitochondrial electron transport chain, molecular oxygen accepts the electrons to reform water. Thus, the entire journey is cyclic and always in motion. This cycle forms the basis for all life on earth.

CONCEPT CHECK ANSWERS #1

The net impact of all three proteins is that compounds A and C is concentrated in the cytosol of this cell.

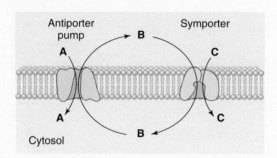

CONCEPT CHECK ANSWERS #2

Scenario 1: If the glut2 failed to reach the cell surface all together, glucose would build up to exceptionally high levels in the intestinal epithelial cells. Assuming the cell doesn't burst from the corresponding influx of water to compensate for the osmotic pressure caused by such an imbalance of glucose across the plasma membrane, this glucose would not be able to enter the bloodstream, and the rest of the cells in the body would rapidly suffer from the loss of glucose. Perhaps another building block (e.g., amino acids) would be properly transported across the epithelial cell membrane, and this might be able to sustain the rest of the cells.

Scenario 2: If the TGN missorted the glut to the apical membrane rather than the basolateral membrane, the cell would suffer no immediate damage: the glucose that entered via the symporter would simply diffuse back out again via the glut, forming a futile cycle of glucose transport that would not contribute anything to the bloodstream on the basolateral side of the epithelial cell. So, while the epithelial cell would likely remain healthy until the glucose in the intestine was depleted, the rest of the cells in the body would suffer same fate as in scenario 1.

<u>Answer 1:</u> We breathe oxygen because it is the sponge that soaks up the last bit of energy in our food and clears the remnants of our food out of the cells, in the form of water. This is important because if we don't clear the cells of this last remnant, it accumulates and slows down the whole process of cellular digestion of food, and it eventually comes to stop. If we stop breathing for several minutes, the accumulated metabolic intermediates cause our cells to die.

<u>Answer 2:</u> Oxygen is the final electron acceptor in the mitochondrial electron transport chain. If our cells do not have access to molecular oxygen, respiratory complex IV cannot be oxidized, and this leads to stable reduction of the upstream respiratory complexes. Once redox stops, no more energy is being extracted from the electrons, so no proton gradient is formed across the mitochondrial inner membrane, and F_1F_o ATPase cannot synthesize ATP. Lactic acid fermentation will not sustain our cells in the absence of oxygen for very long, because it is 18-fold less efficient at oxidizing glucose. The resultant drop in ATP pools causes the Na^+/K^+ ATPase to cease pumping ions, and eventually our cells reach equilibrium with the extracellular environment, and they die.

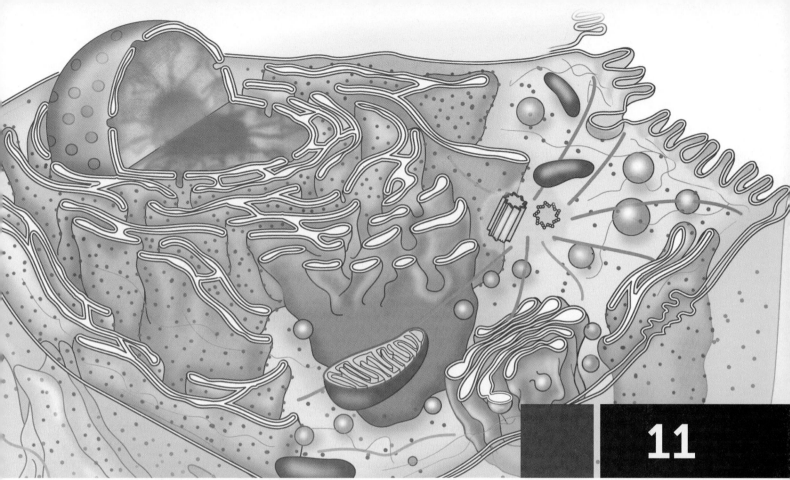

Signal Transduction and Cellular Communication

11.1 The Big Picture

So far we have defined the building blocks of cells (Chapters 1–4) and used them to build (1) a cytoskeleton and extracellular matrix to permit cellular movement and provide structural support inside, outside, and between cells (Chapters 5–6); (2) a mechanism for replicating DNA and properly segregating the copies during cell division (Chapter 7), plus a mechanism for converting DNA information into RNA and proteins that are targeted to their proper cellular locations (Chapters 8–9); and (3) a method for capturing environmental energy and using it to power the chemical reactions underlying all these activities (Chapter 10). To finish the text, let's now focus on linking these stories together to yield an integrated "*really* big picture" of a complete cell.

It's important to point out that, up until now, we've treated the "cell" as a generic concept. But no cell is average, any more than any living person is truly average. By tying the strands of information together, we can bring the picture to life and show how cells can share common properties but be distinct from one another. In short, it's time to start building real, specialized cells.

Our final story concerns how cells achieve this specialization. Specialization is a major decision for cells. To borrow the terminology on college campuses, no cell in our bodies is an "undeclared major": every cell in a multicellular organism has to be especially good at accomplishing some tasks, at the expense of others that it simply cannot do. Specialization affects nearly every aspect of cell structure and function that we've discussed in the previous chapters, from the types and amounts of building blocks needed to the winding state of DNA. Making this decision requires considerable "thought," because in most cases it is irreversible. How cells make these decisions is strikingly similar to how we make decisions in everyday life, and this resemblance forms the basis for cell biology principle #7: signaling networks are the nervous system of a cell.

This means that, just as we receive information about the environment and our own physical state, integrate and process this information, and then choose the appropriate action with which to respond, cells perform the same activities. This decision making takes place constantly, whether we are aware of it or not. While we are busy thinking in the everyday sense, the cells composing our bodies constantly monitor our basic physiological functions, such as blood pressure, temperature, oxygen consumption, and signal small changes in metabolism, breathing rate, and so on to keep us as close to homeostasis as possible (see *The Plasma Membrane Is a Semipermeable Barrier between a Cell and the External Environment* in Chapter 1).

To understand a cellular nervous system, we will address four topics, each covered by a major section in this chapter:

- First, we'll examine the concept of a signal transduction pathway in fairly general terms. While the number of individual components in a pathway may be large, the overall structure of most pathways is quite similar. We'll use this structure as the foundation for building specific pathways later in the chapter.
- Second, we'll examine the kinds of information that cells detect, and how they do so. While every change in a cell's environment could conceivably act as a signal, cells ignore most environmental changes and focus on a select few that have the most immediate impact on their function. The most common device for detecting environmental changes is a specialized class of proteins called receptors.
- Third, we'll examine the molecules most commonly found in signaling pathways. The number of different molecules known to participate in signaling pathways is far too large for us to describe them all, so we'll use categories to group them into a more manageable set of modules.

- Fourth, we'll put the concepts in the first three sections to work by looking at some well-known examples of cell signaling in the final section. These signal transduction pathways will be our tools for linking the first three strands together in the final three chapters (**BOX 11-1**).

11.2 Signaling Molecules Form Communication Networks

KEY CONCEPTS

- Signaling networks relay information from the extracellular environment to the interior of a cell.
- The basic unit of a signaling network is a signal transduction pathway, which carries one specific signal in a single direction from the source (a receptor) to the effector.
- Most signal transduction pathways are composed of several different molecules that activate each other in a carefully controlled sequence of binding interactions.

The anatomy of a cellular signaling network is complicated, just as the human nervous system is. Nonetheless, we can simplify matters considerably by starting with its general properties and then adding details in layers. Let's begin with the basic question: What does a cell signaling network actually do? The simplest answer is that it detects and transmits information from one location to another, just as our nervous system does. In cellular terms, this process is called signal transduction. Information travels in only one direction through nerve cells in the nervous system, and the same is true for cell signaling networks.

The Function of Signaling Networks Is to Convert Extracellular Information into an Appropriate Cellular Response

In Chapters 1–10, the term *information* almost always refers to the *instructions* stored in DNA (like books in a library) and converted into actions by proteins and RNAs. In signaling networks, "information" refers to the *state* of an object. Perhaps this seems equally abstract, but in fact this is how most of us use the word in everyday life. When we ask our friend what time it is, what we literally mean is "what is the state of the clock?" When the state of the clock is transmitted from our friend to us, information is being moved from one place to another. We process this information, perhaps realize we are late for class, and then generate an appropriate response, like hurrying across campus.

When cells transmit information this way, we call it cell signaling. A simple example is when a cell comes into contact with an object in the environment: the contact causes changes in the shape (state) of proteins in the plasma membrane. This information is passed into the cell interior, and the cell responds by either adhering to or releasing the object. Every signal that affects a cell targets at least one effector molecule to generate a suitable response.

▪▪ Signaling Networks Are Composed of Signals, Receptors, Signaling Proteins, and Second Messenger Molecules

The molecules that compose signaling networks include proteins, proteoglycans, peptides, amino acids, lipids, nucleotides, and charged atoms (Ca^{+2}, Na^+, Cl^-, etc.). By convention, these molecules are classified into four groups, according to the *function* they play within the network, rather than their structure. Signals transmit information from one cell to another, typically by traveling through the extracellular space. Proteins, amino acids, lipids, sugars, and nucleotides can act as signals. **Receptors** bind to signals, and are always proteins or proteoglycans (for a review of proteoglycan structure, see *Proteoglycans Provide Hydration to Tissues* in Chapter 6). Once a receptor binds to a signal, it transmits the information to a target in the cell. In some cases, the receptor does this itself, but more commonly it passes the information off to downstream runners that are especially good at moving through the cell interior; many of these runners are **signaling proteins**. In most signaling networks, several signaling proteins pass information from one to another, resembling runners in a relay race. Second messenger molecules also participate in this relay, and are distinguished from signaling proteins by the simple fact that they are not proteins. They are called second messengers because they never start or end a signaling event; they are intermediates by definition (**BOX 11-2**).

The path of information triggered by a single signal is called a **signal transduction pathway**, and it is defined by both the signal that triggers it, as well as the molecules that compose it, as shown in **FIGURE 11-1**. Signal transduction pathways must have one signal, a receptor for the signal, and at least one **effector**, which generates the cellular reaction to the signal. The information always travels in one direction, from the receptor to the final target effector molecule. For example, the pathways triggered by signals called **growth factors** contain protein receptors for the growth factors on the cell surface and typically several copies of about half a dozen different signaling proteins that pass the information from the receptor to the cell interior.

Growth factor signals terminate at several different effectors, generating a number of different responses. Via **branching**, a single signal like this can affect several targets simultaneously; in our nervous system analogy, this is equivalent to receiving a single painful input, such as a burn on our finger, and processing it to generate several complementary responses (e.g., contracting numerous hand and arm muscles to pull away from the heat). In other cases, several different signals can converge on a common target, or they can signal independently, as **FIGURE 11-2** shows. An excellent visual example of how this occurs is the world record for the number of dominoes that fall after a single domino is tipped: by placing different colored dominoes in staggered positions, the effect of a single tipped domino can be quickly amplified and distributed over a wide area, creating beautiful patterns. Some run in parallel, some converge, and others disperse over a wide area. In 2008,

---

BOX 11-2 FAQ ▪

What's the difference between a *protein* that functions as a *signal* and a *signaling protein*? *Signals* are always in the extracellular space; some of these happen to be in the form of proteins. But *signaling proteins* never exit a cell. Instead, they help move information from one location to another in the same cell. Think of *signals* as functioning *outside* cells, while *signaling proteins* function *inside* cells.

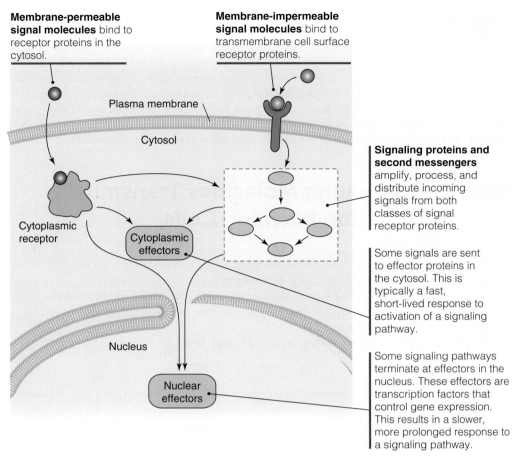

Membrane-permeable signal molecules bind to receptor proteins in the cytosol.

Membrane-impermeable signal molecules bind to transmembrane cell surface receptor proteins.

Plasma membrane

Cytosol

Cytoplasmic receptor

Cytoplasmic effectors

Signaling proteins and second messengers amplify, process, and distribute incoming signals from both classes of signal receptor proteins.

Some signals are sent to effector proteins in the cytosol. This is typically a fast, short-lived response to activation of a signaling pathway.

Nucleus

Nuclear effectors

Some signaling pathways terminate at effectors in the nucleus. These effectors are transcription factors that control gene expression. This results in a slower, more prolonged response to a signaling pathway.

FIGURE 11-1 Simple schematic of signal transduction pathways.

the world record was nearly 4.5 million dominoes, and it took two hours for all of them to fall down. In cells, a single signal molecule can trigger a signal transduction pathway containing about the same number of molecules in a fraction of a second.

The domino example illustrates another critical concept in signal transduction: the more choices a signaling network has, the more sophisticated it can be. A handful of dominoes cannot generate the beautiful patterns created by millions of dominoes, no matter how they are arranged; likewise, a single straight line of dominoes cannot generate these patterns, regardless of how long it is. To create the complexity, one must have the ability to channel the initial trigger into a branched network that contains many different components. By necessity, then, many signaling networks are both long (multiple steps) and complex (multiple branch points). To appreciate how important this complexity is, consider all of the ways our bodies respond to the signal *adrenaline* during a fight-or-flight reaction. Tipping the "adrenaline domino" changes blood flow, breathing, metabolic functions, and a host of other reactions necessary to permit us to rapidly react to danger. From the perspective of cell biology, that too is a beautiful picture (**BOX 11-3**).

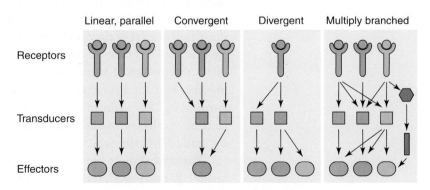

FIGURE 11-2 Signaling pathways use linear, convergent, divergent, and branched signaling pathways to generate complex responses to external signals.

11.3 Cell Signaling Molecules Transmit Information between Cells

KEY CONCEPTS

▸ Signals arise from the extracellular space, and must bind a receptor to be effective.
▸ Most signals are molecules that cannot penetrate the plasma membrane, so they bind to receptor proteins on the cell surface. Those signals can then pass through membranes and are bound by receptors in the cytosol.
▸ Receptors are grouped into six classes, according to their structure, binding partners, and cellular location.

Having described the general properties of cellular signaling networks, let's now follow the path of information through these networks. By definition, a signal transduction pathway begins when a signaling molecule (often called a **ligand**) binds to its target receptor. To do so, it must be released from its original site and travel some distance to its target. The signals that transmit this information are grouped into three classes based on their physical state.

Membrane-impermeable signals bind to cell surface receptors. These signals include proteins, peptides, amino acids, and proteoglycans. A well-known example is **neurotransmitters**, which transmit information between neurons. Because all of these molecules are polar, they cannot penetrate the hydrophobic interior of the plasma membrane and thus cannot transmit information directly into cells. Instead, the receptors change shape after binding to these signals, and this shape change is propagated along the entire length of the receptor so that eventually the cytoplasmic portion of the receptor (commonly called the tail) changes shape as well. This is how the information supplied by the signal is transmitted across the plasma membrane.

Membrane-permeable signals bind to intracellular receptors. These signals include lipids, gasses, and other small nonpolar molecules that can penetrate the plasma membrane. Common examples of this kind of signal are steroid hormones, such as estrogen and testosterone. It is important to note that because of this property, membrane-permeable signals enter *all* cells they come into contact with, but this does not mean that these signals are effective in all cells. To be effective, the signals must bind to a receptor. A clear example of this is the pattern of hair growing on a man's face. For most men, the skin growing near the lips, over the lower jaw, and on a portion of the cheeks responds to testosterone by growing whiskers, while the skin cells at most other locations of the face do not. All of the facial skin cells contain testosterone, but only those that express the testosterone receptor respond to it in this way.

Physical signals are converted into chemical signals. Examples of physical signals include pressure (e.g., touch), temperature, and light. Cells express receptor proteins that specialize in changing shape when these environmental signals change. A simple example is the stretch receptor proteins that change shape in response to physical deformation; this forms the basis of our sense of touch and ability to balance our bodies. Even though most of these signals can penetrate the plasma membrane, most receptors for physical signals are expressed on the cell surface, and their shape change initiates a signal transduction pathway just as the other two classes of signals do.

Intercellular Signals Are Secreted into the Extracellular Space

Membrane-permeable and membrane-impermeable signals are synthesized by cells and released into the extracellular space, where they diffuse to their targets. Membrane-impermeable signals typically are released via exocytosis (See *Exocytosis Is Vesicular Transport of Molecules to the Plasma Membrane and Extracellular Space* in Chapter 9), and those that are membrane permeable are synthesized in the cytosol and enter the extracellular space by diffusing through the plasma membrane. Once they exit the cell, they are transported by the flow of liquid that surrounds all cells. In the human body, the circulatory system is the primary means of distributing signals throughout the body. Cells that are specialized to secrete signals are often concentrated into secretory organs, which lie very close to blood vessels for this reason.

Six Classes of Receptors Are Sufficient to Detect a Vast Array of Environmental Stimuli

Receptors are grouped into six classes based on structural similarity, shown in **FIGURE 11-3**. Some are enzymes that catalyze chemical reactions, while others serve as docking sites for enzymes, and some do both. Variations in the nucleotide sequences in the coding region of receptor genes in the same family give rise to enough different receptor proteins to permit our cells to specifically bind to hundreds of signals. Each individual cell in a multicellular organism expresses only a subset of these receptors, and thus has a limited number of "senses" for their environment.

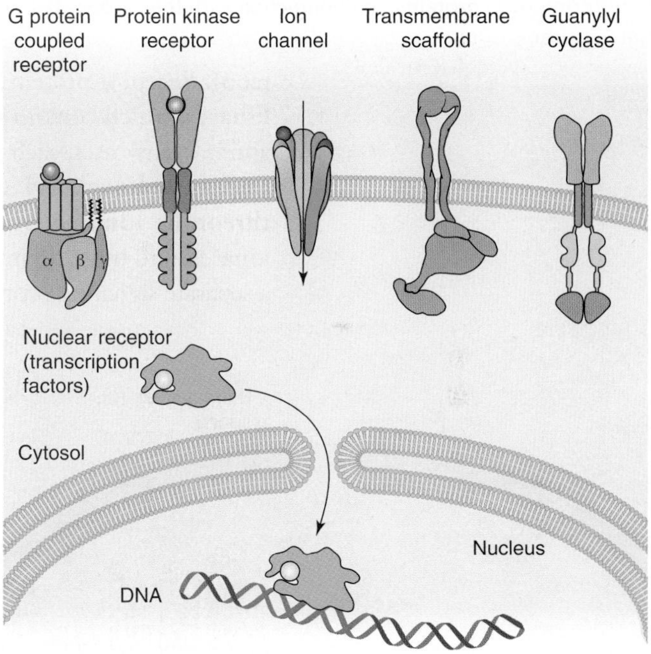

G Protein Coupled Receptors Activate G Proteins

One of the most common receptor types on the cell surface is defined by the fact that it binds to only one class of intracellular signaling protein. These receptors are called **G protein coupled receptors**, often abbreviated GPCRs. Their signaling protein partners are called G proteins because they bind to the nucleotide GTP to regulate their shape and activity. (We'll discuss G proteins in the next section.) GPCRs share other structural features, as shown in **FIGURE 11-4**. These receptors span the membrane seven times (see *Integration of Transmembrane Proteins Requires Specific Amino Acid Sequences* in Chapter 8), using α helices and are often called **seven transmembrane spanning receptors** for this reason. The ligand-binding domain of these receptors lies near the amino terminus, and the cytosolic loop linking the fifth and sixth α helices binds to its G protein partner.

An important property of these receptors is that, when activated, they can trigger the activation of several copies of the same G protein. This occurs because activated G proteins dissociate from the GPCR, leaving the binding site available to bind and activate additional G proteins. The longer the GPCR remains bound to its ligand, the more copies of its G protein it can activate, thereby amplifying the effect of a single signal molecule. GPCRs typically

FIGURE 11-3 Receptors are grouped into six classes based on their structure and cellular location.

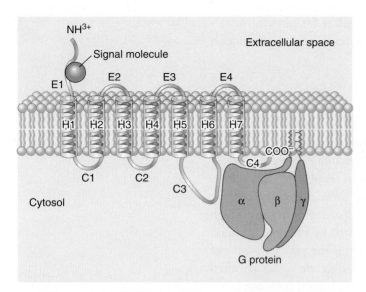

FIGURE 11-4 The general structure of a seven transmembrane receptor. Note that the signal molecules for these receptors bind to the external loops, especially those near the amino terminus. The cytoplasmic loop between H5 and H6 binds to and activates heterotrimeric G proteins, in conjunction with the carboxy tail of the receptor.

bind to protein, peptide, or amino acid-based signals. They are often the target of intracellular signaling proteins that chemically modify them to help control their activity.

■■■ Receptor Protein Kinases Phosphorylate Signaling Proteins

Receptor protein kinases are transmembrane proteins that, unlike GPCRs, are composed of at least two subunits and therefore have a quaternary structure. (Some plants and bacteria contain single polypeptide receptor protein kinases, but we won't discuss these here.) Each of the subunits contains a single α-helical transmembrane domain, so they can diffuse relatively easily in the membrane. This is an important regulatory feature of these receptors: the subunits of inactive receptors dissociate from one another, and only bind to one another when they come into contact with the proper ligand.

A second distinguishing feature of these receptors is that their cytoplasmic tails contain protein kinase domains. A protein kinase is an enzyme that attaches phosphate groups to the side chains of specific amino acids (see *Cells Chemically Modify Proteins to Control Their Shape and Function* in Chapter 3). Receptor protein kinases are grouped into two classes. The first is **receptor tyrosine kinases**, which contain two subunits. Each of the two subunits contains a tyrosine kinase, and in many cases, each phosphorylates tyrosines on the opposite subunit, a behavior called **transautophosphorylation**, shown in **FIGURE 11-5**. The second class is **receptor serine/ threonine kinases**. These are more heterogeneous in structure than receptor tyrosine kinases, and most do not undergo transautophosphorylation; instead, they phosphorylate a separate signaling protein as **FIGURE 11-6** illustrates.

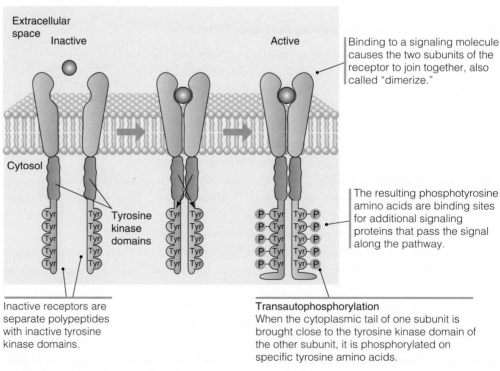

Inactive receptors are separate polypeptides with inactive tyrosine kinase domains.

Binding to a signaling molecule causes the two subunits of the receptor to join together, also called "dimerize."

The resulting phosphotyrosine amino acids are binding sites for additional signaling proteins that pass the signal along the pathway.

Transautophosphorylation
When the cytoplasmic tail of one subunit is brought close to the tyrosine kinase domain of the other subunit, it is phosphorylated on specific tyrosine amino acids.

FIGURE 11-5 Model of growth factor receptor activation.

Phosphoprotein Phosphatases Remove Phosphate Groups from Signaling Proteins

Phosphoprotein phosphatases are enzymes that undo the work of protein kinases by removing phosphate groups from proteins, as **FIGURE 11-7** shows. Most protein phosphatases target either phosphorylated tyrosines or phosphorylated serines and threonines, thereby mimicking the specificity of the kinases. These phosphatases often shut down signal transduction pathways when they dephosphorylate signaling proteins.

Guanylyl Cyclases Produce the Signaling Molecule Cyclic GMP

Receptor guanylyl cyclases are dimeric enzymes that, in humans, bind to a very specific polypeptide signal called atrial natriuretic peptide. Once atrial natriuretic peptide binds, the enzyme portion of the receptors becomes activated and converts GTP into a somewhat unusual form of GMP called cyclic GMP, abbreviated cGMP, shown in **FIGURE 11-8**. The enzyme portion is located in the cytosolic portion of the receptor, so binding of atrial natriuretic peptide to the extracellular portion of the receptor triggers an increase in cGMP on the opposite side of the plasma membrane, and this is how the signal information is transferred to the cell interior. cGMP then binds to other signaling proteins, thereby continuing the signaling cascade.

Ion Channel Receptors Permit Ion Fluxes

Ion channel receptors are not enzymes. They transmit signal information by permitting ions to flow from one side of the plasma membrane to another, just as nonreceptor ion channels do. Their signature feature, as seen in **FIGURE 11-9**, is binding to a signaling ligand, and

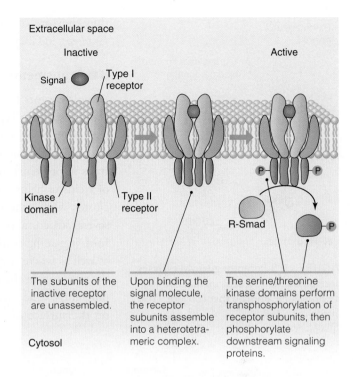

The subunits of the inactive receptor are unassembled.

Upon binding the signal molecule, the receptor subunits assemble into a heterotetrameric complex.

The serine/threonine kinase domains perform transphosphorylation of receptor subunits, then phosphorylate downstream signaling proteins.

FIGURE 11-6 Serine/threonine kinase receptor activation leads to phosphorylation of a signaling protein. This example shows how transforming growth factor β (signal) triggers phosphorylation of its receptor and subsequent phosphorylation of a downstream signaling protein (R-Smad).

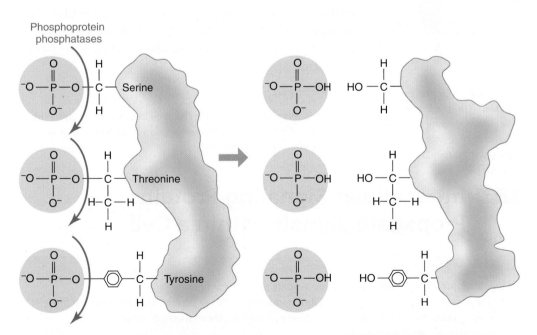

FIGURE 11-7 Protein phosphatases break the phosphoester bond linking phosphate groups to serine, threonine, and tyrosine side chains. Note that the removal of the phosphates from a protein changes its shape, and hence its functional state.

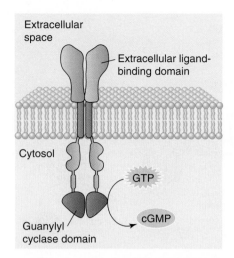

FIGURE 11-8 Receptor guanylyl cyclases are homodimeric receptors that contain a cytoplasmic domain that converts GTP into cyclic GMP.

thus they are often called **ligand-gated channels** (see *Protein Channels Dissipate Gradients* in Chapter 10). Many neurotransmitter signals bind to this class of receptor.

■■ Transmembrane Scaffolds Recruit Intracellular Signaling Proteins

Some receptors function best when they cluster into large aggregates, especially on the cell surface. Typically, these proteins have multiple binding sites on their cytoplasmic tails for many different signaling proteins. The receptors can control which of these sites are accessible at any given time (by controlling their shape, per the three traits of proteins), so they can change the kinds of signals they transmit. Formation of a cluster helps control signaling because some binding sites can be buried in the cluster, while others are exposed. For this strategy to work, the receptors also recruit several structural proteins into the cluster, resulting in what is called a **signaling scaffold**. Signaling scaffolds control which signaling proteins can bind to the complex, as well as where they are positioned in space. By changing the number and shape of the receptors in the complex, cells can assemble different types of scaffolds that signal differently. One excellent example of such a scaffold is the integrin cluster that forms on the surface of cells when they adhere to extracellular matrix proteins, shown in **FIGURE 11-10** (to review integrins, see *Most Integrins Are Receptors for Extracellular Matrix Proteins* in Chapter 6).

■■ Nuclear Receptors Are Transcription Factors

Membrane-permeable signals adhere to receptor proteins in the cytosol, and these receptors typically have very short signal transduction pathways because they move directly into the nucleus once they are bound and activated, as **FIGURE 11-11** shows. They are called **nuclear receptors** for this reason. Many of these receptors bind directly to DNA and control the expression of genes, and this bypassing of the plasma membrane signal transduction machinery permits these signal transduction pathways to skip the complex decision making that takes place in most others. Because they help control the expression of genes, most nuclear receptors are also classified as transcription factors.

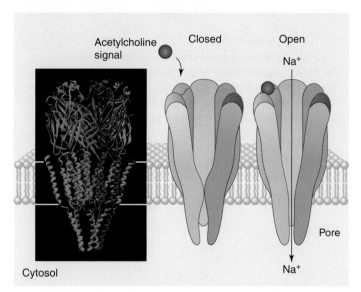

FIGURE 11-9 Ligand-gated channels typically form a central pore that opens when a ligand binds to the receptor. In this example, the neurotransmitter signal acetylcholine is shown binding to its receptor on the cell surface, triggering a shape change that permits Na+ ions to flow through the pore. Structure from Protein Data Bank 2BG9. N. Unwin, J. Mol. Biol. 346 (2005): 967–989.

CONCEPT CHECK #1

Cancer is often associated with changes in cell surface signaling receptors. What kinds of changes would you expect to find in these proteins in cancer cells?

11.4 Intracellular Signaling Proteins Propagate Signals within a Cell

KEY CONCEPTS

▸ Signaling proteins rapidly transmit and amplify signal information.
▸ As information passes through a signal transduction pathway, it often changes physical form.
▸ Signaling proteins are grouped into six classes based on their structure, location, and mechanism of signal transmission.
▸ Second messengers are nonprotein molecules that link signaling proteins together in signal transduction pathways.

Let's now move past the receptors and on to the most numerous players in signal transduction pathways, the **intracellular signaling proteins**. These proteins act as the go-betweens that carry information from the receptors to the effectors. Despite their great structural variations, signaling proteins share two common features:

- First, they are highly mobile in cells. This is important because mobility is a key requirement for transmitting information in any context, even our everyday world. At the molecular level, high mobility means rapid diffusion. Some of these proteins diffuse through the cytosol, and others diffuse in membranes, but they all move very quickly.
- Second, many of these proteins are enzymes capable of catalyzing chemical reactions to rapidly amplify the magnitude of a signal. Those that are not enzymes themselves are capable of binding to enzymes.

Information Changes Form in Signaling Networks

To help visualize how cell surface receptors pass signals on to additional proteins before reaching their

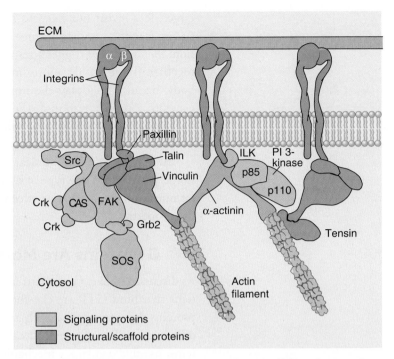

FIGURE 11-10 Integrin receptors form signaling scaffolds. Clusters of integrin receptors recruit many different cytosolic proteins that together form a signaling scaffold. Examples of scaffold and signaling proteins are indicated by different colors.

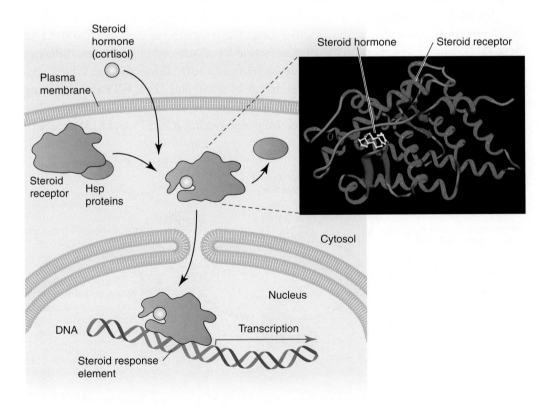

FIGURE 11-11 The steroid receptor binds to steroid hormones when they diffuse into the cytosol. The resulting shape change permits the receptor to pass through a nuclear pore complex and bind to specific DNA sequences called steroid response elements (SREs). Image courtesy of Tim Vickers.

target, let's revisit our nervous system analogy. A simple way of describing the nervous system is that it consists of three cell types: *sensory neurons* receive environmental information, *motor neurons* trigger a response, and *interneurons* form the network that links them together. There are far more interneurons than sensory or motor nerves in the human body, because they play the important role of processing the incoming information to generate a proper response. (For this example, think of the brain as a huge cluster of interneurons.) Intracellular signaling proteins play a similar role in cells. They amplify and distribute signals, then integrate them to trigger the correct response.

Along the way, the physical form of signals often changes repeatedly as the information is passed from one type of signaling protein to another. Intracellular signaling proteins far outnumber receptors or effectors and are grouped into six classes. Let's have a look at the properties of each.

■ ■ G Proteins Are Molecular Switches

As discussed above, G proteins use GTP binding as a regulatory mechanism. Not all proteins that bind GTP are classified as G proteins; the name is used primarily to describe proteins that propagate signals at the plasma membrane and on the Golgi apparatus. Two different families of proteins, called **monomeric G proteins** and heterotrimeric G proteins, use this regulatory mechanism. Monomeric G proteins are single polypeptides that have at least two different binding sites and contain a **GTPase** domain—one binding site is occupied by GTP or GDP, and the other binds to a target protein. When the G protein is bound to GTP, it is in an activated shape that promotes binding to its target. Eventually, the GTPase domain cleaves the terminal phosphate from the bound GTP, generating GDP. In turn, this causes the protein to shift into an inactive shape. At some later point, the protein releases the GDP and binds to another GTP molecule, reactivating it.

This bind-cleave-release cycle (called the GTPase cycle) repeats continuously, shifting the G protein between active and inactive states like a switch, as seen in **FIGURE 11-12 (BOX 11-4)**. Monomeric G proteins are also bound by regulatory proteins that influence their activation state. **Guanine nucleotide exchange factors**, abbreviated as GEFs, activate these G proteins by promoting the exchange of GDP and GTP. **GTPase activating proteins**, abbreviated as GAPs, have the opposite effect: when they bind, they stimulate the GTPase activity in the G protein, effectively inactivating it. GEFs and GAPs are acti-

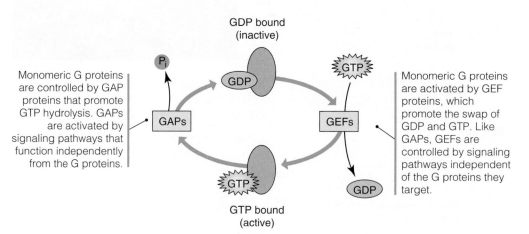

FIGURE 11-12 The GTPase cycle controls the activation state of G proteins. When bound to GDP, these proteins are inactive. When they release GDP and bind to GTP, they become activated. The G protein eventually cleaves the terminal phosphate off GTP, converting it to GDP and inactivating the protein.

vated by separate signal transduction pathways and compete for access to the same monomeric G proteins, thereby integrating pathways. This is an excellent example of a cellular decision-making mechanism.

Heterotrimeric G proteins function in much the same way as their monomeric relatives, except they contain three different polypeptides, and these are called the α, β, and γ subunits. The α subunit resembles a monomeric G protein, in that it binds GTP, cleaves it to form GDP, and binds to a target protein. The β and γ subunits are permanently attached to one another, and are often referred to as a single β/γ subunit for this reason. The primary function of the β/γ subunit is to bind to and stabilize the inactive (GDP-bound) form of the α subunit.

The α and β/γ subunits of heterotrimeric G proteins, as well as monomeric G proteins, are anchored to the cytoplasmic leaflet of a membrane by a lipid added posttranslationally to the C terminus (see *Membrane Proteins Associate with Membranes in Three Different Ways* in Chapter 4). This means they bind to other membrane proteins and the adaptor proteins associated with them. G protein coupled receptors bind to and activate heterotrimeric G proteins, but do not bind monomeric G proteins. A GPCR bound to its ligand triggers a shape change in the heterotrimeric G protein bound to it, separating the α subunit from the β/γ subunit and stimulating the swap of GDP for GTP in the α subunit. The α subunit propagates the signal by binding to its target. It remains attached to its target as long as it is bound to GTP. When it cleaves the GTP to form GDP, the resulting shape change causes it to release its target, shutting off the signaling, as shown in **FIGURE 11-13**. Some β/γ subunits also bind to a target, so that activation of a heterotrimeric G protein can result in two different signals.

Protein Kinases Phosphorylate Downstream Signaling Proteins

Protein kinases attach phosphate groups to the side chains of tyrosine, serine, and threonine amino acids. The receptor protein kinases described above belong to this family of proteins, which also includes a tremendous number of nonreceptor kinases, shown in **FIGURE 11-14**. Most protein kinases that are not receptors are located in the cytosol, where they bind to and phosphorylate other signaling proteins and some effector proteins. In many cases, one protein kinase phosphorylates another kinase, which then phosphorylates yet another kinase, and so on, like tipping dominoes. In these cases, attaching a phosphate activates a protein, but this is not universally true. Some proteins can be phosphorylated on several different amino acids, and some phosphates may be activating while others are inactivating. Several protein kinases have the ability to enter and exit the nucleus, like nuclear receptors; they do not bind to DNA themselves, but instead phosphorylate proteins that do. It is estimated that human cells express over 150 different protein kinases.

Lipid Kinases Phosphorylate Phospholipids

As their name suggests, lipid kinases add phosphate groups to lipids. In particular, these proteins phosphorylate phospholipids in the cytoplasmic leaflet of membranes, especially the

A chemical signal (ligand) binds to the receptor protein, causing the receptor protein to change conformation, allowing it to interact with the heterotrimeric G protein.

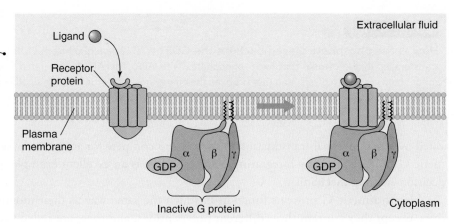

The receptor protein stimulates the exchange of GTP in the G_α subunit, causing the Gα subunit to change conformation. This conformational change triggers the dissociation of the G protein complex, producing $G_\alpha \cdot$ GTP complex and a Gβγ complex.

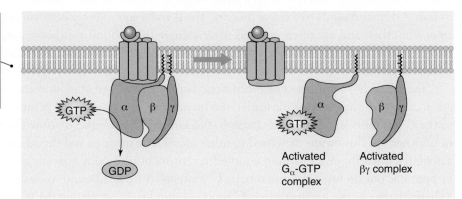

The free Gα subunit interacts with one effector (E1) such as adenylate cyclase while free Gβγ subunit complex interacts with another effector (E2).

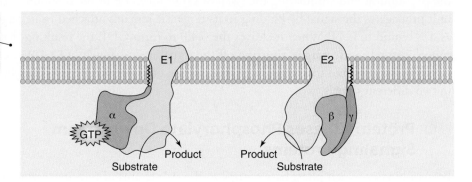

The Gα subunit catalyzes GTP hydrolysis. The resulting $G_\alpha \cdot$ GDP complex binds to the Gβγ complex to re-form the heterotrimeric G complex.

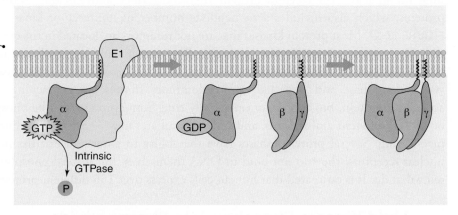

FIGURE 11-13 A heterotrimeric G protein signaling cycle.

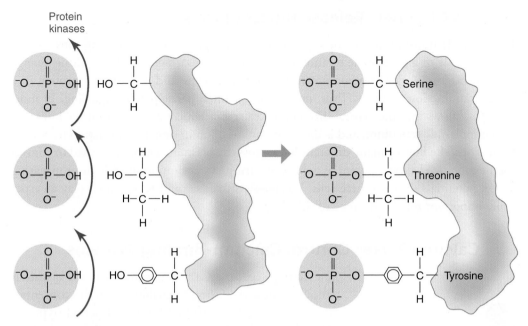

FIGURE 11-14 Protein kinases add phosphate groups to signaling proteins and effectors. One family targets serine and threonine amino acids, and a second targets tyrosine amino acids.

plasma membrane. Typically, the phosphate is added to the polar head group of a phospholipid. This is important because phosphorylation significantly changes the shape of this head group, making it "stand out" amid the enormous pool of other unmodified phospholipids in the same membrane (see *Polar Head Groups Confer Additional Specificity on the Structure of Phospholipids* in Chapter 4). The modified phospholipid then binds to its target protein(s) in the membrane, passing the signal down the pathway. In some cases, several different kinases can phosphorylate the same phospholipid at different sites, generating a multiphosphorylated product, as illustrated in **FIGURE 11-15**. A good example of this is phosphatidylinositol, which can be phosphorylated at three different sites on its head group. (We'll see this in our pathway examples in the last section.)

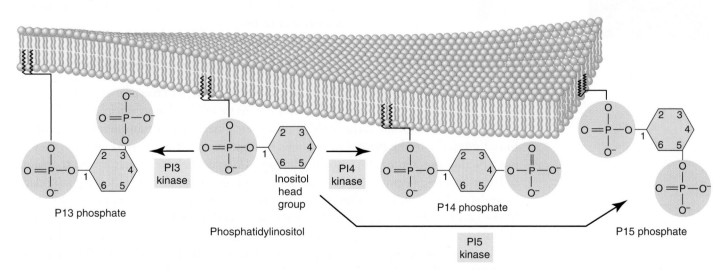

FIGURE 11-15 Lipid kinases add phosphates to phospholipids. In this example, the phospholipid phosphatidylinositol is being modified by three different kinases. PI3 kinase adds a phosphate group to the 3′ carbon on the inositol head group, generating PI3-phosphate. PI4 kinase and PI5 kinase add phosphates to the 4′ and 5′ carbons of inositol, respectively.

Ion Channels Release Bursts of Ions

Ion channels transmit signals in signal transduction pathways by allowing ions to pass from one side of a membrane to another. This qualifies as a signal *if* the resulting change in concentration (or membrane potential) alters the shape and activity of a protein in a signal transduction pathway. Thus, not all ion channels participate in signaling. In most cells, Ca^{+2} flows into the cytosol, from either the extracellular space or from the lumen of the endoplasmic reticulum, and is the most common ion in signal transduction pathways. In neurons, the flow of many different ions across the plasma membrane is used to control the membrane electrical potential, which in turn plays an important role in propagating action potentials through these cells. Voltage-sensitive Na^+ and K^+ channels play an especially important role in this type of signaling.

Calcium Fluxes Control Calcium-Binding Proteins

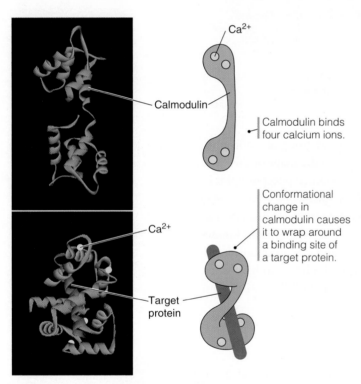

FIGURE 11-16 Calmodulin is an example of a calcium-sensitive signaling protein.

In a "resting" cell, the Ca^{+2} concentration in the cytosol is typically quite low. During a signaling event, this concentration can increase 5- to 10-fold for a brief period (typically <1 to 10 seconds), then return to normal. During the brief burst, some Ca^{+2} ions bind to cytosolic proteins, thereby changing their shape and activating them, as **FIGURE 11-16** shows. One good example is the protein calmodulin. After calmodulin binds calcium, it then binds to and activates other proteins, many of them effectors; calmodulin, therefore, functions near the end of many signal transduction pathways. Calmodulin is inactivated when the cytosolic Ca^{+2} concentration drops back to resting levels. This is caused by calcium ATPase pumps and calcium antiporters (see *Energy-Coupled Carrier Proteins (Pumps) Build Gradients* and *Passive Carrier Proteins Dissipate Gradients* in Chapter 10) in the plasma membrane and endoplasmic reticulum, which actively remove Ca^{+2} ions from the cytosol. An example of calmodulin-mediated signaling is the pathway linking GPCRs to the activation of myosin during smooth muscle contraction.

Adenylyl Cyclases Form Cyclic AMP

Adenylyl cyclases are related to receptor guanylyl cyclases, in that they convert a nucleotide triphosphate (ATP) into a cyclic, monophosphate form (cAMP). These enzymes are also transmembrane proteins found primarily in the plasma membrane, but unlike receptor guanylyl cyclases, they do not function as receptors for signaling ligands. Instead, they are bound by the α subunit of heterotrimeric G proteins. For example, one type of α subunit, called $\boldsymbol{\alpha_s}$, stimulates adenylyl cyclases, while another type, called $\boldsymbol{\alpha_i}$, inhibits them, as shown in **FIGURE 11-17**. These two α subunits are part of different heterotrimeric G proteins that are in turn stimulated by different GPCRs. This situation illustrates another type of cellular decision making during signaling: when two different ligands bind to their GPCRs, they can "compete" at the level of a single adenylyl cyclase, yielding a single response.

cAMP binds to one class of serine/threonine protein kinases, appropriately named **cAMP-dependent protein kinases**. This name is just long enough to invite abbreviation, so most

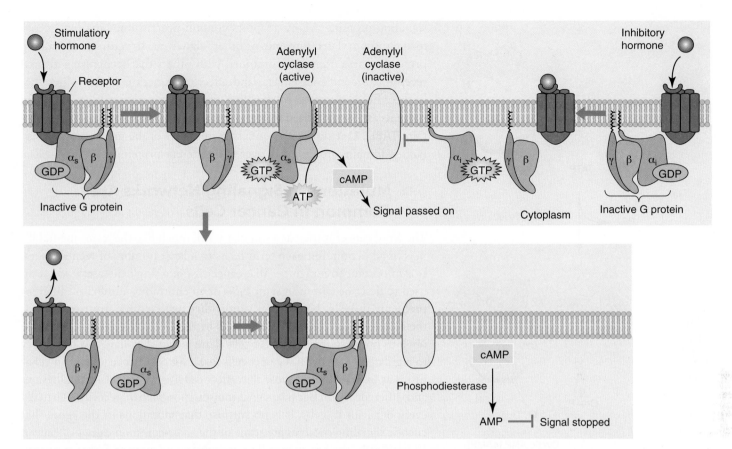

FIGURE 11-17 Adenylyl cyclase is a target of competing regulatory pathways. A stimulatory subunit (G α_s) from one heterotrimeric G protein triggers adenylyl cyclase activity, while an inhibitory subunit, G α_i, from a different heterotrimeric G protein has the opposite effect. Phosphodiesterase converts cAMP into the inactive molecule AMP, shutting off the adenylyl signaling pathway.

cell biologists refer to it as **protein kinase A**, or simply **PKA**. In parallel, the cGMP generated by receptor guanylyl cyclases binds to and activates another serine/threonine kinase named **protein kinase G**, or **PKG**. Both cAMP and cGMP are inactivated by an enzyme called **phosphodiesterase**, which cleaves the phosphodiester bond forming the "cyclic" portion of the molecule, as **FIGURE 11-18** shows.

Second Messengers Are Small, Rapidly Diffusible Forms of Chemical Signals

The nonprotein components of signal transduction pathways are collectively called **second messengers**. These include the Ca^{+2} ions, the cAMP and cGMP molecules mentioned previously, small hydrophobic molecules, and even some gasses. Two key features of these molecules are their small size and ability to rapidly diffuse in the cytosol and/or the hydrophobic interior of a membrane. In addition to rapidly transmitting signals from one location to another, these molecules can also amplify a signal, because the proteins that trigger their release typically release hundreds or thousands of them during one activation cycle.

Adaptors Facilitate Binding of Multiple Signaling Proteins

A key element of nearly all signal transduction pathways is a class of proteins that are neither receptors nor enzymes. These proteins, known as adaptors, are composed exclusively of binding domains and motifs, and serve as links between signal receptors and other

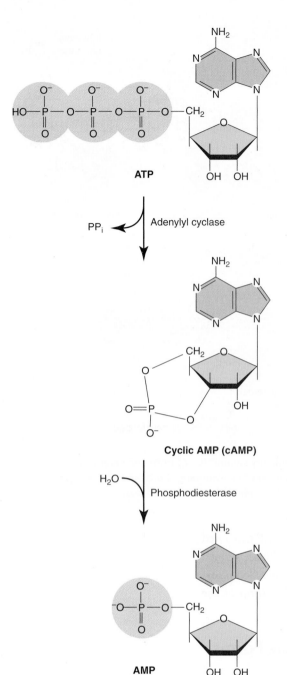

FIGURE 11-18 Adenylyl cyclase converts ATP to cAMP, and phosphodiesterase cleaves the phosphoester bond between the phosphate and the 3′ carbon of ribose, converting cAMP to AMP.

signaling proteins. Many adaptors contain protein domains that recognize phosphorylated tyrosines or other "signature" structures on signaling proteins. These domains, combined with others that recognize activated receptors, second messengers, and effectors, function as the "glue" that literally holds elements of signaling networks together. Often, signaling proteins are characterized by the combination of these domains they contain. **TABLE 11-1** shows a recent estimate that 9 of the most common signaling domains are found in over 600 different proteins in human cells.

Mutations in Signaling Networks Are Common in Cancer Cells

The word cancer is used as a singular term, much like the terms mental illness or addiction, but each term refers to a large number of related issues. It is important to remember that cancer is not a single disease: it is a term used to describe common symptoms of an enormous number of different problems in cells. Nearly every individual who suffers from cancer has their own unique set of dysfunctions in their cells. Of course, the most obvious problem that all cancer cells share is that they have lost the ability to regulate their growth. Cancer cells lose a lot of other regulatory mechanisms as well, including those that affect cell migration, cell–cell adhesion, and differentiation. Because signal transduction pathways control virtually every behavior in cells, it is no surprise that mutations in the genes that encode signaling proteins are some of the most common causes of cancer. Mutations in the genes encoding growth factors, growth factor receptors, hormones, hormone receptors, protein and lipid kinases, G proteins, and even adaptor proteins can cause cancer. Most often, the same cell must develop mutations in several of these genes to actually transform into a cancer cell; this is why the incidence of cancer increases with age: as a person ages, the number of mutations in their cells increases (see *Mutations in DNA Give Rise to Variation in Proteins, Which Are Acted Upon by Natural Selection* in Chapter 2) until some type of threshold is reached in one of these cells and it loses control of its signal transduction pathways.

CONCEPT CHECK #2

Some drugs designed to treat cancer patients target signaling proteins commonly mutated in cancer cells. Many of these have serious side effects in healthy cells. Explain why a highly specific drug that targets a single signaling protein could cause these side effects.

11.5 A Brief Look at Some Common Signaling Pathways

KEY CONCEPTS

▸ Hundreds of different receptors, signaling proteins, and effectors combine into a complex network of interacting pathways within a single cell.

▸ Despite the tremendous complexity of signaling networks, many share common features that help set the standard for our current understanding of how signal transduction pathways function.

▸ Some signal transduction pathways trigger short-term cellular changes via very long and complex sets of signaling interaction, while others contain very few steps and have relatively long-term effects on cells.

TABLE 11-1 Common signaling domains and their prevalence in representative model organisms.

Domain	Human	Mouse	Fruitfly	Roundworm	Yeast
SH3	223	124	113	83	26
WW	91	27	21	40	9
PDZ	234	119	98	106	3
SH2	112	73	33	67	1
PTB	34	14	7	23	0
14-3-3	8	4	4	2	1
BRCT	39	23	28	44	9
FHA	16	9	17	12	13
C2	149	94	51	93	22
Total genes	30,000	30,000	14,000	19,000	6,300

Deciphering signal transduction pathways in cells is a complicated process, somewhat similar to tracing the path of an email sent over the internet. There is no shortage of potential interactions during the transmission, making it difficult to identify the actual players involved. In cell biology, the first step of working out signal transduction is to simply identify the molecules involved, regardless of how they are arranged. This stage is essentially complete for human cells; in our domino analogy, we have discovered just about all of the "signaling" dominoes.

We now face the daunting task of figuring out how they are arranged, both spatially and temporally: we want to know the sequence of dominoes that line up between the first and last to fall. This task is far from complete, and our current picture of most signal transduction pathways is usually a composite assembled from a number of different cell and tissue types. To demonstrate what these composites look like, let's take a brief tour of some of the most common types of signal transduction pathways. Keep in mind that the goal here is to understand the concept of signaling pathways, rather than exhaustively describe them (BOX 11-5).

Protein Tyrosine Kinase Signaling Pathways Control Cell Growth and Migration

As their name suggests, growth factors stimulate the growth of most mammalian cells. Several growth factors are named according to the cells that secrete them. One common

---

BOX 11-5 TIP

As in previous chapters, it is important to focus on the structure and function of these signaling pathways before attempting to memorize them. One way to do this is to substitute the generic class of molecule for the actual name. For example, the "FGF–FGFR–Grb2–Sos–ras–Raf–MEK–Erk" pathway can also be generically described as "signaling protein–receptor–adaptor–GEF–G protein–protein kinase–protein kinase–protein kinase." Doing this allows you to see the relationship between the classes of signaling molecules more easily, and reveals important patterns: for example, GEFs are always placed immediately upstream of G proteins, and several protein kinases can be linked together in a series.

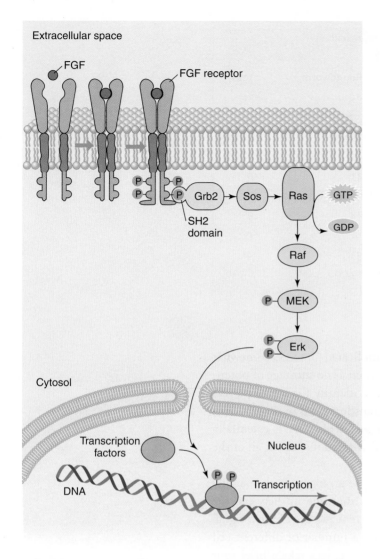

Extracellular space

FGF

FGF receptor

Grb2 → Sos → Ras

SH2 domain

GTP

GDP

Raf

P—MEK

P P Erk

Cytosol

Transcription factors

Nucleus

P P

Transcription

DNA

FIGURE 11-19 A simplified version of an FGF signaling pathway to the nucleus. The receptor dimerizes on the cell surface, undergoes transautophosphorylation, then serves as a docking site for adaptor proteins that link to and activate G proteins and protein kinases. The final kinase enters the nucleus and phosphorylates transcription factors, which then bind to DNA and control gene expression.

example is **fibroblast growth factor**, commonly abbreviated FGF. Twenty-two different forms of FGFs are known in human cells, and they bind to a family of proteins called **FGF receptors** (**FGFRs**). FGFRs are typical homodimeric receptor tyrosine kinases. Each FGFR subunit binds to an FGF ligand, causing a shape change that induces subunit dimerization. A sample of a composite FGF signaling pathway that targets the nucleus is shown in **FIGURE 11-19**.

After the receptor subunits bind to each other, they undergo tyrosine transautophosphorylation. The cytoplasmic tails are then bound by a number of different proteins that contain domains specific for phosphotyrosine. These domains are called **SH2** and PTB. Depending on the cell type, a number of different pathways branch off of these phosphotyrosine-binding proteins. One well-characterized pathway consists of an adaptor called Grb2 (pronounced "grab two"), which binds to phosphotyrosines on the FGFR via an SH2 domain. This then induces Grb2 to change shape, enabling it to bind to a GEF called Sos (pronounced "sauce"). This interaction activates Sos, so it binds to a monomeric G protein called **ras**. Ras responds by releasing GDP and binding GTP, and the activated form of ras binds to and activates a serine/threonine kinase called Raf. Activated Raf phosphorylates a rather unusual protein kinase called MEK; what makes it unusual is that it is capable of phosphorylating tyrosine *and* threonine amino acids. Its substrate is a serine/threonine protein kinase named Erk (**BOX 11-6**).

Once Erk is phosphorylated, it forms a dimer which is capable of phosphorylating other signaling proteins in the cytosol or the nucleus. Some of these dimers control the polymerization of actin filaments at the cell periphery, as well as the activation of myosin light chain kinase (see *Myosins Are a Family of Actin-Binding Motor Proteins* in Chapter 5), and thereby control the rate and direction of cell migration. Phosphorylated Erk dimers can also enter the nucleus to phosphorylate nuclear proteins. Many of the nuclear proteins are transcription factors, and their phosphorylation triggers a cascade of gene expression. The proteins encoded by these genes include some that regulate entry into the S phase of the cell cycle, such as cyclins and cyclin-dependent protein kinases (we will discuss these topics in more detail in Chapter 13). As we will see

---

BOX 11-6 TIP

Capitalization and italicization of names in signal transduction. You might notice throughout this and other texts that some scientific names are capitalized and/or italicized, while others are not. Further complicating matters, numerous conventions exist for using capitalization and italicization, depending on the specific species of organism and type of structure being discussed. Signal transduction suffers from this confusion, too. Rather than trying to memorize the conventions, I suggest you get used to seeing many different forms of writing the same name, and follow whatever convention applies in each situation. In the signal transduction field, most signaling protein names are not capitalized or italicized, except acronyms, which are typically written in all capital letters.

in the next two chapters, the mechanisms controlling these cascades are complex and carefully regulated, so that cells only begin to divide under appropriate conditions.

Growth factor receptors are only one class of receptors that use the so-called MAP kinase signaling pathway. Another important class of cell surface receptor, the integrins, also uses MAP kinases to signal. Recall from Chapter 6 that integrins bind to extracellular matrix proteins and form clusters on the surface called focal complexes (see *Specialized Integrin Clusters Play Distinct Roles in Cells* in Chapter 6). In addition to serving as a physical link between the extracellular matrix and the cytoskeleton, these complexes also serve as signaling scaffolds. Integrins are not enzymes, so to make use of MAP kinase signaling, they must recruit signaling enzymes into the focal complex. One very well-known tyrosine kinase that binds to integrin cytoplasmic tails is focal adhesion kinase (FAK). Clusters of FAK undergo transautophosphorylation, and the phosphotyrosines that result are docking sites for a number of SH2-containing proteins including Grb2, as Figure 11-10 shows. This is how signals from growth factor receptors and integrins can complement one another in the control of cell growth. In fact, growth factor receptors are commonly found in focal complexes.

Heterotrimeric G Protein Signaling Pathways Regulate a Great Variety of Cellular Behaviors

Heterotrimeric G proteins are activated exclusively by GPCRs. Over 800 different GPCRs have been indentified, and they control a wide range of signaling pathways, including those triggered by neurotransmitters, hormones, odorants (molecules we can smell), and photons of light (i.e., vision). One very common pathway used by heterotrimeric G proteins includes adenylyl cyclase and PKA, as shown in **FIGURE 11-20**. Inactive PKA is a tetrameric protein, consisting of two catalytic subunits that are serine/threonine kinases, and two regulatory subunits that hold the catalytic subunits in an inactive conformation. cAMP generated by adenylyl cyclase binds to the regulatory subunits, inducing a shape change that causes them to separate from the catalytic subunits. Once freed, the catalytic subunits bind to and phosphorylate a number of different cytosolic proteins. Like Erk dimers, PKA catalytic subunits can also enter the nucleus to phosphorylate transcription factors. One of the best known substrates of PKA is called **cyclic AMP response element binding protein** (**CREB**). When CREB is phosphorylated, it binds to a protein called CREB binding protein (CBP), and together, the two proteins control the expression of a large number of genes. The CREB/CBP complex is targeted to specific genes, which include a DNA sequence called a cAMP response element that lies upstream of the core promoter, about 100 basepairs from the TATA box (we will discuss this structure in Chapter 12). Once CREB binds to the CRE, CBP interacts with the basal transcription complex to promote transcription

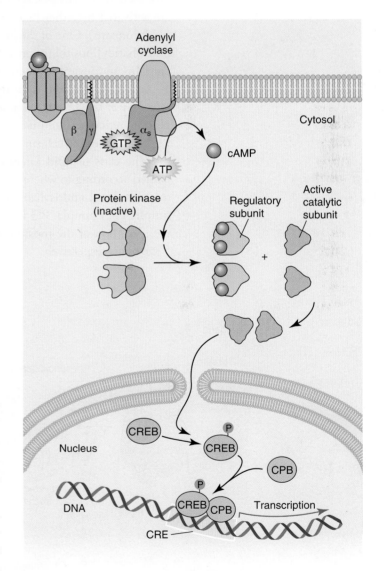

FIGURE 11-20 A sample cAMP signaling pathway. Adenylyl cyclase generates cAMP after activation by a G_s heterotrimeric G protein subunit. The cAMP then binds to regulatory subunits of cAMP-dependent protein kinase, causing the catalytic subunits to dissociate. The catalytic subunits then phosphorylate downstream signaling proteins and effectors.

by RNA polymerase (for a review, see *Transcription Begins After a RNA Polymerase Binds to a Promoter Site on DNA* in Chapter 8). cAMP/PKA/CREB signaling is so ubiquitous in cells that it is not actually known how many different signals use this pathway or the exact number of genes under its control. Mice lacking functional CREB have severe deficits in long-term memory and poorly developed regions of the brain, suggesting this may be one of the most important functions of this pathway in mammals.

Phospholipid Kinase Pathways Work in Cooperation with Protein Kinase and G Protein Pathways

Most **phospholipid kinases** function at the plasma membrane and are activated by signaling proteins, rather than by receptors. Phosphorylated phospholipids are in turn broken down by enzymes called phospholipases, which are also membrane proteins. Phospholipid kinases may be activated by protein kinases or monomeric G proteins, and phospholipases are activated by either heterotrimeric G protein α subunits or by phosphotyrosines (via SH2 domains). One of the best characterized lipid kinase pathways is that used to modify phosphatidylinositol (often abbreviated PI), shown in **FIGURE 11-21**. The head group of PI contains an inositol sugar linked to the glycerol backbone via a phosphate attached to its 1′ carbon. Recall from Chapter 4 (see *The Synthesis of Phosphoglycerides Begins at the Cytosolic Face of the SER Membrane*) that PI is synthesized in the cytosolic leaflet of the endoplasmic reticulum, and it is *not* transported to the exoplasmic leaflet. This means that all PI that reaches the cell surface (via exocytosis) remains in the cytosolic leaflet.

The class of lipid kinases that modifies PI is called PI kinases, and each kinase is named according to where it adds a phosphate to PI. The naming scheme is rife with acronyms for long, multisyllabic names, but don't let that cause confusion. The process is very simple. For example, PI3 kinase (PI3K) replaces the hydroxyl group with a phosphate on the 3′ carbon of the inositol sugar on PI. PI4K and PI5K add phosphates to the 4′ and 5′ carbons, respectively.

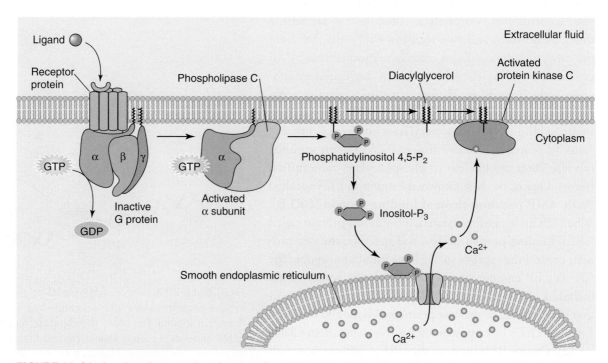

FIGURE 11-21 The phosphoinositol 4,5-bisphosphate (PIP_2) signaling pathway that activates protein kinase C.

Many times, these enzymes act on PI in a defined sequence, shown in Figure 11-15. One commonly used mechanism progresses like this: First, PI4K modifies PI, generating PI 4-phosphate, often called PIP. Next, PI5K kinase adds its phosphate, generating PI 4,5-bisphosphate, abbreviated PIP$_2$. PIP$_2$ functions as a second messenger when it is cleaved by the enzyme phospholipase C. This cleavage occurs between the 1′ phosphate and the glycerol backbone, yielding a molecule of diacylglycerol (DAG) and inositol 1,4, 5 trisphosphate (IP$_3$).

It is important to note that, after this cleavage, the signal initially carried by PIP$_2$ has now been converted into *two* signals with very different chemical properties, as Figure 11-21 shows. We need to keep track of both of them. The IP$_3$ molecule is small and highly charged, and therefore diffuses through the cytosol relatively easily. Eventually it reaches the smooth endoplasmic reticulum, where it binds to and opens a ligand-gated calcium channel. The Ca^{+2} ions stored in the ER spill out into the cytosol and bind to calcium-binding proteins, including calmodulin. Some of these ions also bind to and activate an enzyme called protein kinase C (PKC). Once PKC is activated, it uncovers a binding site for the phospholipid phosphatidylserine at the inner surface of the plasma membrane. This interaction, plus other anchoring proteins (e.g., Receptor for Activated C Kinase, also known as RACK), traps the activated PKC at the plasma membrane.

Meanwhile, the DAG molecule, because it is mostly hydrocarbon, remains in the plasma membrane, but because of its small size it diffuses rapidly in the lipid bilayer. Eventually, it reaches the activated PKC in the membrane, binds to it, and helps keep PKC in its activated conformation. The entire pathway is shut down by PIP$_2$ phosphodiesterase and IP$_3$ phosphatase, which reverse the effects of the phospholipid kinases by removing the phosphate groups from the inositol sugar, as **FIGURE 11-22** shows.

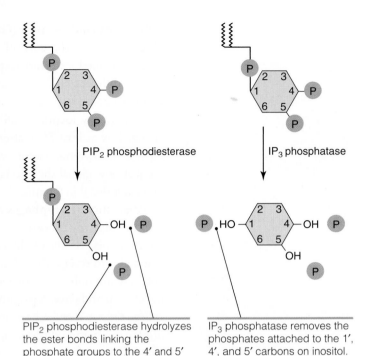

PIP$_2$ phosphodiesterase hydrolyzes the ester bonds linking the phosphate groups to the 4′ and 5′ carbons, yielding PI.

IP$_3$ phosphatase removes the phosphates attached to the 1′, 4′, and 5′ carbons on inositol.

FIGURE 11-22 PIP$_2$ phosphodiesterase and IP$_3$ phosphatase inhibit PIP$_2$ signaling. Both enzymes dephosphorylate second messengers in the PIP$_2$ signaling pathway.

■■ Steroid Hormones Control Long-Term Cell Behavior by Altering Gene Expression

One steroid hormone–mediated signal transduction pathway most college students can readily identify with is that associated with exposure to long-term stress. The steroid hormone **cortisol** is one of the most well-known stress signals. Cortisol is released by the outer layer (cortex) of the adrenal glands and is a potent anti-inflammatory signal. Localized application of cortisol is commonly prescribed for a range of maladies, including bee stings and eczema. These treatments are typically not used for long, perhaps a few times a day for a week or less, but chronic, full-body exposure to cortisol—over the length of a school year, for example—can have some very debilitating effects. Ample experimental evidence supports the hypothesis that cortisol links the psychological impacts of stress to many of the physical ailments associated with it. How the brain processes the complex notion of stress into a cortisol signal is still not well understood, but there is no doubt it happens. Some of the most familiar responses to chronic cortisol exposure include weight gain, lethargy, sleeplessness, and a compromised immune system.

How does this happen? Because cortisol is a steroid, it enters every cell in the body, and many different cell types express the cortisol receptor, more commonly known as a **glucocorticoid receptor** (**GR**) (because it also binds other glucocorticoid-class hormones). Once bound to the cortisol, the GR enters the nucleus and binds to a gene regulatory sequence called a **steroid response element** (**SRE**). Several different variations of the SRE exist in different cell, tissue, and animal types, making it difficult to determine exactly how many genes are responsive to cortisol, but at least 30 have been identified in humans so far. A recent study identified 28 different proteins in a single cell type where expression was changed by at least 75% after only 3 days of constant exposure to cortisol.[1] These proteins include transcription factors, which in turn control the expression of other genes, demonstrating just how global the effects of cortisol exposure are. (We will discuss transcription factors in greater detail in Chapter 12.) The cells used in this study are called **monocytes**, which differentiate into **macrophages** and are a critical component of the immune system; this may help explain why exposure to chronic stress increases the likelihood of getting sick. Because the molecular details of the cortisol response are not yet thoroughly understood, products that claim to reverse the effects of stress by inhibiting cortisol or reversing its impact on the body are not only ineffective, they may even be dangerous. Until scientists better understand the exact path between psychological stress and its effects on the body, the best treatment for college fatigue is still a healthy diet, regular exercise, and plenty of sleep. No pill, powder, or drink can get around this simple fact.

CONCEPT CHECK #3

Many students attempt to combat the effects of long-term stress by adding stimulants, like caffeine, to their diet, or by smoking tobacco to calm their nerves. Given that caffeine and nicotine bind to GPCRs, whereas cortisol binds to GR, propose an explanation for why these "treatments" are not effective over a long period.

11.6 Chapter Summary

Signal transduction is one of the most complicated cellular behaviors. Nonetheless, we can see patterns in the signal transduction machinery that give us important clues as to how information is passed from the extracellular environment through the plasma membrane and into the cell interior. One of the most obvious properties of signal transduction pathways is that they are composed of at least three elements: a signal, a signal receptor, and an effector. In most cases, the link between the receptor and effector is formed by a series of several signaling proteins and second messenger molecules that interact in a linear sequence. Information flows in only one direction through these pathways.

Classification is an important tool in deciphering the vast number of signaling molecules cells use. For example, signal molecules are classified as either membrane-permeable or membrane-impermeable, and their receptors are classified as cell surface proteins or cytosolic proteins. The cell surface receptors are further classified according to the kinds of signal molecules they bind, the signaling proteins they interact with, and their molecular composition. Many receptors are enzymes, and those that are not frequently bind to enzymes as part of signal transduction. This helps amplify the effect of a signal, making it easier for the appropriate effector proteins to respond. Finally, the signaling proteins that link receptors to their effectors are grouped into six classes, based on their mechanism of activation.

Most cell surface receptors are linked to several different effectors in the same cell. Some of these lie in the cytosol, and the impact on them is relatively short lived. Other

---

[1] Billing AM, Fack F, Renaut J, Olinger CM, Schote AB, Turner JD, Muller CP. Proteomic analysis of the cortisol-mediated stress response in THP-1 monocytes using DIGE technology. J Mass Spectrom. 2007 Nov;42(11):1433–44.

effectors lie in the nucleus, and their impact is much longer lasting, because they influence the transcription of genes and thereby alter a cell's phenotype. Most membrane-permeable signals, such as steroid hormones, have very short signaling pathways that target gene transcription and are therefore long lasting.

CONCEPT CHECK ANSWERS #1

Cancers are good examples of cells that have lost the ability to control themselves, and so we would expect to find changes in these receptors that prevent them from being controlled. A common problem is mutation of receptors such that they are locked in one conformation. Receptors locked in an active conformation will signal even if they are not bound to ligands, while those locked in an inactive state can never be activated even if a ligand binds to them. The lesson here is that not being able to turn on and shut off signaling pathways can be extremely dangerous for a cell.

CONCEPT CHECK ANSWERS #2

The signaling proteins that mutate in cancer cells are not there just to cause cancer—the normal form of these proteins plays an important role in regulating cell behavior. Because signaling proteins undergo complex interactions with other signaling proteins in signaling networks, inhibiting even one protein with a highly specific drug has the potential to impact a wide range of signaling pathways. While this may reverse the effect of the aberrant signaling in the cancer cell, it is quite possible that the same inhibition will disrupt a healthy cell that uses the same protein. Cancer cells are very similar to normal cells in most respects, so disabling a cancer cell has the potential to impact healthy cells at the same time. Very little about cancer cell behavior is so specific that it is not shared by at least some other cells in the same patient. This is still one of the major hurdles facing cancer patients, physicians, and researchers.

CONCEPT CHECK ANSWERS #3

GPCRs are receptors for several different neurotransmitters that transmit information between nerve cells. They target numerous effectors in cells, including the pathways our nerve cells use to process thoughts. Cortisol, on the other hand, bypasses all of these pathways by binding the GR and moving directly to the nucleus to change the transcription pattern in our cells. Ingesting caffeine and nicotine interferes with the normal thought processes we use to manage stress, but does not significantly impact the mechanisms driving the stress response. We can't eat/drink/smoke our way out of a stressful situation, for this reason.

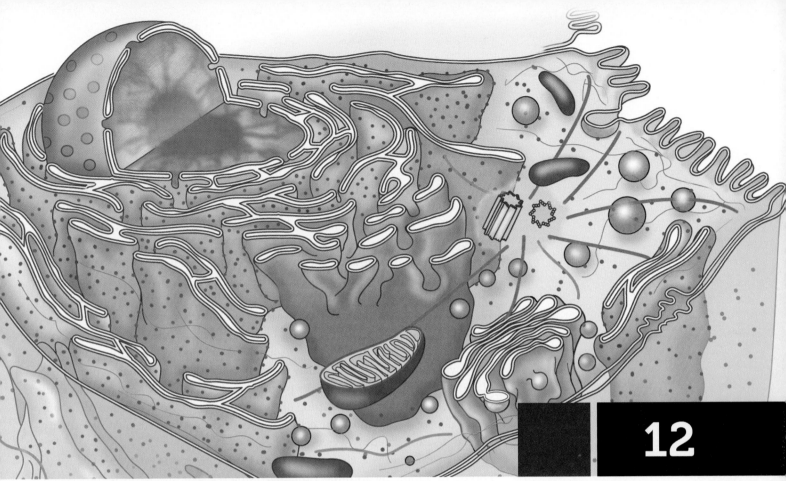

Control of Gene Expression

12.1 The Big Picture

In this chapter we continue the story of cell specialization that began in Chapter 11. Here, we will focus on the fraction of cellular signaling pathways that enter the nucleus, target nuclear effectors, and thereby control gene expression. (We will not address control of gene expression in prokaryotes.) We pick up the story where the common signaling pathways discussed in the previous chapter enter the nucleus, as shown in **FIGURE 12-1**. For historical reasons, a great deal is known about cytosolic signaling proteins and DNA-binding proteins, but the signaling events that link them are not as well understood. We'll look at what is known about these links to keep the story moving. This chapter contains three major sections, covering the following subjects:

- First, we'll examine how the key nuclear signaling proteins from Chapter 11 enter the nucleus. The picture of how most cytosolic proteins that control gene expression accomplish this is a bit blurry, and likely contains significant gaps. Focusing on a few well-known examples will be sufficient for our purposes here.

- Second, we will examine how DNA-binding proteins (the *effectors* of nuclear signaling pathways) function to control gene expression. A tremendous amount is known about this subject, which falls squarely within the domain of the field of molecular biology. Rather than plumb the depths of this field here, we will sample key elements while keeping our focus on the theme of cell specialization. As we do, we'll encounter our cell biology principle #8: protein complexes are cellular decision-making devices.

  What this means is that cells make complex decisions, such as which phenotype to adopt, by forming protein complexes that specialize in processing tremendous amounts of information and generating a few critical outputs. A common analogy for this process is how groups of people make decisions by forming committees that receive multiple sources of information and process it

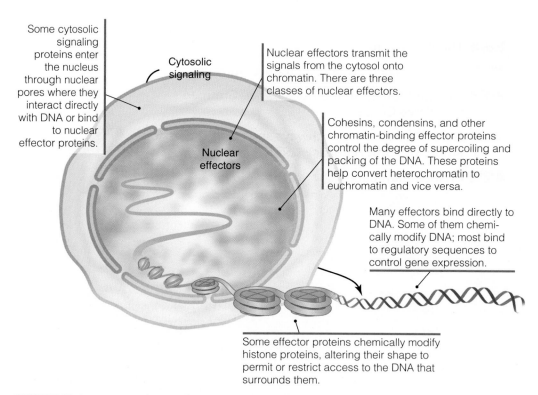

Some cytosolic signaling proteins enter the nucleus through nuclear pores where they interact directly with DNA or bind to nuclear effector proteins.

Cytosolic signaling

Nuclear effectors transmit the signals from the cytosol onto chromatin. There are three classes of nuclear effectors.

Cohesins, condensins, and other chromatin-binding effector proteins control the degree of supercoiling and packing of the DNA. These proteins help convert heterochromatin to euchromatin and vice versa.

Nuclear effectors

Many effectors bind directly to DNA. Some of them chemically modify DNA; most bind to regulatory sequences to control gene expression.

Some effector proteins chemically modify histone proteins, altering their shape to permit or restrict access to the DNA that surrounds them.

FIGURE 12-1 How cytosolic signaling impacts the nucleus.

to reach a consensus. Signaling proteins can form large clusters like this, and so can transcription factors. Using this principle, we can simplify complicated regulatory mechanisms by looking for cases where large complexes of proteins process multiple inputs to yield a single activity, such as initiation of transcription or the unwinding of a stretch of DNA. By looking for these telltale decision makers, we can avoid getting caught up in the minutiae and keep our focus on one of the key features of biological diversity: differential gene expression.

- Third, we'll have a look at how changes in gene expression result in changes in signaling pathways. New proteins arising from altered gene expression often impact the signaling pathways that triggered their synthesis. The concept of a product affecting the pathway that created it is called **feedback**, and we'll look at a few examples of feedback that include signaling and transcriptional control. This will set the stage for examining the mechanisms controlling cell division and programmed cell death in the next chapter.

Before we begin, remember that our overall goal is to integrate the previous chapters into a coherent picture of an actual living cell that does many things at once. For this reason, we'll rely heavily on material in previous chapters, and I've made numerous references to earlier sections in the book. This is a relatively short chapter, and I strongly suggest reviewing these earlier sections as we make our way through this chapter (**BOX 12-1**).

12.2 Many Signaling Proteins Enter the Nucleus

KEY CONCEPTS

▸ Numerous classes of cytosolic signaling proteins routinely pass into and out of the nucleus.
▸ Proteins that bind to membrane-soluble signaling molecules move directly into the nucleus and function as transcription factors.
▸ Some plasma membrane–associated signaling proteins and protein fragments enter the nucleus, but their function in the nucleus is not well understood.
▸ Nuclear signaling proteins that do not bind DNA regulate proteins that do.
▸ The same regulatory mechanisms that control cytosolic signaling proteins control nuclear signaling proteins; phosphorylation/dephosphorylation is the best understood nuclear regulatory mechanism.

The tremendous complexity of cytosolic signaling networks discussed in Chapter 11 extends into the nucleus for those signaling molecules that pass through the nuclear pore complexes. The total number of signaling molecules that enter the nucleus as part of their signal transduction function is not known, but scientists recently discovered that several proteins previously thought to be restricted only to the cytosol have the ability to cycle into and out of the nucleus, suggesting a new paradigm in cell signaling may be on the horizon. The signaling proteins include heterotrimeric G proteins, adenylyl cyclase isoforms, phospholipases, protein kinase A (PKA), and even some G protein-coupled receptors. In some cases, the function of this nuclear cycling is quite clear: cytosolic signaling proteins enter the nucleus to bind to and modify the structure and function of DNA, RNAs, and nuclear proteins. In other cases, the reasons for nuclear translocation are still not known. As researchers learn more about the functions of these signaling molecules, some surprising new models for how information is transported into the nucleus have appeared.

Many cytosolic signaling proteins contain nuclear localization sequences (see *Nuclear Localization Sequences (NLS) and Nuclear Export Signals (NES) Are Amino Acid Sequences Recognized by NLS and NES Receptors, Respectively* in Chapter 8) that enable them to enter the nucleus through nuclear pore complexes and bind to their targets. Other signaling proteins enter the nucleus via mechanisms that are not well understood. Because we cannot describe all of the proteins that carry signals into the nucleus, let's focus on the classes of receptors and cytosolic signaling proteins we discussed in Chapter 11 for examples illustrating the general properties of nuclear signaling pathways.

Nuclear Receptors Translocate from the Cytosol to the Nucleus during Signaling

As we discussed in Chapter 11, some receptors for extracellular signaling molecules are not located at or in the plasma membrane. Those that bind membrane-permeable signals, such as steroid hormones and nitric oxide, are located in the cytosol. Upon binding their ligand, many of these receptors move into the nucleus. These nuclear receptors are among the best known examples of signal transduction from the cytosol to the nucleus, and we need not expand on the discussion in Chapter 11 to illustrate their significance.

Notch Is a Transmembrane Scaffold Receptor That Enters the Nucleus

The protein Notch is a well-characterized example of a cell surface receptor moving into the nucleus. Notch influences cell differentiation by inhibiting cells from choosing certain differentiation fates. At the cell surface, Notch is activated when it binds to a different receptor protein on a neighboring cell. The best known of these is a protein called Delta. After it binds Delta, Notch undergoes a shape change that permits two proteinases to break the peptide backbone, generating three different pieces of Notch. The cytosolic portion of the receptor, also called the Notch intracellular domain (NICD), separates from the plasma membrane and enters the nucleus, where it binds to a protein called CSL, as **FIGURE 12-2** shows. CSL binds to DNA and suppresses transcription. After NICD binds to CSL, the suppressive effect is removed, and gene transcription proceeds. The genes suppressed by CSL include several inhibitory transcription factors, thus, the net effect of Notch activation is to block the expression of many genes, effectively limiting the number and types of proteins a cell can create.

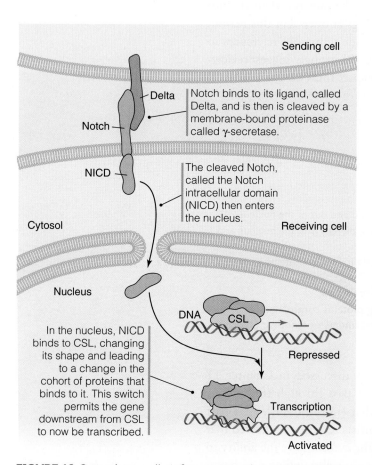

Notch binds to its ligand, called Delta, and is then is cleaved by a membrane-bound proteinase called γ-secretase.

The cleaved Notch, called the Notch intracellular domain (NICD) then enters the nucleus.

In the nucleus, NICD binds to CSL, changing its shape and leading to a change in the cohort of proteins that binds to it. This switch permits the gene downstream from CSL to now be transcribed.

FIGURE 12-2 Notch is a cell surface receptor that signals in the nucleus.

G Protein Coupled Receptors and GPCR Fragments Signal in the Nucleus

Most GPCRs bind to extracellular ligands on the cell surface and transmit the signal to heterotrimeric G proteins, triggering a range of signaling pathways. Recent experiments have identified additional GPCRs that associate with the Golgi apparatus or endoplasmic reticulum rather than the plasma membrane, and over 20 different GPCRs that actually move into the nuclear envelope or nucleoplasm (see **TABLE 12-1**). The route into the nucleus for most of these receptors is unknown, but one obvious possibility is that they diffuse through the nuclear pore complexes directly from the endoplasmic reticulum membrane, because it is continuous with the nuclear envelope (see *The Nucleus Is the Central Storehouse of Genetic Information* in Chapter 1). Some of these receptors contain putative (possible but unproven) nuclear localization sequences that could facilitate this transport.

Some GPCRs are proteolytically cleaved before they enter the nucleus, similar to Notch. **FIGURE 12-3** illustrates an example of a GPCR fragment signaling in the nucleus of nerve cells. A GPCR called D-Frizzled 2 (DFZ2) is endocytosed after it binds to its ligand (called Wingless), is cleaved by a mechanism not yet understood, and the carboxy-terminal portion enters the nucleus and initiates gene transcription. Other GPCRs leave the cell surface upon activation and participate in more conventional signaling pathways

TABLE 12-1 GPCRs found in the nucleus.

GPCR Subfamilies	Cell/Tissue	Function
Family 1 mAChR	Cornea, corneal epithelial/endothelial cells	↑ DNA and RNA Poll activity
APJ	Cerebellum, hypothalamus	Unknown
α-AR	Neonatal ventricular myocytes, adult	α AR; ERK activation, proliferation
β-AR	Adult & neonatal ventricular myocytes	↑ Nuclear cAMP, G protein coupling
AT R	Vascular endothelial cells (ECs), VSMCs	↑ Nuclear (Ca^{2+}), ERK & p38 activation
CysLT$_1$	Colon tissue, colorectal carcinomas	↑ Nuclear (Ca^{2+}), ERK activation
CXCR4	Hepatoma, colorectal cancer	Tumorilgenesis?
EP receptors	Brain, myometrium, liver	↑ Nuclear (Ca^{2+}), ERK & PKB activation
ETR	Liver, VSMCs, adult ventricular myocytes	↑ Nuclear (Ca^{2+}), phosphorylation
LPA$_1$R	Hepatocytes, cerebral microvessel ECs	↑ Nuclear (Ca^{2+}), ERK & PKB activation
MT2	Placental chorlocarcinona cells	Unknown
NK$_1$R	Dorsal root ganglia neural cells	Unknown
OPR1/2	NG 108-15 neurohybrid cells (OPR1)	G protein coupling, PKC activation
PAFR	Liver, cerebral microvascular ECs, brain	↑ Nuclear DAG, nuclear (Ca^{2+})
Y$_1$R	Pituitary gland, pituitary cells, endocardial	Unknown
Family 2 PTH$_1$R	Kidney, liver, gut, uterus, ovary	DNA synthesis, mitosis?
Family 3 mGluR5	Germline, intestine	↑ Nuclear (Ca^{2+})
GnRHR		Unknown

mAChR, muscarinic acetylcholine receptor; APJ, apelin receptor; α-AR, α-adrenergic receptor; β-AR, β-adrenergic receptor; AT1R, angiotensin II type 1 receptor; CysLT1, cysteinylleukotriene type 1 receptor; CXCR4, CXC chemokine receptor 4; EP receptors, prostaglandin type E receptor; ETR, endothelin receptor; GLRN-R, ghrelin receptor; LPAIR, lysophosphatidic acid receptor type-1; MC2R, melanocortin-2 receptor; MT2, melatonin receptor 2; NK1R, tachykinin receptor 1; OPR1/2, opiold peptide receptor 1/2; PAFR, Platelet-activating factor receptor; PRLR, prolactin receptor; VPAC, vasoactive intestinal peptide receptor; Y1R, neuropeptide Y receptor 1; PTH1R, Type I parathyroid hormone receptor; mGluR5, metabotropic glutamate receptor 5; GnRHR, gonadotropin releasing hormone receptor; ECs, endothelial cells; PKB, protein kinase B; PKC, protein kinase C; VSMCs, vascular smooth muscle cells

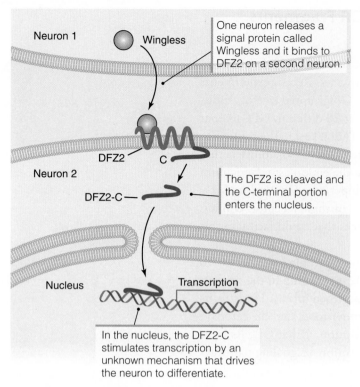

Neuron 1

Wingless

One neuron releases a
signal protein called
Wingless and it binds to
DFZ2 on a second neuron.

DFZ2   C

Neuron 2

DFZ2-C

The DFZ2 is cleaved and
the C-terminal portion
enters the nucleus.

Nucleus   Transcription

In the nucleus, the DFZ2-C
stimulates transcription by an
unknown mechanism that drives
the neuron to differentiate.

FIGURE 12-3 D-Frizzled (DFZ2) is a GPCR that signals in the nucleus.

by binding to cytosolic signaling proteins. In some cases, a GPCR enters the nucleus as a complex with its ligand and its associated heterotrimeric G protein.

Heterotrimeric G Proteins Target Many Cellular Compartments, Including the Nucleus

The traditional paradigm describing the structure and function of heterotrimeric G proteins places them exclusively at the plasma membrane, where they associate with GPCRs and other membrane-associated signaling proteins, as described in Chapter 11. But considerable evidence suggests that, as for GPCRs, this is not the complete picture. For example, intact heterotrimers are present in the nucleus as well as the Golgi apparatus and endoplasmic reticulum membranes, and some are associated with the cytoskeleton. G proteins can also translocate between various cellular locations in response to hormonal stimulation. In some cells, the number of α subunits greatly exceeds the number of β/γ subunits, suggesting that at least some α subunits may function independent of β/γ, negating the need to form heterotrimers. Some G proteins associate with receptor tyrosine kinases, and thereby offer the potential for considerable crosstalk between signaling pathways triggered by these receptors and GPCRs.

Several Elements of Phosphatidylinositol Signaling Pathways Are Present in the Nucleus

The conventional phosphatidylinositol-based signaling discussed in Chapter 11 has recently been augmented by some surprising evidence that a similar pathway operates inside the nucleus. For example, PIP_2 can be detected in the membrane of isolated nuclei, along with the enzymes responsible for synthesizing and metabolizing PIP_2 and other phosphatidylinositols. This implies that nuclear phospholipase C could generate diacylglycerol and IP_3 directly in the nucleus, thereby activating nuclear forms of protein kinase C. Also, several PIP receptors have been found in the nucleus, including nuclear hormone receptors, histones, and chromatin remodeling complexes. It is still not clear if this nuclear pathway is isolated from that taking place at the plasma membrane, or if it represents nuclear translocation of cytosolic and plasma membrane proteins and phospholipids.

Receptor Protein Tyrosine Kinases Signal in the Nucleus

Just as traditional GPCRs remain on the cell surface when they signal, so too do most receptor protein kinases; however, some receptors can enter the nucleus by a mechanism that bears striking similarity to that mentioned previously for GPCRs. **FIGURE 12-4** shows how a receptor tyrosine kinase called Ryk binds its ligand (Wnt), then is cleaved by two different proteases to generate a cytosolic fragment that enters the nucleus to control gene transcription. Other receptor tyrosine kinases, including the insulin receptor and a member of the epidermal growth factor receptor family called ErbB4, are likewise cleaved by proteases, and their cytoplasmic tails can enter the nucleus. In most cases, the proteases responsible for this cleavage are not known, nor is it clear how the receptor fragments enter the nucleus, since they lack the traditional nuclear localization sequences we discussed in Chapter 9. Another receptor, called ErbB2, can enter the nucleus after it has been endocytosed. The intact receptor is recognized by nuclear pore proteins and members of the importin family of nuclear transport proteins, suggesting it has a nuclear localization sequence, though it shares only partial similarity with traditional NLS sequences. Amazingly, the ErbB2 passes through the nuclear pore complex in a membrane vesicle, suggesting that vesicular trafficking networks, such as the one we discussed in Chapter 9 (see *Vesicles Shuttle Material between Organelles in the Endomembrane System*), may include the nucleus as a target organelle. In at least one case, an extracellular ligand (fibroblast growth factor) *and* its receptor tyrosine kinase are translocated together from the plasma membrane to the nucleus, perhaps in similar vesicles.

Some Protein Kinases Phosphorylate Nuclear Proteins

Most cytosolic protein kinases are incapable of entering the nucleus, but a few cases of nuclear translocation are known. One is the MAP kinase family of serine/threonine kinases. **FIGURE 12-5** illustrates an example of a nuclear signaling network triggered by p38 and ERK1/2, two members of this family. ERK1/2 is held inactive in the cytosol by a MAP kinase called MEK1/2. Once MEK1/2 is activated, it phosphorylates ERK1/2, and this both activates it and separates it from MEK1/2. ERK1/2 then enters the nucleus by at least two different mechanisms: monomers of ERK1/2 pass through nuclear pore complexes by passive diffusion, and dimers of ERK1/2 are actively transported using the Ran-dependent mechanism outlined in Chapter 8 (see *The Direction of Nuclear Transport Is Controlled by the G Protein Ran*). Both p38 and ERK1/2 target nuclear protein kinases called mitogen- and stress-activated protein kinases, or MSKs. MSKs are somewhat unusual kinases, in that they are permanent residents of the nucleus. Once activated, MSKs phosphorylate a number of different proteins, including the transcription factor CREB (cAMP response element binding protein) and the chromatin proteins histone H3 and HMG-N1 (high-mobility group protein N1). Their large number of substrates permits MSKs to control the expression of many different genes in different cell types.

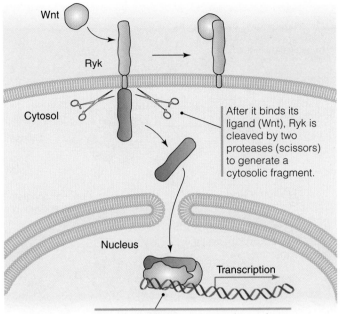

After it binds its ligand (Wnt), Ryk is cleaved by two proteases (scissors) to generate a cytosolic fragment.

The cleaved fragment enters the nucleus and interacts with other unknown proteins to control transcription.

FIGURE 12-4 Ryk is a tyrosine kinase receptor that signals in the nucleus.

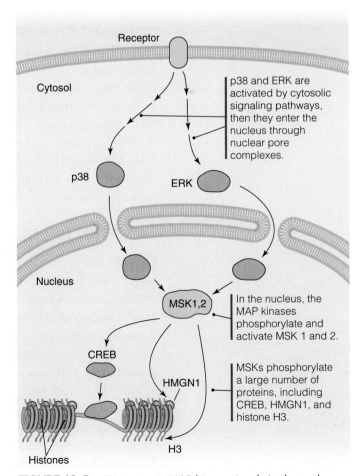

p38 and ERK are activated by cytosolic signaling pathways, then they enter the nucleus through nuclear pore complexes.

In the nucleus, the MAP kinases phosphorylate and activate MSK 1 and 2.

MSKs phosphorylate a large number of proteins, including CREB, HMGN1, and histone H3.

FIGURE 12-5 MSKs transmit MAP kinase signals in the nucleus.

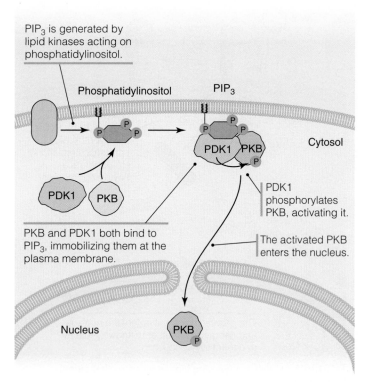

PIP$_3$ is generated by lipid kinases acting on phosphatidylinositol.

Phosphatidylinositol

PIP$_3$

Cytosol

PDK1 PKB

PDK1 PKB

PKB and PDK1 both bind to PIP$_3$, immobilizing them at the plasma membrane.

PDK1 phosphorylates PKB, activating it.

The activated PKB enters the nucleus.

Nucleus

PKB

FIGURE 12-6 Protein kinase B signals in the nucleus.

Protein kinase B (also called PKB or Akt) is a cytosolic kinase activated by another cytosolic kinase called PDK1. This typically occurs at the plasma membrane, as **FIGURE 12-6** shows. PKB and PDK1 are attached to the cytosolic surface of the plasma membrane by binding to a phosphorylated phospholipid called phosphatidylinositol-3,4,5 trisphosphate (PIP$_3$). PIP$_3$ is created from phosphatidylinositol by the action of three lipid kinases, PI3 kinase, PI4 kinase, and PI5 kinase (see Figure 11-15). PDK1 (and most likely a second, related kinase called PDK2) phosphorylate one serine and one threonine in PKB, and this induces a change in shape that detaches it from PIP$_3$. It then passes through the nuclear pore complex and phosphorylates a large number of proteins, including many transcription factors that control a great variety of cellular behaviors including growth, differentiation, and survival. Mutations in PKB and the kinases that activate it are thought to contribute to many diseases, including cancer. Recent experiments identifying PIP$_3$ inside the nucleus suggest that PKB activity can be controlled directly in the nucleus.

Protein kinase C (PKC) is another well-known kinase that, somewhat surprisingly, appears to signal in both the cytosol and nucleus. As discussed in Chapter 11 (see *Phospholipid Kinase Pathways Work in Cooperation with Protein Kinase and G Protein Pathways*), PKC isoforms activated by IP$_3$ and diacyl glycerol move from the cytosol to the plasma membrane. This makes it difficult to explain how active PKC could accumulate in the nucleus, yet several lines of evidence strongly suggest that PKC binds to the interior of the nuclear envelope. Some PKC isoforms are able to bind to the nuclear matrix, immobilizing them in the nucleoplasm. Once inside the nucleus, PKC phosphorylates a variety of proteins (**TABLE 12-2**) that control a wide range of cellular behaviors, including proliferation and migration; for example, several forms of PKC phosphorylate nuclear lamins (see *Lamins Form a Strong Cage Inside the Nucleus* in Chapter 5), thereby helping initiate the onset of mitosis. Nuclear PKA is also involved in the onset of some cancers.

PTEN Is a Nuclear Phosphatase

Considerable protein and lipid phosphorylation takes place in the nucleus, and additional proteins and lipids previously phosphorylated in the cytosol enter the nucleus as well. As we discussed in Chapter 11, phosphorylation is a powerful mechanism for controlling protein and lipid function, especially in signaling networks. However, this is only true if the impact of phosphorylation can be reversed by detaching the phosphates at some point. This is the job of phosphatases. One of the best known nuclear phosphatases is named **phosphatase and tensin homolog**, commonly abbreviated as **PTEN**. PTEN is located in both the cytosol and the nucleus, where it targets different phosphorylated substrates. PTEN is a dual specificity phosphatase, meaning it can dephosphorylate both proteins and lipids. PTEN can enter and exit the nucleus by several means, including simple diffusion, signal sequence–dependent mechanisms, and phosphorylation-dependent mechanisms. In the nucleus, PTEN suppresses cell growth, most likely by inactivating transcription factors and the kinases that phosphorylate them.

TABLE 12-2 Examples of proteins phosphorylated by PKC in the nucleus.

Phosphorylated Protein	Data Source (In Vitro/In Vivo)	PKC Isoform	Effect of Phosphorylation
Histone H1	In vitro	All classes	n.d.
Histone H2B	In vitro/in vivo	B	Apoptotic DNA fragmentation (?)
Histone H3	In vitro/in vivo	A	Enhanced gene transcription
Lamin A	In vitro/in vivo	n.d.	Lamin disassembly/nuclear targeting
Lamin B	In vitro/in vivo	A, BII	Mitotic/apoptotic lamin disassembly
DNA topoisomerase I	In vitro	n.d.	Enzyme activation
DNA topoisomerase IIa	In vitro	n.d.	Enzyme activation
DNA polymerase B	In vitro	n.d.	Enzyme inactivation
Vitamin D3 receptor	In vitro/in vivo	B	Transcriptional activation
RNA polymerase II	In vitro	n.d.	Increased RNA sysnthesis rate
CREB	In vitro	n.d.	n.d.
PI-PLC B	In vitro/in vivo	A	Enzyme inactivation
HMG proteins	In vitro	A, B	DNA-binding affinity reduction
P53	In vitro	n.d.	Increased DNA-binding affinity

An ATP-Binding Calcium Ion Channel Is Present in the Plasma Membrane and Nuclear Envelope in Some Neurons

Ion channel proteins are widely distributed throughout organelles in eukaryotic cells, where they control the passage of ions from one side of the membrane to the other (see *Protein Channels Dissipate Gradients* in Chapter 10). Many of these are ligand gated, meaning their open or closed configuration is controlled by direct binding of signaling molecules. Most ligand-gated channels remain in place, meaning they do not move from one membrane to another; however, at least one ATP-gated calcium channel has been identified in both the plasma membrane and nuclear envelope of some neurons, raising the possibility that these channels may circulate between the two compartments. (ATP is sometimes used as an extracellular signaling molecule.) The nuclear envelope contains many other ion channels that regulate the flow of Ca^{+2} and Cl^- ions from the perinuclear space (the area between the inner and outer nuclear membranes) and the cytosol. Some evidence suggests that these ion fluxes help control the shape (and function) of nuclear pore complexes.

An Adenylyl Cyclase Present in the Nucleus

Classical adenylyl cyclases are transmembrane proteins that convert ATP to cAMP at the cytosolic face of the plasma membrane, but there are at least 10 different forms of adenylyl cyclases known in mammals, and one of these is found in the nucleus. Coupled with the localization of PKA catalytic subunits and PKA-binding proteins, this suggests that a PKA signaling pathway may function entirely in the nucleus. The function of such a pathway is not yet known.

CONCEPT CHECK #1

The functions of some nuclear signaling proteins are quite clear, while the functions of others are still hard to define. Based on what is known about how they function at the plasma membrane and in the cytosol, propose three different ways that nuclear GPCRs could trigger phosphorylation of nuclear proteins.

12.3 | Effector Proteins in the Nucleus Are Grouped into Three Classes

KEY CONCEPTS

▸ Cohesins hold two DNA strands together by forming a ring-shaped complex around them.

▸ Condensins form a ring-shaped structure that gathers loops of DNA together.

▸ Histones are phosphorylated by a number of protein kinases; histone phosphatases remove them. This modification of histones is critical for permitting gene expression.

▸ Histone acetyltransferases add acetyl groups to histones, thereby loosening their interaction with the DNA surrounding nucleosomes. This loosening activates the neighboring DNA by promoting expression of genes in this area. Histone deactylases remove the acetyl groups.

▸ Histone methylation creates a binding site for heterochromatin proteins and promotes DNA silencing.

▸ Histone ubiquitination can either promote transcription or trigger histone degradation, depending on the number of ubiquitin proteins added to each histone.

▸ General transcription factors assemble at the core promoter of a gene and serve as the foundation for RNA polymerase activation.

▸ Activator and repressor proteins are transcription factors that bind to and control the activity of the general transcription complex. Some activators bind to enhancer sequences located hundreds of base pairs away from the core promoter.

Effectors are the endpoints of signaling pathways. The nucleus contains a wealth of effector proteins, all ultimately concerned with regulating the organization and usage of DNA. These are grouped into three classes, based on the kinds of molecules they target.

Cohesins and Condensins Help Control the Packaging State of Chromatin

Cohesins and condensins are members of a family of proteins called SMC (structural maintenance of chromosome) proteins. Both cohesins and condensins form large complexes with other SMC proteins that are responsible for controlling the condensation of chromosomes during prophase (see *Prophase Prepares the Cell for Division* in Chapter 7) and the proper alignment of the sister chromatids during mitosis, as well as regulation of gene expression, DNA repair, and centromere organization. **FIGURE 12-7** shows models of how they might accomplish this. Both condensins and cohesions are composed of two different

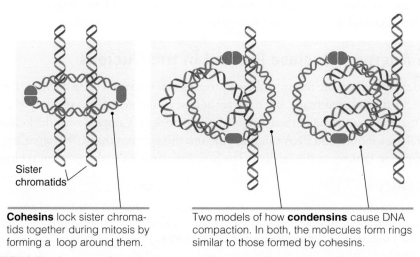

Sister chromatids

Cohesins lock sister chromatids together during mitosis by forming a loop around them.

Two models of how **condensins** cause DNA compaction. In both, the molecules form rings similar to those formed by cohesins.

FIGURE 12-7 Cohesins and condensins control the spatial arrangment of chromatin.

SMC subunits, each composed of two polypeptides that twist around one another to form a coiled-coil structure. The heterodimeric proteins form a complex, aligned "head-to-head" at one end, with DNA- and ATP-binding domains at the other. Cohesins are phosphorylated and activated by protein kinases called polo-like kinases; activation of the polo-like kinases can, in turn, be traced back to tyrosine kinase signaling pathways in the cytosol. Cohesins bind two strands of DNA together by forming a ring-shaped structure that encloses the DNA, and condensins are thought to condense DNA by forming similar ring-shaped structures enclosing loops of DNA. While it is known that DNA condensation requires ATP energy, it is still not entirely clear exactly how that ATP is used. It is possible these proteins hydrolyze ATP to gather the DNA into these rings.

Histone Modifiers Control the Structure of Nucleosomes

Chromatin remodeling, illustrated in **FIGURE 12-8**, is the term used to describe the process of displacing histones to control access to DNA. In some cases, this means nucleosomes are simply slid from one position in chromatin to another, which can change the spacing between nucleosomes. In other cases, entire nucleosomes are temporarily removed from a particular region of DNA. Large ATP-consuming complexes are primarily responsible for performing these remodeling activities. Humans have at least eight different complexes, and best known complex is called SWI/SNF.

One of the most important aspects of chromatin remodeling is structural (chemical) modification of histones, summarized in **FIGURE 12-9**. As we discussed in Chapter 2 (see Figure 2-17), histones can be modified in at least four different ways. First, they can be phosphorylated. Serine and threonine amino acids in the amino-terminal "tails" of core

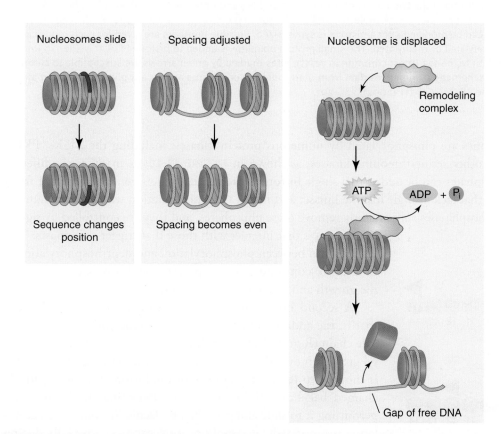

FIGURE 12-8 Chromatin remodeling is modification of histones to permit their removal or sliding along a piece of DNA. Opening up gaps between nucleosomes permits RNA polymerases and transcription factors to bind to promoters.

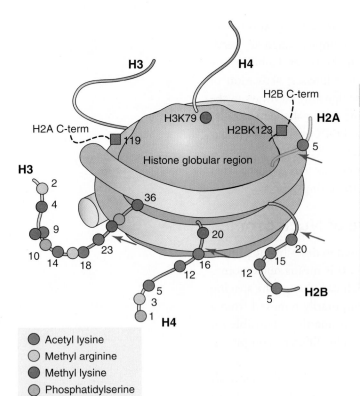

Histone	Site	Modification	Function
H3	K-4	Methylation	Transcription activation
H3	K-9	Methylation	Chromatin condensation
	K-9	Methylation	Required for DNA methylation
	K-9	Acetylation	Transcription activation
H3	S-10	Phosphorylation	Transcription activation
H3	K-14	Acetylation	Prevents methylation at Lys-9
H3	K-79	Methylation	Telomeric silencing
H4	R-3	Methylation	Transcriptional activation
H4	K-5	Acetylation	Nucleosome assembly
H4	K-12	Acetylation	Nucleosome assembly
H4	K-16	Acetylation	Nucleosome assembly
	K-16	Acetylation	Transcriptional activation

Legend:
- ● Acetyl lysine
- ○ Methyl arginine
- ● Methyl lysine
- ○ Phosphatidylserine
- ■ Ubiquinated lysine

FIGURE 12-9 Histone modifications on the nucleosome core particle. The graphic represents a schematic of a nucleosome core particle. For simplicity, modifications are shown on only one of the two copies of histones H3 and H4 and only one N-terminal tail is shown for histones H2A and H2B. The C-terminal tails of one histone H2A molecule and one histone H2B molecule are also shown (dashed lines). Colored symbols indicate histone sites that can be modified after a histone is synthesized. Residue numbers are given for each modification. Lysine-9 in histone H3 can be either acetylated or methylated. The small protein ubiquitin is added to histone H2A at lysine-119 (most common in mammals) and to histone H2B at lysine-123 (most common in yeast). Sites marked by green arrows are susceptible to cutting by a proteinase in intact nucleosomes. The schematic summarizes data from many different organisms and so any given organism may lack particular modifications. Adapted from Turner, B. M., *Cell* 111 (2002): 285–291.

histones are phosphorylated by numerous protein kinases, including the MSKs, PKCs, and other serine/threonine kinases, as shown in **FIGURE 12-10**; as many as six different phosphates can be added to a single histone protein. Researchers generated mutant fruitflies that lack a single histone kinase, and found that this mutation was lethal. A number of phosphatases subsequently remove these phosphates, and they are controlled by signaling pathways that interact with those that trigger the kinases. This competition between phosphorylation and deprrhosphorylation is thus key to controlling gene expression, and plays a critical role in the growth and differentiation of cells.

A second important modification of histones is acetylation, that is, the addition of an acetyl group to the amino group in the side chain (R group) of the amino acid lysine. Addition of the acetyl group removes the positive charge in the amino group, and this is thought to reduce the affinity of the histone for the negatively charged backbone of DNA, thereby loosening the histone and permitting it to slide along a DNA molecule. In some cases, acetylation promotes the assembly of nucleosomes, especially during DNA replication. Acetylation is also associated with the activation of the nearby DNA (i.e., it promotes local unfolding of

Sites of modification in H3

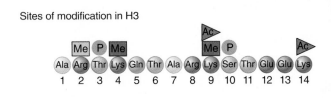

Sites of modification in H4

FIGURE 12-10 The N-terminal tails of histones H3 and H4 can be acetylated, methylated, or phosphorylated at several positions.

DNA and transcription of the genes in that region of a chromosome), as **FIGURE 12-11** shows. Enzymes called histone acetyltransferases add the acetyl groups, and histone deactylases (also called HDACs) remove them. One important histone acetyltransferase is p300/CREB binding protein, also called CBP. As its name implies, CBP binds to activated CREB and acts as a coactivator that links CREB to the basal transcription complex. CREB is a common target of PKA, and CREB phosphorylation promotes CBP binding, histone acetylation, and subsequent gene transcription. This is how many cAMP-associated signaling pathways help control transcription (for a review of cAMP signaling, see *Heterotrimeric G Protein Signaling Pathways Regulate a Great Variety of Cellular Behaviors* in Chapter 11).

The third important histone modification is methylation of lysine and arginine amino acids. Methylation is associated with chromatin condensation, though the direct link between the two events is still not clear. One important family of proteins that bind methylated histones is the heterochromatin proteins. After histone H3 is methylated on lysine-9, this creates a binding site for a heterochromatin protein that subsequently attracts an enzyme that methylates the surrounding DNA. This coupling of two kinds of methylation is important for the silencing of DNA (see *DNA Is "Silenced" in Heterochromatin* in Chapter 2). Histone demethylase enzymes remove the methyl groups, permitting heterochromatin to be partially unwound to form euchromatin.

Finally, histones can be ubiquitinated. Recall from Chapter 3 (see Figure 3-15) that ubiquitin is a small (67 amino acid) protein that is commonly attached to other proteins by enzymes called **ubiquitin ligases**. In many cases, serial addition of several ubiquitins (called polyubiquitination) on cytosolic and nuclear proteins triggers their degradation by nucleosomes. But when only a few ubiquitins are added to an amino acid, the effect is quite different. For example, when two or three ubiquitins are added to lysine in histone H2B, this triggers (and in yeast, is necessary for) methylation of lysines 4 and 79 in histone H3. This, in turn, promotes the recruitment of basal transcription factors to the core promoter of some genes, and thereby helps regulate the initiation and elongation stages of transcription. Curiously, the ubiquitins on H2B must be removed by one member of a family of **ubiquitin proteases** in order for RNA polymerase II to be properly activated. It is possible that several rounds of ubiquitination and deubiquitination may take place during the elongation phase of transcription. Ubiquitination of histone 2A interferes with elongation.

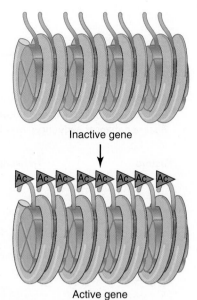

Inactive gene

Active gene

FIGURE 12-11 Acetylation associated with gene activation occurs by directly modifying specific sites on histones that are already incorporated into nucleosomes.

Transcription Factors Promote the Expression of Genes

The term *transcription factor* refers to a large number of proteins that promote gene transcription. By some measures, nearly every protein in the nucleus has some impact on genes, but the term is specifically applied to proteins that assist RNA polymerases to initiate transcription at the promoter region of genes. These are further divided into proteins that bind directly to DNA sequences and those that do not. Members of both types are effectors of signaling pathways. Let's merge signaling with the DNA information pathway story we began in Chapter 8 (see *Transcription Begins After a RNA Polymerase Binds to a Promoter Site on DNA*), by focusing on the level of DNA sequences and the proteins that bind to them.

General Transcription Factors Assemble at the Core Promoter

FIGURE 12-12 shows the general organization of DNA regions upstream of the initiation start site for a gene that is transcribed by RNA polymerase II. The region stretching about 200 base pairs immediately upstream of the start site is called the promoter. The promoter contains several distinct DNA sequences that play different functions in

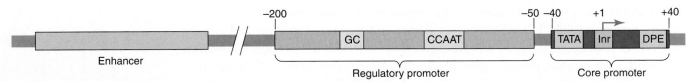

FIGURE 12-12 Regulatory regions controlling a gene transcribed by RNA polymerase II. The core promoter contains the TATA box, initiator, and in some cases a downstream promoter element. The regulatory promoter lies just upstream of the core promoter. More distant enhancer sequences may be hundreds or thousands of base pairs away, either upstream or downstream of the promoter.

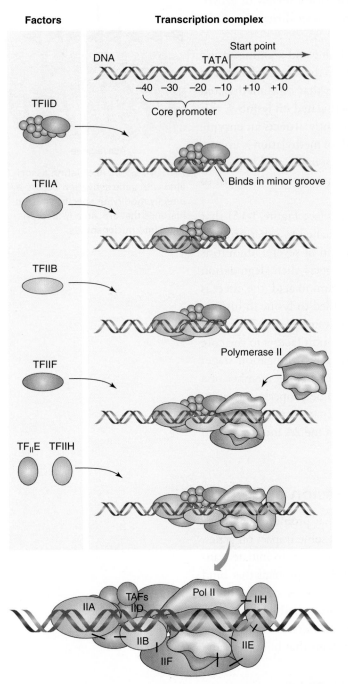

FIGURE 12-13 The sequential assembly of the general transcription complex for a gene trasncribed by RNA pol II. TFIID binds to the TATA box within the core promoter, and the rest of the general transcription factors bind in sequence. Eventually RNA pol II binds to the complex. Known binding interactions are shown with lines. Adapted from R. G. Roeder, Nat. Med. 9 (2003): 1239-1244.

assembling the transcription complex for a gene. The region between +40 and about −40 base pairs is called the core promoter. Within the core promoter, a region called the TATA box (because it contains the sequence TATAXAXA, where X is either T or A) is found in many mRNA-encoding genes. The core promoter also contains a sequence called the **TFIIB recognition element** (**BRE**). Those mRNA-encoding genes that lack a TATA box instead have a five-base-pair sequence called a **downstream core promoter element** (**DPE**) in the +28 to +32 region of the gene.

Together, the BRE and either the TATA box or DPE serve as binding sites for elements of the general (also known as basal) transcription complex. **FIGURE 12-13** shows the binding sequence of proteins that form the general transcription complex at the TATA box in yeast. TFIID, a multi-subunit protein, starts the sequence by binding to the TATA box. Once RNA polymerase attaches to this complex, it is called the preinitiation complex. This complex is minimally active, meaning that in a test tube, it can transcribe a gene, but at very low efficiency. In cells, this complex virtually never works alone; it interacts with other transcription factors to control the rate of transcription. This raises an important question: How do these additional transcription factors find and bind to this particular region of DNA?

The answer to this question is somewhat complicated, but it illustrates an important theme in cell and molecular biology, so let's take some time to go through it carefully. The key is to pay close attention to the three traits of proteins we discussed in Chapter 3 (for a review, see *Proteins Are Polymers of Amino Acids That Possess Three Important Traits* in Chapter 3). There are four steps to this answer:

- First, let's remember the different levels of protein structure. DNA-binding transcription factors contain secondary structures called *motifs,* organized into several binding *domains.* These particular motifs share a common property: they all contain α helices that fit into the grooves of a DNA double helix. There are three especially common motifs in DNA binding proteins, including transcription factors; for example, many DNA-binding proteins contain multiple copies of a 23-amino-acid-long loop called a **zinc finger** motif, illustrated in **FIGURE 12-14**. Zinc fingers

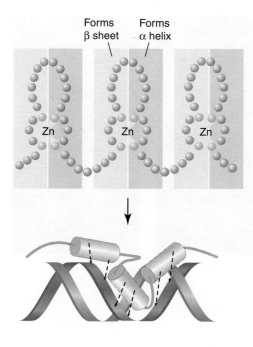

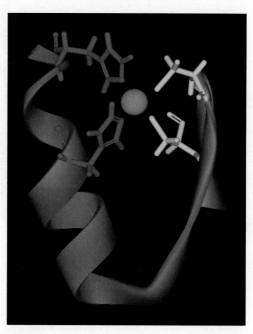

FIGURE 12-14 Two views of a zinc finger DNA-binding motif. The left diagram shows how the alpha helix for each zinc finger fits into the grooves on a DNA double helix. The right diagram shows the three dimensional structure of a zinc finger. The zinc ion is magenta, and the four amino acids bound to it are yellow and blue. Protein Data Bank ID: 1ZNF. Lee, M. S., et al., Science 245 (1989): 635-637.

require a series of four noncovalent bonds to a zinc ion (Zn^{+2}) to stabilize the secondary structure. The carboxyl end of the motif forms an α helix that fits into the major and minor grooves of the DNA double helix, while the amino end forms a β sheet. Most proteins that contain zinc fingers have several of them in succession.

Another common motif is called **helix-turn-helix (HTH)** because it consists of a stretch of 40 to 50 amino acids, organized into two α helices separated by a short loop. Proteins that contain HTH domains often form homo- and heterodimers with other HTH-containing proteins. The third common motif is called a **leucine zipper**, shown in **FIGURE 12-15**. It consists of two α helices that contain several leucine amino acids, plus positively and negatively charged amino acids. A leucine zipper is so-named because of its very specific arrangement of the side chains on the α helices. The positive and negative charges of these side chains are distributed so that they complement one another, drawing the two helices together into a slightly twisted coiled coil, somewhat similar to how the teeth in a zipper interlock. As Figure 12-15 shows, these two helices fit into the major grooves of DNA. Like HTH proteins, transcription factors that contain leucine zippers often form homo- and heterodimers.

■ Second, every different transcription factor has its own unique shape. (For the sake of simplicity, let's imagine that these three motifs are the only protein structures that stably bind DNA.) This means different proteins have different versions of these DNA-binding motifs. For example, two proteins that contain zinc fingers can have different amino acids in their zinc fingers (the exact sequence of amino acids that results in the formation of a zinc finger is still not known), which means that their zinc fingers are slightly different in shape. When they

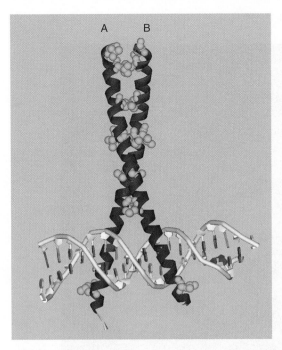
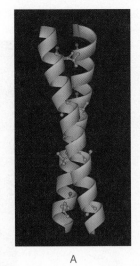
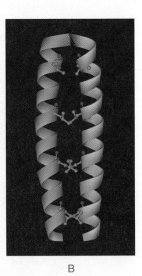

FIGURE 12-15 Two views of a leucine zipper. The left diagram shows a model of a leucine zipper attached to the major groove in DNA. The right diagram shows two images of a leucine zipper to emphasize (A) their coiled-coil nature, and (B) the interactions among the leucine side chains, respectively. Structure on left from Protein Data Bank 1YSA. T. E. Ellenberger, et al., Cell 71 (1992): 1223-1237. Prepared by B. E. Tropp. Structures on right from Protein Data Bank 2ZTA. E. K. O'Shea, et al., Science 254 (1991): 539–544. Prepared by B. E. Tropp.

bind to a DNA double helix, the α helices fit into the grooves, allowing them to slide back and forth along a DNA sequence. But because their shapes are not identical, the α helices fit into the grooves differently. This feature, plus the difference in overall tertiary (or even quaternary) structure of every different transcription factor, gives each one its unique stable shape.

■ Third, proteins with unique shapes can recognize different sequences of nucleotides in DNA because each base pair has its own size and shape; every unique DNA-binding protein has its preferred "shape" (sequence) of DNA. This is how transcription factors find the TATA box, CAAT box, and so on.

■ Fourth, when any protein binds to its target, it changes shape. This is fundamental to understanding how almost every activity in a cell is controlled. In many cases, the new shape becomes a target for yet other proteins to bind. This can start a chain reaction of binding interactions that result in the formation of huge multiprotein complexes, such as the general transcription complex in Figure 12-13. As each new protein is added to the complex, the entire complex changes shape; therefore, there are many, many different overall shapes possible for any cluster of proteins. Also, according to the three traits of proteins, every different shape has its own functional state, which helps explain why so many proteins collect at or near the promoter of a gene: rather than simply turn a gene on or off (two choices), cells can carefully adjust the activity of a gene (i.e., the rate of its transcription) by making subtle changes (multiple choices) in the protein complex that controls it.

As shown, the key to understanding how a gene is controlled is understanding the proteins that bind to the core promoter, plus all of the other proteins that "pile on" the general transcription complex. Fortunately, these proteins are classified according to the kind of DNA sequence they bind to, and/or the impact they have on the functional state of the general transcription complex. Let's have a look at these important proteins next.

Activators and Repressors Control the General Transcription Factors

As their name suggests, **activators** promote the transcription of a gene by helping to activate RNA polymerase II. Alternatively, **repressors** are transcription factors that decrease transcription of a gene by either preventing activating proteins from binding to the transcription complex, or by binding to members of the complex and preventing them from adopting an active conformation; they accomplish the second method by binding to distinct regions of DNA called **regulatory sequences**. One of these, called the **regulatory promoter**, lies just upstream of the core promoter, seen in Figure 12-12. It contains two characteristic sequences. One of these is GGCCAATCT, and it is commonly called the **CAAT box**, and the other is called the **GC box** because its sequence is GGGCGG. The regulatory promoter binds to two classes of transcription factors. Because activators and repressors bind so close to the core promoter, they can often bind directly to the general transcription complex. In some cases, proteins called **coactivators** and **corepressors** bind to their respective activators or repressors to strengthen their impact; some of them serve as bridges between the activators or repressors and the general transcription complex. One example of a repressor is CAAT enhancer binding protein; when it binds to the CAAT box, it inhibits expression of genes required for cell growth, and is considered a tumor suppressor protein for this reason. Repressors attract enzymes called DNA methyltransferases (see Figure 2-18), which add methyl groups to cytosine bases, thereby silencing the associated genes.

Other sequences are shared by regulatory promoters in many different genes. One class of these sequences is called **response elements.** For example, the cAMP response element (CRE) is found in many genes that are sensitive to signaling pathways using cAMP as a second messenger. The CRE is bound by an activator called CRE binding protein, or CREB. CREB is activated when it is phosphorylated by protein kinase A. The glucorticoid response element (GRE) is bound by an activated glucocorticoid receptor, which binds to the GRE with zinc fingers.

Enhancer Sequences Bind Activators at a Distance from the Promoter

Enhancer sequences share similarities with regulatory promoters, in that they bind to the same kind of activators. In fact, they contain a higher concentration of regulatory sequences than the regulatory promoters, which means they are assembly sites for large activator complexes. They are also located hundreds or even thousands of base pairs away from the initiation start site of a gene. This presents something of a problem, however: If this complex forms so far away from the promoter, how does it control said promoter? The answer is quite simple, once we remember that DNA is not a linear molecule. So often we draw DNA as a straight line, for the sake of convenience and simplicity, but this has the potential to give the impression that, at the level of individual genes, DNA is simply "flat." By returning to the concept that the backbone of DNA is flexible because of the single covalent bonds that hold it together (see Figure 2-12), we can see that DNA can loop back on itself, as **FIGURE 12-16** shows. Two important proteins make this possible. One is responsible for inducing bending in the DNA backbone and then holding the bent DNA in place. The second is called a mediator; in

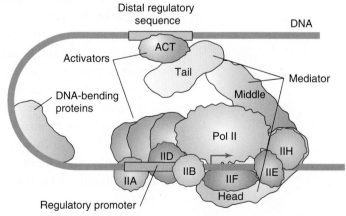

FIGURE 12-16 Mediator helps form an enormous protein complex near the transcription start site. DNA-bending proteins induce a loop to form, drawing the distant regulatory sequence close to the core promoter. Activators bound to the proximal promoter and distal regulatory sequence are held on top of the general transcription factors by the mediator complex. Adapted from S. Björklund and C. M. Gustafsson, *Trends Biochem. Sci.* 30 (2005): 240-244.

yeast, it contains 20 subunits that collectively span the distance between the cluster of activators bound to an enhancer and the general transcription complex. Together, the six general transcription factors at the core promoter, bound to additional transcription factors at the regulatory promoter and enhancer sequences and held together by the mediator, form the enormous protein complex that gives eukaryote cells such careful control over the expression of their genes.

12.4 ■ Signal Transduction Pathways and Gene Expression Programs Form Feedback Loops

Once a signaling pathway enters the nucleus and triggers a change in gene expression, the impact of this change is only felt once the expressed genes begin functioning. Several different types of RNAs do not need to be translated to begin functioning; snRNAs, microRNAs, and others begin work in the nucleus as soon as they are made. rRNAs must travel to the nucleolus, where they combine with ribosomal proteins to form ribosome subunits. tRNAs and mRNAs must be transported into the cytosol to begin the process of translation (for a review, see *Translation Occurs in Three Stages* in Chapter 8).

Because proteins are the key elements in signal transduction and gene expression, signaling pathways that trigger the expression of protein-coding genes can have a major impact on how these pathways function. This is known as feedback, and the proteins linked together this way form what is called a **feedback loop**; **FIGURE 12-17** illustrates examples of both negative and positive feedback loops in cells. A negative feedback loop is formed between MAP kinases and MAPK phosphatase (MKP). Phosphorylated, active MAPK enters the nucleus and initiates transcription of the MKP gene; MKP is synthesized and removes the phosphates from MAKP, inactivating it. Positive feedback occurs when a receptor called TLR1 activates a kinase called Jun kinase 1, which activates transcription of more TLR1 mRNA.

Cell biologists and molecular biologists have learned much about how individual proteins organize into pathways and feedback loops, but the full complexity of signal integration has yet to be fully understood. To help put the concept of cellular decision making in perspective, let's consider how one decision maker, a MAP kinase called ERK1/2, helps a cell decide whether to grow or differentiate. **FIGURE 12-18** shows three different types of signaling pathways that converge on ERK. GPCRs, receptor tyrosine kinases, and integrins all trigger ERK activation, but through different intermediates. On paper, it looks like all three of these pathways to ERK activation are equivalent, and should generate the same outcome, but experimental evidence suggests that *how* ERK is activated is just as important as its activation state. Clearly, our comprehension of how ERK functions is incomplete. Considering that ERK is better understood than most of the thousands of known signaling molecules, unraveling this web will keep us busy for years to come.

CONCEPT CHECK #2

As we will see in the next chapter, the mechanisms controlling cell growth are complex and require expression of several hundred different proteins, all acting at different times and locations in the cell. Yet some cells can be triggered to divide after binding to just one signal molecule at the surface. How can you explain this?

Negative feedback

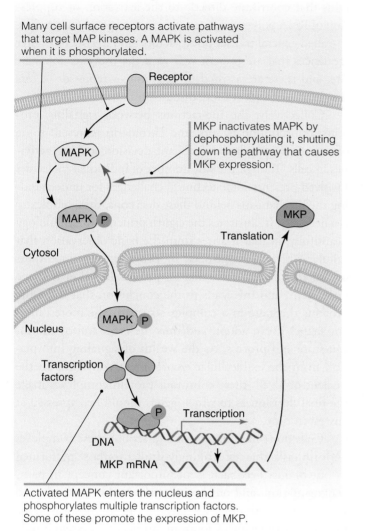

Many cell surface receptors activate pathways that target MAP kinases. A MAPK is activated when it is phosphorylated.

Receptor

MAPK

MKP inactivates MAPK by dephosphorylating it, shutting down the pathway that causes MKP expression.

MAPK P

MKP

Translation

Cytosol

MAPK P

Nucleus

Transcription factors

P

Transcription

DNA

MKP mRNA

Activated MAPK enters the nucleus and phosphorylates multiple transcription factors. Some of these promote the expression of MKP.

Positive feedback

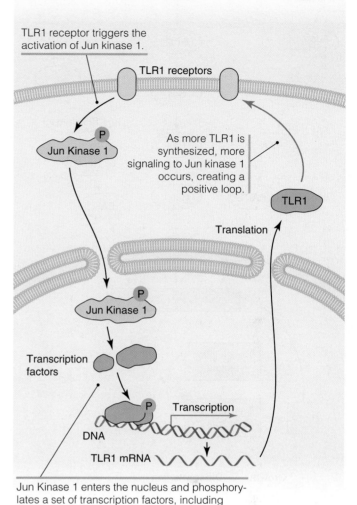

TLR1 receptor triggers the activation of Jun kinase 1.

TLR1 receptors

P

Jun Kinase 1

As more TLR1 is synthesized, more signaling to Jun kinase 1 occurs, creating a positive loop.

TLR1

Translation

P

Jun Kinase 1

Transcription factors

P

Transcription

DNA

TLR1 mRNA

Jun Kinase 1 enters the nucleus and phosphorylates a set of transcription factors, including those that promote the expression of TLR1.

FIGURE 12-17 Examples of positive and negative feedback loops in signal/gene expression pathways.

12.5 Chapter Summary

The purpose of this chapter is to illustrate how some signaling pathways that originate in the cytosol can impact the expression of genes in the nucleus. While a great deal is known about signaling molecules that function in the cytosol, relatively few that enter the nucleus are well understood. Recent evidence suggests that numerous signaling proteins and second messengers, previously thought to function only at the plasma membrane or in the cytosol, may be active in the nucleus as well. At this stage, we are still trying to unravel how many of these molecules might function in the nucleus. The diversity of signaling molecules discovered in the nucleus presents compelling, but incomplete, evidence that they might form nucleus-specific pathways. Once this issue is settled, it is likely that our current paradigms of signal transduction in the nucleus will be significantly revised.

The nuclear regulatory proteins that are best understood are those that bind directly to chromatin. Some simply bind to specific sequences of DNA without changing them, while others are enzymes that add or remove chemical modifications of histones and DNA. Phosphorylation, acetylation, methylation, and ubiquitination are the most common modifications, and these either promote or suppress expression of genes near the modifications.

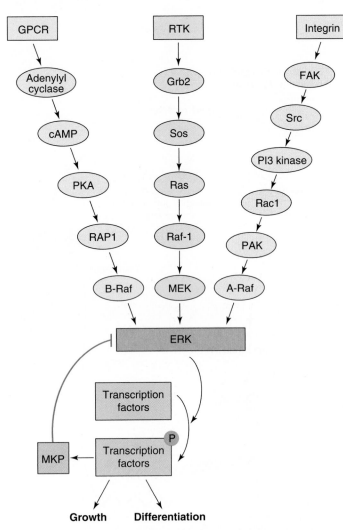

FIGURE 12-18 ERK is a decision-making protein. Three different signaling pathways converge and activate ERK, which then phosphorylates transcription factors leading to cell growth or cell differentiation. MKP provides negative feedback for ERK activity.

Most of these modifications take place on histones. The proteins that contribute directly to the activation or suppression of RNA polymerase are called transcription factors. The so-called general transcription factors bind to core promoter sequences that are present very near the transcription start site, and these are assisted by activator or repressor factors that bind to DNA at more distant sites.

Collectively, the interactions between signaling proteins, transcription factors, and chromatin represent some of the most complex and elegant decision making activities in cells. Though the vast number of individual molecules involved presents some technical challenges for understanding the mechanisms behind these decisions, this complexity is a necessity. It illustrates the eighth principle of cell biology beautifully, in that it arises from the basic observation that cellular activity is controlled by proteins. The three traits of proteins demonstrate that all protein function is shape dependent, and this leads to the conclusion that the more proteins there are in a complex structure, the more different shapes it can adopt, and thus the more information it can store and process. As the wealth of signaling information from the extracellular environment converges on the nucleus of a cell, these enormous protein complexes make the final decision as to which genes should be expressed at any given time.

Cells don't have brains, of course, but these complexes perform tasks that are strikingly similar to those performed in our brains. Feedback is an important concept in pathway regulation, and our brains perform this routinely and with such dexterity that we rarely notice. In most cases, the importance of feedback loops in our nervous system is only apparent when it goes awry, such as when we trip over our words when speaking, or if we temporarily lose our balance. Cells, too, are remarkably well balanced. In the next chapter, we'll examine two of the most important decisions these complexes have to make: (1) when to halt nearly every routine cellular activity, and (2) when to trigger the processes of cell replication or cell death.

CONCEPT CHECK ANSWERS #1

Phosphorylation of nuclear proteins requires activation of protein kinases. Since GPCRs are not kinases, they control kinase activity indirectly via signaling proteins and second messengers. Plasma membrane GPCRs can activate at least three different protein kinases by different pathways. One uses heterotrimeric G proteins to activate a phospholipase that generates IP_3 and diacyl glycerol. IP_3 could bind to a calcium channel, releasing calcium from the intermembrane space. These molecules, in turn, stimulate protein kinase C. All elements of this pathway have been found in the nucleus, though it is not clear which GPCRs and their associated G protein subunits might trigger it in any single cell type. Table 12-1 lists candidate targets of PKC in the nucleus.

A second possible pathway activates PKB/Akt by generation of PIP_3. PIP_3 is created by phosphatidylinositol kinases, and at least one of them, PI3 kinase, is a target of a heterotrimeric G protein. As with the PKC pathway, all of the necessary elements have been identified in the nucleus, but the exact identities of the players involved has not been identified for a specific cell type.

A third possible pathway could activate PKA. GPCRs can activate PKA by triggering adenylyl cyclase activity via heterotrimeric G proteins. All of the elements necessary for this pathway to exist have been identified in nuclei, but it is still not clear if they are all present in the same nucleus. Unraveling nucleus-specific versions of any of these pathways is seriously complicated by the fact that many of these molecules are present and activated in the cytosol as well, leaving the possibility that the cytosolic and possible nuclear pathways overlap in the nucleus. This means it is very difficult, for example, to determine whether a cAMP molecule is generated directly in the nucleus or diffuses in after being made in the cytosol.

CONCEPT CHECK ANSWERS #2

The first issue to address is how a single molecule can affect an entire cell. An important feature of most signaling pathways is signal amplification: binding to a single signal could trigger the activation of hundreds, thousands, or even millions of downstream signaling molecules. However, regardless of how much amplification takes place, the products of a single signaling pathway are limited. To increase the diversity of the signal, we can invoke the branching network that links signaling pathways together; with branching, the original signal can be converted into several different signals. Finally, we have to explain how hundreds of proteins can be expressed in response to a single initial stimulant. To help explain this, we can use response elements, such as CRE or GRE, which are found in the promoters of many different genes. Finally, we know that at least some of the genes being triggered by the initial signal may encode transcription factors, which can then trigger a second (and third, and so on) wave of gene expression. Each new wave is driven by the transcription factors activated by the previous round of expression. We will use this wave-after-wave concept in the next chapter.

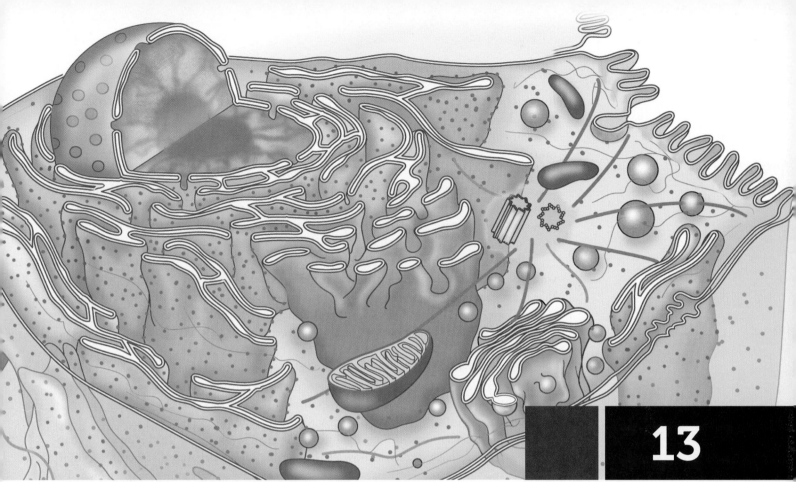

The Birth and Death of Cells

13.1 The Big Picture

In this chapter we continue integrating the subjects discussed in previous chapters to create a coherent, top-down view of how and why cells behave. Recall that one of the missions of this book is to discuss cells while avoiding inert facts; to be meaningful, a fact must have a rationale associated with it—a reason to exist, in effect—that helps answer an important question. So, we will begin this chapter with the most important question a cell ever has to answer: whether to continue living or die. We will access virtually all of the subject matter in previous chapters to help us understand how cells make this decision. This strategy gives rise to cell biology principle #9: progression through the cell cycle is the most vulnerable period in a cell's life.

The reason for this is clear: faithful duplication of its entire structure, culminating with cell division, leaves a cell little room for error, and the nature of inheritance ensures that most genetic errors are passed on to the progeny. A corollary to this principle is that in most cases, a healthy cell would rather die than pass serious damage on to its daughter cells. Principle 3—DNA integrity is the top priority for all cells (Chapter 7)—is especially applicable here.

As they embark on their journey of replication and division, cells are constantly checking their progress and actively deciding whether to continue on, or give up and die, if they make a mistake. Let's look at some examples of when this decision making occurs in multicellular organisms, to illustrate this vulnerability:

- First, cells must consider the physical challenges they face. In the previous chapters, we have discussed how cells assemble highly organized structures that are specialized to help balance outside forces and generate some of their own. For example, cytoskeletal proteins form filaments that regulate cell shape, internal membrane trafficking, and resistance to mechanical forces (see Chapters 5 and 9). In multicellular organisms, cell–cell and cell–ECM interactions play important roles in defining where and how a cell contributes to the overall function of the organism (see Chapter 6). But cell division via mitosis is a solo act; cells must decide to release all of their attachments to other cells, and most of their attachments to the ECM, so they can round up and form the contractile ring. Losing these structures makes cells especially vulnerable to physical forces. In effect, most of the beneficial traits of multicellularity fail to help a cell when it is dividing, and some may even be dangerous; the fluid shear caused by circulating blood can injure or rupture dividing cells, for example.

- Second, recall from Chapter 7 that DNA is completely replicated, and then tightly compacted during cell division (see *Prophase Prepares the Cell for Division*). This means that cells must decide to make most of the information stored in DNA inaccessible during cell division. This limits the number and type of responses a cell can generate to challenges at that time.

- Finally, the faithful replication of an entire cell requires innumerable enzymatic reactions to occur in a carefully choreographed sequence, so innumerable complications could arise if a cell fails to follow the sequence properly. The energy transduction and signal transduction pathways covered in Chapters 10 and 11 are good examples of the complex machinery that must be reproduced with tremendous accuracy. Many anticancer drugs capitalize on this principle, in that they target one or more of these vulnerabilities.

To illustrate cell biology principle #9, we'll cover two major topics, each discussed in a separate section:

- The first explores how cells make the decision to begin moving through the stages of replication, and why some cells never make this journey. Principle #8 from Chapter 12 is especially applicable here, as the focus is on decision-making protein complexes that decide when a cell begins its journey. These decision-making complexes are arranged in hierarchical order and in a specific temporal sequence in the cell, and most of this first section is devoted to how these decisions are made. To help keep them in order, they are grouped into 10 numbered and named steps. This section is intended to be read more than once. I suggest that during the first reading, skim through this section to get the overall idea, and then, once the connections between the major ideas (i.e., the minor subject headings) are clear, go through it again to add the facts. And, as before, focus on *concepts before names*. To keep the material in perspective, we'll follow a storyline, much like the one we used in Chapter 10. If at any time the material seems confusing (or inert), return to the basic question: How is this helping a cell decide whether to live or die?

■ The second section concerns how cells decide to die. It is much shorter than the first section, for a reason: rather than probe the depths of cell death at the same level of detail, I prefer that we focus on how cells decide to live (and grow), then view death as the alternate choice (when conditions dictate) for the decision-making complexes we discuss in the first section. By the time we reach the end of the chapter, my goal is for the reader to be able to randomly flip to any page in the earlier chapters and generate a rationale for how the subject on that page pertains to some aspect of this chapter. The focus questions in this chapter will provide some examples to help get us started.

13.2 New Cells Arise from Parental Cells That Complete the Cell Cycle

KEY CONCEPTS

▸ Cells divide by following a carefully scripted program of molecular events collectively called the cell cycle.

▸ The cell cycle is subdivided into five phases named G_1, S, G_2, M, and G_0. Cells not actively dividing reside in G_1 or G_0 phase.

▸ Progression through the cell cycle is under the control of proteins that form checkpoints to monitor whether the proper sequence of events is taking place. Cells halt at these checkpoints until they complete the necessary steps to continue.

▸ The G_1/S checkpoint, called the restriction point or start point, is where cells commit to completing cell division. Proteins called cyclins play an important role in advancing cells through checkpoints.

▸ Cell division takes place in 10 steps.

One of the first challenges facing the early cell biologists was determining where cells come from. Given what we know today, this question may seem trivial, but the answer was far from obvious to scientists that had yet to understand how sperm and eggs gave rise to complex organisms. Nearly 200 years passed between the discovery of cells by Robert Hooke and the concept that cells arose from preexisting cells, formally proposed by Rudolf Virchow. It took nearly 100 more years to develop plausible explanations for how the very first cell formed on earth, though this question is still not conclusively answered. We have yet to find any means for converting a combination of simple molecules into a complete living organism *de novo* ("from scratch") (**BOX 13-1**).

Leaving aside the question of the origin of the first cell, we now know a great deal about how one parental cell can replicate to form two daughter cells. But we are far from understanding the complete picture: biologists estimate that about 10 million different organisms currently exist on earth, and many of these are composed of several different cell types (for example, humans contain over 200 different cell types). It is now clear that the mechanisms controlling the division of these cells are not identical. Nonetheless, the similarities between cell types are sufficient to allow us to build an integrated, albeit simplified, model of cell division. Consistent with their taxonomic classifications, prokaryotes use mechanisms that are quite different from eukaryotes, and differences within eukaryotes likewise reflect their evolutionary history. Multicellular organisms use mechanisms not found in unicellular organisms, and plants use mechanisms not found in animals. Even different cells within the same organism use their own unique machinery to control when and how they divide (**BOX 13-2**).

Faced with this vast amount of information, we must tackle a problem similar to that confronting the first modern cell biologists: What are the *fundamentals* of cell division? Rather than simply listing the most common features of cell division, let's restate the question: What is the machinery cells use to decide how and when to divide? This allows us to filter the plethora of data, such that we look for extensions of the signaling and gene expression mechanisms discussed in the previous two chapters.

The Cell Cycle Is Divided into Five Phases

In the most simplified model of cell division, cells are assumed to live forever. Their life cycle, therefore, spans the period from their appearance (resulting from mitosis and cytokinesis [see *Cytokinesis Completes Mitosis by Partitioning the Cytoplasm to Form Two New Daughter Cells* in Chapter 7] of their parental cell) to their division, when they generate two daughter cells. Because the daughter cells possess the same abilities to grow and divide, biologists view the lifespan of a cell as a *repeating* cycle.

The formal name for the events that control the growth and division of a eukaryotic cell is the cell cycle. By convention, the cell cycle is represented by a circle that lists essential events in a clockwise, chronological order, as shown in **FIGURE 13-1**. The circle is divided into sections called **phases**, and the events in one phase cannot begin until the events in the phase preceding it are complete. One turn of the cycle begins at the start of the G_1 phase, and ends with the completion of M phase. The G_1, S, and G_2 phases are sometimes collectively referred to as **interphase**, to distinguish them from the radical structural changes that occur in most cells during M phase.

Note that many of the cellular activities we have covered in previous chapters are represented in Figure 13-1: cell growth, including synthesis of new membranes (Chapter 4) and new proteins (Chapter 8) necessary for cell division, takes place in the G_1 and G_2 phases; DNA replication (Chapter 7) occurs in S phase; and mitosis and cytokinesis (Chapter 7) define the M phase. Accordingly, we will not discuss these subjects in detail here, but will reference specific sections from other chapters when appropriate.

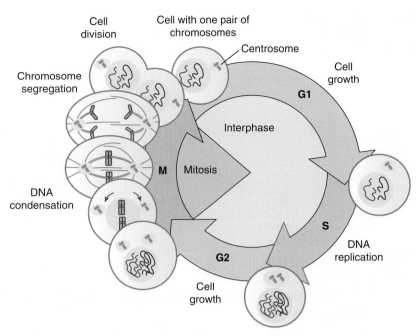

FIGURE 13-1 The cell cycle. Cells reside in G_1 (or G_0, not shown) until stimulated to grow, then progress through the cycle in clockwise order. Phases G_0, G_1, S, and G_2 are collectively called interphase.

■■ "Resting" Cells Reside in G_0 or G_1 Phase

The fifth phase of the cell cycle, called G_0, is technically not part of the loop, as it represents a state of arrest. Some cells enter G_0 under conditions of environmental stress (e.g., nutrient deprivation) and then return to G_1 once conditions are favorable, while other cells enter G_0 as a protective measure to prevent them from even attempting to divide. For example, most fully differentiated neurons and skeletal muscle cells would lose their ability to function if they retracted their elaborate cytoskeleton, rounded up, and divided, so they reside in G_0. Cells in G_0 are assumed to live in a condition resembling a permanent G_1 state. Other names for cells in G_0 include *senescent* (from the Latin word for "growing old"), *quiescent* (Latin for "quiet"), and *resting*. While these names may apply to some G_0 cells, they should be used with caution, as they suggest the possibility that all cells in G_0 must be metabolically inactive, unresponsive to signals, or otherwise nonfunctional. Keep in mind how much work nerves and skeletal muscles do before drawing this conclusion.

■ Several Checkpoints Define Critical Decision-Making Events in the Cell Cycle

Cells that progress through the cell cycle must perform a daunting number of tasks, as we will discuss below. Many of these tasks must be performed in a strict order: for example, DNA must be *fully* replicated before it can be condensed into the tightly bundled chromosomes that move on the mitotic spindle. To help ensure that the critical steps are done properly before progressing, cells use clusters of proteins that validate successful completion of these steps, then actively promote the steps immediately after them. These protein complexes can also halt progression of the cycle if problems are detected, so, for example, if errors in DNA that arise during replication (S phase) are not repaired, these proteins will prevent the onset of DNA condensation (G_2 phase) until the errors are corrected.

The periods in the cell cycle where this quality control takes place are called **checkpoints**. Dozens of checkpoints have been identified so far, but not all of them are well understood. (It is often easier to observe a halt in the cell cycle than to understand what causes it.) Many are finely tuned to detect very specific problems: different checkpoint proteins examine single- and double-stranded DNA breaks in S and G_2, for example. One easy way to keep track of them is to simply classify checkpoints according to where in the cell cycle they function. **FIGURE 13-2** illustrates examples of major checkpoints in each phase of an actively growing cell.

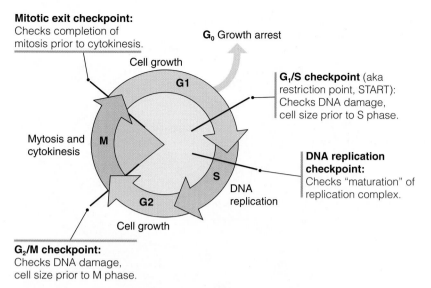

Mitotic exit checkpoint:
Checks completion of
mitosis prior to cytokinesis.

G₂/M checkpoint:
Checks DNA damage,
cell size prior to M phase.

G_0 Growth arrest

G₁/S checkpoint (aka
restriction point, START):
Checks DNA damage,
cell size prior to S phase.

**DNA replication
checkpoint:**
Checks "maturation" of
replication complex.

FIGURE 13-2 Checkpoints control progression through the cell cycle. Some of the major checkpoints are shown.

The G₁/S Checkpoint Is the Point of No Return

Not all checkpoints have the same amount of influence on the cell cycle. While all checkpoints have the ability to halt it, most can only do so until their specific "problem" is fixed. However, one checkpoint near the G_1/S boundary, called the **restriction point** in multicellular organisms and the **START point** in yeast, has ultimate control over whether cells will complete the cycle.

The power of this checkpoint can be easily demonstrated by doing an experiment with yeast cells. In low-nutrient conditions, or in the presence of signals called mating factors, yeast cells stop progressing through the cell cycle, but only if they are in early or mid-G_1 phase. Under these conditions, they can easily remain in G_1 for several days. If they are in any other part of the cycle, they will continue through until they reach G_1, and then stop. If nutrients are added and/or the mating factors removed, the cells will resume cell cycle progression. As long as they remain in early to mid-G_1, swapping these environmental conditions will start and stop the cells no matter how many times the conditions change. This behavior continues until the cells reach a specific point in late G_1; at this point, removing nutrients and/or adding mating factors no longer stops them, and the cells progress through the complete cell cycle, stopping again only after the cells cycle back to G_1.

This experiment demonstrates that there is a single "point of no return" in the cell cycle, and this lies in late G_1. Once cells cross this point, they are committed to completing the entire cycle. Any interruptions triggered by other checkpoints are temporary, and persist only as long as it takes to fix the problems that activated them. Resting at these checkpoints is not an option (**BOX 13-3**).

The G₂/M Checkpoint Is the Trigger for Large-Scale Rearrangement of Cellular Architecture

The first checkpoint that was discovered controls the progression from G_2 to M phase, because the transition (remodeling of the cytoskeleton, dissolution of the nuclear envelope, and so on, that takes place during mitosis [see Chapter 7]) is easy to see with a light microscope. Put simply, because it is fairly easy to see the onset of M phase, it is equally easy to identify a window of time when the transition to M is controlled. Over the course of several years, scientists learned that the decision-making complex for entering M phase was composed of proteins that were only active during a specific "window" in the cell cycle. These proteins were initially called **mitosis promoting factor**, or **MPF**.

FIGURE 13-3 illustrates two of the early experiments characterizing MPF. In one, triggering mitosis in one cell (an immature frog egg) with the hormone progesterone causes a change in the cytosol, such that simply injecting a small amount of that cytosol into a second, unstimulated cell induces mitosis of that cell, too. What is especially remarkable is

BOX 13-3 TIP

A simple way to remember that cells cannot rest at the temporary checkpoints after G_1 is to create a "cell cycle law": *no healthy cells ever stop in S, G_2, or M phase.*

that cytosol collected from the second cell when it is in M phase can induce mitosis in a third cell, and so on. This demonstrates that a cytosolic factor (MPF) is capable of inducing mitosis. Importantly, the observation that this effect is not reduced by dilution through serial injection into other cells suggests that MPF initiates a change in the cytosol of the second cell that can be passed on to other cells. We'll see how this occurs in the sections below. The second experiment shows how the activity of MPF in cells varies over the length of a cell cycle, and peaks only during mitosis. This cyclic activity pattern is characteristic of the key proteins that regulate progression through the cell cycle.

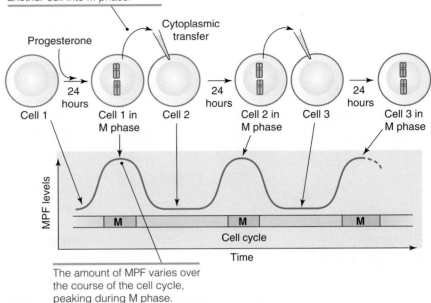

Injection of cytosol from one cell in M phase is sufficient to drive another cell into M phase.

The amount of MPF varies over the course of the cell cycle, peaking during M phase.

FIGURE 13-3 Early experiments characterizing the activity of mitosis promoting factor (MPF).

Activation of Cyclin-CDK Complexes Begins in G_1 Phase

Later experiments focused on isolating MPF and similar protein complexes that control phases of the cell cycle. Scientists learned that MPF is composed of two proteins that only function when they are bound to each other and activated by the appropriate signaling proteins. One of these proteins is called a **cyclin**, because its expression level changes dramatically (i.e., cycles) through the cell cycle, peaking only when it is required, as **FIGURE 13-4** shows. While a single cyclin is sufficient to control the entire cell cycle in some yeasts, mammalian cells contain many different cyclins that regulate distinct events. Cyclins bind to **cyclin-dependent kinases** (**CDKs**) that phosphorylate numerous other proteins, thereby controlling their activity. Mammalian cyclin-CDK complexes also contain two additional polypeptides to form a **quaternary complex**.

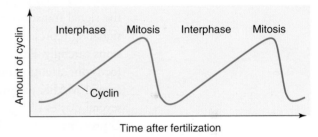

FIGURE 13-4 Scientists discovered the first cyclins when they noted that high cyclin levels were associated with the onset of mitosis in embryos. Cyclin levels dropped sharply after mitosis.

Humans have at least 12 different cyclin-CDK complexes, resulting in a somewhat complicated naming scheme that reflects the fact that many different scientists studied these proteins independently before realizing they were related. For example, MPF has been renamed cyclin B/CDK1, in accordance with the current standard whereby cyclins are assigned letter names (when necessary, subdivided by numbers) such as A1, A2, B1, C, and so on, and the CDKs are assigned numbers. In yeast, CDKs are also commonly referred to as **cell division cycle** (**cdc**) proteins, so most CDKs have a cdc name as well. CDK1 is also called cdc2, for example. For simplicity, most scientists now use a generic naming scheme: cyclins A1, A2, B, C, and so on, and CDK1, CDK2, CDK3, and so on. In this scheme, MPF is called cyclin B/CDK1 (**BOX 13-4**).

BOX 13-4 TIP

The names given to cyclin-CDK complexes are an excellent example of *not* following the naming scheme we discussed in BOX 3-4 Despite the fact that neither cyclins nor CDKs can function in isolation (and are therefore *subunits*), they are nonetheless called proteins.

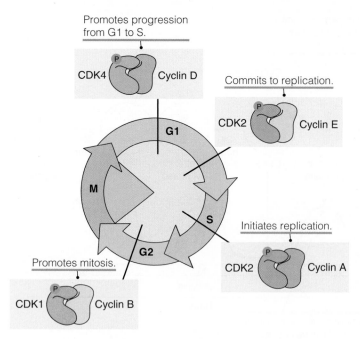

Promotes progression from G1 to S.

CDK4 — Cyclin D

Commits to replication.

CDK2 — Cyclin E

Initiates replication.

CDK2 — Cyclin A

Promotes mitosis.

CDK1 — Cyclin B

FIGURE 13-5 Distinct cyclin-CDK complexes control progression through cell cycle checkpoints.

FIGURE 13-5 highlights some of the key activities in the cell cycle that are controlled by cyclin-CDK complexes. These proteins are only part of an intricate, multilayered mechanism for defining checkpoints, sensing conditions inside and outside of the cell, and orchestrating the reorganization of cells as they move through each phase of the cycle. To navigate through this complexity, we will follow cyclin-CDKs in a temporal sequence of 10 steps, tracing progrowth stimuli in the extracellular environment to the cell interior, then examining the sequence of events that carry a cell from G_1 phase through the complete cell cycle. Along the way, we will review details covered in previous chapters, and use the cell cycle as a theme for linking these concepts together.

■■ Cell Cycle Step 1: Signal Transduction Initiates Cell Cycle Progression

Once a cell passes the restriction/START checkpoint, a global replication program takes charge. The best understood signals that initiate this program are the growth factor proteins and hormones we discussed in Chapter 11 (see *Protein Tyrosine Kinase Signaling Pathways Control Cell Growth and Migration* and *Steroid Hormones Control Long-Term Cell Behavior by Altering Gene Expression*); in fact, most of the signal transduction we covered in Chapter 11 takes place during G_1. Remember that signaling represents a form of cellular decision making, and healthy cells plan their divisions carefully. Rather than follow all of the signaling pathways in Chapter 11 here, we will focus our attention on those containing MAP kinases (MAPKs). Mammalian cells contain several MAPK pathways, as shown in **FIGURE 13-6**. Each pathway contains members of several classes of signaling proteins, including receptor tyrosine kinases, adaptor proteins, G proteins, and protein kinases.

FIGURE 13-7 shows a simplified version of a MAPK signaling pathway triggered by platelet derived growth factor (PDGF). PDGF binds to a receptor tyrosine kinase and activates a signaling pathway very similar to that shown in Figure 11-19. To respond to PDGF and similar peptide growth factors, most healthy cells must maintain contact with an ECM substrate, such as the focal complexes discussed in Chapter 6 (see *Specialized Integrin Clusters Play Distinct Roles in Cells*). These complexes serve as signaling scaffolds, as shown in Figure 11-10. Integrin receptors in focal complexes recruit signaling proteins capable of activating several pathways, including those containing MAPKs. In many cases, the receptor tyrosine kinases are contained in focal complexes, so both the ECM and growth factor signaling pathways form a single, cooperative decision-making complex on the cell surface.

■■ Cell Cycle Step 2: Changes in Gene Expression Are Required for Progression through the Restriction Point

Progression through the cell cycle requires expression of a large number of genes; in budding yeast, approximately 15% of the entire genome (corresponding to approximately 800 genes) is activated during the cell cycle, including an average of 300 genes during G_1/S transition. While the specific set of genes necessary varies from one cell type to another, the pattern initially observed for MPF applies to the restriction point as well. Figure 13-5 shows that progression through the restriction point in mammalian cells requires activation of at least

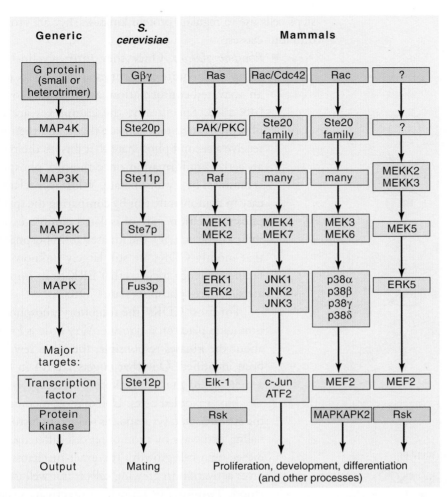

Generic	S. cerevisiae	Mammals			
G protein (small or heterotrimer)	Gβγ	Ras	Rac/Cdc42	Rac	?
MAP4K	Ste20p	PAK/PKC	Ste20 family	Ste20 family	?
MAP3K	Ste11p	Raf	many	many	MEKK2 MEKK3
MAP2K	Ste7p	MEK1 MEK2	MEK4 MEK7	MEK3 MEK6	MEK5
MAPK	Fus3p	ERK1 ERK2	JNK1 JNK2 JNK3	p38α p38β p38γ p38δ	ERK5
Major targets: Transcription factor	Ste12p	Elk-1	c-Jun ATF2	MEF2	MEF2
Protein kinase		Rsk		MAPKAPK2	Rsk
Output	Mating	Proliferation, development, differentiation (and other processes)			

FIGURE 13-6 The family of mitogen-activated protein kinases (MAPKs) and their upstream regulatory proteins. This extensive network controls expression of hundreds of genes in yeast (*S. cerevisiae*) and mammals. The generic names of each level of control are shown at left. MAPK2 = MAP kinase kinase; MAPK3 = MAP kinase kinase kinase, and so on.

two cyclin-CDK complexes: cyclin D1 / CDK4 (or CDK6) and cyclin E / CDK2. The expression of most CDKs does not vary much throughout the cycle, but without their corresponding cyclins, they are not functional. Cyclin E is normally expressed in very low amounts in quiescent cells, but numerous progrowth signals, including ERK activation by receptor tyrosine kinases and integrin-based complexes, trigger its transcription. As such, the signaling necessary to begin the cell cycle must pass into the nucleus and interact with the transcription regulatory proteins we discussed in Chapter 12. Figure 13-7 summarizes this for a MAPK pathway, with a single arrow indicating phosphorylation of the transcription factor Jun. For a more detailed view of MAPK nuclear signaling, see Figure 12-5.

■■ Cell Cycle Step 3: Pro- and Antigrowth Signaling Networks Converge at the G₁/S Cyclin-CDK Complexes

As cyclins D, E, and A accumulate in the cytosol, they bind to their respective CDKs. This dimerization changes the shape of the kinases, allowing access to their substrate(s), while not being sufficient to activate them. Given the enormous consequences, the decision to progress past the restriction point cannot be this simple and, in fact, the mechanisms controlling activity of restriction point CDKs represent some of the most complicated decision-making

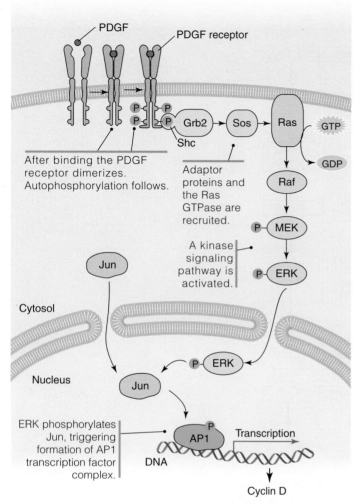

After binding the PDGF receptor dimerizes. Autophosphorylation follows.

Adaptor proteins and the Ras GTPase are recruited.

A kinase signaling pathway is activated.

ERK phosphorylates Jun, triggering formation of AP1 transcription factor complex.

FIGURE 13-7 A simplified MAP kinase signaling pathway. MAP kinase enters the nucleus, phosphorylates transcription factors, and thereby controls gene expression.

steps cells use to regulate protein kinases. They are grouped into four classes:

- *Phosphorylation.* Once they form, cyclin-CDK complexes require covalent modification to adopt an activated conformation, as illustrated in **FIGURE 13-8**. Specifically, the complexes are first phosphorylated to inactivate them, and then they receive a second phosphate that pushes them into an active conformation once the first phosphate is removed by a phosphatase. While it is relatively easy to demonstrate this by comparing the specific phosphorylation state of isolated complexes with their kinase activity, the kinases and phosphatases that modify CDKs are still largely unknown. So far, no evidence suggests that ERK or its immediate substrates phosphorylate CDKs.

  For most CDKs, the inhibitory phosphorylation takes place on a tyrosine. Very little is known about the kinases responsible, though a few have been identified. One that targets CDK4 in G_1 is a member of the src family of nonreceptor tyrosine kinases, called c-yes. Unlike most members of the src kinases, c-yes responds to growth factor signaling pathways by promoting cell differentiation rather than cell growth. The exact mechanism of c-yes activation in growing cells is not well understood. Two kinases known to inactivate cyclin-CDK complexes at the G_2/M checkpoint are called Wee1 and Chk1. They phosphorylate the equivalent tyrosine on CDK2, halting mitosis. So far, only a handful of phosphatases are known to remove the inhibitory phosphate from the tyrosine on CDKs.

  The best known activating kinases are the aptly named **cyclin activating kinases (CAKs)**. Budding yeast (*Saccharomyces cerevisiae*), fission yeast (*Schizosaccharomyces pombe*), and mammals express one structurally distinct CAK each. In vitro, CAK phosphorylates a threonine amino acid in the kinase activation loop near the active site of several CDKs, including CDK1, 2, 3, 4, and 6, when they are bound to their respective cyclins. This causes a conformational change in the target cyclin/CDK, permitting easy substrate access to the active site. Experimental evidence that CAKs do this *in vivo* is still not conclusive.

- *Binding by kinase inhibitory proteins.* These proteins bind to individual CDKs or cyclin-CDK complexes as shown in Figure 13-8, and inhibit them in a number of different ways. They are named primarily by their apparent size in kilodaltons, followed by acronyms of their specific targets or functions. Examples include the INhibitors of cdKinases (INKs, e.g., p15[INK4B], p16[INK4A], p18[INK4C], p19[INK4D]), the Cdk Interacting Proteins (CIPs, e.g., p21[CIP1]), and the cyclin-dependent Kinase Inhibitor Proteins (KIPs, e.g., p27[Kip1] and p57[Kip2]).

  The known INKs target CDK4 and CDK6 complexes specifically, and use at least two means to inhibit their activation: first, in vitro, p16[INK4A] competes with cyclin D for binding to CDK4 and CDK6, preventing the formation of the

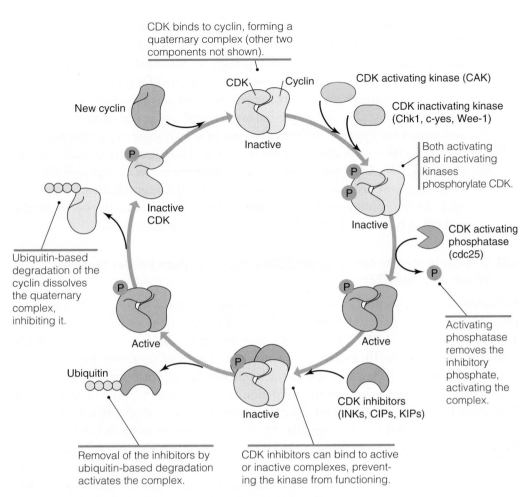

FIGURE 13-8 Summary of the cyclin/CDK activation-inactivation cycle.

quaternary complex and serving as one of the first mechanisms cells could use to arrest progression through the G_1/S checkpoint; and second, the INKs also bind to intact complexes, inhibiting their kinase activity.

Curiously, p21^{CIP1} plays a dual role in controlling CDK activity. One copy of the p21^{CIP1} protein is part of the quaternary complex formed by most cyclin/CDKs, and it serves to *activate* the complex. Later in G_1, as more p21^{CIP1} accumulates in the cell, additional copies of the protein attach to the quaternary complex, *inhibiting* it. The KIPs bind only to fully assembled quaternary complexes, and physically prevent CAK from activating the CDKs.

- *Subcellular location.* One straightforward means of preventing CDK activation is to physically separate cyclins from their CDK partners. Cyclins and CDKs are synthesized in the cytosol and enter the nucleus through nuclear pores. A few cyclins routinely pass into and out of the nucleus when they are not actively engaged in cell cycle activities, though it is not clear how. By mechanisms that are not fully understood, this shuttling comes to halt at the proper time for the appropriate cyclin-CDK complex to form in the nucleus. The best understood example of this occurs at the G_2/M checkpoint, when phosphorylation of cyclin B causes it to accumulate in the nucleus and form a complex with CDK1.

- *Protein degradation.* Another fairly simple means of preventing CDK activitation is to degrade components of the quaternary complex, as shown in Figure 13-8. Because these proteins are synthesized in the cytosol, they are subject to degradation by proteasomes (see *Proteins in the Cytosol and Nucleus Are Broken Down in the Proteasome* in Chapter 3). Proteasomes are most commonly found in the cytosol, and are also located in the nucleus, though the mechanism(s) governing their nuclear import are not well understood. To be targeted for proteasomal degradation, a protein must have several copies of the small protein ubiquitin attached to it. The mechanism for identifying and tagging proteins for degradation is complex and carried out by many different proteins; in at least some cases, ubiquitination of proteins takes place directly in the nucleus, and is triggered by cytosolic signals. Cyclins, CDKs, CDK inhibitors, and assembly proteins are all degraded this way, and the rate of degradation varies over time for different regulators of the cell cycle. In fact, proteasomal degradation is critical for maintaining the cyclic rise and fall of cyclin and CDK inhibitor levels in actively growing cells.

▪▪ ■ Cell Cycle Step 4: Active Cyclin/CDKs Phosphorylate Pocket Proteins, Which Activate E2Fs

Following assembly, removal of the inhibitory phosphate, and addition of the activating phosphate, the cyclin D / CDK4 (or cyclin D / CDK6) quaternary complex becomes fully active. As shown in **FIGURE 13-9**, one of the most important substrates for this complex is a protein called Rb. The Rb protein is a member of a family of regulatory proteins called **pocket proteins**, and is classified as a tumor suppressor because its primary function is to stop unregulated passage through the restriction point, as often happens in tumor cells. (The name Rb is an abbreviation for the cancer retinoblastoma, which often results when this protein is mutated.) In resting cells, Rb and other pocket proteins bind to a family of regulatory transcription factors called E2Fs, inactivating them. Phosphorylation by CDK causes pocket proteins to change shape, releasing the E2Fs. Mammals contain eight E2F genes, giving rise to nine different E2F proteins.

E2Fs are some of the most critical decision-making proteins in the cell cycle. They are sometimes referred to as the "tipping point" of the cell cycle, because their activation can promote either *growth* (progression through the restriction point) or *death*. (We will pursue cell death later in this chapter.) Consequently, their activities are very tightly regulated, by a number of different mechanisms. Three of the most important are:

- *Dimerization.* E2Fs must form dimers to bind DNA. Most (E2F1–6) can form heterodimers with a family of proteins called **dimerization partners** (**DP**, numbered 1–4). Thus, the relative abundance and type of dimerization partners at any given time can affect what complexes E2Fs form, and consequently which genes they will impact. The dimers bind to specific DNA sequences in the regulatory promoters of hundreds of genes, activating expression of some and inhibiting expression of others.

- *Positive feedback loops.* **FIGURE 13-10** shows examples of some of the positive feedback loops triggered by E2Fs. In particular, these dimers can

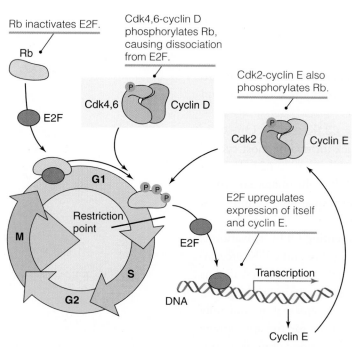

FIGURE 13-9 The transcription factor E2F is inactivated by Rb binding. When Rb is phosphorylated, it cannot bind to E2F, thus allowing E2F to upregulate expression of numerous genes including itself and cyclin E. CDK2 / cyclin E can then phosphorylate more Rb, resulting in an amplification loop.

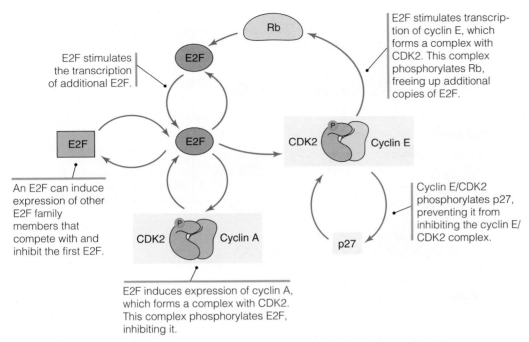

E2F stimulates the transcription of additional E2F.

E2F stimulates transcription of cyclin E, which forms a complex with CDK2. This complex phosphorylates Rb, freeing up additional copies of E2F.

An E2F can induce expression of other E2F family members that compete with and inhibit the first E2F.

Cyclin E/CDK2 phosphorylates p27, preventing it from inhibiting the cyclin E/CDK2 complex.

E2F induces expression of cyclin A, which forms a complex with CDK2. This complex phosphorylates E2F, inhibiting it.

FIGURE 13-10 Examples of positive (green) and negative (red) feedback loops controlling E2F function.

promote the expression of E2F genes, creating more E2F proteins. In addition, E2F complexes can induce expression of cyclin E, which complexes with CDK2 and leads to more phosphorylation of pocket proteins and CDK inhibitory proteins (e.g., p27^{Kip1}), further increasing the pool of active E2Fs. These positive feedback loops accelerate E2F expression.

■ *Negative feedback loops.* E2Fs can also induce expression of cyclin A, as shown in Figure 13-10. When it binds CDK2, the resulting complex can bind to an E2F, then phosphorylate the DP it is attached to, thereby preventing the heterodimer from binding DNA efficiently.

Exactly how E2Fs enhance expression of some genes while suppressing expression of others is still not clear. One model, shown in **FIGURE 13-11**, illustrates how E2F complexes can repress and activate expression of the same gene. According to this model, during early G_1, an E2F dimer attached to a pocket protein binds to the regulatory promoter and forms a binding site for repressor transcription factors; this, in turn, binds to the basal transcription complex, preventing activation of RNA polymerase (see *Activators and Repressors Control the General Transcription Factors* in Chapter 12). Another means of repression is thought to occur when E2Fs recruit histone deacetylases, histone methyltransferases, and DNA methyltransferases to the promoter region, covalently modifying and silencing the histones and/or DNA associated with the genes (see Figure 2-18 and *Histone Modifiers Control the Structure of Nucleosomes* in Chapter 12). Upon activation of cyclin-CDK complexes and phosphorylation of the pocket protein, the E2F complex releases the repressor and leaves the regulatory promoter, to be replaced by a different E2F complex that binds to an activator and thereby enhances expression of the gene. Note that in this model, the key difference between repression and activation of a gene is simply the type of E2F dimer that happens to bind the regulatory promoter. This is why regulating the relative abundance of E2F proteins is so important for controlling progression through the G_1/S checkpoint.

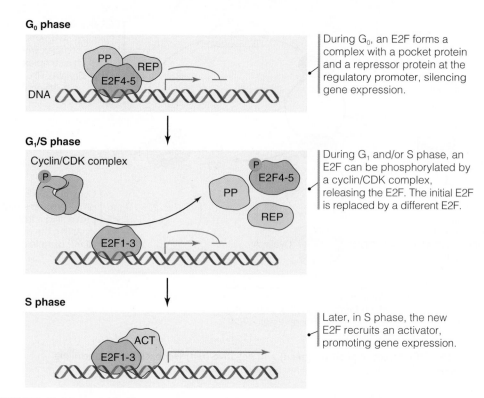

G$_0$ phase

DNA

During G$_0$, an E2F forms a complex with a pocket protein and a repressor protein at the regulatory promoter, silencing gene expression.

G$_1$/S phase

Cyclin/CDK complex

E2F1-3

During G$_1$ and/or S phase, an E2F can be phosphorylated by a cyclin/CDK complex, releasing the E2F. The initial E2F is replaced by a different E2F.

S phase

ACT

E2F1-3

Later, in S phase, the new E2F recruits an activator, promoting gene expression.

FIGURE 13-11 A model of how E2F transcription factors can suppress or activate gene transcription.

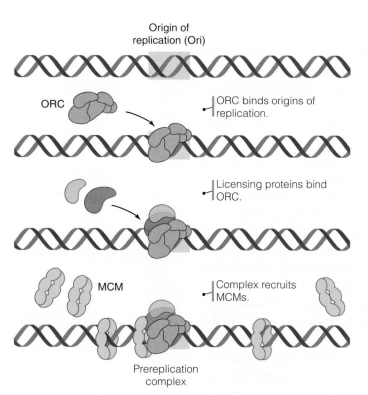

Origin of replication (Ori)

ORC

ORC binds origins of replication.

Licensing proteins bind ORC.

MCM

Complex recruits MCMs.

Prereplication complex

FIGURE 13-12 Assembly of the prereplication complex. ORC binds replication origins throughout the cell cycle, then licensing proteins bind to ORC in G1. MCM hexamers bind to the ORC/licensing proteins, completing prereplication.

DNA Replication Occurs in S Phase

At some point, the biochemical and gene expression changes initiated by G$_1$ cyclin/CDKs drive a cell past the point of no return, and the cycle enters S phase. The function of S phase is to copy every nucleotide of every chromosome with as few errors as possible. We discussed many of the details of DNA replication in Chapter 7, and will therefore focus here on the control points rather than the replication events themselves.

Cell Cycle Step 5: The DNA Replication Machinery Is Activated by Protein Kinases

One specific requirement for moving into S phase is the G$_1$ phase assembly of a **prereplication complex** at each origin of replication on a chromosome, illustrated in **FIGURE 13-12**. A prereplication complex is composed of four structures: the origin recognition complex, which binds to the DNA sequence that defines the replication origin (see *DNA Replication Begins at Sites on Chromosomes Called Origins of Replication* in Chapter 7); the **minichromosome maintenance (MCM)** complex; and two additional licensing proteins that ensure that each replication origin is used only once per cell cycle.

Once S phase begins, the conversion of the prereplication complex into a full replication complex requires the binding of several additional proteins. This occurs in three steps, as seen in **FIGURE 13-13**:

- First, a cyclin A / CDK2 quaternary complex phosphorylates one of the licensing proteins, targeting it for ubiquitin-mediated proteolysis.
- Second, another protein kinase complex, called DDK, phosphorylates the MCM complex, activating it. MCM is a helicase, and thus begins unwinding the DNA. Several MCMs can be activated in an active replication complex, and these move outward from the origin in both directions, thereby exposing single DNA strands and generating replication forks (see *During Replication, Specialized Proteins Unwind and Separate the Two Strands to Form a Replication Fork* in Chapter 7).
- Third, DNA polymerases bind to the replication complexes and begin synthesizing DNA.

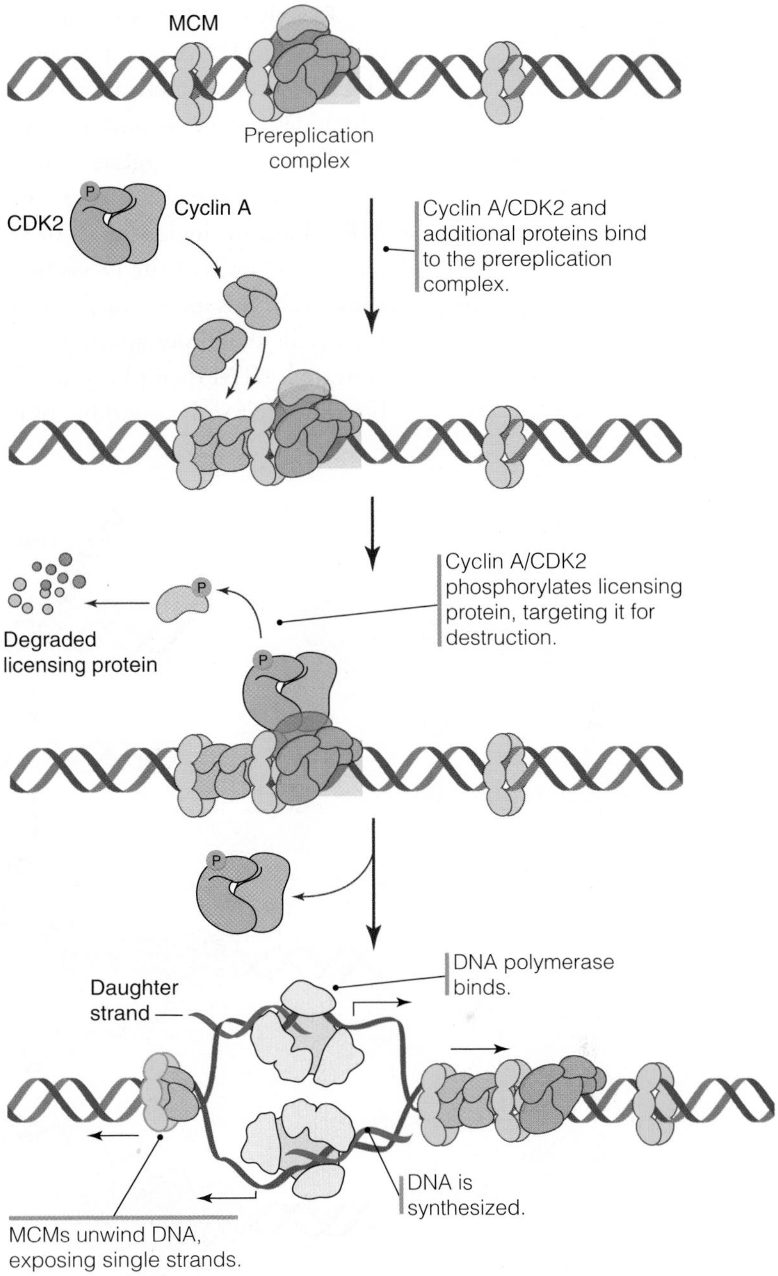

■■■ Cell Cycle Step 6: DNA Integrity Is Ensured by the G_1/S, S/G_2, and G_2/M Checkpoints

DNA is constantly being damaged by environmental sources (e.g., ultraviolet radiation, environmental toxins) and metabolic products (e.g., oxygen radicals). Yet, once cells commit to replicating their DNA in S phase, the DNA must be as "healthy" as possible, so cells expend considerable effort to detect and repair DNA damage before S phase is complete. As we discussed in Chapter 7 (see *Cells Have Two Main DNA Repair Mechanisms*), cells use multiple mechanisms to proofread their newly synthesized DNA and correct any errors. These efforts are complemented by an equally complex network of at least 700 proteins that search for uncorrected errors and halt progression through the G_1/S, S/G_2, and G_2/M checkpoints until the errors are corrected. Sorting out how these proteins interact to perform this function is a daunting task, similar to the challenges faced by those studying signal transduction and the regulation of gene expression.

One very helpful clue is that repair of double- and single-stranded breaks in DNA, respectively, are triggered by only two protein kinases, called **ATM (ataxiatelangiectasia mutated)** and **ATR (ataxiatelangiectasia and Rad-3-related)**. These two kinases are responsible for phosphorylating every one of the proteins known to control DNA repair and maintenance; this suggests a hierarchical organization of the DNA quality-checking system, with ATM and ATR at the top. As their names suggest, these kinases were first discovered in individuals suffering from the genetic disease ataxiatelangiectasia, which is a rare autosomal recessive disorder caused by mutations in the ATM gene. Inactive ATM and ATR proteins form homodimers that separate upon activation.

FIGURE 13-13 Activation of the replication complex.

Exactly how DNA damage is detected is still largely unknown, but analysis of mutant cells lacking DNA repair functions has identified some proteins that interact with ATM and ATR immediately after DNA is damaged. One of these is called the **MRN complex**, which detects single- and double-stranded breaks in DNA. One current model, illustrated in **FIGURE 13-14**, suggests how the MRN complex interacts with ATM to detect breaks in DNA and signal the repair machinery to the injury site. Two proteins present in the MRN bind to single strands of DNA; when DNA is damaged, it breaks apart, generating loose (free) ends of the DNA backbone at the breakpoints. If the MRN complex binds to these ends, this induces a shape change that permits a dimer of ATM to bind the complex, then pulls the dimer apart, permitting each monomer to autophosphorylate and become activated. ATM then phosphorylates a form of histone 2A called H2X, and this serves as a binding site for the assembly of the repair machinery.

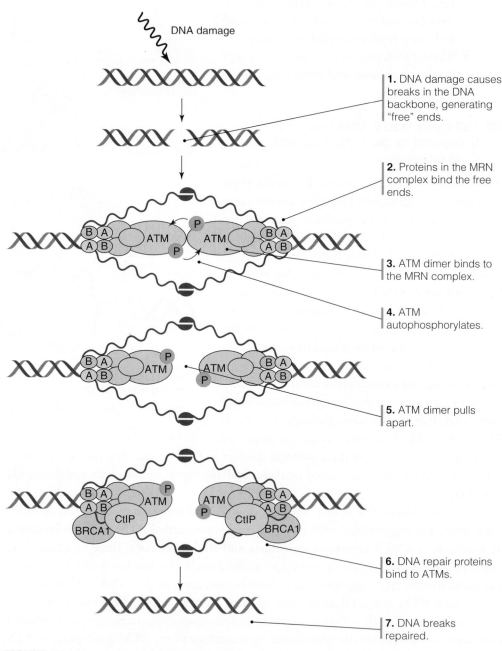

FIGURE 13-14 A current model for DNA repair.

While repair is taking place, it is critical for cells to halt their cell cycle. Two important substrates of activated ATM and ATR and serine/threonine kinases are called Chk2 and Chk1, respectively. They halt progression into the M cycle until DNA repair is complete. In turn, these kinases phosphorylate numerous targets. Some of the best understood are illustrated in **FIGURE 13-15**:

- *The G_2/M cyclin B / CDK1 complex.* Phosphorylation by Chk1 inactivates the complex.
- *Cdc25, the activating phosphatase for the G_2/M cyclin B / CDK1 complex.* Phosphorylation of cdc25 by Chk1 immediately inactivates it, and also targets cdc25 for ubiquitin-mediated degradation, thus preventing progression through the G_2/M checkpoint. (As we will see later, phosphorylation of cdc25 can have opposite effects, depending on which kinases phosphorylate it.)
- *Wee1, the inhibitory kinase for CDK1 at the G_2/M checkpoint.* Phosphorylation by Chk1 increases Wee1 activity, complementing the effect of cdc25 degradation.
- *p53, a tumor suppressor protein.* In healthy, cycling cells, p53 is normally bound to a protein called mdm2. This association targets p53 for ubiquitination and degradation. When p53 is phosphorylated by Chk2 (as well as by ATM and ATR), it is released from mdm2, where it then forms a tetramer that binds to a p53 response element in the regulatory promoter of over 100 genes, as shown in **FIGURE 13-16**. The genes targeted by p53 are responsible for a broad range of cellular behaviors, from cell cycle arrest to differentiation and cell death, though how (or if) they interact is still not clear. Likewise, how p53 generates a specific response to Chk2 phosphorylation is also not known. One of the best known targets of p53 is p21^{CIP1}. When DNA is damaged, p53 can induce expression of this CDK

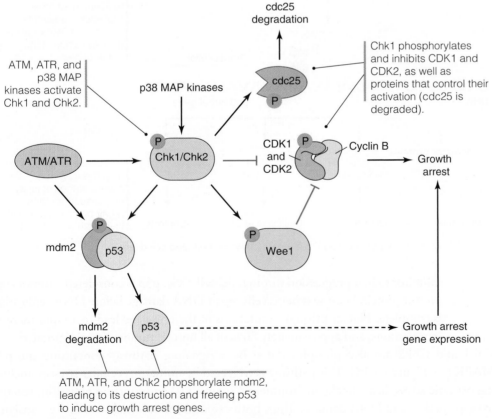

FIGURE 13-15 Growth arrest induced by Chk1 and Chk2.

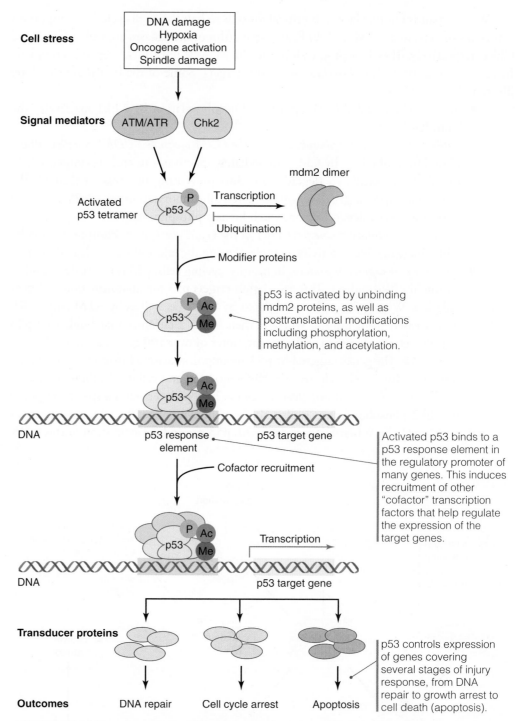

Cell stress

DNA damage
Hypoxia
Oncogene activation
Spindle damage

Signal mediators ATM/ATR Chk2

mdm2 dimer

Activated
p53 tetramer

P p53 Transcription
Ubiquitination

Modifier proteins

P Ac p53 Me

p53 is activated by unbinding
mdm2 proteins, as well as
posttranslational modifications
including phosphorylation,
methylation, and acetylation.

P Ac p53 Me

DNA

p53 response element   p53 target gene

Cofactor recruitment

Activated p53 binds to a
p53 response element in
the regulatory promoter of
many genes. This induces
recruitment of other
"cofactor" transcription
factors that help regulate
the expression of the
target genes.

P Ac p53 Me

Transcription

DNA

p53 target gene

Transducer proteins

p53 controls expression
of genes covering
several stages of injury
response, from DNA
repair to growth arrest to
cell death (apoptosis).

Outcomes DNA repair   Cell cycle arrest   Apoptosis

FIGURE 13-16 How p53 controls DNA repair, growth arrest, and cell death.

inhibitor to halt progression through the cell cycle. p53 is considered a tumor suppressor protein because it helps cells repair DNA damage before DNA replication is complete. Not surprisingly, mutations in the p53 gene lead to a rapid increase in mutations, and approximately 50% of all human tumors contain mutated p53. Chk1 and Chk2 are also phosphorylated by a signaling pathway containing the p38 MAPK (see Figure 13-15). This pathway is triggered by numerous cellular stresses, including osmotic stress, heat shock, environmental toxins, UV and ionizing radiation, reactive oxygen species, and DNA damage. Thus, both external and internal stresses are translated into checkpoint arrest by these kinases.

G₂ Phase Prepares Cells for Mitosis

Much of what takes place in G₂ phase is difficult to visualize with microscopes, but considerable biochemical activity takes place, most of it dedicated to two important tasks. First, G₂ represents the last-chance stage for cells to correct any errors in DNA structure before mitosis begins. The DNA repair machinery we discussed earlier remains active in G₂, and helps ensure that prior to compaction during prophase, replicated chromosomes are as structurally sound as possible.

Cell Cycle Step 7: Cells Increase in Size During G₂ Phase

Second, G₂ is also the last chance for cells to literally bulk up before dividing. At first, this may seem somewhat trivial, because a cell can generally undergo mitosis even if its size varies. But discovery of Wee1, one of the most important regulatory kinases in G₂, helped illustrate how important cell volume is during the cell cycle. As stated above, Wee1 is the inhibitory kinase for cyclin B / CDK1. Wee1 was first discovered in mutant yeast that divided at half the normal cell volume, as shown in **FIGURE 13-17**. This phenotype in fission yeast was first observed, by accident, in a laboratory in Scotland; *wee* is the Scottish word for "small," and hence these cells were called "wee ones."

Cell Cycle Step 8: Cyclin B / CDK1 Activation Drives Cells through the G₂/M Checkpoint

Transition from G₂ to M is literally determined by cell size, yet how most cells sense their volume is largely unknown. One of the best understood mechanisms is used by fission yeast and is composed of a pathway of protein kinases that collectively suppress growth when they are in contact with each other. As the cells grow, the kinases are pulled apart from one another, shutting off the inhibition; a current model for how this occurs is shown in **FIGURE 13-18**. Fission yeast cells are oval shaped immediately after mitosis, and elongate into a cylindrical shape as they grow. The pathway begins at the tips of the cell, where a kinase called Pom1 is held onto the membrane by a lipid-anchored membrane protein called Mod5. In small cells, Pom1 has access to kinases called Cdr1 and Cdr2, which are members of a different complex that lies at the middle of the cell. When Pom1 phosphorylates proteins in this complex (perhaps Cdr1 and Cdr2 directly), the kinases are inactive. These kinases are, in turn, able to phosphorylate and inhibit Wee1. This model thus contains three successive rounds of inhibition: Pom1 inhibits Cdr1 and Cd2, which inhibit Wee1, which inhibits CDK1. This presents a bit of a brain teaser for most of us, so it may be easiest to follow if we start at the end of the pathway and work backward:

- When cyclin B / CDK1 is active, cells *can* enter mitosis.
- When Wee1 inhibits CDK1, cells *do not* enter mitosis.

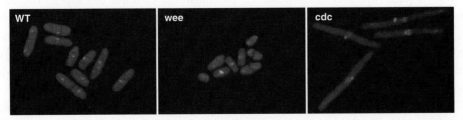

FIGURE 13-17 Wee1 mutation affects cell size. Compared to normal ("wild-type," WT) yeast, Wee1 mutants grow to half normal size before dividing. In contrast, cells containing mutant cdc25 grow larger than normal. Photo courtesy of K. Adam Bohnert and Kathleen L. Gould, Vanderbilt University Medical Center.

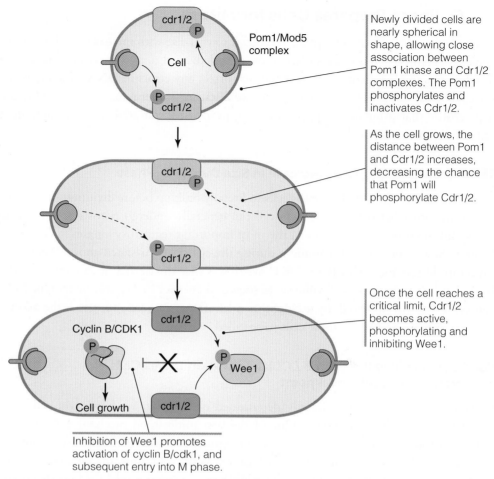

Newly divided cells are nearly spherical in shape, allowing close association between Pom1 kinase and Cdr1/2 complexes. The Pom1 phosphorylates and inactivates Cdr1/2.

As the cell grows, the distance between Pom1 and Cdr1/2 increases, decreasing the chance that Pom1 will phosphorylate Cdr1/2.

Once the cell reaches a critical limit, Cdr1/2 becomes active, phosphorylating and inhibiting Wee1.

Inhibition of Wee1 promotes activation of cyclin B/cdk1, and subsequent entry into M phase.

FIGURE 13-18 A model for cell size control of cell growth in yeast in G2.

- Cdr1 and Cdr2 inhibit Wee1. This is a double negative (inhibition of an inhibitor), so when Cdr1 and Cdr2 are active, cells *can* enter mitosis.
- Pom1 inhibits Cdr1 and Cdr2. This is a triple negative (inhibition of an inhibitor of an inhibitor), so cells *do not* enter mitosis when Pom1 is active.

To understand how cells switch from G_2 to M, we follow the pathway forward: As the cells grow in size through G_2, the two poles move away from the medial band, lowering the concentration of Pom1 kinase in the middle of the cell. As Pom1 kinase activity decreases at the band, Cdr1 and Cdr2 kinase activity increases, which in turn inhibits Wee1 and thus allows activation of CDK1. The larger the cell grows, the less Pom1 influences Cdr1 and Cdr2. At some point, a threshold size is reached where enough CDK1 becomes active to drive the onset of mitosis.

Similar pathways may control the G_2/M transition in other cell types. The Mod5-Pom1 complex at the tips of yeast also contains members of a MAP kinase signaling pathway, suggesting MAP kinases may regulate G_2/M progression in these cells. In yeast, this pathway is associated with cellular stress, providing a means for cells to temporarily halt the cycle in G_2 until the stress is removed. Addition of MAPK-activating growth factors to mammalian cells in G_2 causes a delay in mitosis, perhaps via a similar mechanism.

The rate-limiting step in CDK1 activation is removal of two inhibitory phosphates placed in CDK1. In addition to phosphorylation of tyrosine 15 by Wee1, a second kinase called Myt1 phosphorylates threonine 14 (and tyrosine 15, if necessary). These phosphates are removed by cdc25 phosphatase. We can trace the activation of cdc25 all the way back to the focal complexes formed by cells as they spread and migrate on extracellular matrix proteins, as shown in **FIGURE 13-19**. A serine/threonine kinase associated with focal complexes, called Ste20-like kinase (Slk), is activated by focal adhesion kinase (see Figure 11-10), and phosphorylates another serine/threnonine kinase, called polo-like kinase1 (plk1). Plk1 plays three important roles in promoting G_2/M progression:

- First, it also phosphorylates Wee1 and Myt1 (targeting their ubiquitination and degradation).
- Second, it phosphorylates cyclin B. How this phosphorylation of cyclin B promotes CDK1 activity is not clear.
- Third, it phosphorylates and *activates* cdc25. Note that this yields the exact opposite result from phosphorylation by Chk1, which we discussed earlier. It is likely that these two effects are caused by phosphorylation of different amino acids in cdc25, generating different shape changes. This is somewhat analogous to the activating and inhibiting phosphorylation of CDKs.

Once cyclin B / CDK1 is activated, it initiates a positive feedback loop, similar to the G_1 phase loop discussed above (see **FIGURE 13-20**). In this loop, CDK1 phosphorylates cdc25, further activating it, and phosphorylates Wee1, inhibiting it. The resulting surge in CDK1 activity triggers phosphorylation of numerous other proteins (estimated to be approximately 180 different proteins in yeast), pushing cells into M phase.

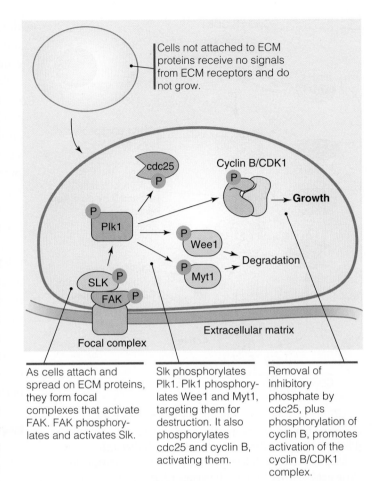

As cells attach and spread on ECM proteins, they form focal complexes that activate FAK. FAK phosphorylates and activates Slk.

Slk phosphorylates Plk1. Plk1 phosphorylates Wee1 and Myt1, targeting them for destruction. It also phosphorylates cdc25 and cyclin B, activating them.

Removal of inhibitory phosphate by cdc25, plus phosphorylation of cyclin B, promotes activation of the cyclin B/CDK1 complex.

FIGURE 13-19 A model for how adhesion to ECM promotes cell growth in mammalian cells.

Mitosis and Cytokinesis Occur in M Phase

Cell Cycle Step 9: Chromosome Alignment Is Ensured by the Mitotic Spindle Assembly Checkpoint

(See **BOX 13-5**.) One of the most important stages during mitosis is the attachment of kinetochores on duplicated chromosomes to kinetochore microtubules in the mitotic spindle. Until all chromosomes are properly attached to the spindle and aligned at the metaphase plate, progression to anaphase is actively inhibited by a mechanism known collectively as the **spindle assembly checkpoint**. This checkpoint helps ensure that the replicated chromosomes are properly separated; in mutant cells incapable of triggering it, separation of chromosomes is imbalanced, resulting in a condition known as **aneuploidy**. Aneupolid cells are often quite defective, and many die or become cancerous.

Many details for specific organisms and cells are still not known, making it difficult to construct a consensus model for how the spindle assembly checkpoint functions. The one constant in nearly all organisms is the presence of a large protein complex called

BOX 13-5 TIP

The mechanisms of mitosis and cytokinesis are discussed in Chapter 7, and you are encouraged to review them before reading this section.

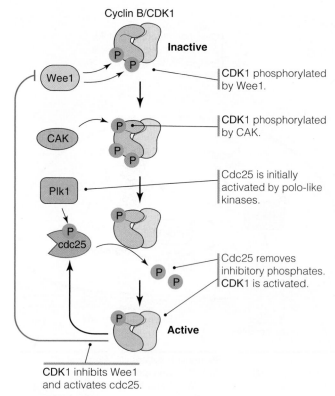

Cyclin B/CDK1

Inactive

CDK1 phosphorylated by Wee1.

CDK1 phosphorylated by CAK.

Cdc25 is initially activated by polo-like kinases.

Cdc25 removes inhibitory phosphates. CDK1 is activated.

Active

CDK1 inhibits Wee1 and activates cdc25.

FIGURE 13-20 Phosphorylation of CDK1 primes it for activation but also keeps it in an inactive state. Polo-like kinases (Plks) activate the cdc25 phosphatase, which then activates a small amount of CDK1 by removing its inhibitory phosphate. Once activated, CDK1 phosphorylates cdc25 and augments its activity. In addition, once some CDK1 is activated, it can phosphorylate and inactivate Wee1. This autoamplification loop results in precipitous activation of CDK1.

the anaphase-promoting complex/cyclosome (APC/C). In one model, illustrated in **FIGURE 13-21**, a **mitotic checkpoint complex** (**MCC**), containing an essential APC/C component called cdc20, binds to kinetochore microtubules that have yet to engage a kinetochore. The cdc20 is inhibited by being in the MCC, as is APC/C. As chromosomes attach to the kinetochore microtubules and are aligned at the metaphase plate, tension in microtubules increases, causing some cdc20 inhibitor proteins in the MCC to leave the complex, allowing cdc20 to bind to and activate the APC/C. Other proteins also bind to the APC/C, permitting it to bind to distinct sets of substrates. Plk1 also phosphorylates APC/C, further activating it. APC/C is an ubiquitin ligase, and it targets several proteins, including plk1, cyclin B, and securin. These proteins are ubiquitinated in a specific order to ensure that the sequence of events during mitosis is carefully maintained. Upon degradation of securin, the enzyme separase is free to digest cohesin, the protein that holds sister chromatids together (see *The Onset of Anaphase Requires Dissolving the Connections between Sister Chromatids* in Chapter 7). Collectively, this turnover of APC/C target proteins permits cells to complete mitosis.

■■ Cell Cycle Step 10: Onset of Cytokinesis Is Timed to Begin Only after Mitosis Is Complete

Cytokinesis requires the contraction of the contractile ring that lies just beneath the plasma membrane, perpendicular to the long axis of the mitotic spindle. It is important that the myosin motors in the ring not activate until mitosis, including reconstitution of the nuclear membrane, is complete. While most of the events controlling the onset and progression of cytokinesis are not well understood, a few key patterns appear to be conserved across many different species, from yeast to mammals. One is that the transition from telophase to cytokinesis is dependent upon the *loss* of protein phosphorylation, especially for cyclin B / CDK1 substrates. Whether this results from ubiquitination of cyclins, increase in the activity of phosphatases, or both, is still being debated. Cells expressing modified cyclins that cannot be ubiquitinated enter cytokinesis after a considerable delay, demonstrating that loss of cyclins is not strictly required for cytokinesis, but the loss appears to control the timing of cytokinesis. Second, the dephosphorylation of CDK substrates occurs in a specific order, suggesting that cytokinesis follows a pattern of dephosphorylation similar to that found for the phosphorylation cascades that push cells through cycle checkpoints.

CONCEPT CHECK #1

For each of the following events, explain whether they *activate* or *inhibit* cell cycle progression, and propose a mechanism to explain how they might accomplish this.
- Addition of PDGF to cells, in mid-G_1
- Addition of PDGF to cells, in early G_2
- Removal of nutrients from cells, in S
- Removal of nutrients from cells, in M
- Addition of a drug that inhibits Pom1, in G_2
- Loss-of-function mutation for cyclin B, in any phase

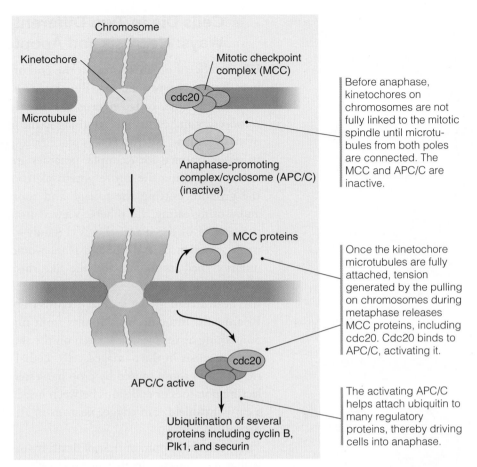

Chromosome

Kinetochore

Mitotic checkpoint complex (MCC)

cdc20

Microtubule

Anaphase-promoting complex/cyclosome (APC/C) (inactive)

Before anaphase, kinetochores on chromosomes are not fully linked to the mitotic spindle until microtubules from both poles are connected. The MCC and APC/C are inactive.

MCC proteins

Once the kinetochore microtubules are fully attached, tension generated by the pulling on chromosomes during metaphase releases MCC proteins, including cdc20. Cdc20 binds to APC/C, activating it.

cdc20

APC/C active

Ubiquitination of several proteins including cyclin B, Plk1, and securin

The activating APC/C helps attach ubiquitin to many regulatory proteins, thereby driving cells into anaphase.

FIGURE 13-21 A model for anaphase promotion by APC/C.

13.3 Multicellular Organisms Contain a Cell Self-Destruct Program That Keeps Them Healthy

KEY CONCEPTS

▸ Cells die either by traumatic injury (necrosis) or by a self-destruct program called apoptosis.

▸ Apoptosis begins through at least two molecular mechanisms, called intrinsic and extrinsic pathways.

▸ The family of proteins called caspases includes proteases that promote the degradation of organelles and cytosolic proteins during apoptosis.

Death is, of course, a natural part of life on Earth. Discussing cell death is a bit tricky, however, because since we rarely experience death at the level of individual cells in our everyday lives, very little of this process is intuitive. As tempting as it may be to compare the birth and death of single cells to the birth and death of organisms more familiar to us, this type of analogy is often more misleading than beneficial. Instead, let's consider the seemingly paradoxical view that cell death can serve as a strategy for keeping multicellular organisms alive and healthy. Some biologists even consider it a form of altruism.

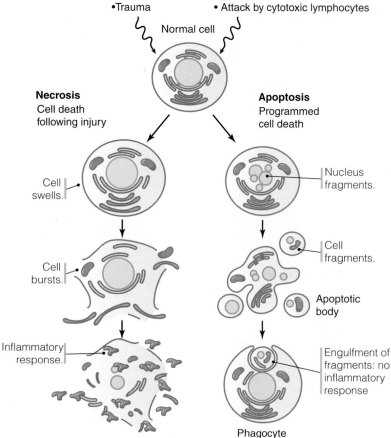

- Trauma

- DNA damage
- Withdrawal of essential growth factors or nutrients
- Detachment from substrate
- Attack by cytotoxic lymphocytes

Normal cell

Necrosis
Cell death
following injury

Cell
swells.

Cell
bursts.

Inflammatory
response.

Apoptosis
Programmed
cell death

Nucleus
fragments.

Cell
fragments.

Apoptotic
body

Engulfment of
fragments: no
inflammatory
response

Phagocyte

FIGURE 13-22 Cellular damage can result in necrosis, which has a different appearance than apoptosis, as organelles swell and the plasma membrane ruptures, without chromatin condensation.

Cells Die in Two Different Ways: Necrosis and Apoptosis

Ultimately, all organisms reach the end of their lives by one of two means, illustrated in **FIGURE 13-22**. One, called **necrosis**, is considered *involuntary* and contributes nothing to the health of an organism. Necrosis takes place in response to severe trauma; for example, potatoes are collected by pulling the entire potato plant out of the ground, cutting the foliage and roots off, and storing/eating the tubers. Viewed from the perspective of the plant, this is certainly a traumatic experience, as no potato plant would voluntarily shed its foliage and die like this. Any environmental insult (e.g., dehydration, excessive heat, freezing) can induce necrosis in plants. In animals, most necrosis results from injuries. Something as simple as a scratch on the skin can cause necrosis in the damaged cells. Lethal injuries, such as those caused by crop harvesters collecting potatoes or collisions between insects and a car windshield, can trigger necrosis in all of an organism's cells.

A second important trait of death-by-necrosis is that it is "messy," in that necrotic cells rupture as they are dying, releasing their contents into the extracellular space. This typically interferes with the function of neighboring cells, and requires cleanup to restore the tissue to its normal functional state. In animals, cleanup is typically performed by cells of the immune system.

Apoptosis Is a Property of All Animal Cells and Some Plant Cells

Because cells have very little control over necrosis, most cell biologists focus instead on understanding the second means of dying. In almost all ways, this form, called apoptosis, is the exact opposite of necrosis (**BOX 13-6**). Some of the key differences, which we will examine in greater detail later in this chapter, are:

- *Apoptosis is voluntary.* One useful way of placing apoptosis into context is to consider it an alternative to G_0 for exiting the cell cycle. Just as in G_0, cells actively elect to halt cycling, and they use proteins to direct the cell into an apoptotic state. Unlike G_0,

---

BOX 13-6 FAQ

How is *apoptosis* pronounced in English? The word *apoptosis* is taken directly from ancient Greek, where the prefix *apo* (away, off) is joined to the word *ptosis* (falling) to mean "falling off," as with tree leaves or flower petals. In Greek, the p in *ptosis* is silent, so one pronunciation is a-po-TOW-sis, with the emphasis on the third syllable. The second pronunciation is a-pop-TOW-sis, where the p in *ptosis* is spoken in the second syllable. This pronunciation results from the application of modern English grammatical rules to the word, and is often the first choice of most native English speakers. But there is no formal consensus in the scientific community, so be prepared to hear both.

however, apoptosis is absolutely irreversible for every cell that adopts it, because it results in the death of the cell.

Let's briefly consider why cells would *choose* to die rather than continue living. Perhaps the most obvious reason is extreme damage, when the amount of effort required to repair the damage is greater than the cost of simply killing off the cell and replacing it (via division of a neighboring cell). This occurs, for example, in burned skin: the damaged skin is triggered to die, and is replaced by aggressive growth of the underlying healthy cells. The damage can be caused by environmental hazards, toxins, or even by the cells themselves: many of the clinical symptoms of Parkinson's disease result from apoptosis of neurons in the brain, though it is still not understood what causes the damage that triggers their death.

Another reason for choosing to die is that a cell has served its function in the organism and is no longer needed. This argument could only apply to multicellular organisms, of course, and indeed apoptosis is not found in any single-celled organisms. Apoptosis is especially useful during plant and animal development, when gross morphological changes occur prior to maturity. One very visual example is the removal of the cells in the webbing that lies between the digits in the hands and feet of humans and in the paws of mice and some other mammals, as shown in **FIGURE 13-23**, and another is the loss of the tail when a tadpole develops into a frog. Apoptosis also occurs in many plants, ranging from developing onion root tips to cacti. Loss of these cells is necessary for the organism to function properly as an adult, and thus contributes to the health of the adult. Because it occurs at a predetermined time during development, this form of apoptosis is called **programmed cell death**. Note that programmed cell death is not simply another name for apoptosis, though the two terms are commonly used interchangeably; rather, it is a very specific example of the general mechanism of apoptosis.

- *Apoptosis is carefully orchestrated.* While necrosis can happen at any moment to any cell, considerable deliberation can take place before cells enter apoptosis. Much of this decision making is performed by proteins controlling the cell cycle checkpoints we discussed earlier in this chapter. Just as progression through cell cycle checkpoints occurs in carefully organized steps, apoptosis too must take place in a defined sequence of events. The rationale is the same as for checkpoints: this helps ensure quality control and a predictable outcome. While death by necrosis can be instantaneous, apoptosis takes several hours to complete.

- *Apoptosis is "clean."* The desired outcome of apoptosis is a collection of small, membrane-enclosed fragments of cytoplasm that can be internalized by neighboring cells via endocytosis, as shown in Figure 13-22. The organelles have all been digested, and most proteins and nucleic acids have been broken down into small fragments that can be easily used by the endocytosing cells. This is considered a "clean" death, because virtually the entire cell is recycled and used by other cells without "spoiling" the surrounding environment.

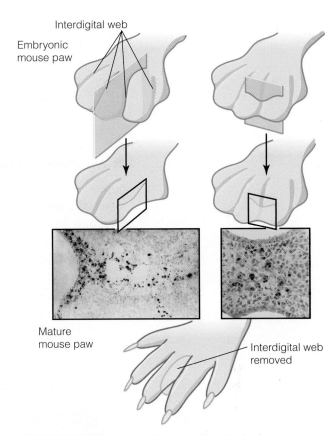

FIGURE 13-23 Sections of the interdigital web show cell death (dark-staining nuclei). This cell death has the characteristics of apoptosis. Photos reprinted from Curr. Biol., vol. 9, M. Chautan, et al., Interdigital cell death can occur through…, pp. 967-970, Copyright (1999) with permission from Elsevier [http://www.sciencedirect.com/science/journal/09609822]. Photos courtesy of Pierre Golstein, Centre d' Immunologie.

- *Apoptosis is a form of self-defense.* Most infectious agents, such as viruses and some bacteria, enter cells as part of their life cycle. One clever way for an organism to combat infection is to kill the infected cells before the infection can spread. This occurs for example in tomato plants infected with a fungus called Alternaria, and in animals cells infected with viruses. In humans, lung macrophage cells apoptose during their battle with the bacterium *Streptococcus pneumonia*, which causes pneumonia. Many cancer cells can also be cleared from the body by apoptosis.

██■ Apoptosis Is Induced via at Least Two Different Pathways

Apoptosis responses are grouped into two classes, based on the location of the triggering source. The extrinsic pathway, summarized in **FIGURE 13-24**, begins at the cell surface, in response to extracellular signals. In many ways it resembles the signaling pathways we discussed in Chapter 11, in that cells pass this apoptosis "information" forward from a cell surface receptor to cytosolic proteins that are activated in a defined order. The receptor for this pathway is called a **death receptor**; in vertebrates, the death receptors belong to a family of proteins called **tumor necrosis factor receptors**. The receptors form a homotrimer

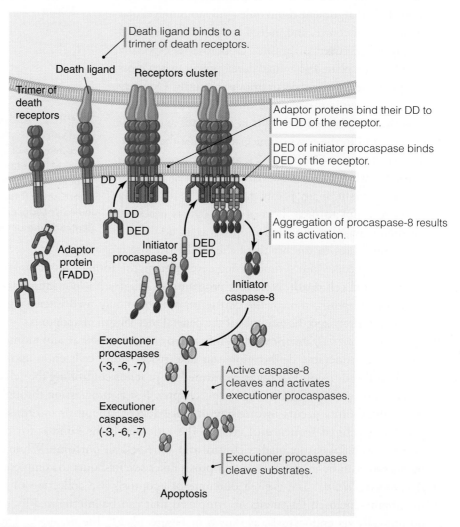

FIGURE 13-24 Ligation of death receptors causes the recruitment of the adaptor protein FADD to the intracellular region of the death receptor, via death domain (DD)–(DD) interactions. Caspase-8 is then recruited to FADD via death effector domain (DED)–(DED) interactions. The dimerization of caspase-8 activates it through induced proximity. The active caspase-8 can cleave and activate executioner caspases to cause apoptosis.

that binds to clusters of extracellular **death ligands**, and then change their shape to expose a **death domain** on the cytosolic portion of each receptor. The death domain binds to an adaptor protein, analogous to how SH2 and SH3 domains interact with phosphotyrosines and proline-rich regions in other proteins, respectively (see Figure 11-19). Next, the adaptor protein binds to a protease through a **death effector domain** (**DED**).

One of the first death ligands discovered was called Fas, and the adaptor protein that binds to the death domain in most death receptors is called **Fas-associated protein with death domain** (**FADD**) for this reason. Once bound to the receptor, FADDs recruit the first of a series of proteases that begin digesting the cell interior. We'll discuss these proteases in the next section.

The **intrinsic pathway** originates at the outer membrane of the mitochondria, and is triggered primarily by intracellular signals that transmit information about the state of the health of the cell. Many of the intrinsic signals arise from severe DNA damage, oxygen radicals, membrane disruption, or the presence of toxic substances that have penetrated the plasma membrane. These triggers activate transcription of genes belonging to the bcl-2 family of proteins. The links between the actual triggers and this transcription are still not well understood, but several observations strongly suggest that, like passage through the restriction point in the cell cycle, activation of apoptosis is controlled by a host of regulatory proteins.

One of the best characterized mechanisms of activating the intrinsic pathway is that triggered by DNA damage. Overall, it appears that the decision to halt the cell cycle and repair the DNA or surrender to apoptosis is based on a careful balancing act between pro- and antiapoptosis signals. Recall from earlier in this chapter that the MRN complex detects breaks in DNA and recruits a host of proteins to repair the damage. Downstream of MRN, the transcription factors p53 and E2F control expression of genes necessary to execute the repair. It turns out that p52 and E2F, as well as several other transcription factors, can also initiate apoptosis by selectively promoting expression of apoptosis genes rather than DNA repair genes.

While it is still unclear how this critical decision is reached, the proapoptosis transcription factor E2F1 and the tumor suppressor p53 and play a prominent role, at least in some cell types. For example, activation of E2F1 via phosphorylation of Rb occurs in at least four ways:

- First, E2F1 can be preferentially activated in response to DNA damage by Rb acetylation. Rb interacts with E2F1 at two sites: the primary binding site that binds many E2Fs, and a secondary site that is specific to E2F1. This second site can be acetylated (see Figure 3-15) in response to DNA damage, and when this occurs, Rb releases E2F1 even if Rb is never phosphorylated. This means that Rb can remain bound to other E2Fs while releasing only E2F1. This tips the balance in favor of E2F1, and promotes apoptosis.
- Second, E2F1 is directly phosphorylated by ATM and Chk2, while E2F2 and E2F3 are not, and this too selectively activates E2F1.
- Third, E2F1 can become acetylated after DNA damage, and this biases it toward proapoptotic genes.
- Fourth, E2F1 promotes the expression of two proteins that bind p53 and direct it toward proapoptotic genes.

The effects of this proapoptotic signaling are counterbalanced by prosurvival signaling pathways. The protein kinase B/Akt signaling pathway, which uses phosphorylated phosphatidylinositol as a second messenger (see *Phospholipid Kinase Pathways Work in Cooperation with Protein Kinase and G Protein Pathways* in Chapter 11) suppresses apoptosis, as does the ras-MAPK signaling pathway (see Figure 11-19). **FIGURE 13-25** shows how

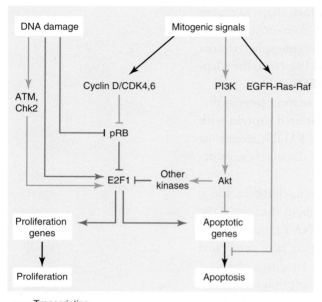

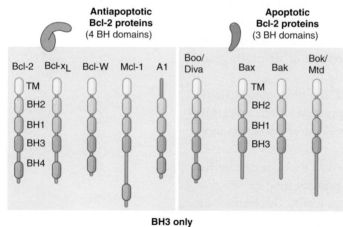

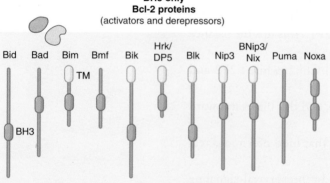

FIGURE 13-25 E2F1 lies at the heart of the growth versus death decision-making system. It receives both stimulatory and inhibitory signals, and the balance between these determines what types of genes E2F1 will induce.

FIGURE 13-26 The Bcl-2 family proteins share up to four Bcl-2 homology domains (BH) and can be antiapoptotic or proapoptotic. The proapoptotic proteins include multidomain proteins and the BH3-only proteins.

E2F1 lies at the center of this balancing act. Note that, combined with the material we discussed in Chapter 11, we can now trace a nearly continuous path from the cell surface to specific genes that control cell life and death. We'll use this information in Chapter 14, when we discuss cancer.

Some of the most important targets of pro- and antiapoptotic transcription factors are members of the bcl-2 family of proteins, illustrated in **FIGURE 13-26**. The family is divided into three groups, depending on the number and type of **bcl homology (BH) domains** each protein has. Members of the BH3-only group contain one or two copies of the BH3 domain, and these are the proteins targeted by proapoptosis transcription factors. These proteins compete with members of the antiapoptotioc group to access the apoptotic group in an elaborate hierarchy, shown in **FIGURE 13-27**. At the top, antiapoptotic proteins, such as bcl-2, bind to and inhibit the action of the apoptotic proteins Bak and Bax, all of which takes place in the cytosol. When BH3-only proteins are added to the mix, they bind to the antiapoptotic proteins, preventing them from binding Bak and Bax. As we saw with DNA repair, this is another example of double inhibition: BH3-only proteins inhibit antiapoptotic proteins that normally inhibit Bak and Bax.

The result of this double inhibition (a.k.a. activation) is that Bak and Bax move from the cytosol into the outer mitochondrial membrane, where they form a channel-like structure that permits proteins in the intermembrane space of the mitochondrion to leak out into the cytosol. This event, called **mitochondrial outer membrane permeabilization (MOMP)** is the defining moment for the onset of apoptosis via the intrinsic pathway. It is still not clear whether Bak and/or Bax alone form these channels, or whether they recruit other proteins in the mitochondrial outer membrane to form them. (It is interesting to note that members of the BH3-only family that possess a hydrophobic tail can insert themselves into the outer mitochondrial membrane, keeping them closely associated with the channels. These membrane-bound forms are the most effective at inducing apoptosis.) Once MOMP occurs, apoptosis proceeds as illustrated in **FIGURE 13-28**. One of the first events following MOMP is the binding of cytochrome c (a critical player in the electron transport chain: see Stage 3: *The Electron Transport Chain Uses High-Energy Electrons from NADH and FADH₂ to Build a Proton Gradient across the Inner Mitochondrial Membrane* in Chapter 10) to a cytosolic protein called **apoptotic protease activating factor-1 (APAF-1)**. The binding to cytochrome c causes a shape change that permits an APAF-1 to bind to a molecule of dATP, resulting in another shape change that exposes an **oligomerization domain**. Seven of these APAF-1 proteins bind together to form a complex called an **apoptosome** (meaning *an apoptosis-causing body*), the center of which binds to and activates the first stage of proteolytic enzymes that begin the process of digesting the cell interior. Note that an apoptososme is not a proteasome (see *Proteins in the Cytosol and Nucleus Are Broken Down in the Proteasome* in Chapter 3).

■■ Apoptosis Triggers the Activation of Special Proteases: The Caspases

Thus far in this book we have viewed proteolytic enzymes primarily as cleanup agents that digest broken or damaged proteins (e.g., in the proteasome), breakdown complex cargo (e.g., in the lysosome; see Figure 9-17), or remodel the extracellular matrix (see *Proteins in the Extracellular Space Are Digested by Proteinases* in Chapter 3). In each of these cases, digestion represents the end of the story. But the enzymes that digest cells during apoptosis are quite different. Rather than playing supporting roles opposite the rest of the cell proteins, these proteases take center stage: they are the most important proteins for successively executing apoptosis. Many are called cell executioners, for this reason.

These proteases belong to a family of proteins called cysteine aspartate–specific proteases, or more simply **caspases**, illustrated in **FIGURE 13-29**. The structural organization of caspases is quite straightforward: they contain three domains (large and small catalytic domains, plus a prodomain) separated by cleavage sites that can be cut by other proteases. The fully intact forms of the enzymes are inactive, and are called zymogens (derived from an archaic phrase, "to produce by fermentation"—a modern equivalent would be "to activate by digestion"). They are activated when the prodomain is released via severing the cleavage site separating it from the cata-

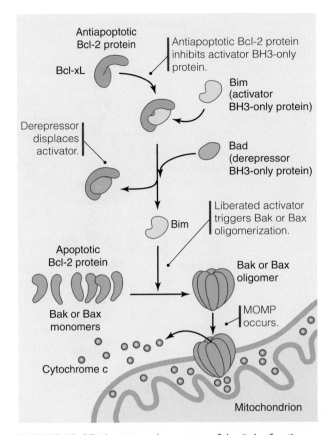

FIGURE 13-27 The BH3-only proteins of the Bcl-2 family function either to directly activate Bax and Bak (activators) or to interfere with the inhibitory functions of the antiapoptotic Bcl-2 family proteins.

lytic domains. This illustrates an important trait of caspases: they can activate each other by digesting away a portion of their structure. This is analogous to transautophosphorylation by receptor protein kinases, but instead of phosphorylating each other, caspases perform **autoproteolysis**, though the result is functionally similar (activation). One important difference, however, is that phosphorylation is reversible, while caspase cleavage is not. Once activated, caspases are always active until they, too, are digested.

Caspases are divided into three groups, based on their function. **Initiator caspases** are activated first, and this requires dimerization. In the extrinsic pathway (Figure 13-24), FADD adaptors bind to death domains on caspase-8 monomers, creating a cluster. This cluster undergoes autoproteolysis, and the cleaved dimers are released into the cytosol, where they begin digesting their substrates. Some of their primary targets are the **executioner caspases** (-3, -6, and -7), which digest hundreds of different proteins inside cells. Like initiators, executors are inactive until the prodomains are removed. Executioner caspases can also cleave the prodomain off other executioner caspases, establishing a positive feedback loop that leads to the rapid, inevitable death of a cell.

In the intrinsic pathway, the apoptosome recruits the monomers of inactive initiator procaspase-9. Caspase-9 undergoes induced dimerization, then autoproteolysis, and is released, targeting the executioner caspases. Note that despite the significant differences between the extrinsic and intrinsic pathways, ultimately they converge on the same executioners. In fact, the two pathways are sometimes linked, in that the initiator caspase-8 activated by the extrinsic pathway can activate a BH3-only protein called bid. This stimulates MOMP and thereby activates the entire mitochondrion-based apoptosis pathway.

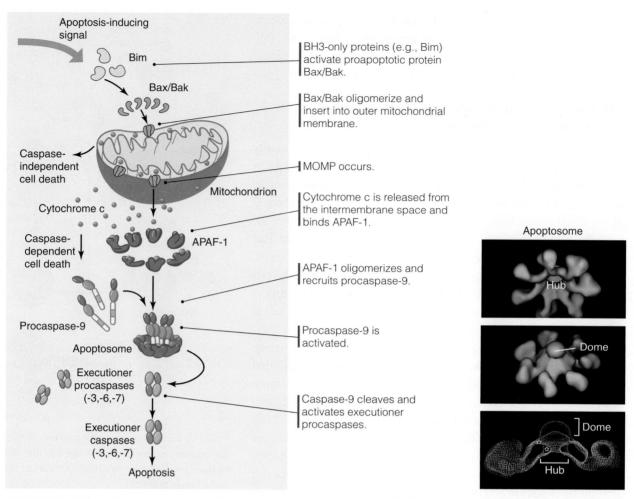

Apoptosis-inducing
signal

Bim

Bax/Bak

Caspase-
independent
cell death

Cytochrome c

Caspase-
dependent
cell death

APAF-1

Procaspase-9

Apoptosome

Executioner
procaspases
(-3,-6,-7)

Executioner
caspases
(-3,-6,-7)

Apoptosis

Mitochondrion

BH3-only proteins (e.g., Bim)
activate proapoptotic protein
Bax/Bak.

Bax/Bak oligomerize and
insert into outer mitochondrial
membrane.

MOMP occurs.

Cytochrome c is released from
the intermembrane space and
binds APAF-1.

APAF-1 oligomerizes and
recruits procaspase-9.

Procaspase-9 is
activated.

Caspase-9 cleaves and
activates executioner
procaspases.

Apoptosome

Hub

Dome

Dome

Hub

FIGURE 13-28 Signals for the induction of apoptosis trigger changes in the Bcl-2 family proteins, which function to inhibit (antiapoptotic proteins) or promote (proapoptotic proteins) apoptosis. As a result, the proapoptotic multidomain proteins of the Bcl-2 family may be activated and, if so, cause permeabilization of the outer membranes of all mitochondria in the cell. Mitochondrial outer membrane permeabilization (MOMP) allows proteins of the intermembrane space to diffuse into the cytosol, including cytochrome c, which activates APAF-1. This leads to recruitment and activation of the initiator caspase-9, which cleaves and activates executioner caspases to cause apoptosis. A model of apoptosome structure, including hub and dome regions, is shown at right. Photos reprinted from Mol. Cell, vol. 9, D. Acehan, et al., Three-dimensional structure of the apoptosome..., pp. 423-432, Copyright (2002) with permission from Elsevier [http://www.sciencedirect.com/science/journal/10972765]. Photos courtesy of Christopher W. Akey, Boston University School of Medicine.

The third class of caspases, called **inflammatory caspases**, appears to contribute little or nothing to apoptosis.

Stereotypical Morphological Changes Take Place during Apoptosis

Once the executioner caspases have been activated, the cell gradually changes its structure. One of the earliest observable changes, triggered by a mitochondrial protein that escapes during MOMP, is condensation of chromatin within the nucleus. This correlates with cleavage of the DNA into fragments that form a characteristic "ladder" pattern when separated by gel electrophoresis. This cleavage is performed by substrates of caspases-3 and -6, called caspase-activated DNase and DNA fragmentation factor. As this occurs, DNA repair proteins are cleaved as well, preventing any slowdown of the program. Finally, nuclear lamins and other proteins in the nuclear scaffold (see *DNA Is Bound to a Protein/RNA Scaffold* in Chapter 2) are degraded, causing the nuclear membrane to break into fragments that resemble small bubbles within the cell. The generation of these fragments is called **karyorrhexis** (from Greek and Latin, meaning "to break the nucleus").

In the cytosol, the cytoskeleton collapses, causing deflation of filopodia and lamellopodia (see *Migrating Cells Produce Three Characteristic Forms of Actin Filaments: Filopodia, Lamellopodia, and Contractile Filaments* in Chapter 5) and shrinkage of the cell. The phospholipid composition of the plasma membrane changes, such that phosphatidylserine now appears in abundance in the exoplasmic face (see *Lipid Bilayers Are Asymmetrical* in Chapter 4). It is still not clear how this occurs, but it is a hallmark of apoptosis. As the cell shrinks, the plasma membrane bulges in some regions, forming blisters called **blebs**. In most cells, the organelles persist for the duration of apoptosis, and are slowly enclosed in plasma membrane fragments to form **apoptotic bodies**.

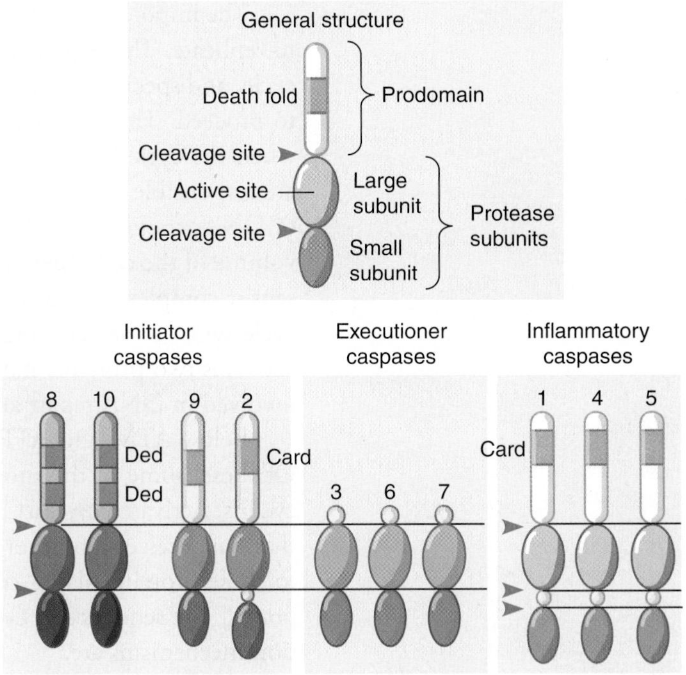

■■ Apoptotic Cells Are Cleared by Phagocytosis

Phagocytosis is a form of internalizing foreign objects that is performed by many cells, ranging from fibroblasts and epithelial cells to so-called "professional" phagocytic cells (e.g., macrophages) specialized to engulf large foreign objects. The principle differences between endocytosis (see *Endocytosis Begins at the Plasma Membrane* in Chapter 9) and phagocytosis are (1) endocytosis is always receptor mediated, while phagocytosis is not, and (2) endocytosis internalizes soluble molecules (proteins, lipids, etc.), while phagocytosis is targeted to solid objects (bacteria, apoptotic bodies, etc.). Most studies in this area thus far have focused on the phagocytosis of cells that have undergone apoptosis due to infection by bacteria or viruses. In these cases, expression of foreign particles (fragments of the infectious agent) on the surface of the apoptotic cell triggers phagocytosis, mediated by blood-borne mediators of the innate immune response. Some evidence is emerging that receptors for cell surface phosphatidylserine are important for this, and studies with inhibitory drugs implicate several members of the classical signaling pathways (e.g., monomeric G proteins, PI3K, MAP kinases) as well, though no universal consensus has emerged. It is likely that the precise mechanism of phagocytosis of apoptotic cells is specific to the cell types involved.

FIGURE 13-29 Different types of vertebrate caspases are shown schematically. Note the prodomains and protein–protein interaction regions of the initiator and inflammatory caspases. DED = Death effector domain.

CONCEPT CHECK #2

Possessing the ability to trigger apoptosis is inherently dangerous to individual cells, as the ability to trigger death without sufficient protection against accidental triggering is an open invitation to cellular suicide. Yet such a partial apoptosis mechanism likely existed in our early ancestors if apoptosis arose by natural selection. (A common argument made by creationists is that evolution offers no reason for "half" of a complex mechanism, such as apoptosis, to exist.) What evidence can you cite to support the argument that apoptosis evolved from ancestral, prokaryotic proteins? Be creative, but follow the principles of evolution.

13.4 Chapter Summary

To live or die is the most important decision a cell makes, and most cells make this decision almost constantly, as they endure various forms of injury. The decision is made after very careful monitoring of a multitude of cellular functions. This monitoring, in turn, requires extensive intercommunication between regulatory proteins, analogous to the metabolic and signaling pathways that keep healthy cells functional.

The importance of the live-or-die decision is especially evident when cells are triggered to replicate. The sequence of events necessary for full replication of a cell is called the cell cycle, and specific points in this cycle mark periods when cells carefully deliberate whether to proceed. These points are called checkpoints, and the master checkpoint that defines when the cycle begins is called the restriction point. The proteins enforcing checkpoints are responsible for ensuring proper base pairing, replication, compaction, and segregation of DNA strands, as well as the even distribution of other cellular structures and even the volume of the cell. These proteins are organized in a hierarchical fashion. A DNA damage sensor complex, called the MRN complex, lies near the top of the list, and halts the cell cycle when it detects single- or double-stranded breaks in DNA; in turn, this complex activates two proteins, named ATM and ATR, that collectively control all other proteins involved in DNA repair and maintenance.

Below ATM and ATR lie a multitude of signaling proteins and regulatory proteins. Of these, some of the most critical are the cyclin-dependent kinases and their associated cyclins. Activated cyclin/CDK complexes can, by themselves, trigger division of a naïve cell that has received no other stimulus to divide. Addition and removal of phosphate groups to these proteins play a central role in controlling progression through cell cycle checkpoints. The sequence of events is often grouped into pathways, as classical signal transduction mechanisms are.

The decision to die, or undergo apoptosis, can be made at almost any point in the cell cycle. While some apoptosis is carefully planned (e.g., programmed cell death), most apoptosis occurs when a cell is too damaged to repair. Two inputs to the apoptosis machinery are receptors at the cell surface and cytosolic proteins that trigger mitochondrial outer membrane permeabilization; these two routes are called the extrinsic and intrinsic pathways, respectively. Apoptosis, unlike signal transduction, is irreversible because it is mediated by protease enzymes rather than by short-term protein modifiers such as kinases, G proteins, and phosphatases. As one wave of proteases is activated, they cleave additional "pro" forms of other proteases to activate them, propagating the destructive wave. The end product of apoptosis is a collection of small apoptotic bodies containing the partially digested remains of the cell interior. These bodies are internalized by neighboring cells through phagocytosis.

CONCEPT CHECK ANSWERS #1

▶ Addition of PDGF to cells, in mid-G1. This will likely help drive the cell through the restriction point, by a pathway similar to that shown in Figure 13-7. Phosphorylation of Jun will enable it to bind to the transcription factor fos, forming the AP1 heterodimer that binds to the regulatory site on several genes, including cyclin D. The cyclin D protein will bind to CDK4 or CDK6, forming a complex that can be activated, driving cells into S phase. Once past the restriction point, the cell should complete the entire cell cycle.

▶ Addition of PDGF to cells, in early G2. This will possibly halt progression into M phase by the mechanism shown in Figure 13-18. Receptor tyrosine kinases can activate MAPK pathways, and the Pom1 kinase is activated by MAPK signaling. Activation of Pom1 will inhibit Cdr1 and Cdr2, thereby allowing Wee1 to inhibit cyclin B / CDK1.

▶ Removal of nutrients from cells, in S. This will likely have no impact on S phase events, and, based on experiments with yeast, it is likely that the cells will not be sensitive to the removal of the nutrients until they reach G_1 phase. At that point, the cells will arrest (enter G_0) until the nutrients are restored.

▶ Removal of nutrients from cells, in M. This will likely have the same effect as removal of nutrients in S phase: the cells will arrest in G_0 once they complete M phase.

▶ Addition of a drug that inhibits Pom1, in G2. This would likely trigger M phase and completion of the cell cycle. Inhibition of Pom1 would introduce a *fourth* layer of inhibition controlling cyclin B / CDK1: no Pom1 activity→Cdr1 and Cdr2 are active→Wee1 is inactive→cyclin B / CDK1 is active.

▶ Loss-of-function mutation for cyclin B, in any phase. Cyclin B controls the activity of CDK1 in G_2 phase. Without it, CDK1 cannot be activated, thus cells should not be able to escape G_2, regardless of where they were in the cycle when the mutation occurred.

When one examines individual elements of the apoptotic mechanism, apoptosis requires nothing new to evolve. For example:

▸ The executioner caspases are simply proteases, and every cell expresses proteases as part of its normal housekeeping duties. Mutation of proteases can easily explain how an ancestral protease mutated to form the first caspase.

▸ Caspases are inhibited by their own tails, in a "pro" form. Pro-forms of proteins are quite common, and a powerful way to regulate protein function; the collagens and elastin are examples we have discussed in this book (see Chapter 6).

▸ The oligomerization of Bax, Bak, APAF-1, and so on, occurs by altering the shape of proteins to expose binding sites, in complete agreement with the three traits of proteins. The same is true for the inhibitory proteins.

▸ Insertion of Bax and Bac into the outer mitochondrial membrane is quite similar to how other transmembrane proteins enter mitochondria, and the fact that they form a channel is all but expected, since channel formation is one of the most likely outcomes of a complex made of multiple copies of the same polypeptides. The nuclear pore complex makes the Bax/Bak channel look simple, by comparison.

▸ The death receptor functions the same way most signal transduction receptors do: it forms a multimer, binds to a ligand in the extracellular space, and recruits a cytosolic adaptor protein to bind its tails as a result. All protein–protein associations in this pathway are mediated by specific binding domains, just as we saw for signal transduction.

So, how can one explain "half" of an apoptosis mechanism in a cell? The death receptor may have evolved from a signal transduction pathway that became less relevant over time, such that mutation would be tolerated rather than prove immediately fatal. By itself, it does no harm. The same argument applies for nearly every other component of the intrinsic and extrinsic pathways—the Bax/Bak channel may have provided permeability to another molecule before it mutated to permit cytochrome c to escape, and so on. Even the caspases, without a mechanism for activating them, would have been harmless. Since we cannot observe every organism that ever existed on Earth, we cannot conclusively prove this idea, just as we cannot experimentally prove major steps in early evolution. But the fact that proteins closely related to apoptosis proteins are used by modern prokaryotes adds great strength to this argument.

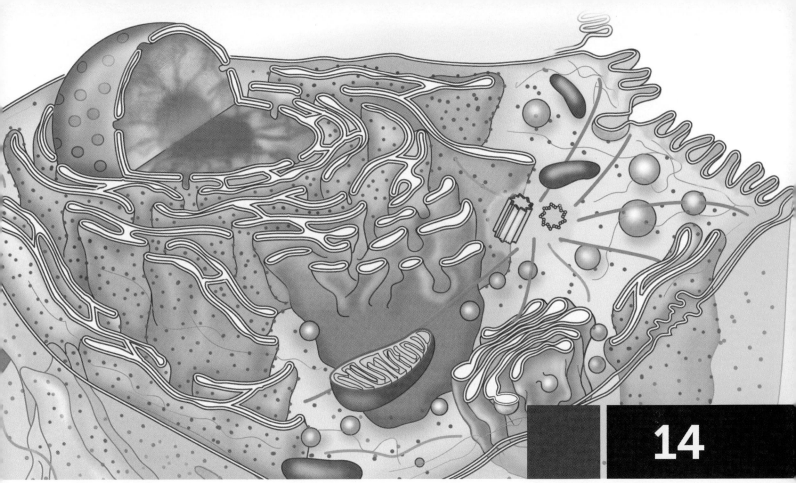

Tissues

14.1 The Big Picture

The first 13 chapters of this book have collectively addressed the fundamental structure and function of cells, especially those in multicellular animals. For many students, the next highest level of biological organization is covered in courses in anatomy and physiology, which address how groups of cells work as teams to perform functions that individual cells simply cannot. This suggests that the cellular teams may have developed these abilities *de novo,* independent of their cellular ancestry. But natural selection, currently the model accepted by virtually all scientists of how life on Earth evolved, argues strongly against this notion. Just as we have used natural selection to explain how complex eukaryotes arose from less complex prokaryotes, we can also apply natural selection to explain how specialized teams of cells arose as well. The key concept is that every evolutionary "advance" arises from variations of existing biological organization. Thus, anatomy and physiology are built on the principles of cell biology.

In most cases, courses in anatomy and physiology begin with this as a given fact, leaving the students to generate the bridges between cell biology and higher levels of biological organization themselves. The goal of this chapter is to help identify and explain some of these bridges. To guide us, we'll use the last principle in this book, cell biology principle #10: tissues form macroscopic equivalents of individual cells.

This means that most of the concepts we have discussed in this book, including the previous nine principles, can be scaled up to the next level of cellular organization (in cell biology, the word *macroscopic* is commonly used to describe a structure that can be seen with the unaided eye, in contrast with *microscopic*). Using this reasoning, a tissue can be viewed as simply the next largest functional unit of multicellular organisms, so this chapter also serves as a review of the previous principles and shows how they apply to tissues and organs as well. For example, tissues are composed of extracellular material surrounding cells, just as cells are composed of phospholipid membranes and the molecules they enclose. Numerous other examples can be found in texts discussing cell physiology.

This chapter is organized into four major sections, each focused on one of the four tissue types in animals. Keep in mind that the goal of these sections is to illustrate how cell biology merges into anatomy and physiology, without comprehensively covering the structure or function of these tissues. As with previous chapters, we will look for patterns, but this time consider them in a broader context, using the principles as a guide. For example, what is most valuable at a different level of biological organization, and how is it protected (principle #3)? How are decisions made (principle #8)? What is the most vulnerable aspect of life at this other level, and is there any way to correct errors (principle #9)? These questions apply to tissues, organs, individuals, populations, and even ecosystems. Once we see how these levels are connected, learning more biology could become much simpler, and that, ultimately, is what every biology text is striving for.

14.2 Epithelial Tissues Form Protective, Semipermeable Barriers between Compartments

KEY CONCEPTS

▸ Epithelial tissues share important properties with cellular membranes: they possess structural polarity and protect the material they enclose, forming a semipermeable barrier.
▸ Epithelial cells contain specialized structures that help them form strong bonds to each other and the extracellular matrix.
▸ Epithelial tissues are specialized to perform specific functions, including protection, secretion, transport, and absorption.

During gastrulation, a developmental stage found in nearly all metazoans, the embryo changes from a hollow ball to a three-layered structure containing the three **germ layers** that give rise to all other cells in the organism. These layers are called the **ectoderm** (outer layer), **mesoderm** (middle layer), and **endoderm** (inner layer), as shown in **FIGURE 14-1**. Epithelial tissues are derived from all three of these layers; for example, epithelia in the skin, ovaries, and lungs arise from the ectoderm, mesoderm, and endoderm, respectively. In general, the word *epithelium* (plural, *epithelia*) is roughly equivalent to the more familiar word *membrane.* By applying principle #10, we can classify epithelial tissues as the macroscopic equivalent of the plasma membrane in cells. (A similar structure called an *endothelium* arises from endoderm and lines the cavities of blood and lymphatic vessels. We will group them together with epithelia due to their structural and functional similarities.) Just as a plasma membrane serves as a semipermeable boundary in a cell, a membrane is a selective boundary that protects the cells it covers; in addition, each epithelium is also specialized to provide additional functions specific to the needs of those cells, just as the

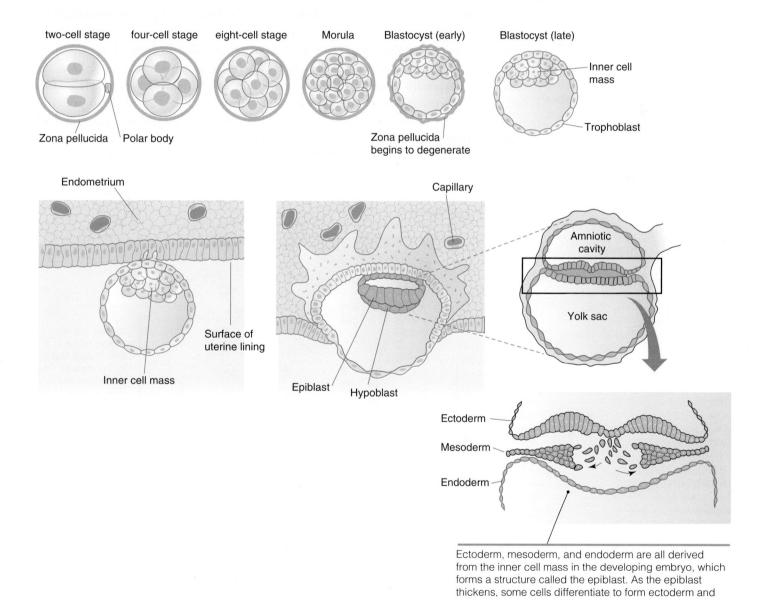

FIGURE 14-1 Origin of three embryonic germ layers in mammals.

plasma membrane of different cell types is specialized. Let's begin by discussing some of the common features shared by epithelia, and then turn our attention to some examples of different epithelial tissues to illustrate how these specializations create new functions in multicellular organisms.

Epithelial Cells Have Structural Polarity

The compartments that most epithelia protect are hollow and shaped like tubes or pouches, very similar to the structures formed by phospholipid bilayers. The hollow space is generally called a lumen, though some have more specific names (e.g., *alveolar space* in lungs). In addition to protecting the underlying tissue from the contents of the lumen, epithelia maintain chemically distinct compartments on either side of the membrane. This, too, is quite similar to the function of a cellular membrane, but it occurs on a much larger scale. **FIGURE 14-2** shows a comparison between these two membrane types. While the concept of selective transport applies to both membranes, the types of materials selectively transported by epithelia are, in general, much larger than those transported across phospholipid bilayers.

A second important function of epithelia is to protect underlying tissues from physical trauma. Whereas cellular membranes are held together primarily by weak noncovalent chemical bonds between individual phospholipids, lipids, and proteins, the cells in an epithelium are typically joined by an enormous number of bonds between cellular and extracellular proteins to form a very sturdy sheet. The exact makeup of these protein–protein bonds varies from one epithelium to another, but is grounded in a fundamental property of all epithelial cells: they possess structural polarity. As we discussed in Chapter 5

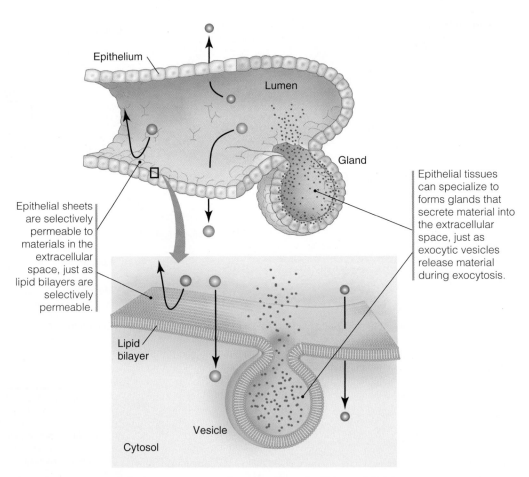

Epithelial sheets are selectively permeable to materials in the extracellular space, just as lipid bilayers are selectively permeable.

Epithelial tissues can specialize to forms glands that secrete material into the extracellular space, just as exocytic vesicles release material during exocytosis.

FIGURE 14-2 Epithelial tissues are macroscopic analogs of cellular membranes.

(see Box 5-8), the term *polarity* has many different meanings in biology, and in this case it refers to a specific arrangement of organelles and other structures in a cell. Polarized cells have at least two distinct "sides" that are structurally and functionally distinct. (For a review, see *Tight Junctions Form Selectively Permeable Barriers between Cells* in Chapter 6.) In epithelial cells, structural polarity is represented by apical and basolateral domains on the plasma membrane. These domains are defined by the junctional complex we discussed in Chapter 6, which is responsible for the mechanical strength of epithelial tissues.

■■ Epithelial Tissues Are Classified According to Cellular Structure and Function

FIGURE 14-3 shows some examples of different epithelial tissues. **Simple epithelia** are composed of a single layer of epithelial cells directly in contact with the basal lamina (see *The Basal Lamina Is a Specialized Extracellular Matrix* in Chapter 6). **Simple squamous epithelia** contain flattened epithelial cells, and are found in tissues where gas diffusion across the epithelium is most important, such as in capillaries and the smallest chambers of the lung. **Simple cuboidal** and **simple columnar** epithelia contain cube- and column-shaped epithelial cells, respectively, and are found in locations where the epithelial cells secrete and/or absorb material in the lumen (e.g., ovary, kidney nephrons, intestinal lining). **Stratified epithelia** contain two or more layers of epithelial cells and are further classified according to the shape of these cells (e.g., stratified squamous epithelia); because of the multiple layers, not all cells in stratified epithelia are connected to the basal lamina. **Transitional epithelia** contain several layers of different types of epithelial cells, and are typically found in organs that routinely swell and shrink, such as the bladder.

■■ Some Epithelial Tissues Are Optimized for Protection: The Epidermis

The term *organ* refers to groups of tissues that function together, and therefore represents the next highest order of organization above tissues. Skin is considered an organ, because it is composed of several tissues, as **FIGURE 14-4** shows. The outermost tissue of skin is called the **epidermis** and is an example of a **keratinized stratified squamous epithelium**, consisting of a layer of epithelial cells connected to the basal lamina, covered by several layers of squamous cells. One distinctive feature of keratinized epithelium is that the topmost cells undergo a specialized form of programmed cell death called **cornification**, so the surface of skin is actually made of dead cells. This top layer is called the keratinized layer because the cells express tremendous amounts of the intermediate filament protein keratin and incorporate it into desmosomes (see *Desmosomes Are Intermediate Filament-Based Cell Adhesion Complexes* in Chapter 6) linking neighboring cells. As the top cells die and dry out, most of the cellular contents are shed, leaving behind a network of intermediate filaments that provide tremendous strength to the epidermis.

Some regions of epithelium are further specialized to form pocket-shaped **glands** that secrete substances. According to principle #10, these represent the scaling up of individual pump

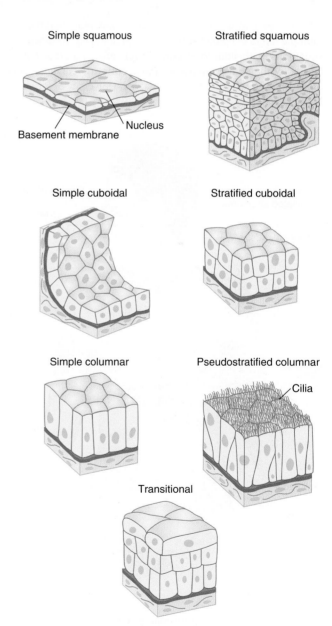

FIGURE 14-3 Examples of different types of epithelial tissues.

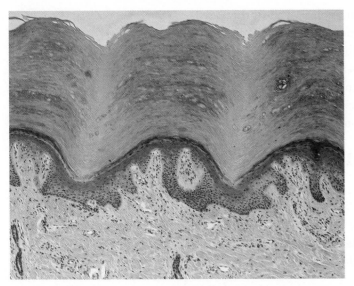

FIGURE 14-4 Structural organization of skin. Note that the multiple layers of epithelial cells in the keratinized, stratified squamous epithelium (dark purple, top) serve as a protective barrier for the underlying tissues (pink, bottom). © Donna Beer Stolz, Ph.D., Center for Biologic Imaging, University of Pittsburgh Medical School.

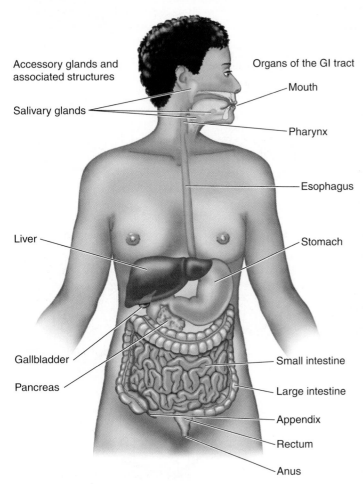

FIGURE 14-5 The gastrointestinal system. © leonello calvetti/ ShutterStock, Inc.

proteins on the cell surface (see *Cells Use a Combination of Channel, Carrier, and Pump Proteins to Transport Small Molecules across Membranes* in Chapter 10). The epidermis of all terrestrial mammals contains sebaceous (oil) and eccrine (sweat) glands. **Sebaceous glands** secrete **sebum**, the oily substance that confers water resistance to skin. Infection of sebaceous glands by *Propionibacterium acnes* bacteria triggers an inflammatory immune response that often leads to the appearance of the visible lesions characteristic of acne vulgaris, or common acne. Recent studies suggest that the epithelial cells lining sebaceous glands play many important roles in the maintenance of skin health, including control of the immune system, production of hormones, and delivery of antioxidants to the skin surface.

Some Epithelial Tissues Are Optimized for Absorption: The Gastrointestinal System

The gastrointestinal system, illustrated in **FIGURE 14-5**, is a series of organs responsible for digesting and absorbing food and water. Most of the lining of the mouth, including the gums, is composed of keratinized stratified squamous epithelium. Salivary glands secrete saliva to aid in the physical and chemical breakdown of food, and these glands contain stratified columnar epithelial cells. Nonkeratinized stratified squamous epithelium, which lacks the dead, cornified layer found in keratinized epithelia, is present in the esophagus. As food passes from the mouth through the esophagus, it encounters a simple columnar epithelium in the stomach. The stomach contains at least four different types of epithelial cells, and is organized into deep folds called **gastric pits** that secrete acid and a digestive enzyme called gastrin. To help prevent the gastrin from digesting the proteins in the epithelia (i.e., the stomach would digest itself), these cells are covered by a thick layer of mucous, as **FIGURE 14-6** shows. (Note how similar this is to the glycocalyx that lines lysosomes, illustrating principle #10; see Figure 9-17). This mucous is composed of heavily glycosylated glycoproteins that are delivered to the apical surface by exocytosis. Effective delivery of the mucous requires these epithelial cells to be properly polarized; delivery to the basal surface would not protect the cells because the acid and gastrin are present in the lumen. Thus, the health of an entire organ is heavily dependent on the organization of the cytoskeleton, motor proteins, and organelles in the epithelial cells lining said organ.

After passing through the stomach, partially digested food enters the small intestine, where digestion is completed by enzymes delivered from the pancreas. This is a very tightly orchestrated process, because the body must be careful to

The figure labels for FIGURE 14-5:

Accessory glands and associated structures
- Salivary glands
- Liver
- Gallbladder
- Pancreas

Organs of the GI tract
- Mouth
- Pharynx
- Esophagus
- Stomach
- Small intestine
- Large intestine
- Appendix
- Rectum
- Anus

digest the food without damaging its own organs, which are of course built from the same fundamental building blocks, as we discussed in Chapters 1–4. As the food is digested in the small intestine, it is absorbed by active transport (see *Energy-Coupled Carrier Proteins (Pumps) Build Gradients* in Chapter 10). The absorption of glucose we discussed in Chapter 10 (see Figure 10-15) is a good example of how this absorption takes place. Water absorption occurs by osmosis, as salts (especially sodium) are actively transported from the intestinal lumen into the epithelium; this is the same mechanism individual cells use to absorb water, and again illustrates our principle #10. As the bulk of a meal moves from the small to the large intestine, absorption of nutrients is completed and the remaining waste is concentrated for excretion.

Some Epithelial Tissues Are Optimized for Transport: The Kidney

The function of the kidney is to filter and concentrate wastes in the extracellular fluid for excretion by the urinary system. The overall strategy for accomplishing this is illustrated in **FIGURE 14-7**. A small artery (called an **arteriole**) carries blood into a receiving chamber called Bowman's capsule, where the fluid portion of blood is filtered through capillaries; all of the blood cells remain in the exiting arteriole, while the bulk of the fluid enters a long tube called a **nephron**. This filtering is not very selective, and a lot of important materials in blood depart the capillaries, and so one of the first tasks facing a nephron is recapturing these valuable materials and transporting them back into the blood vessel. The nephron is lined by a simple cuboidal epithelium, and the portion immediately downstream of the Bowman's capsule, called the **proximal tubule**, is where most of the reabsorption takes place; **FIGURE 14-8** shows how this is accomplished. Because the cells lining the nephron lumen are epithelial, they are linked by junctional complexes. Tight junctions in the junctional complexes limit the diffusion of materials through the spaces between cells, so these cells must transport most of the valuable materials out of the lumen. Some of these materials are membrane soluble (e.g., steroid hormones) and can diffuse freely across the cell membranes, but the rest must be moved by active transport. Unlike in the small intestine, the flow rate through the nephron is quite high, so the cells in the tubule have a small window of time to recapture these materials. In cases where the concentration of a molecule is unusually high in the blood (e.g., exceptionally high glucose levels in diabetics), some of the material is not recaptured and instead is excreted in the urine; this is one of the reasons why a urinalysis is a common means of testing for diabetes.

Carcinomas Are Cancers Originating from Epithelial Cells

Because they cover and protect other tissues, most epithelial cells are exposed to a variety of traumas. Part of the body's defense permits epithelial cells to be scraped off or undergo programmed cell death quite

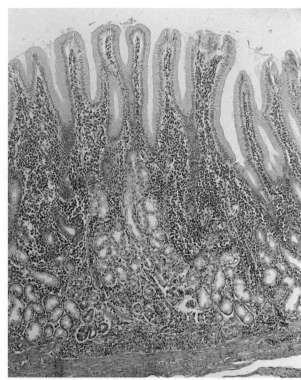

FIGURE 14-6 Cross section of the stomach. Note the deep folds (gastric pits). The protective mucous layer is stained light pink (top) and covers the epithelial cells in this image. © Donna Beer Stolz, Ph.D., Center for Biologic Imaging, University of Pittsburgh Medical School.

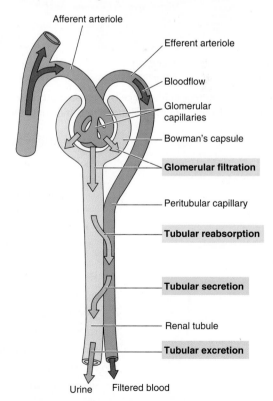

FIGURE 14-7 Summary of kidney function. The nephron collects wastes filtered from the blood in Bowman's capsule and converts them into urine for excretion.

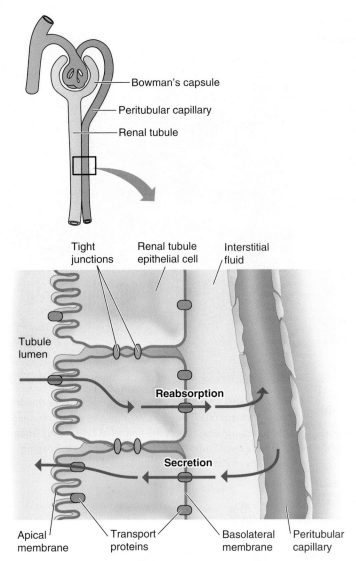

Tight junctions

Renal tubule epithelial cell

Interstitial fluid

Tubule lumen

Reabsorption

Secretion

Apical membrane

Transport proteins

Basolateral membrane

Peritubular capillary

Bowman's capsule

Peritubular capillary

Renal tubule

FIGURE 14-8 Kidney epithelial cells use transport proteins (pumps, carriers, and channels) to move material into and out of the bloodstream.

frequently, so there is a high turnover of epithelial cells in healthy animals. This creates a need for many replacement cells, so epithelial cells progress through the cell cycle more frequently than most other cells (note that this does not mean that they move through the cycle *faster;* they simply spend less time between divisions). Combined with the effects of exposure to damaging agents, the high frequency of division means that these cells are more likely to mistakenly pass mutations on to their progeny cells. For this reason, tumors arise most frequently in epithelial tissues. Recall from Chapter 13 (see *Cell Cycle Step 6: DNA Integrity Is Ensured by the G_1/S, S/G_2, and G_2/M Checkpoints*) that mutations in cell cycle regulators, especially checkpoint proteins, can greatly diminish the cell's ability to correct genetic errors. If tumors spread to other regions in the body, they are classified as cancers. The most common cancers (e.g., lung, breast, colon, skin) are all derived from epithelial tissues in the affected organs. Epithelial-derived cancers are called **carcinomas**.

CONCEPT CHECK #1

Because they resist mechanical force and environmental trauma, most common injuries take place in epithelial tissues. Apply cell biology principle #10 to propose an analogy for how an individual cell heals its wounds.

14.3 Nervous Tissues Store and Transmit Information as Electrical Charge

KEY CONCEPTS

▸ Nervous tissues are composed of neurons and supportive cells called glial cells.

▸ Neurons transmit information in the form of an electrical current that is passed between neurons. The network of interconnected neurons in an organism is functionally analogous to the signaling network in a single cell.

▸ The electrical current is conducted through individual neurons by a coordinated transport of ions across the plasma membrane. This transport is collectively called an action potential.

Recall that in Chapter 2 we used a library as an analogy for the nucleus, because it stores nearly all the information necessary for a cell to function properly, in the form of DNA (see Box 2-2). Let's now extend that analogy a bit: a library cannot keep a community functional unless there are individuals able to translate the library's information into activities outside the library. As we discussed in Chapters 8 and 9, cells use elaborate strategies for translating DNA information in the nucleus into activities outside the nucleus. When we apply this same analogy to tissues, a similar pattern emerges: nervous tissues store critical information an individual needs to remain alive, but this information must be translated into actions by other tissues. By applying principle #10, we can classify nervous tissues as the macroscopic equivalent of the information transfer network—composed of

DNA, signal transduction, and regulation of transcription and translation—in cells; this is illustrated in **FIGURE 14-9**. The key difference between these two systems is the way in which information is stored. In nervous tissue, the information responsible for controlling the actions of other tissues is stored as electrochemical gradients across cellular membranes, not as DNA. Note that gradients of charged atoms like this first appeared in the energy storage mechanisms we discussed in Chapter 10; natural selection simply allowed some cells to adapt this mechanism for a new purpose, and electrical signaling became an important communication system in multicellular organisms.

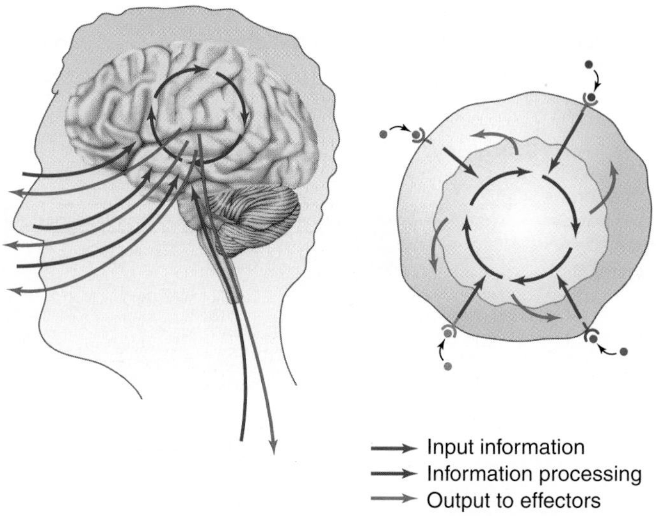

Just as we discussed for epithelial tissues, different nervous tissues are specialized to perform a subset of the tasks required to store and transmit information. By analogy, some nervous tissues serve as information storage and processing centers—much like DNA in a nucleus—while others control the transmission of information into and out of these centers, similar to signaling pathways that enter the nucleus and trigger changes in gene expression resulting in protein synthesis. This link between nervous tissues and cellular signaling was introduced in Chapter 11 as cell biology principle #7. To better understand this analogy, let's have a closer look at how nervous tissues store and transmit information.

→ Input information
→ Information processing
→ Output to effectors

FIGURE 14-9 The nervous system is a macroscopic analog of cellular signal transduction. Incoming information is carried by neurons (left) or signaling molecules (right), processed, then translated into outgoing information that targets tissue effectors (e.g., muscles, glands) or molecular effectors (e.g., motor proteins).

Nervous Tissues Are Composed of at Least Two Different Cell Types

Nervous tissues contain two types of cells: neurons, which store and transmit electrical information, and glial cells, which provide support to the neurons but do not actually manipulate the information. This is very similar to how genetic information is stored; recall from Chapters 7 and 8 that the most stable form of genetic information is chromatin, which is composed of DNA *and* supporting proteins, just as nervous tissue is nerves *and* supporting cells. Both are needed to keep the information safe and accessible.

Neurons Transmit Signals via Action Potentials

When nervous tissues send electrical signals, the signals must pass from one cell to another in a reliable, reproducible fashion. Neurons have solved this problem by adapting a property all cells share, the chemical disequilibrium across their plasma membrane. As cells pump ions across the plasma membrane, the charge separation creates an electrical potential, as we discussed in Chapter 10 (see *The Amount of Potential Energy Stored in an Ion Gradient Can Be Expressed as an Electrical Potential*). All mammalian cells have a resting potential of approximately −60 to −80 milliVolts. Neurons use this potential to store and transmit neural information. To do so, they generate a wave of electric current that flows from one end of the cell to another; this wave is called an action potential. Action potentials are controlled by signal transduction mechanisms that first appeared in prokaryotes to aid with the selective opening and closing of ion channels. Discussing action potentials gives us an excellent opportunity to review the concepts introduced in Chapters 10 and 11 in a very different context.

FIGURE 14-10 illustrates an example of how an action potential is generated. It begins when a **ligand-gated ion channel** in the plasma membrane binds to its ligand (called a

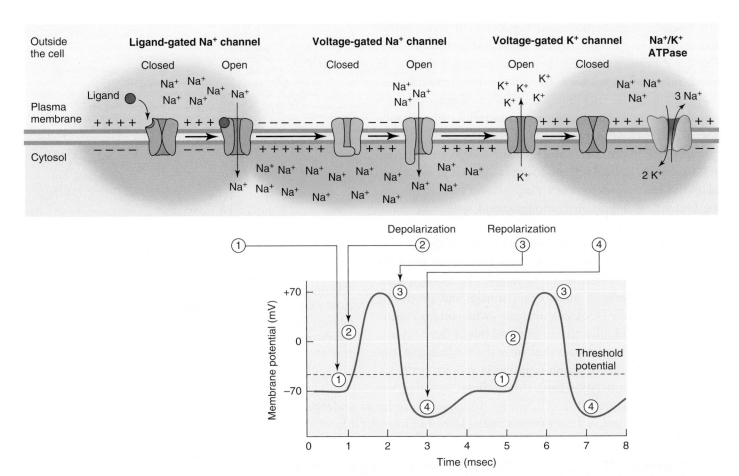

FIGURE 14-10 Summary of action potential generation in a neuron. At step 1, a neurotransmitter binds to a ligand-gated Na⁺ channel, opening it. Na⁺ ions move into the cytosol, partially depolarizing the plasma membrane. Once the threshold potential has been reached, step 2 begins and voltage-gated Na⁺ channels open, permitting more Na⁺ ions to enter the cytosol and further depolarize the membrane. At step 3, voltage-gated K⁺ channels open, permitting K⁺ ions to cross from the cytosol to the extracellular space, and the membrane begins repolarizing. At step 4, membrane polarization is restored, and K⁺ channels close. During this period, voltage-gated Na⁺ channels cannot open (refractory period). After the refractory period ends, another action potential can begin, repeating steps 1–4.

neurotransmitter), resulting in a shape change in the channel protein that permits ions to pass through it. This is known as a *fast response*, because the same protein that binds the signal also initiates the depolarization. In other cases, the receptor activates a heterotrimeric G protein signaling pathway to reach the ion channel; these are called *slow responses*, and they can result in opening or closing of ion channels. Regardless of their speed, nearly all neurotransmitter receptors are located in dendrites. While many different ion channels can impact an action potential, by convention most physiologists concentrate on Na⁺ channels when describing the onset of an action potential. Once the Na⁺ channels open, they permit Na⁺ to flow through the channels from outside the cell to the cytosol. Within a few milliseconds, a "cloud" of Na⁺ ions forms near the cytosolic face of the plasma membrane around the open channels; this is called depolarization of the membrane, because the polarity of the electrical potential reverses in this region of the membrane. (Note how this use of the word *polarity* differs from that used to describe epithelial cells earlier.) One way to visualize the depolarization is to imagine that the molecules in the cytosol are normally colored red, while the Na⁺ ions outside the cell are colored blue; depolarization occurs when enough blue ions enter the cytosol to create a blue-green patch near the membrane.

Within this depolarized zone, **voltage-gated Na⁺ channels** are triggered to open once a **threshold potential** (about 20 milliVolts more positive than the resting potential) is reached, allowing still more Na⁺ ions to enter the cell. The process repeats as more of the membrane is depolarized, spreading the zone of depolarization. Thus, opening the first Na⁺ channels starts a chain reaction that permits Na⁺ ions to enter the cell in an outwardly spreading circle. In our color visualization, we would see a blue-green patch spread along the cytoplasmic side of the plasma membrane like a wave spreads across a body of water when a pebble is dropped in. A key feature of action potentials is that they spread spontaneously, and at extremely high speed.

This rapid propagation is only useful, however, if neurons can manage it. They do so by placing nearly all of their ligand-gated channels in one region of the cell, called the **dendrite**. Dendrites are located at the extreme edge of neurons, as **FIGURE 14-11** shows. With respect to the direction of travel of an action potential, neurons are highly polarized cells. (Note that we are once again shifting the meaning of the word *polarity*, back to cellular architecture as we discussed with epithelial cells.) Action potentials that begin in the dendrite regions of neurons flow "forward" through the **soma** (body) and into the **axon hillock**. Every neuron has only a single axon hillock, so all action potentials that pass through the soma ultimately arrive at the same destination regardless of where they began. The function of the axon hillock is to integrate all of the excitatory (and inhibitory) signals delivered by the dendrites, and thereby generate a single response, which is passed as an action potential to the **axon**. The axon, which may be branched, carries the action potential to the portion(s) of the plasma membrane closest to the target(s) of the neuron (remember that neurons work in teams, so they pass action potentials from one to another in a sequence). This near-neighbor region is called the **synapse**, and we will discuss it in more detail later in the chapter.

Before we pass the action potential forward to another cell, let's back up and discuss how action potentials end. Let's return to Figure 14-10, and focus on what happens to the Na⁺ channels over time. The ligand-gated channels, which started the entire reaction, are inactivated when their ligand no longer binds to them (their concentration at the receptor typically drops over time until a new nerve impulse triggers more to be released). Closing the voltage-gated channels, however, is a bit more complex. Once enough of these channels open to create a membrane potential of approximately +60 milliVolts, the electrical

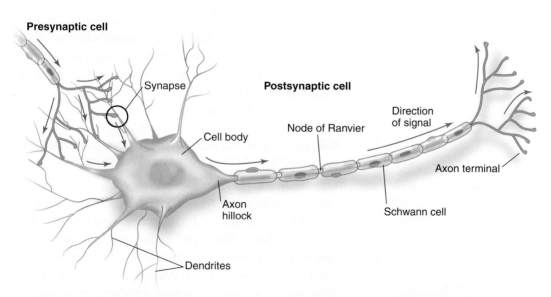

FIGURE 14-11 Structural organization of nerve cells. Nerves generate action potentials at dendrites, and these propagate through the soma to the axon hillock, then on to the axon terminal, where they trigger release of neurotransmitters.

environment surrounding these proteins induces another shape change that closes them (see *Proteins Change Shape in Response to Changes in Their Environment* in Chapter 3). This halts the chain reaction, and the channels then enter what is known as a **refractory period**—during this time, they cannot open, even if the electrical potential is favorable for opening. This refractory period ensures that two action potentials do not overlap at the axon hillock or the synapse.

Opening and closing the Na⁺ channels leaves the cell with a new problem: it is still depolarized. To solve it, neurons express **voltage-gated K⁺ channels** that open when the plasma membrane potential approaches +60 milliVolts. This permits K⁺ ions to move down their chemical gradient and out of the cell (recall that at resting potential, the concentration of K⁺ is much higher inside a cell than outside; see *The Na⁺/K⁺ ATPase Maintains the "Resting Potential" across the Plasma Membrane* in Chapter 10). The outflow of K⁺ ions rapidly restores the negative potential in the cell—in fact, it briefly becomes too negative (also called hyperpolarized). At this point, the membrane depolarization state has been restored, but at a cost: the cell has allowed Na⁺ ions to enter and K⁺ ions to leave, and these must be returned to their proper locations to maintain homeostasis. This final job is performed by the Na⁺/K⁺ ATPase. Notice that an action potential is simply a discrete sequence of ion transport events, grounded by the same pump protein responsible for maintaining ion gradients in all cells.

Why don't all cells generate action potentials? Only those that express the specific proteins discussed here can, and these proteins are derivatives of ion transport proteins that have been present since the very first cells evolved. According to the current theory of evolution, the variation introduced by mutations caused some cells to express "mutant" channel proteins that responded to changes in voltage, and this provided an advantage that, due to natural selection, spread to other organisms over time. Thus, evolution occurs at the level of individual proteins, as we discussed in Chapter 2 (see *Mutations in DNA Give Rise to Variation in Proteins, Which Are Acted Upon by Natural Selection*).

▪▪ Glial Cells Support Neurons and Increase the Speed of Action Potential Transmission

Glial cells constitute approximately 90% of the cells in nerve tissue, but they are not directly involved in transmission of electrical signals. There are five types of glial cells in the human body. Two types, known as **oligodendrocytes** and **Schwann cells**, participate in accelerating the transmission of action potentials through neurons by wrapping their plasma membrane around the axons of neurons, as **FIGURE 14-12** shows. These regions of the plasma membrane are highly specialized to act as electrical insulators that inhibit ion exchange, and are called **myelin sheaths**.

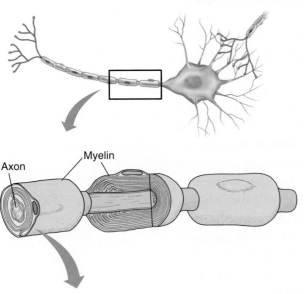

Axon Myelin

Neurolemmocyte Cytoplasm Neurolemma Myelin sheath

Axon

Nucleus

(a) PNS

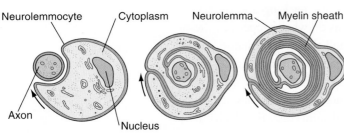

Axon

Oligodendrocyte

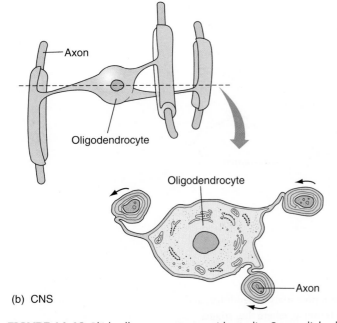

Oligodendrocyte

Axon

(b) CNS

FIGURE 14-12 Glial cells wrap neurons with myelin. Some glial cells wrap a portion of only one axon, while others can wrap many axons.

Special gaps in the myelin sheaths, called **Nodes of Ranvier**, permit ion exchange across the neuron plasma membrane. By limiting ion exchange to these areas, myelin accelerates action potential transmission. One way to visualize this is to consider myelinated portions of an axon "invisible" during an action potential, because almost no ion exchange takes place there. From this perspective, the depolarization caused by influx of Na^+ ions at one node is not propagated as a wave, but rather as a series of short "jumps" from one node to the next. The same sequence of ion transport events takes place at each node, but because of the myelin sheaths, the action potential moves faster down axons that are myelinated. Not all neurons have a myelinated axon.

The insulating properties of myelin are due to the exceptionally high lipid concentration (approximately 80%) in the plasma membrane that forms it. These lipids repel water and ions, and include standard membrane phospholipids as well as derivatives of a lipid called **sphinganine**, as shown in **FIGURE 14-13**. When a sphinganine is linked to a fatty acid and modified to create a double bond, the structure is called **ceremide**. Ceremide resembles diacyl glycerol, the foundation of phospholipids, in that it contains two long chain lipids linked by a hydrophilic bridge (see Figure 4-1). Most ceremide in myelin is further modified to form two different **glycolipids**. One of these, galactocerebroside,

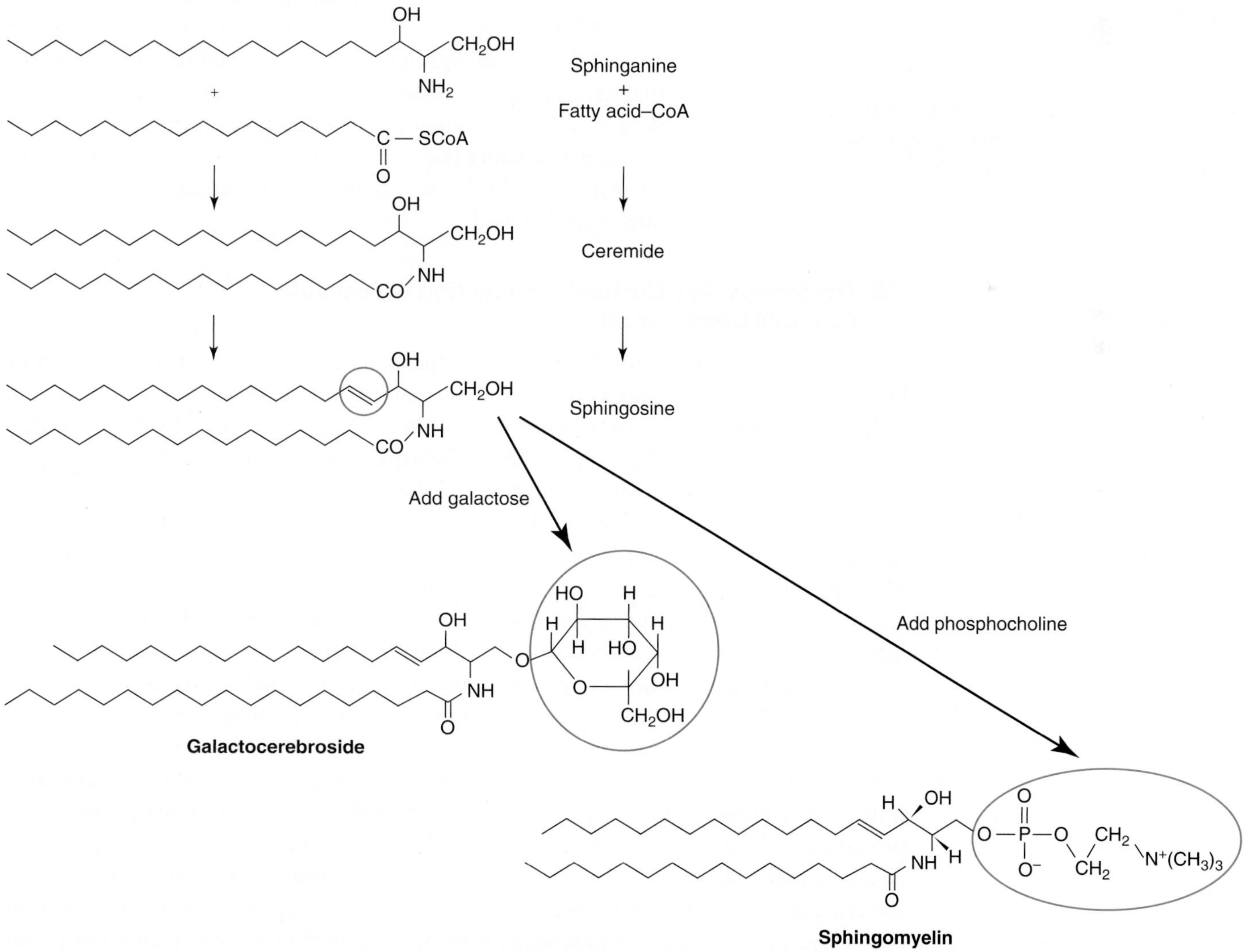

FIGURE 14-13 Synthesis of two lipids enriched in myelin. These and other lipids insulate the neuron plasma membrane.

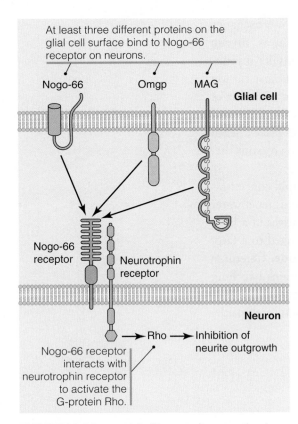

At least three different proteins on the glial cell surface bind to Nogo-66 receptor on neurons.

Nogo-66   Omgp   MAG

Glial cell

Nogo-66 receptor

Neurotrophin receptor

Neuron

Rho → Inhibition of neurite outgrowth

Nogo-66 receptor interacts with neurotrophin receptor to activate the G-protein Rho.

FIGURE 14-14 A model of how myelin-associated glycoprotein (MAG) and Nogo-66 receptor participate in suppression of neurite outgrowth in neurons.

contains a single galactose sugar attached to the hydrophilic bridge; when a phosphocholine is attached to the bridge, the molecule is called sphingomyelin. Note that galactose and phosphocholine are the structural analogs of polar head groups in phospholipids. Both galactocerebroside and sphingomyelin are assembled on the surface of the endoplasmic reticulum then shipped to the plasma membrane via the endomembrane system, as we discussed for phospholipids in Chapter 4 (see *Additional Membrane Lipids Are Synthesized in the Endoplasmic Reticulum and Golgi Apparatus* and Figure 4-15).

Contact between the neuron plasma membrane and glial cell membrane is established by two transmembrane proteins: Nogo-66 receptor in the neuron, and myelin-associated glycoprotein (MAG) in the glial cell, which binds to sialic acid sugars on the Nogo-66 receptor. The sialic acid sugars are added to the receptor as it passes through the Golgi apparatus en route to the plasma membrane (see *Posttranslational Modification of Proteins in the Golgi Is a Stepwise Process* in Chapter 9). The Nogo-66 receptor was given this name because when it attaches to several proteins in glial cells, including MAG, neurons are not able to extend any dendrite or axon progenitors (called **neurites**); this is illustrated in **FIGURE 14-14**. Thus, when nervous tissue is injured, the presence of MAG / Nogo-66 receptor complexes on the surrounding cells prevents them from sending out new dendrites or axons to bridge the damaged area, and the tissue loses function. If this damage occurs in areas of the brain controlling response inhibition (during so-called Go/Nogo tasks), performance is severely impaired. Overcoming this inhibition is a major goal for those seeking to help damaged brains restore functionality as they heal.

■■ The Synapse Is a Customized Junction to Facilitate Cell–Cell Communication

Once an action potential reaches the end of the axon, the information it encodes must be passed downstream to the next cell in the sequence. Nearly all neurons use a structure called a **chemical synapse** to transmit this information, as shown in **FIGURE 14-15**. Perhaps the most distinctive feature of a chemical synapse is the tiny space, called the **synaptic cleft**, between the axon plasma membrane and the membrane of the target cell. When an action potential reaches the end of the axon (also called an axon terminal), the depolarization phase of the action potential triggers the opening of voltage-gated Ca^{+2} channels. These permit Ca^{+2} ions to enter the cytosol from the extracellular space; once inside the cell, they bind to membrane proteins in special secretory vesicles called synaptic vesicles. This triggers fusion of the vesicles with the plasma membrane, completing exocytosis.

Synaptic vesicles are an example of regulated secretion, which we discussed in Chapter 9 (see *Cells Use Two Mechanisms for Controlling the Final Steps of Exocytosis,*). These vesicles contain neurotransmitter molecules, which act as ligands for receptor proteins in the plasma membrane of the target cell. Unlike most exocytosed molecules, neurotransmitters are synthesized in the cytosol, then transported into the vesicle interior by carrier proteins (see *Passive Carrier Proteins Dissipate Gradients* in Chapter 10), and stored until an action potential triggers their exocytosis. The neurotransmitter molecules diffuse across the synaptic cleft and bind to their receptors, thereby passing the action potential from one cell to the next. The synaptic vesicles are recycled by endocytosis, then refilled with neurotransmitter for the next round of exocytosis.

If a nerve cell in culture was fed a medium containing a nonhydrolyzable analog of ATP (e.g., ATP with sulfur atoms replacing oxygen atoms in the covalent bonds linking the phosphates together), it would no longer generate action potentials. Explain why.

14.4 Muscle Tissues Convert Chemical Signals into Mechanical Force

KEY CONCEPTS

▸ Muscles are effectors targeted by the nervous system. They are composed primarily of muscle tissue.

▸ Muscle tissues are classified into three types: skeletal, cardiac, and smooth muscle; each contains a different type of muscle cell. Collectively, muscle and bone tissues form the functional analog of the cytoskeleton in tissues.

▸ Skeletal muscle cells are multinucleated, terminally differentiated cells composed primarily of parallel actin and myosin bundles called sarcomeres. They are stimulated to contract by the electrical current passing from neurons to the muscle cells.

▸ Cardiac muscles are structurally similar to skeletal muscle cells, except they contract autonomously and are found only in the heart.

▸ Smooth muscles lack sarcomeres, and are capable of contracting in several directions. This enables them to line tubes such as blood vessels and airways and control their diameter. Smooth muscle cells are also capable of sustaining contractions much longer than skeletal or cardiac muscle cells.

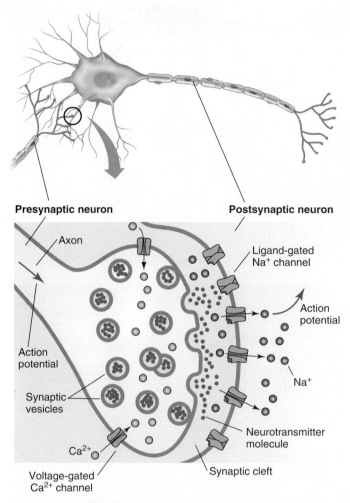

FIGURE 14-15 Transmission of information at a chemical synapse. At rest, a neurotransmitter is stored in secretory vesicles near the membrane of the axon terminal. As the action potential reaches the axon terminal, voltage-gated Ca^{+2} channels open, triggering exocytosis of secretory vesicles. These bind to receptors on the target cell, initiating a new action potential.

Nerve cells communicate with three classes of target cells: other neurons, endocrine cells (which secrete hormones), and muscle cells. Just as signaling pathways target effectors to generate a response, muscles act as effectors by translating neural information into motion; muscles, therefore, implement many of the decisions made by the nervous system. If we view the nervous system as an elaboration of a signaling network, by analogy muscles are the actin/myosin and microtubule-associated motor systems that generate motion in a cell. And, like the different cytoskeletal networks committed to different types of motion, muscle tissues are subdivided into three types, shown in **FIGURE 14-16**, each responsible for a different type of motion in a multicellular animal.

Skeletal Muscle Cells Are Multinucleated, Highly Specialized Cells

The most abundant muscle type in humans is skeletal muscle. As its name implies, skeletal muscle is responsible for movement of the skeleton, the functional analog of the cytoskeleton in individual cells.

From the perspective of scale, skeletal muscle is simply a collection of actin/myosin complexes large enough to generate the force necessary to move the skeleton. The structural

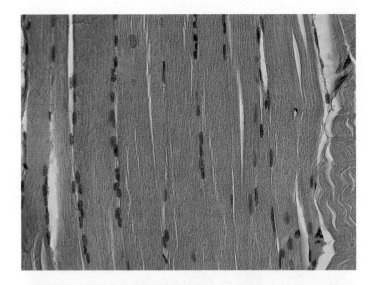

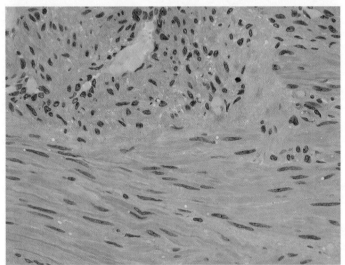

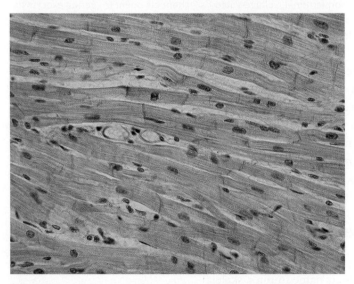

FIGURE 14.16 Three classes of muscle tissue: skeletal (top), smooth (middle) and cardiac (bottom). Note that both skeletal and cardiac muscle form large, multinucleated cells called fibers. Images courtesy of © Donna Beer Stolz, PhD, Center for Biologic Imaging, University of Pittsburgh Medical School.

organization of skeletal muscle reflects this single-minded purpose, and is one of the best examples of how the single-cell contractile machinery has scaled up.

■■ Sarcomeres Are Combined to Form Muscle Fibrils

To generate force, myosin motor proteins must bind to and move actin filaments in a carefully coordinated fashion. We first discussed this concept in Chapter 5 (see *Striated Muscle Contraction Is a Well-Studied Example of Cell Movement*) when we introduced the sarcomere; now, let's revisit the sarcomere to illustrate how it is scaled up to form skeletal muscle tissue. Notice in **FIGURE 14-17** that several hundreds or even thousands of sarcomeres can be aligned end-to-end to form a **muscle fibril**, which spans the entire length of a muscle cell. Hundreds or even thousands of fibrils are bundled together in parallel in a single muscle cell, so that the force generated by each contracting sarcomere is oriented in the same direction. When viewed under a microscope, the aligned sarcomeres look like stripes, so the term ***striated*** (from Latin, referring to a furrow) is often used as well.

If we look more closely at the structure of muscle fibers, the scaling effect becomes quite clear. Notice in Figure 14-16 that these fibers are surrounded by a single plasma membrane, but have several nuclei. This arrangement is somewhat unusual in animal cells, and is a direct result of fusion between hundreds or thousands of individual muscle precursor cells. As animals became larger over the course of evolution, they reached a size where a single, mono-nucleated cell could no longer span the distance between two muscle attachment points on the skeleton, so several muscle cells teamed up to form a single unit. (This strategy is so effective that now mononucleated skeletal muscle cells only appear during development, when muscles are first built.) These fused muscle cells are considered to be **terminally differentiated**. As we discussed in Chapter 13 (see *"Resting" Cells Reside in G_0 or G_1 Phase*), they enter G_0 phase for the remainder of their lives. This is the practical solution to the problem of a single cell having several nuclei.

■■ Action Potentials Stimulate Skeletal Muscle Contraction

As **FIGURE 14-18** shows, some regions of the muscle fiber plasma membrane are specialized to form the receiving side of a synapse. These regions are called motor end plates, and when one of these is paired with the axon terminal of a nerve, it forms a special synapse called the neuromuscular junction. This means that the plasma membrane of skeletal muscle cells can conduct action potentials in response to neurotransmitters released by the neuron. (The mechanism of propagating an action potential in a muscle cell is very similar to that used by neurons.) Another distinctive feature of skeletal muscle cells is the transverse tubule, often called the T tubule, formed by deep folds of the plasma membrane. As action potentials travel outward from the neuromuscular junction and along the plasma membrane, they also travel down the T tubules, where they depolarize portions of the smooth endoplasmic reticulum (called the **sarcoplasmic reticulum** in striated muscle cells), triggering the release of Ca^{+2} ions from the ER lumen. The presence of Ca^{+2} ions in the cytosol activates muscle contraction, as shown in Figure 14-18.

▪▪▪ Skeletal Muscle Contraction Is Driven by the Crossbridge Cycle

At rest, myosin is not bound to actin within the sarcomere. A three-protein complex wrapped around actin filaments covers the myosin binding site, and initiation of contraction occurs when this complex changes shape after binding to Ca^{+2} ions. Once this complex slides away from the binding site, myosin latches on and immediately begins pulling on the myosin head. The orientation of the myosin heads in the sarcomeres ensures that the actin filaments slide toward the center, shortening the length of the sarcomere. Because they are linked to one another in series, the shortening of each sarcomere is cumulative for an entire myofibril. Adjacent myofibrils can slide past one another if necessary, but typically most myofibrils in the same cell contract at nearly the same time, so an entire muscle fiber will commit to a contraction. So long as action potentials continue to arrive at the neuromuscular junction, Ca^{+2} ions will remain in the cytosol and myosin will continue to pull. This is how sustained skeletal muscle contractions are maintained.

A muscle remains contracted as long as myosin continues to exert force on actin. The mechanism controlling myosin activity in skeletal muscle is called the **crossbridge cycle**, illustrated in **FIGURE 14-19**. It consists of five steps:

1. The cycle begins when myosin makes contact with actin, initiating sliding of the actin. Myosin is also bound to ADP at this step.
2. The myosin completes a power stroke, sliding the actin approximately 80 Angstroms (8 nm) toward the center of the sarcomeres. The organization of the sarcomeres ensures that several myosins can bind to a single actin. In one experiment, a single actin filament moved more than 600 Angstroms after one round of the crossbridge cycle, suggesting that a myosin may "hand off" its actin to another myosin to achieve maximum sliding.
3. Once the power stroke is complete, myosin remains attached to actin until an ATP molecule binds to it. This condition is called **rigor**, because it is where the cycle stops when ATP is depleted in the muscle cell.
4. When ATP binds to myosin, it releases the actin filament. This also causes the release of ADP.
5. Finally, the myosin uses the energy released by ATP cleavage to reset. This is often called "cocking," a reference to the cocking of a firearm prior to its firing.

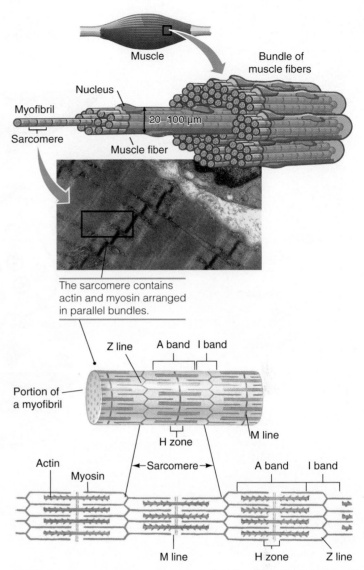

FIGURE 14-17 Structural organization of skeletal muscle. Courtesy of Louisa Howard, Dartmouth College, Electron Microscope Facility.

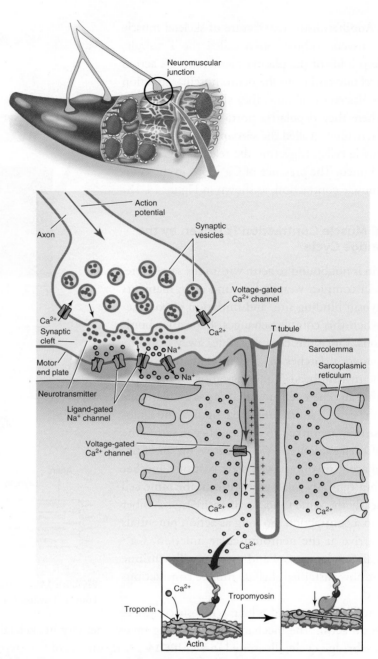

FIGURE 14-18 How an action potential triggers skeletal muscle contraction.

Cardiac Muscle Is Responsible for Pumping Blood

Cardiac muscle is another form of striated muscle, but is found only in the cardiovascular system, where it forms the contractile tissue (e.g., heart muscle) that pumps blood. In many ways, cardiac muscle cells resemble skeletal muscle cells in that they are multinucleated, terminally differentiated, and their actin-myosin contractile machinery is aligned in sarcomeres. One important difference, however, is that cardiac muscle cells are branched rather than linear. This is important, because it allows these cells to generate force in more than one direction at the same time. Whereas skeletal muscle has one main function—to pull two bones closer to each other—cardiac muscle has the more complex job of squeezing blood through a series of tubes (blood vessels).

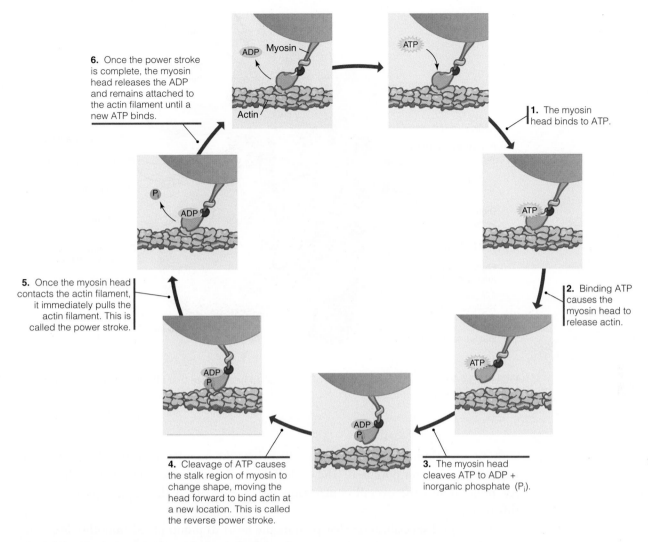

6. Once the power stroke is complete, the myosin head releases the ADP and remains attached to the actin filament until a new ATP binds.

ADP Myosin

ATP

Actin

1. The myosin head binds to ATP.

2. Binding ATP causes the myosin head to release actin.

5. Once the myosin head contacts the actin filament, it immediately pulls the actin filament. This is called the power stroke.

4. Cleavage of ATP causes the stalk region of myosin to change shape, moving the head forward to bind actin at a new location. This is called the reverse power stroke.

3. The myosin head cleaves ATP to ADP + inorganic phosphate (P$_i$).

FIGURE 14-19 The crossbridge cycle for skeletal muscle contraction.

The job of pumping blood is quite complicated, especially in a multichambered heart such as a human heart, as **FIGURE 14-20** shows. The reason is that the contractions of the heart muscle must be carefully timed to ensure unidirectional flow. This can be easily visualized by squeezing a water balloon: random squeezing shifts the water in several directions, disrupting any uniform flow, which can only happen when one end of the balloon is squeezed in a wave-like fashion. In a four-chambered heart, four such waves have to take place each time the heart beats. To complicate matters further, the blood flow changes directions as it passes through the heart.

▇■ The Pacemaker Is a Specialized Form of Cardiac Muscle

This choreography is so critical that the heart can control its beating pattern even when it is disconnected from the rest of the body. Three specializations make this possible. First, some heart muscle cells have evolved so that they no longer contract, but instead create action potentials on their own, without the need of nerve input. These are called **pacemaker cells**, and they are concentrated in two areas of the heart called the sinoatrial (SA) node and the atrioventricular (AV) node. These cells generate their own action potentials by slowly letting ions move across the plasma membrane

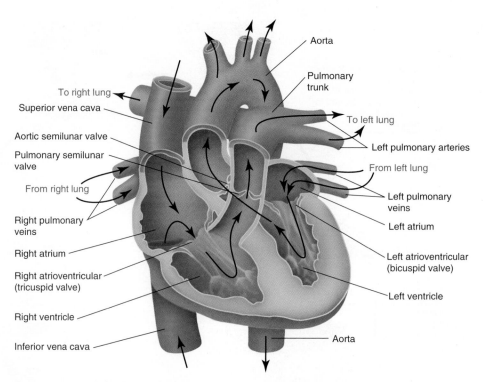

FIGURE 14-20 Structural organization of mammalian heart. Note that blood flow changes direction in the left and right ventricles.

Labels in figure:
- Aorta
- Pulmonary trunk
- To right lung
- Superior vena cava
- To left lung
- Aortic semilunar valve
- Left pulmonary arteries
- Pulmonary semilunar valve
- From left lung
- From right lung
- Left pulmonary veins
- Right pulmonary veins
- Left atrium
- Right atrium
- Left atrioventricular (bicuspid valve)
- Right atrioventricular (tricuspid valve)
- Left ventricle
- Right ventricle
- Aorta
- Inferior vena cava

without any stimulus, and this eventually brings the membrane to threshold potential, triggering an action potential. The cycle time in the SA node is faster than that in the AV node; it sets the overall contraction rate and is called the cardiac pacemaker for this reason.

The second specialization that permits the heart to pump blood smoothly is a group of modified cardiac muscle cells that conduct the action potential away from the SA node. These cells, collectively called **conduction fibers**, ensure that the action potential reaches each region of the heart in the proper sequence. They conduct action potentials much faster than the rest of the cardiac muscle cells.

■■ Cardiac Muscle Cells Are Electrically Coupled

The third specialization solves one of the most pressing problems in the heart, namely how the initial action potential generated by the SA node is spread to the rest of the contractile cells. If they communicated the way most neurons do, each muscle cell would have to build chemical synapses with its neighbor cells, then convert the action potential into neurotransmitter release, which would then have to be converted into a new action potential in the recipient cell, and so on. Instead, cardiac muscle cells are joined by collections of gap junctions called **intercalated disks**, shown in **FIGURE 14-21**. Because gap junctions permit ion exchange between cells (see *Gap Junctions Allow Direct Transfer of Molecules between Adjacent Cells* in Chapter 6), an action potential can be passed from one cell to another without any conversion whatsoever. The use of electrical synapses ensures that action potentials generated by the SA node pass through the first chamber of the heart (the right atrium) in a wave, thereby maintaining consistent blood flow. The contraction of the other three chambers is triggered by the conduction fibers, which are also linked by gap junctions.

Cardiac Muscle Contraction Is Determined by Intrinsic and Extrinsic Controls

In addition to the pacemaker in the SA node, which sets the contraction rate of the heart on its own (intrinsic control), cardiac contraction can also be affected by the nervous system (extrinsic control). Specifically, nerves form chemical synapses with pacemaker cells in the AV node, and then release neurotransmitters that activate signaling pathways in the pacemaker cells. These pathways trigger ion channels in the plasma membrane, and thereby control the rate of membrane depolarization, which in turn controls the rate of action potential generation, and ultimately heart rate and blood flow. This helps explain how meditation and other relaxation techniques can help slow heart rate.

Smooth Muscle Cells Generate Force in Three Dimensions

Smooth muscle differs significantly from striated muscle, as Figure 14-16 shows. One of the most visible differences is that it is not striated, which indicates that the actin/myosin complexes mediating contraction of smooth muscle are not aligned in neat rows, but are spread out. Actin filaments collect at regions called **dense bodies**, somewhat analogous to the Z discs in sarcomeres. This does not mean that smooth muscle is less effective at contracting, or that it generates less force because its contractile machinery is disoriented. Instead, smooth muscle cells possess a trait that neither skeletal nor cardiac muscle cells have: they can

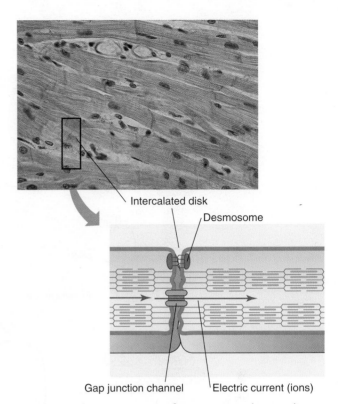

FIGURE 14-21 Transmission of action potentials in cardiac muscle. Note that ions pass directly from the cytosol of one cell to the next, carrying electric current. No neurotransmitter is needed to carry information between cells. © Donna Beer Stolz, PhD, Center for Biologic Imaging, University of Pittsburgh Medical School.

simultaneously contract in all three dimensions. This makes smooth muscle cells especially effective for controlling the diameter of tubes, and most tubular structures in the human body are lined by smooth muscle. The actin cytoskeleton of smooth muscle cells is also dynamic, and can rapidly remodel in response to the load it bears during contraction.

A second key difference between smooth muscle and striated muscle is that smooth muscle cells can maintain continuous contraction for prolonged periods (hours), while striated muscle can maintain contraction for minutes at most. The reason for this is reflected in the mechanism controlling myosin activation, shown in **FIGURE 14-22**. Instead of using Ca^{+2} ions to expose myosin binding sites on actin, this mechanism uses Ca^{+2} ions as part of a calmodulin-based signal transduction pathway (see *Calcium Fluxes Control Calcium-Binding Proteins* in Chapter 11). This pathway stimulates **myosin light chain kinase** to phosphorylate a serine in the regulatory light chain subunit on myosin, locking it in an activated conformation until the phosphate is removed by **myosin light chain phosphatase.** This allows myosin in smooth muscle to remain active for much longer periods than in striated muscle.

Because contraction of smooth muscle can persist for long periods of time, the decisions to contract and relax must be made carefully. By applying cell biology principle #8—protein complexes are cellular decision-making devices—we can predict that the machinery upstream of myosin light chain kinase and myosin light chain phosphatase should include at least one protein complex. In fact, the machinery is so intricate that the total number of proteins involved on regulating smooth muscle contraction is still not known. Figure 14-22 shows some of the signaling pathways involved. All six classes of signaling molecules discussed in Chapter 11 (see *Intracellular Signaling Proteins Propagate Signals within a Cell*) are represented.

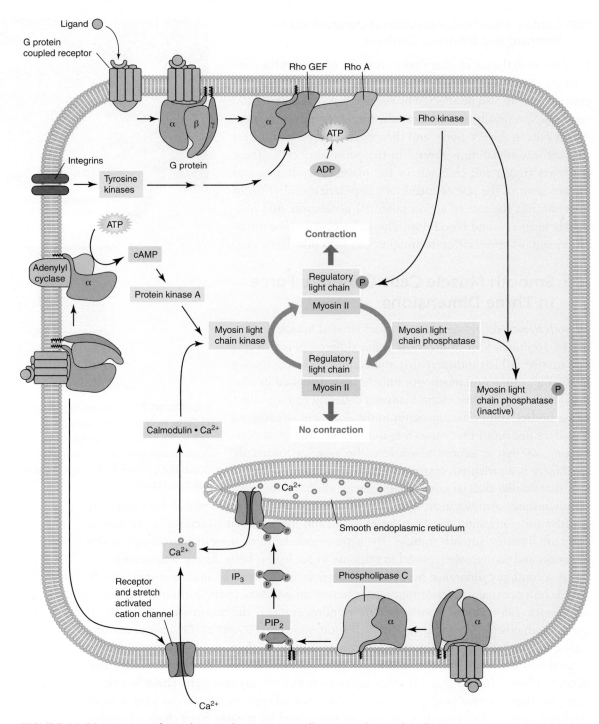

FIGURE 14-22 Summary of signaling mechanisms controlling smooth muscle cell contraction. Note that all converge on myosin II regulatory light chain phosphorylation.

![Icon] **Vascular Smooth Muscle Cells Control the Diameter of Blood Vessels**

Vascular smooth muscle forms a layer called the **media** around all types of blood vessels that are larger than capillaries; this list includes the aorta and other arteries, arterioles, veins, and venules, as shown in **FIGURE 14-23**. The main function on this layer is to control the flow of blood through the vessels, and thereby control blood pressure. **FIGURE 14-24** illustrates that at rest, vascular smooth muscle is partially contracted, so that it can respond to signals and contract further or relax to raise or lower blood pressure, respectively. These signals include hormones and other signaling molecules released into the bloodstream (e.g., adrenaline) that bind to cell surface receptors, and direct stimulation by nerves. By expressing different receptors, smooth muscle surrounding different blood vessels can respond differently to the same signal. Release of adrenaline from the adrenal glands causes some blood vessels to constrict (e.g., in the intestine), while others relax (in skeletal muscles), thereby shunting blood to the most critical organs during a fight-or-flight response. Chronic constriction of some vascular smooth muscle contributes to high blood pressure, kidney failure, and other debilitating conditions, and is the target of many hypertension-reducing drugs.

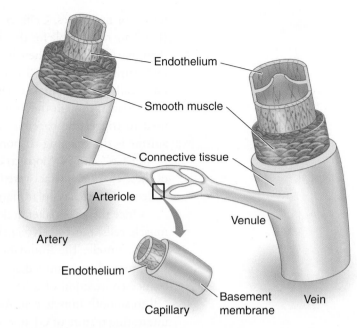

FIGURE 14-23 Smooth muscle forms a contractile layer around all but the smallest blood vessels.

■■■ Airway Smooth Muscles Control Airflow

Like vascular smooth muscle, airway smooth muscle forms a layer around all but the smallest airways in the trachea and lungs. In healthy individuals, partial constriction of this layer helps keep airways open by providing the necessary tension, often called tone, to maintain an oval shape of the airway wall. In addition, constriction is important during development of the airway and during healing of injuries. In rare cases, severe constriction can close off a branch of the airway to prevent inhaled particles from migrating into and damaging the extremely fragile terminal sacs, called **alveoli**. However, over constriction is largely associated with asthma, one of the most common respiratory illnesses. Asthma may be triggered by a number of irritants, ranging from smoke and other pollutants to allergens (which trigger an inflammatory response in the lungs). The signaling mechanisms that control constriction of airway smooth muscle are as complex as those for other types of smooth muscle, so finding the exact cause of an individual's asthma is extremely difficult. Current treatments induce dilation of airway smooth muscle, but this must be used with caution as they are rarely selective for airways, and may interfere with other smooth muscle types, including those in the vascular system.

■■■ Smooth Muscle Is Responsible for Peristalsis in the Gut

Smooth muscle cells that line the gastrointestinal (GI) tract, the long tube connecting the mouth with the anus, exhibit a behavior not seen in most other smooth muscle types. To ensure proper digestion of food and absorption of nutrients, these cells contract in carefully timed waves that propel the food mass forward. This contraction is called peristalsis (from the Greek words meaning *wrapped around an area,* or *constricted*). One easy way to experience this behavior is to feel the contraction of the throat during a swallow. The act of swallowing triggers a peristaltic wave that moves through the esophagus.

For well over 100 years, biologists have known that coordinated smooth muscle contraction was responsible for the peristaltic wave in the GI tract, but even the most recent studies have yet to fully explain how peristalsis is controlled. Most current models suggest two levels of regulation. The first establishes a baseline rate of contraction called the slow wave. This regulation is called intrinsic or enteric, because it occurs independent of any outside signals, such as from

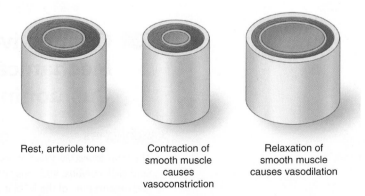

Rest, arteriole tone

Contraction of smooth muscle causes vasoconstriction

Relaxation of smooth muscle causes vasodilation

FIGURE 14-24 Vascular smooth muscle can contract and relax to control the diameter of blood vessels.

nerves or hormones. Cells in the smooth muscle cell layer called Interstitial Cells of Cajal (ICC) are responsible for the slow wave, which is caused by generating action potentials and then passing them to their neighboring smooth muscle cells, which contract. This is strikingly similar to how the SA and AV nodes trigger cardiac muscle contraction, and ICCs are often called GI pacemaker cells for this reason. ICCs are present throughout the length of the GI tract, though they vary considerably in structure, and the exact mechanisms ICCs used to trigger the pacemaker action potential vary by location. Central to most mechanisms is the opening of nonselective cation channels on the plasma membrane, which permit Ca^{+2} and Na^+ ions to enter the cytosol, depolarizing the plasma membrane. Some of these channels are activated by outward stretching of the GI tube, implicating the cytoskeleton and extracellular matrix as sensors of a food mass (often called a bolus) in the gut.

Once the pacemaker action potential occurs, it is distributed to neighboring smooth muscle cells, most likely via clusters of gap junctions resembling the intercalated disks of cardiac muscle. The abundance of these structures varies in different ICCs, suggesting that some may play an immediate role in triggering the contraction of many cells, while others induce contraction of a few cells and let the contractile wave pass from smooth muscle cell to smooth muscle cell. All GI smooth muscle cells are linked via gap junctions. One interesting feature of GI smooth muscle is that the ion primarily responsible for the depolarization phase of an action potential is Ca^{+2} rather than Na^+. Most of the Ca^{+2} that enters the cytosol in this phase comes through voltage-gated calcium channels on the cell surface, rather than from the sarcoplasmic reticulum.

The second level of smooth muscle cell contraction is extrinsic, and occurs primarily through synapses formed between nerve cells and ICCs. The type of neurotransmitters released by these nerve cells varies according to location within the GI tract, as well as by the kind of stimulus delivered by the nerve. Some stimuli increase membrane permeability to cations, accelerating the frequency and increasing the strength of peristaltic contractions (relative to the intrinsic pacemaker), while others have the opposite effects, by hyperpolarizing the ICC membrane. Similarly, hormones released by endocrine cells lining the GI tract can bind to cell surface receptors on ICCs and smooth muscle cells, activating signaling pathways that regulate membrane permeability to ions and/or phosphorylation of myosin light chain.

CONCEPT CHECK #3

If a heart is surgically removed from an animal, such as a frog, it will continue beating for several minutes in the dissecting tray if kept wet with a solution containing NaCl. However, if a solution containing potassium chloride (KCl) is poured onto the heart, it will stop beating. Bathing a stopped heart in NaCl solution again will restore its beating, and this start-stop cycle can be repeated several times before the heart dies. Apply your understanding of action potential and cardiac muscle contraction to explain these observations.

▪ 14.5 Connective Tissues Provide Mechanical Strength and Cushioning to the Animal Body

KEY CONCEPTS

▸ The term *connective tissue* refers to a wide variety of tissues that collectively fill spaces between epithelial, nervous, and muscle tissues. In most cases, connective tissue is characterized by the type and organization of the ECM it contains. It is analogous to the ECM surrounding individual cells.

▸ Most connective tissues are classified into five types. Each type is populated by different cells that secrete and organize the ECM.

Unlike the three other tissue types, most connective tissue is characterized by the type and organization of its ECM, rather than by the cells it contains. The reason is fairly straightforward: well-organized ECM structures are stronger (and often bigger) than individual cells, so they contribute most to the mechanical properties of connective tissue. The specific function of connective tissues varies by location, but all share one common property: they provide the "glue" that holds cells and other tissues together. Recall from Chapter 6 that multicellular organisms exhibit three behaviors that distinguish them from unicellular forms of life (see *The Extracellular Matrix Is a Complex Network of Molecules That Fills the Spaces between Cells in a Multicellular Organism*), and that the ECM plays a critical role in all three:

- Form stable bonds between neighboring cells
- Govern the location and behavior of the cells
- Provide a means for intercellular communication

The exact same criteria apply to connective tissues. Thus, to scale up from individual cells to connective tissues is simply a matter of considering connective tissues as large collections of ECM molecules, such that they dominate the behavior of this tissue type. By extension, most descriptions of connective tissue focus on the composition and organization of the ECM, with one clear exception we will discuss below.

Most connective tissues fall into one of five classes , as shown in **FIGURE 14-25**, loose connective, dense connective, elastic, reticular, and adipose. One convenient means of understanding how many of these classes work is to examine a long bone such as a femur (thigh bone). The hardest part of the bone (sometimes called bone proper) is an example of dense connective tissue, because it contains a high density of ECM molecules (primarily type I collagen; see *Collagen Provides Structural Support to Tissues* in Chapter 6) and relatively few cells. The cells that secrete the ECM and mineralize it to form bone are called **osteoblasts** and **osteocytes**. The thin, translucent layer of tissue wrapped around the outer surface of the bone is called the periosteum, and it is an example of dense, *irregular* connective tissue; it is classified as irregular because its collagen fibers are not aligned in any specific direction or order. By comparison, the collagen fibers in bone proper are neatly aligned in parallel sheets stacked perpendicular to each other, similar to the structure of plywood.

The ends of long bones, which form joints, are lined by another form of dense connective tissue, cartilage, which is synthesized from cells called chrondrocytes. Cartilage is subdivided into three types: hyaline, elastic, and fibrocartilage. Hyaline cartilage is composed primarily of type II collagen and forms a strong, yet slippery surface that resists the eroding effects of bones rubbing against each other in a joint. Cartilage is also so well organized that no manmade substance, including Teflon and Kevlar, can match its strength and slickness on a per-weight basis. Thus, wherever the body is normally exposed to physical trauma, cartilaginous connective tissue helps absorb the force. It is perhaps not surprising that cartilage contains relatively few blood vessels, because they could be damaged by the physical trauma.

Two types of **fibrous connective tissues** connect to the outer surface of long bones. One is ligament, a flexible, cord-like material that holds bones together, especially across joints (e.g., femur to pelvis). Ligament is highly organized, with type I collagen fibers aligned along the long axis. The cells that secrete the collagen and align it in ligaments are called **fibroblasts**, because they build fibers. This arrangement means that a ligament resists tensile (pulling) forces along its long axis very well, but is comparatively weak when pulled in any other direction. This kind of alignment is called **anisotropic**, which means it cannot resist forces equally in all directions; note how the anisotropic organization of a ligament reflects a similar structure in individual type I collagen fibers (see Figure 6-3), providing another example of how cellular structure scales up to tissues. In contrast, the cartilage lining the ends of the bone is far more isotropic, because it resists force from

Cartilage, a dense connective tissue, is assembled by chondrocytes.

Tendon is a fibrous connective tissue that connects bones and skeletal muscles. It is assembled by tenocytes.

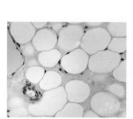

Adipose tissue is found in the marrow of some bones. An adipocyte stores lipid in a droplet that constitutes the bulk of the cell volume. A small layer of cytosol surrounds the droplet.

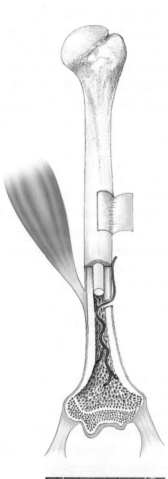

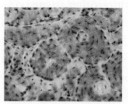

Bone, a dense connective tissue, is assembled by osteoblasts and osteocytes (small dots). Bone matrix surrounds blood vessels (large cavities).

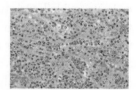

Reticular connective tissue (dark lines) is formed in the bone marrow by fibroblasts. The majority of cells in the marrow are blood cells.

Periosteum is an irregular dense connective tissue that lines the outer surface of bones.

Ligament is a fibrous connective tissue that connects bones. It is composed primarily of parallel bundles of collagen, and is assembled by fibroblasts.

FIGURE 14-25 Examples of connective tissue types found in and attached to bone. Cartilage © Astrid & Hanns-Frieder Michler/Photo Researchers, Inc. Bone © Donna Beer Stolz, PhD, Center for Biologic Imaging, University of Pittsburgh Medical School. Periosteum © Biophoto Associates/Photo Researchers, Inc. Reticular connective tissue © Carolina Biological Supply Co./Visuals Unlimited, Inc. Adipose tissue © Donna Beer Stolz, PhD, Center for Biologic Imaging, University of Pittsburgh Medical School. Tendon © Donna Beer Stolz, PhD, Center for Biologic Imaging, University of Pittsburgh Medical School. Ligament © Donna Beer Stolz, PhD, Center for Biologic Imaging, University of Pittsburgh Medical School.

many different directions (though no biological material is truly isotropic). As with cartilage, ligament is poorly vascularized, to reduce the risk of blood vessel damage caused by the mechanical forces to which ligaments are exposed.

The second fibrous connective tissue connected to bone is tendon, which links bone with skeletal muscle. Like ligament, tendon contains many type I collagen fibers aligned along the long axis, and is, in many ways, structurally similar to ligament. One important difference between tendon and ligament is that tendon contains a much higher proportion of elastin fibers, so it is more elastic and can absorb greater loads for short periods of time. The cells that build tendons are called tenocytes (meaning *cells that build tendons*), and like ligament and cartilage, tendon contains relatively few blood vessels, for the same reasons.

If we look inside the femur, three more types of connective tissue are visible. The first is **reticular connective tissue**, which forms a network of filaments in the bone marrow. These filaments are composed primarily of type III collagen fibers, arranged into a net-like structure. Lying between the fibers is **adipose tissue**, which is composed almost entirely of adipocytes (*fat cells*). In addition to serving as energy stores (in the form of triacyl glycerol and other long chain lipids stored in lipid vesicles), adipose tissue can also serve as cushioning and thermal insulation (especially under the skin). Adipose tissue is the only common connective tissue dominated by cells rather than ECM. **Blood plasma** (i.e., the liquid fraction of blood) is a special form of connective tissue, in that it supports and protects circulating blood cells. In healthy individuals, plasma constitutes approximately 55% of blood volume, and cells account for 45%. Thus, even blood is mostly connective tissue.

Finally, the tissue that contains the highest proportion of elastin is **elastic connective tissue**, and its primary function is to protect the surrounding cells and tissues by absorbing physical trauma and springing back to its original shape. There is no formal definition of elastic connective tissue; most biologists simply consider it to be areas of connective tissue that also contain high amounts of elastin. Good examples include body parts containing elastic cartilage, such as the ears and nose, and another is skin: elastic fibers run throughout the dermal layer of skin, and are responsible for its elastic behavior. As we grow older, the elastic fibers wear out and our skin sags, creating wrinkles.

CONCEPT CHECK #4

Explain why knee injuries are so common and take so long to heal. Include a discussion of connective tissue in your answer.

14.6 Chapter Summary

Tissues are collections of cells that work cooperatively to perform functions that individual cells cannot. While the number of distinct cell collections in an animal could be innumerable, they collectively form only four different tissue types. Because multicellular organisms arose from unicellular organisms, the structure and function of tissues obeys the same principles that govern cell biology, scaled appropriately to include multiple cells. This chapter serves as a brief introduction to the study of tissues, and is written to illustrate how the first nine principles discussed in the previous chapters apply to cell clusters.

Epithelial tissues are functional analogs of cellular membranes. They are composed of epithelial cells, which contain numerous specialized structures that enable them to form strong attachments to adjacent cells and the ECM. Epithelia are subdivided into classes, each specialized to perform a different set of functions. Some epithelia (e.g., in the skin) are especially good at forming protective barriers around other tissues, and possess several traits that make them especially resistant to environmental hazards. Others specialize in secretion (e.g., in glands), transport (e.g., in the kidney), and absorption (e.g., in the intestine).

Nervous tissues are responsible for the conduction of neural signals from one location in the body to another, and for processing these signals to trigger appropriate responses. These tissues are formed by two cell types: neurons, which carry the signals in the form of an electrical current, and glial cells, which play a supportive role by protecting and nurturing neurons. The molecular details of how this current is conducted are somewhat elaborate, but central to our understanding of how nerves function. In this chapter, we highlight some of the most important details of signal transmission in neurons, including the major steps in generating and transmitting an action potential. Networks formed by nervous tissues are functionally analogous to signaling networks within individual cells.

Muscle tissues are effectors of nerve signals. The three types of muscle tissue (skeletal, cardiac, and smooth) share one essential feature: their muscle cells use actin and myosin to generate contractile force. Skeletal muscles apply this force to move the skeleton in animals, while cardiac muscle contracts to pump blood. The contractile unit in skeletal and cardiac muscle is called the sarcomere, and is the best understood contractile mechanism in muscle. Smooth muscle cells do not organize their actin and myosin filaments into sarcomeres; instead, they distribute these filaments in all directions, and this enables them to contract in any direction. This increase in directionality comes at a cost, however: while smooth muscle cells can sustain contractions for much longer than skeletal or cardiac muscle cells, they cannot generate as much force as sarcomere-containing muscle cells. Collectively, the three types of muscle tissue are functionally analogous to the cytoskeleton of an individual cell.

Connective tissue is the most diverse tissue type. In general, the function of connective tissue is to provide mechanical strength to surrounding tissues, and to act as the adhesive material that holds these tissues in place. Some forms of connective tissue are highly specialized and form readily recognizable structures such as bone, cartilage, tendons, hooves, horns, nails, and teeth. For nearly all connective tissues, the primary function is determined largely by the amount, type, and organization of ECM within it. Multiple cell types secrete ECM in connective tissue; one of the most abundant in mammals is called a fibroblast. Because ECM plays such a prominent role in connective tissue function, it is analogous to how the ECM functions at the level of individual cells.

CONCEPT CHECK ANSWERS #1

"Wound healing" at the level of single cells occurs when damaged membranes reseal, damaged organelles are replaced, and damaged molecules are removed and replaced by newly synthesized molecules. Analogous to the border function of epithelia, the plasma membrane is the first barrier to cellular damage, and like the cell–cell and cell–ECM junctions formed by epithelial cells to reinforce the entire epithelial tissue, the plasma membrane can be reinforced by linking membrane proteins to the cytoskeleton. When damaged, the plasma membrane can spontaneously reseal due to the amphipathic property of membrane phospholipids, and damaged membrane proteins can be removed and replaced. Because the cell is fundamental unit of life (see Chapter 1), this healing must occur rapidly to preserve the chemical disequilibrium essential to keeping the cell alive. Thus, principle #10 applies to individual cells in this context because membrane phospholipids and proteins are replaced to maintain structural integrity, just as cells and ECM are replaced in damaged epithelia.

CONCEPT CHECK ANSWERS #2

Not being able to hydrolyze ATP means that the energy stored in the bond linking the second and third (β and γ) phosphates is not available to do work. Action potentials require gradients of Na^+ and K^+ ions across the plasma membrane, and these gradients are generated by the pump protein Na^+/K^+ ATPase. If the nerve cell cannot hydrolyze ATP, it cannot maintain these gradients, will rapidly dissipate their existing gradients, and then be unable to generate any new action potentials. If the cell is kept in this medium long enough, it would die because it would not be able to cleave ATP and would effectively starve to death.

CONCEPT CHECK ANSWERS #3

The dissected heart maintained in NaCl solution will continue beating until it has exhausted its energy supply or depleted its intracellular K^+ and Ca^{+2} stores. The NaCl in the solution is critical because it replaces the natural extracellular fluid, which is also rich in NaCl relative to the cell interior. The NaCl fluid helps the heart muscle cells maintain their Na^+ gradient, and is critical for initiating and sustaining action potentials.

When the solution is changed to a high-KCl medium, this reverses the chemical gradients of Na^+ and K^+ across the heart muscle cell plasma membranes: Na^+ is now more abundant inside the cells than outside, and K^+ is now more abundant outside the cells than inside. Under these conditions, the chemical gradients drive Na^+ *out* of the cells and K^+ *into* the cells. Because action potentials begin with a rush of Na^+ ions into the cytosol from the extracellular fluid, this rapid departure of Na^+ from the cytosol would hyperpolarize the plasma membrane, rather than depolarize it, so no action potential is generated. (This hyperpolarization prevents K^+ channels from opening as well.) Without an action potential on the plasma membrane, the remaining downstream events, including release of Ca^{+2} from the sarcoplasmic reticulum, would not occur, so no muscle contraction would occur.

Replacing the KCl solution with the NaCl solution restores the proper direction of the Na^+ and K^+ gradients, and thus the heart begins beating again, due to the pacemaker cells that initiate action potentials in the heart. The proper gradients allow the action potentials to be transmitted from cell to cell and into the T tubules, leading to Ca^{+2} release and muscle contraction. Replacing the NaCl medium with KCl medium will again stop the heart from beating. The solution swap experiment can be repeated for about 30 minutes before the heart exhausts its energy stores and dies.

CONCEPT CHECK ANSWERS #4

Aside from permitting our legs to bend, the function of knees is to absorb compressive force, as evidenced by the abundance of cartilage in the knee joint, and to resist tensile (stretching) forces, as evidenced by the ligaments holding the bones together. The switch from walking on all four limbs to walking on only our hind limbs is a relatively new event in evolutionary terms, and many physiologists and anthropologists propose that our knees are not structured to properly resist these forces for extended periods of time. Thus, sprinting, kicking, leaping, and lifting heavy objects puts tremendous strain on the connective tissues, and they eventually fail. The reason knee injuries heal relatively slowly is that the connective tissues in the knee have relatively few cells or the blood vessels that feed them, so replacing damaged cells and ECM takes much longer than, for example, in the skin (which has a much higher cell density and relative proportion of blood vessels).

Glossary

1° (primary) structure The linear sequence of amino acids in a polypeptide.

2° (secondary) structure The three-dimensional organization of a polypeptide's primary structure (e.g., alpha helix).

3° (tertiary) structure The three-dimensional organization of a polypeptide's secondary structure.

30-nm fiber The structure formed in eukaryotes when nucleosomes are packed together; this represents the condensed form of the beads-on-a-string nucleosome arrangement.

4° (quaternary) structure The three-dimensional organization of polypeptide subunits in a multi-subunit protein.

α-catenin A linking protein between cadherins and the actin-containing filaments of the cytoskeleton.

α-helix A common structural motif in the secondary structure of proteins in which a linear sequence of amino acids folds into a right-handed coil or spiral and is stabilized by internal hydrogen bonding between backbone atoms.

A site The site on a ribosome where charged amino acyl tRNA first binds to mRNA.

acceptor compartment During vesicle transport, the compartment targeted by the vesicle; when transport is complete, the vesicle fuses with the acceptor compartment.

actin-binding proteins Proteins that bind to monomeric actin or actin filaments and alter the function of the actin.

actin filaments Polymers of monomeric actin protein; also called F actin.

action potential A short event in which the electrical membrane potential of a cell rapidly rises and falls.

activators In general, molecules that bind to proteins and promote their folding into a functionally active conformation; in molecular biology, activators are transcription factors that promote the transcription of genes.

active site The location in an enzyme that binds to a substrate and facilitates the resulting chemical reaction.

acyl transferases Enzymes that covalently attach an acyl group to a substrate.

adaptor proteins In signal transduction, proteins that promote the formation of signaling complexes by binding to signaling proteins in a signaling pathway.

adenylyl cyclases Enzymes that convert ATP into cAMP during signal transduction.

adherens junction A cell–cell junction found in the junctional complex of epithelial cells, attached to the actin cytoskeleton.

adipose tissue Tissue composed primarily of connective tissue cells called adipocytes; commonly called fat tissue.

allosteric Modifying the activity of an enzyme by binding to a site other than the active site. This additional site is commonly called an allosteric site.

alternative splicing Joining different combination of exons from one gene to generate different RNAs. Commonly used to explain how a single gene can encode many different polypeptides.

alveoli Generally, small cavities; in the lung, tiny air sacs where the exchange of oxygen and carbon dioxide takes place.

amide plane The plane formed by the atoms that form a peptide bond.

amino acid side chain (or "R" group) The atoms attached to the α-carbon of an amino acid that differentiate it from all other amino acids. Amino acids are commonly grouped into four types, according to the types of R groups they contain.

amino terminus The end of a polypeptide that contains an un-bound amino group.

aminoacyl tRNA Molecule containing an amino acid covalently attached to a tRNA; also called charged tRNA.

aminoacyl tRNA synthetase The enzyme that synthesizes aminoacyl tRNAs by forming a covalent bond between an amino acid and a tRNA.

Amperes or Amps A measure of electrical current.

anaerobic metabolism The chemical reactions cells use to extract energy from food molecules in the absence of atmospheric oxygen.

anaphase (A and B) The period during mitosis when sister chromatids are separated and then pulled to opposite poles of the mitotic spindle.

anaphase-promoting complex, or APC A complex of proteins that initiates the onset of anaphase by targeting some proteins for degradation. Also called a cyclosome.

aneuploidy An abnormal number of chromosomes, commonly caused by errors in DNA replication, DNA repair, or mitosis.

anisotropic Having material properties that are direction dependent. For example, the materials that compose a tendon are capable of resisting stretching forces along the length of the tendon, but offer very little resistance to forces perpendicular to the length of the tendon.

anterograde transport In vesicle transport, the movement of material from the endoplasmic reticulum through the compartments of the Golgi apparatus, and onto their final compartments (e.g., the plasma membrane, lysosomes). Movement in the opposite direction is called retrograde transport.

anticodons Three-nucleotide sequences in tRNA molecules that align with the three-nucleotide codon sequences in mRNA molecules.

antiport Codependent transport of molecules to opposite sides of a membrane by a pump protein or a carrier protein. One example is the sodium–potassium ATPase pump.

apical surface The surface of epithelial cells that faces the lumen.

apoptosis The process of programmed cell death.

apoptosome The large quaternary protein structure formed in the process of apoptosis.

apoptotic bodies Cell fragments produced as a result of apoptosis.

apoptotic protease activating factor-1 (APAF-1) A cytoplasmic protein which forms the central hubs in the apoptosis regulatory network.

aquaporins Proteins embedded in the cell membrane that regulate the flow of water.

arteriole A small-diameter blood vessel that extends and branches from an artery and leads to the capillaries.

astral microtubules Microtubules that originate from the centrosome that only exist during and immediately before mitosis.

ATM (ataxiatelangiectasia mutated) A serine–theonine protein kinase that is recruited and activated by DNA double-strand breaks initiating the activation of the DNA damage checkpoint, leading to cell cycle arrest, DNA repair, or apoptosis.

ATR (ataxiatelangiectasia and Rad-3 related) An enzyme encoded by the ATR gene that is a serine/threonine-specific protein kinase involved in sensing DNA damage and activating the DNA damage checkpoint, leading to cell cycle arrest.

autophagy A catabolic process involving the degradation of a cell's own components through the lysosomal machinery.

autoproteolysis The degradation of a protein by intramolecular digestion.

autotrophic cells Cells that produce complex organic compounds from simple inorganic molecules using energy from light by photosynthesis or by chemosynthesis.

auxilin A protein that is a cofactor for uncoating clathrin-coated vesicles by the chaperone Hsc70.

axon The long slender projection of a neuron that conducts electrical impulses away from the neuron's soma.

axon hillock A specialized part of a neuron's soma that connects to the axon.

axoneme The cytoskeletal structure of cilia or eukaryotic flagella.

β-barrel A large β-sheet that twists and coils to form a closed structure in which the first strand is hydrogen bonded to the last, typically arranged in an antiparallel fashion.

β-pleated sheets A β-sheet is pleated when sequentially neighboring carbon atoms are alternatively above and below the plane of the sheet.

β-sheet A folding pattern found in the secondary structure of proteins that consists of beta strands connected laterally by at least two or three backbone hydrogen bonds.

backbone A series of covalently bonded atoms that together create the continuous chain of the molecule.

basal A term of anatomical location for features associated with the base of the structure.

basal body An organelle formed from a centriole and a short cylindrical array of microtubules, which serves as a nucleation site for the growth of the axoneme microtubules.

basal lamina A layer of extracellular matrix secreted by epithelial cells on which the epithelium sits.

basal surface An anatomical term of location for features associated with the base of a structure.

basal transcription complex A complex of RNA polymerase and multiple transcription factors assembled around the start site on the core promoter that initiates nuclear transcription.

basal transcription factors Protein transcription factors that are important for the transcription of genes into mRNA templates.

base A substance that can accept hydrogen ions or, more generally, donate electron pairs.

base pairs Two nucleotides on opposite complementary DNA or RNA strands that are connected via hydrogen bonds.

basement membrane A thin sheet of fibers that underlies the epithelium, lining the cavities and surfaces of organs such as skin, or the endothelium.

battery One or more electrochemical cells that convert stored chemical energy into electrical energy.

bcl homology (BH) domains Conserved motifs (BH1, BH2, BH3, and BH4) that are characteristics homologs present in the bcl-2 family.

bilayer A double layer of closely packed molecules.

bilayer thickness model A model, based on the varying thickness of the lipid bilayer, to explain how Golgi-resident proteins are sorted to their correct locations.

binding site The region on a protein, DNA, or RNA to which specific molecules or ions bind.

biochemistry The study of chemical processes in living organisms.

biofilms An aggregate of micro-organisms in which cells adhere to each other on a surface.

bleb An irregular bulge in the plasma membrane of a cell caused by a localized decoupling of the cytoskeleton from the plasma membrane.

blood plasma The liquid component of blood in which blood cells are normally suspended.

branching The replacement of a substituent by another covalently bonded chain of that polymer.

bulk transport The trapping and transport of nonspecific cargo.

CAAT box A distinct pattern of nucleotides with GGCCAATCT consensus sequence that occurs upstream to the initial transcription site box that signals the binding site for the RNA transcription factor.

cadherins A class of type-1 transmembrane proteins that play an important role in cell–cell adhesion that are dependent on calcium ions to function.

calcium channel An ion channel that is selectively permeable to calcium ions.

Calvin cycle A series of biochemical redox reactions that take place in the stroma of chloroplasts in photosynthetic organisms.

cAMP-dependent protein kinases An enzyme that functions in the regulation glycogen, sugar, and lipid metabolism and whose activity is dependent on the level of cyclic AMP.

capsule A tough membrane that encloses something. For example, a gelatinous layer formed outside the cell wall of some bacteria is a capsule.

carbohydrates A general term for sugars and related compounds with the general formula $C_m(H_2O)_n$. Can be divided into four groupings: monosaccharides, disaccharides, oligosaccharides, and polysaccharides.

carboxy terminus The end of the amino acid chain terminated by a carboxyl acid group.

carcinomas A tumor tissue derived from putative epithelial cells whose genome has become altered or damaged to such an extent that cells become transformed and begin to exhibit abnormal malignant properties.

cardiac muscle A type of involuntary striated muscle found in the walls and histologic foundation of the heart, specifically the myocardium.

cardiomyocytes Mononuclear cells that compose cardiac muscle.

cargo The cellular products and waste transported in vesicles.

carrier proteins Integral membrane proteins involved in the movement of ions, small molecules, or macromolecules across a biological membrane.

caspases A family of cysteine proteases that plays essential roles in apoptosis, necrosis, and inflammation.

catabolic pathways A metabolic pathways in which a chemical is broken down by a series of chemical reactions.

catalytic domain The site is part of an enzyme where substrates bind and undergo a chemical reaction.

catastrophe The switch from the growth phase to shrinking phase of a microtubule is called a catastrophe.

catenins Proteins found in complexes with cadherin cell adhesion molecule.

cell The smallest functional unit of life from which living organisms are made.

cell–cell junctions A structural connection composed of cell adhesion molecules that exists between cells facilitating cellular interactions.

cell cycle A series of events that takes place in a cell leading to its division and duplication.

cell–ECM junctions A structural connection between cells and the extracellular matrix.

cell division cycle (cdc) A term commonly used to name cell cycle proteins in yeast (e.g., cdc1, cdc2).

cell–matrix junctions Used interchangeably with cell–ECM junctions.

cell surface receptor proteins Specialized integral membrane proteins that participate in communication and signal transduction between the cells and environment surrounding the cells.

cell wall A tough layer located outside the cell membrane that surrounds some types of cells. It provides structural support and protection.

cell-to-cell transport Transport of materials between cells via intercellular connections.

central dogma of molecular biology It states that information cannot be transferred back from protein to either protein or nucleic acid. In other words, the process of producing proteins is irreversible: a protein cannot be used to create DNA.

central nervous system The part of the nervous system that integrates the information that it receives, coordinating the activity of the body. It consists of the brain and the spinal cord and has a fundamental role in the control of behavior.

centralized network A network of components that are focused around a node.

centromere The constricted region of a mitotic chromosome that holds sister chromatids together. It is also the site on the DNA where the kinetochore forms and then captures microtubules from the mitotic spindle.

ceremide A major lipid molecule in the lipid bilayer of the cell membrane that is composed of sphingosine and a fatty acid.

CGN-resident proteins Proteins found in the cis-Golgi network that modify other proteins by adding and subtracting sugars to/from their oligosaccharides.

channels Pore-forming proteins on the plasma membrane of cells.

chaperone proteins or chaperonins A large class of molecules that assist protein folding.

checkpoints A control mechanism that ensures a process has been accurately completed before progress into the next phase begins.

chemical synapse Specialized junctions through which neurons signal via neurotransmitters to other neurons and to nonneuronal cells.

chlorophyll A green pigment found in almost all plants, algae, and cyanobacteria that is critical in photosynthesis.

chloroplast Organelles found in plant cells and other eukaryotic organisms that conduct photosynthesis. Chloroplasts capture light energy to conserve free energy in the form of ATP and reduce NADP to NADPH.

chloroplast stroma The material within the chloroplast is called the stroma. Within the stroma are stacks of thylakoids that are the site of photosynthesis.

cholesterol A waxy steroid of fat that is manufactured in the liver or intestines that is used to produce hormones and cell membranes.

chromatids One of the two identical copies of DNA making up a duplicated chromosome, which are joined at their centromeres, for the process of cell division.

chromatin A combination of DNA and proteins (histones) that make up the contents of the nucleus of a cell. The primary functions of chromatin are to package DNA into a smaller volume to fit in the cell, to strengthen the DNA to allow mitosis and meiosis and prevent DNA damage, and to control gene expression and DNA replication.

chromatin remodeling The enzyme-assisted movement of nucleosomes on DNA that allows proteins like transcription factors to access DNA.

chromosome A single piece of tightly coiled DNA containing genes, regulatory elements, and other nucleotide sequences.

chromosome territories Chromatin that has organized itself into discrete individual patches.

cilia Slender protuberances that project from the much larger cell body. Two types of cilia exist: motile cilia and nonmotile cilia (sensory).

cis configuration Orientation of the functional groups on the same side of a molecule.

cis face of the Golgi apparatus The side of the organelle where substances enter from the endoplasmic reticulum for processing.

cis fatty acids Fatty acids whose hydrogen atoms are on the same side of the double bond.

cis-Golgi network, or CGN The cisternae stack nearest the endoplasmic reticulum on the cis face of the Golgi apparatus to which vesicles fuse.

cistern (plural, cisternae) A flattened membrane disk that makes up the Golgi apparatus and carries Golgi enzymes to help or to modify cargo proteins.

claudins Important proteins of tight junctions that establish the paracellular barrier that controls the flow of molecules in the intercellular space between the cells of an epithelium.

coactivators A protein that increases gene expression by binding to a transcription factor that contains a DNA-binding domain. The coactivator is unable to bind DNA by itself.

coat proteins An ADP ribosylation factor (ARF)-dependent adaptor protein involved in membrane retrograde traffic COPI traffic from the cis-Golgi to the rough endoplasmic reticulum.

coated pit A special type of lipid raft in the plasma membrane, rich in proteins and lipids and that has several functions in signal transduction.

coding sequence The portion of a gene's DNA or RNA, composed of exons, that codes for protein. It is bound nearer the 5′ end by a start codon and nearer the 3′ end with a stop codon.

codon A specific sequence of three nucleotides in mRNA that constitutes the genetic code for a particular amino acid.

cohesin A protein complex that regulates the separation of sister chromatids during cell division, either mitosis or meiosis.

coiled coil A structural motif in proteins, in which alpha helices are coiled together like the strands of a rope.

collagen A fibrous protein rich in glycine and proline that is a major component of the extracellular matrix and connective tissues. It exists in many forms: type I, the most common, is found in skin, tendon, and bone; type II is found in cartilage; type IV is found in basal laminae.

complementary base Two nucleotides on opposite complementary DNA or RNA strands that are connected via hydrogen bonds.

condensin Large protein complexes that play a central role in chromosome assembly and segregation in eukaryotic cells.

conduction fibers Fibers that transmit action potentials.

congression The impaired alignment of chromosomes at the equator of the mitotic spindle.

connective tissue A fibrous tissue found in tendons, blood, cartilage, bone, adipose tissue, and lymphatic tissue.

connexin Gap junction proteins that assemble to form vertebrate gap junctions.

connexons An assembly of six proteins called connexins that can be a part of a gap junction channel between the cytoplasm of two adjacent cells.

constitutive secretion Continuous secretion of proteins from the cell regardless of environmental factors.

contractile cycle The mechanisms of force generation in muscle.

contractile ring A ring made of nonmuscle myosin II and actin filaments that assembles in the middle of the cell adjacent to the cell membrane.

cooperative interaction An interaction of protein for a single cause.

COPII complex A vesicle coat protein that transports proteins from the rough endoplasmic reticulum to the Golgi apparatus.

core oligosaccharide A short chain of sugar residues within Gram-negative lipopolysaccharide.

core particle A nucleosome with a DNA link region.

core promoter The minimal portion of the promoter required to properly initiate gene transcription that contains a binding site for RNA polymerase (RNA polymerase I, RNA polymerase II, or RNA polymerase III).

corepressors A substance that inhibits the expression of genes by indirectly interacting with a gene promotor through a direct interaction with a repressor protein.

cornification The process of forming an epidermal barrier in stratified squamous epithelial tissue.

cortical-actin network A dense concentration of actin filaments that lies just under the plasma membrane.

cortisol A glucocorticoid produced by the adrenal gland that is released in response to stress and a low level of blood glucocorticoids.

costamere A structural–functional component of striated muscle cells that connects the sarcomere of the muscle to the cell.

cotranslational Involving the delivery of a protein that occurs during the process of translation.

covalent bonds A chemical bond characterized by the sharing of pairs of electrons between atoms.

critical concentration The minimum concentration of units needed before a biological polymer will form.

cross-linking proteins Proteins used to bond to polymers and other proteins together.

crossbridge cycle The process by which muscle fibers generate tension through the action of actin and myosin.

cyclic AMP response element binding protein (CREB) A cellular transcription factor that binds to certain DNA sequences called cAMP response elements (CRE), thereby increasing or decreasing the transcription of the downstream genes.

cyclin Protein that controls the progression of cells through the cell cycle by activating cyclin-dependent kinase (Cdk) enzymes.

cyclin activating kinases (CAKs) A complex that activates CDKs by phosphorylating threonine residue in the CDK activation loop.

cyclin-dependent kinases (CDKs) A family of protein kinases that play a role in regulating the cell cycle, transcription mRNA processing, and the differentiation of nerve cells.

cytokinesis The process in which the cytoplasm of a single eukaryotic cell is divided to form two daughter cells.

cytoplasmic leaflet The inner membrane of a lipid bilayer composed of phosphatidylethanolamine, phosphatidylserine, and phosphatidylinositol and its phosphorylated derivatives.

cytoskeleton A cellular scaffold contained within the cytoplasm that is made out of protein and plays a role in intracellular transport and cellular division.

dark reactions / carbon assimilation reactions A series of enzymatic reactions in mitochondria involving oxidative metabolism of sugars made during photosynthesis to produce high-energy phosphate compounds.

death domain A protein interaction module that signals apoptosis.

death effector domain (DED) A protein interaction domain found to regulate a variety of cellular signaling pathways including caspase activation in the apoptosis cascade.

death ligands Proapoptotic signaling molecules.

death receptor A cytokine receptor that binds tumor necrosis factors.

dehydration (or condensation) reaction A chemical reaction in which two molecules combine to form one single molecule, together with the loss of a small molecule (water).

dendrites The branched projections of a neuron that act to conduct the electrochemical stimulation received from other neural cells to the soma of the neuron from which the dendrites project.

dense bodies Amorphous bodies scattered throughout the cytoplasm of smooth muscle fibers that act as attachment sites for myofilaments.

deoxynucleoside Sugars that have had a hydroxyl group replaced with a hydrogen atom.

deoxynucleotides Components of DNA, containing the phosphate, sugar, and organic base.

deoxyribonucleotide A monomer of DNA. Each deoxyribonucleotide comprises three parts: a nitrogenous base, a deoxyribose sugar, and one or more phosphate groups.

deoxyribose A monosaccharide that is derived from the sugar ribose by loss of an oxygen atom.

depolarization A change in a cell's membrane potential, making it more positive or less negative.

depolymerizing proteins Proteins involved in the breakdown of polymers to monomers.

dermal tissue The tissue that covers the outside of the plant and functions to protect the plant from injury and water loss.

desmosome A cell structure specialized for cell-to-cell adhesion which are localized spot-like adhesions randomly arranged on the lateral sides of plasma membranes.

detergents A surfactant.

development The process of growing to maturity.

differentiation The process by which a less specialized cell becomes a more specialized cell type.

dimer (adjective, dimeric) A chemical entity consisting of two structurally similar subunits called monomers.

dimerization partners The proteins that together form dimers.

dipeptide A molecule consisting of two amino acids joined by a single peptide bond.

direct active transport The use of energy to transport molecules across a membrane.

disaccharides The carbohydrate formed when two monosaccharides undergo a condensation reaction that involves the elimination of a small molecule, such as water, from the functional groups only.

discontinuous cell–cell adhesion Leukocytes expression of selectin ligands such that they will bind only to endothelial cells that express the E- and P-selectins preventing them from adhering to vessel walls.

disulfide bonds A covalent linkage between the sulfur atoms on two cysteine residues in different polypeptides in different parts of the same polypeptide.

DNA A nucleic acid that contains the genetic instructions used in the development and functioning of all known living organisms.

DNA decatenation A process that uses ATP energy to sort the replicated sister chromatid DNA into two discrete forms.

DNA helicase An enzyme that catalyzes the energy-dependent unwinding of the DNA double helix during DNA replication.

DNA methylation The addition of a methyl group to DNA.

DNA replication factories The site of ongoing DNA replication.

DNAse II An enzyme that hydrolyzes deoxyribonucleotide linkages in native and denatured DNA, yielding products with 3′ phosphates.

docking A process in which cargo is delivered to the rim of the nuclear pore complex.

docking complex A complex on the cytoplasmic surface of the peroxisome where cargo is delivered.

dolichol phosphate A fatty alcohol.

domain A distinct region of a protein's three-dimensional structure.

donor compartment During vesicle transport, the compartment from which material is transported.

downstream core promoter element (DPE) A core promoter sequence that is within the transcribed portion of a gene.

dynamic instability The coexistence of assembly and disassembly at the (+) end of a microtubule. The microtubule can switch between the growing and shrinking phases dynamically at this region.

dynamin A GTPase responsible for endocytosis in the eukaryotic cell.

dynein A motor protein in cells that converts the chemical energy contained in ATP into the mechanical energy of movement. Dynein transports cargo by "walking" along cytoskeletal microtubules toward the (−) end of the microtubule.

E site The binding site of ribosomes that binds a free tRNA before it exits the ribosome.

ectoderm The outer layer of germ cells that differentiates to form the nervous system.

EF-G-GTP A prokaryotic elongation factor that catalyzes the translocation of tRNA and mRNA down the ribosome.

EF-Tu-GTP A prokaryotic elongation factor that mediates the entry of the aminoacyl-tRNA into a free site of the ribosome.

effector A molecule that binds to a protein and thereby alters the activity of that protein.

elastic connective tissue Bundles of elastin produced by fibroblasts and smooth muscle cells in arteries. These fibers can stretch up to 1.5 times their length, and snap back to their original length when relaxed.

elastin An elastic protein in connective tissue that is important in load-bearing tissue and is used in places where mechanical energy is required to be stored.

electrical charge A physical property of matter, either positive or negative, that causes it to experience a force when near other electrically charged matter.

electrical current The flow of electric charge through a medium.

electrical potential The energy associated with the separation of positive and negative charges.

electricity A general term encompassing a variety of phenomena resulting from the presence and flow of electric charge.

electrochemical gradient A spatial variation of both electrical potential and chemical concentration across a membrane.

electron transport chain (ETC) An event during photosynthesis that couples electron transfer between an electron donor (such as NADH) and an electron acceptor (such as O_2) with the transfer of H^+ ions (protons) across a membrane.

elongation (transcription) The process of using a strand of DNA to serve as the template for RNA synthesis.

elongation (translation) The process of using a strand of mRNA to serve as the template for peptide synthesis.

elongation factors A set of proteins that facilitate the events of translational elongation.

endocytic pathway A process by which cells absorb molecules by engulfing them.

endocytic vesicles Membrane-bound compartments that deliver molecules from the plasma membrane to early endosomes.

endocytosis The process by which cells absorb molecules (such as proteins) by engulfing them.

endoderm The innermost layer of three germ layers formed during embryogenesis, which gives rise to the gut and most of the respiratory tract.

endomembrane system A system composed of the different membranes that are suspended in the cytoplasm that divide the cell into functional and structural compartments, or organelles.

endonucleases Enzymes that cleave the phosphodiester bond within a polynucleotide chain.

endoplasmic reticulum A eukaryotic organelle that forms an interconnected network of tubules, vesicles, and cisternae within cells that functions to synthesize proteins (rough) or lipids (smooth).

endosome A membrane-bound compartment inside eukaryotic cells that transport pathway from the plasma membrane to the lysosome.

endosymbiotic theory A theory that certain organelles originated as free-living bacteria and were taken inside another cell.

energy carriers A substance that can be used to produce mechanical work or heat or to operate chemical or physical processes.

energy storage Storing some form of energy to perform some useful operation at a later time.

enhancer sequences A short region of DNA that can bind with proteins to enhance transcription levels of genes.

entropy A thermodynamic property that can be used to determine the energy available for useful work.

enzymes Proteins that catalyze chemical reactions.

epidermal cells Cells of the outer layer of skin.

epidermis A stratified squamous epithelium that forms the outer layer of the skin.

epigenetics The study of changes produced in gene expression caused by mechanisms other than changes in the underlying DNA sequence.

equilibrative nucleoside transporter (ENT) Transport proteins that transport nucleoside substrates like adenosine into cells.

ER export signals Proteins that contain domains that regulate endoplasmic reticulum exporting.

ER export sites The portion of the endoplasmic reticulum that regulates exportation in the endoplasmic reticulum.

ER retention signal Short amino acid sequence on a protein that prevents it moving out of the endoplasmic reticulum.

ER signal sequence N-terminal signal sequence that directs proteins to enter the endoplasmic reticulum.

euchromatin A lightly packed form of chromatin.

excision repair systems A mechanism that repairs damaged DNA.

executioner caspases Enzymes essential for apoptosis.

exocytic pathway The process by which a cell directs the contents of secretory vesicles out of the cell membrane.

exocytic vesicles Membrane-bound compartments that fuse with the plasma membrane and release intracellular molecules.

exocytosis The release of intracellular molecules contained within vesicles by fusion of the vesicle with the plasma membrane.

exon A nucleic acid sequence in the mature form of an RNA molecule after the introns have been removed.

exon junctional complex, or EJC A complex that plays a major role in posttranscriptional regulation of mRNA.

exoplasmic or ectoplasmic leaflet The layer of the plasma membrane that faces the cell exterior.

exportins Proteins involved in transporting molecules from the nucleus to the cytoplasm.

extracellular matrix A complex mesh-work of proteins and polysaccharides secreted by cells into the space between them that provides structural support and aids in biochemical function.

extracellular space The space outside the cell.

extravasation The leakage of a fluid out of its container. For example, in the case of malignant cancer metastasis, it refers to cancer cells exiting the capillaries and entering organs.

facilitated diffusion The spontaneous passage of molecules or ions across a biological membrane via specific transmembrane integral proteins.

farnesyl and geranylgeranyl lipids Lipids that are covalently attached to the sulfhydryl group of cysteine amino acids on some membrane proteins.

Fas-associated protein with death domain (FADD) An adaptor molecule that bridges the Fas-receptor and other death receptors to caspase-8 during apoptosis.

fat A category of lipids important for both structural and metabolic functions.

fatty acid A carboxylic acid usually derived from triglycerides or phospholipids that contains a long unbranched aliphatic tail that is either saturated or unsaturated.

fatty acid synthase An enzyme that composes a whole enzymatic system that plays a key role in fatty acid synthesis.

feedback The output signal from an event in the past that will influence an event in the future.

feedback loop The path that leads from the initial generation of the feedback signal to the subsequent modification of the future event.

FGF receptors (FGFRs) Receptors that bind to members of the fibroblast growth factor family of proteins.

fibripositors Elongated membrane-bound compartments that cluster at or near the plasma membrane, acting as the location of collagen fibrillogenesis.

fibroblast A type of cell that synthesizes extracellular matrix proteins and plays a critical role in wound healing.

fibroblast growth factor A family of growth factors that are heparin-binding proteins involved in angiogenesis, wound healing, and embryonic development.

fibronectin A high-molecular-weight glycoprotein of the extracellular matrix that binds to integrins, collagen, fibrin, and heparan sulfate proteoglycans and plays a major role in cell adhesion, growth, migration, differentiation, wound healing, and embryonic development.

fibronectin repeats A repeating domain of fibronectin that is classified into three groups, named type I, II, and III.

fibrous connective tissues Dense connective tissue that forms strong rope-like structures such as tendons and ligaments that contain fibers as the main matrix element, generally collagen type I.

fibrous proteins Proteins that are practically insoluble and involved in protection and support forming connective tissue, tendons, bone matrices, and muscle fiber.

filopodium (plural, filopodia) Slender cytoplasmic projections that extend beyond the leading edge of lamellipodia in migrating cells that contain actin filaments cross linked into bundles by actin-binding proteins.

flagellum (plural, flagella) A tail-like projection that protrudes from the cell body of certain prokaryotic and eukaryotic cells, and functions in locomotion.

flippases Transmembrane lipid transporter enzymes responsible for aiding the outward-in movement of phospholipid molecules between the two leaflets that compose a cell's membrane.

floppases Transmembrane lipid transporter enzymes responsible for aiding the inward-out movement of phospholipid molecules between the two leaflets that compose a cell's membrane.

fluid–mosaic model A model stating that biological membranes can be considered as a two-dimensional liquid where all lipid and protein molecules diffuse more or less easily.

focal adhesion Large integrin assemblies through which both mechanical force and regulatory signals are transmitted, mediating the regulatory effects of extracellular matrix adhesion on cell behavior.

focal contacts Integrin clusters that form at the leading edge of migrating cells, and are induced by stimulation of actin filament growth near the plasma membrane.

functional groups Specific groups of atoms within molecules that are responsible for the characteristic chemical reactions of those molecules.

G protein Proteins involved in transmitting chemical signals outside the cell causing changes inside the cell; regulating metabolic enzymes, ion channels, transporters, and other parts of the cell machinery; controlling transcription, motility, contractility, and secretion, which in turn regulate systemic functions such as embryonic development, learning and memory, and homeostasis.

G protein coupled receptors A large protein family of transmembrane receptors that sense molecules outside the cell and activate inside signal transduction pathways and, ultimately, cellular responses.

gamma-(γ) tubulin Globular proteins important in the nucleation and polar orientation of microtubules.

gamma-tubulin ring complex (γTuRC) A combination of several γ-tubulin and other protein molecules that chemically mimic the (+) end of a microtubule and thus allow microtubules to bind.

gap junction A specialized intercellular connection directly connecting the cytoplasm of two cells, which allows various molecules and ions to pass freely between cells.

gap junction channel A connection between intercellular space composed of two connexons.

gastric pits Indentations dotting the surface of the lining epithelium of the stomach which denote entrances to the tubular shaped gastric glands.

gating The opening (activation) or closing (deactivation) of ion channels.

GC box A sequence of contiguous nucleotides: guanine, guanine, cytosine, and guanine, in that order, along a DNA strand often found in the promoter region upstream from the transcription start site.

gel filtration A chromatographic method that uses an aqueous solution to transport the sample through a column to separate molecules by their size, not by molecular weight.

gene A name given to some stretches of DNA and RNA that code for a type of protein or for an RNA chain that has a function in the organism.

genetics A discipline of biology that deals with the molecular structure and function of genes, with gene behavior in the context of a cell or organism with patterns of inheritance from parent to offspring, and with gene distribution, variation, and change in populations.

genome The entirety of an organism's hereditary information encoded either in DNA or, for many types of virus, in RNA. The genome includes both the genes and the noncoding sequences of the DNA and RNA.

germ cells Any biological cell that gives rise to the gametes of an organism that reproduces sexually.

germ layers A group of cells, formed during embryogenesis, that eventually give rise to all of an animal's tissues and organs.

glands An organ in an animal's body that synthesizes a substance for release often into the bloodstream (endocrine gland) or into cavities inside the body or its outer surface (exocrine gland).

glial cell Nonneuronal cells that maintain homeostasis, form myelin, and provide support and protection for the brain's neurons.

globular proteins Globe-like proteins that are more or less soluble in aqueous solutions that act as enzymes, transporters, and messengers.

glucocorticoid receptor (GR) The receptor to which cortisol and other glucocorticoids bind participating in the regulation genes controlling the development, metabolism, and immune response.

glucose A monosaccharide and an important carbohydrate that is used as the primary source of energy and a metabolic intermediate. Glucose is one of the main products of photosynthesis and starts cellular respiration.

glucose transporter (GLUT) Integral membrane proteins that transport glucose and related hexoses.

glycerol A colorless, odorless, viscous liquid that has three hydrophilic hydroxyl groups that are responsible for its solubility in water and its hygroscopic nature. The glycerol backbone is central to all lipids known as triglycerides.

glycerol phosphate shuttle A mechanism that regenerates NAD+ from NADH, a byproduct of glycolysis.

glycerol-3 phosphate An organophosphate derived from the reaction catalyzed by glycerol kinase.

glycocalyx A general term referring to extracellular polymeric material consisting of glycolipids and glycoproteins produced by some bacteria, epithelia, and other cells.

glycogen A molecule that functions as the secondary long-term energy storage in animal and fungal cells, with the primary energy stores being held in adipose tissue.

glycolipid Any lipid to which a short carbohydrate chain is covalently linked, commonly found in the plasma membrane, whose role is to provide energy and also to serve as markers for cellular recognition.

glycolysis The metabolic pathway that converts glucose into pyruvate, releasing free energy in the process and forming the high-energy compounds ATP and NADH.

glycophosphatidylinositol (GPI) A glycolipid that can be attached to the C terminus of a protein during posttranslational modification.

glycosaminoglycan (GAGs) Long, unbranched, highly charged polysaccharides consisting of a repeating disaccharide unit.

glycosidase A common enzyme that catalyzes the hydrolysis of the glycosidic linkage to release smaller sugars.

glycosidic bond The covalent linkage between two monosaccharide residues formed when a carbon atom in one sugar reacts with a hydroxyl group on a second sugar with the net release of a water molecule.

glycosylation The enzymatic process that attaches glycans to proteins, lipids, or other organic molecules. Glycosylation is a form of cotranslational and posttranslational modification.

glycosyltransferase An enzyme that acts as a catalyst for the transfer of a monosaccharide unit from an activated nucleotide sugar to a glycosyl acceptor molecule, resulting in a carbohydrate, glycoside, oligosaccharide, or polysaccharide.

Golgi apparatus An organelle in eukaryotic cells made of stacks of flattened, interconnected cisternae that functions in processing and sorting proteins and lipids destined for other cellular compartments or for secretion.

Golgi retention sequences An amino acid sequence that acts as a specialized retention signal enabling the protein to be retained by the Golgi.

ground tissue The tissue found in plants that has developed from meristem and consists of three simple tissues: perenchyma, collenchyma, and sclerenchyma.

growth factors A naturally occurring substance, usually a protein or steroid hormone, capable of stimulating cellular growth, proliferation, and differentiation.

GTP hydrolysis The unidirectional change of the GTPase from the active, GTP-bound form to the inactive, GDP-bound form by hydrolysis of the GTP through intrinsic GTPase activity, effectively switching the GTPase off.

GTPase-activating proteins A family of regulatory proteins that bind to activated G proteins and stimulate their GTPase activity.

guanine nucleotide exchange factors Activate monomeric GTPases by stimulating the release of guanosine diphosphate (GDP) to allow binding of guanosine triphosphate (GTP).

guard cells Specialized cells located in the leaf epidermis of plants that act in pairs surrounding tiny stomatal airway pores.

gyrase An enzyme that relieves strain while double-stranded DNA is being unwound by helicase, which causes the supercoiling of the DNA.

hairpin loop A pattern that can occur in single-stranded DNA or, more commonly, in RNA. It occurs when two regions of the same strand form a double helix that ends in an unpaired loop.

head group The portion of a phospholipid bilayer that can alter the surface chemistry.

heat energy A form of energy that is transferred by a difference in temperature.

heat of vaporization The energy required to transform a given quantity of a substance into a gas at a given pressure.

heat shock protein 70 (Hsp70) A protein important to the cell's machinery for protein folding, that helps to protect cells from stress.

helix-turn-helix (HTH) A major structural motif capable of binding DNA. It is composed of two α helices joined by a short strand of amino acids and is found in many proteins that regulate gene expression.

hemidesmosomes Very small structures on the inner basal surface of keratinocytes in the epidermis of skin that attach one cell to the extracellular matrix.

heterochromatin A tightly packed form of DNA.

heterogeneous nuclear ribonucleoprotein particle (hnRNP) Complexes of RNA and protein present in the cell nucleus during gene transcription and subsequent posttranscriptional modification of the pre-mRNA.

heterophilic binding The binding of an adhesion molecule in one cell to a nonidentical adhesion molecule in an adjacent cell.

hetero-oligomeric Describes an oligomer consisting of a few different monomers.

histology The study of the microscopic anatomy of cells and tissues of plants and animals. It is performed by examining a thin section of tissue under a light microscope or electron microscope.

histone remodeling The enzyme-assisted movement of nucleosomes on DNA.

homeostasis The maintenance of stable, constant conditions.

homo-oligomeric Describes an oligomer consisting of a few of the same monomers.

homophilic binding The binding of an adhesion molecule in one cell to an identical adhesion molecule in an adjacent cell.

hyaluronan An anionic, nonsulfated glycosaminoglycan distributed widely throughout connective, epithelial, and neural tissues. Hyaluronan contributes significantly to cell proliferation and migration, and may also be involved in the progression of some malignant tumors.

hydration shell See water shell.

hydrogen bonding The interaction of a hydrogen atom with an electronegative atom, such as nitrogen, oxygen, or fluorine, that comes from another molecule or chemical group.

hydrolysis A chemical reaction during which molecules of water (H_2O) are split into protons (H^+) and hydroxide anions (OH^-). This reaction is used to break down certain polymers.

hydrolytic enzymes, or hydrolases An enzyme that catalyzes the hydrolysis of a chemical bond.

importins A type of protein that moves other protein molecules into the nucleus by binding to the nuclear localization signal.

indirect active transport Transport in which one gradient supplies the energy needed to create another gradient causing the transporting of two different proteins.

inflammatory caspases Caspases associated with the immune response.

initiation (transcription) The upstream assembly of the RNA polymerase complex at the promoter region of a DNA template.

initiation (translation) The assembly of cellular and enzymatic reactions that instigate the process by which polypeptides are synthesized.

initiation factors Proteins that bind to the small subunit of the ribosome during the initiation of translation.

initiation site The codon at which peptide synthesis begins.

initiator caspases Cleave inactive pro-forms of effector caspases, thereby activating them.

initiator tRNA A special type of transfer ribonucleic acid (RNA) that initiates protein synthesis by binding to the amino acid methionine and delivering it to the small ribosomal subunit.

inner dense plaque The location on the cytoplasmic side of the plasma membrane where desmoplakin attaches to the intermediate filaments of the cell.

insulator A genetic boundary element that plays two distinct roles in gene expression, either as an enhancer-blocking element, or more rarely as a barrier against condensed chromatin proteins spreading onto active chromatin.

integral membrane proteins A protein molecule that is permanently attached to the biological membrane.

integrin Receptor that mediates attachment between a cell and other cells or the ECM surrounding it. They play a role in cell signaling and thereby regulate cellular shape, motility, and the cell cycle.

integrin complexes A protein complex composed of an α and β subunit.

intercalated disks An undulating double membrane separating adjacent cells in cardiac muscle fibers. Intercalated discs support synchronized contraction of cardiac tissue.

intermediate filaments (IFs) A family of related proteins that share common structural and sequence features. Termed *intermediate* because their average diameter is between those of narrower microfilaments (actin) and wider myosin filaments.

interphase The phase of the cell cycle in which the cell spends the majority of its time and performs the majority of its purposes including preparation for cell division.

intracellular signaling proteins Peptides and proteins found inside the cell that serve to transmit extracellular signals to intracellular effectors.

intrinsic pathway A chemical pathway that occurs inside a cell.

introns Any nucleotide sequence within a gene that is removed by RNA splicing to generate the final mature RNA product of a gene.

invagination An inward fold.

ion channel receptors Receptors bound in the cell membrane that act through synaptic signaling on electrically excitable cells.

ionic bonds A bond formed through an electrostatic attraction between two oppositely charged ions.

isopeptide bond A bond that forms between a side-chain carboxyl group and amino group.

junctional adhesion molecule (JAM) A membrane glycoprotein localized to the intercellular junction of polarized endothelial and epithelial cells.

junctional complex The attachment zone between epithelial cells, typically consisting of the zonula occludens, the zonula adherens, and the macula adherens (desmosome).

karyorrhexis The irregular distribution of chromatin throughout the cytoplasm resulting from the destructive fragmentation of the nucleus of a dying cell.

keratinized stratified squamous epithelium Squamous (flattened) epithelial cells arranged in layers upon a basement membrane whose apical surfaces are protected from abrasion by keratin and kept hydrated and protected from dehydration by glycolipids produced in the stratum granulosum.

kin recognition model A model based on the ability to recognize kin. For example, the aggregation of proteins that recognize each other's membrane-spanning domains. As these proteins are carried through the Golgi apparatus by anterograde vesicles, they encounter other proteins with similar transmembrane domains and bind to them. Eventually, by a mechanism still not fully understood, this complex no longer enters vesicles and is therefore trapped in a specific cistern.

kinesin A motor protein found in eukaryotic cells that performs anterograde transport powered by the hydrolysis of ATP along microtubule filaments.

kinesin-related motor proteins A class of motor proteins that are able to move along microtubules.

kinetic energy The energy an object possesses due to its motion.

kinetochore The compact protein complexes that attach to the chromosomes on either side of the centromere at the onset of prometaphase and act as an attachment site for kinetochore microtubules.

kinetochore microtubules A collection of microtubules that are directly attached to kinetochores on the chromosomes.

Krebs cycle Part of the metabolic pathway involved in the chemical conversion of carbohydrates, fats, and proteins into carbon dioxide and water to generate a form of usable energy, essential for all living cells, especially those that use oxygen as part of cellular respiration. In eukaryotic cells, the Krebs cycle occurs in the matrix of the mitochondria.

lactic acid A carboxylic acid that plays a role in various biochemical processes.

lactic acid transporter protein Proteins involved in the transport of monocarboxylic acids such as lactic acid and pyruvic acid across cellular membranes.

lagging strand The DNA strand formed at the replication fork as short continuous fragments that are synthesized in the 5′ to 3′ direction and later joined.

lamellopodium (plural, lamellopodia) A cytoskeletal protein actin projection on the mobile edge of the cell that contains an actin mesh to propel the cell across a substrate.

laminin A major matrix protein active in the basal lamina influencing cell differentiation, migration, adhesion, and survival.

late endosomes Endosomes that are distinguished by the time it takes for endocytosed material to reach them Late endosomes receive materials from early endosomes and deliver it to lysosomes.

leading strand The template strand of DNA that is replicated continuously in the 5′ to 3′ direction toward the replication fork.

leaflets One of the two single molecular layers that compose the lipid bilayer.

lectins Sugar-binding proteins that play a role in biological recognition phenomena involving cells and proteins.

leucine zipper A common coiled-coil structural motif in proteins usually found as part of a DNA-binding domain in various transcription factors.

ligand A substance that forms a complex with a biomolecule to serve a biological purpose. It is a signal-triggering molecule that binds to a site on a target protein.

ligand-gated ion channel Transmembrane ion channels that are opened or closed in response to the binding of a chemical messenger (i.e., a ligand).

ligases An enzyme that catalyzes the joining of two large molecules by forming a new chemical bond, usually with accompanying hydrolysis of a small chemical group.

light and heavy chains An antibody is a "Y"-shaped molecule that consists of four polypeptide chains; two identical heavy chains and two identical light chains connected by disulfide bonds.

light-harvesting complex A complex of proteins associated with pigment molecules that captures light energy and transfers it to reaction-center pigments in a photosystem.

light reactions / energy transduction reactions The stage of photosynthesis during which energy from light is used for the production of ATP.

linker DNA Double-stranded DNA between two nucleosome cores that, in association with histone, holds the cores together.

lipid-binding proteins Carrier proteins for fatty acids and other lipophilic substances that facilitate the transfer between extra- and intracellular membranes.

lipid raft hypothesis A hypothesis that suggests changes in membrane-dependent functions depend upon the cell-membrane fluidity.

lipids Naturally occurring amphiphilic small molecules that includes fats, waxes, sterols, monoglycerides, diglycerides, phospholipids, and others. The main biological functions of lipids include energy storage, as structural components of cell membranes, and as important signaling molecules

lipoproteins A biochemical assembly that contains both proteins and lipids water-bound to the proteins. Many enzymes, transporters, structural proteins, antigens, adhesins, and toxins are lipoproteins.

liposome Artificially prepared vesicles made of a lipid bilayer.

loop domains A configuration useful for packaging a massive chromosome into a small space so that it is readily accessible.

lumen The inside space of a tubular structure, such as an artery or intestine.

lysosomal associated membrane proteins (LAMPs) A membrane glycoprotein of the lysosome that is involved in the transport of materials across the membrane.

lysosomal integral membrane proteins (LIMPs) A membrane glycoprotein involved in the transport of newly synthesized hydrolases to the lysosome.

lysosome Cellular organelles that contain acid hydrolase enzymes to break down waste materials and cellular debris.

macrophages White blood cells produced by the differentiation of monocytes in tissues that function in defense mechanisms by engulfing and digesting cellular debris and pathogens.

major groove A structure within the double helix of DNA that allows proteins to bind to DNA and is wider than the minor groove.

mannose-6 phosphate A molecule bound by lectin that acts as a targeting signal tagging proteins from the cis-Golgi to the trans-Golgi.

mannose-6 phosphate receptors Proteins that bind newly synthesized lysosomal hydrolases in the trans-Golgi network and deliver them to prelysosomal compartments.

matrix The material between animal or plant cells, in which more specialized structures are embedded.

mature elastin Proteins that have been fully assembled into a polymer.

medial-Golgi stacks One of the four cisternae stack functional regions.

membrane A selectively permeable barrier that separates the interior of all cells from the outside environment.

membrane domains The portion of the protein that interacts with the membrane.

membrane-impermeable signals Signals that bind to cell-surface receptors.

membrane-permeable signals Signals that bind to intracellular receptors.

membrane-spanning (or transmembrane) proteins Proteins that span from one side of a membrane through to the other side of the membrane and function as gateways to deny or permit the transport of specific substances across the membrane.

membrane trafficking This transportation of vesicles between organelles.

membrane transport proteins A membrane protein involved in the movement of ions, small molecules, or macromolecules across a biological membrane.

mesoderm The middle of three primary cell layers found in embryogenesis, lying between the ectoderm and the endoderm, that gives rise to connective tissue, blood , and other tissues.

messenger ribonucleoprotein particle (mRNP) Nucleoprotein that contains RNA and is implicated in pre-mRNA splicing and is among the main components of the nucleolus.

metaphase plate An imaginary line that is equidistant from the two centrosome poles.

micelle An spherical aggregate of phospholipids or other amphipathic molecules that form spontaneously in an aqueous solution.

microfiber (or microfibrillar) sheath A tough coating surrounding the elastic fibers of elastin.

microfilament The thinnest filaments of the cytoskeleton responsible for cell movement and changes in cell shape.

microtubule-capping proteins Proteins that cap the end of microtubules after assembly.

microtubule organizing center (MTOC) A structure found in eukaryotic cells from which microtubules emerge that function to organize flagella and cilia and the organization of the mitotic and meiotic spindle.

microtubules Cytoskeletal fibers that are formed by polymerization of α- and β-tubulin monomers and exhibit structural and functional polarity.

midbody A transient structure found in mammalian cells present near the end of cytokinesis just prior to the complete separation of the dividing cells. The midbody structure contains bundles of microtubules derived from the mitotic spindle that are important for completing the final stages of cytokinesis, a process called abscission.

milliamp A unit of electric current, equivalent to 1/1,000 of an amp.

minichromosome maintenance (MCM) A complex that has a role in both the initiation and the elongation phases of eukaryotic DNA replication, specifically the formation and elongation of the replication fork.

minor groove A structure within the double helix of DNA that is narrower than the major groove.

mismatch repair A system for recognizing and repairing erroneous insertion, deletion, and incorrect incorporation of bases that can arise during DNA replication and recombination.

mitochondria A membrane-enclosed organelle that functions as the "cellular power plant" generating most of the cell's supply of ATP.

mitochondrial matrix A matrix that contains soluble enzymes that catalyze the oxidation of pyruvate and other small organic molecules.

mitochondrial outer membrane permeabilization (MOMP) The defining moment for the onset of apoptosis that happens when Bak and Bax move from the cytosol into the outer mitochondrial membrane, where they form a channel-like structure that permits proteins in the intermembrane space of the mitochondrion to leak out into the cytosol.

mitochondrial signal sequences Short signal sequences at or near their amino in cytosolic proteins targeted to organelles.

mitosis promoting factor, or MPF A heterodimeric protein that stimulates the mitotic and meiotic cell cycles.

mitotic checkpoint complex (MCC) A checkpoint involved in the regulation of mitosis in which the complex binds to kinetochore microtubules that have yet to engage a kinetochore.

mitotic spindles Structures that separate the chromosomes into the daughter cells during mitosis.

model organisms A nonhuman species that is extensively studied to understand particular biological phenomena, with the expectation that discoveries made in the organism model will provide insight into the workings of other organisms.

molecular gradients A gradient caused by a difference in molecular concentration.

monocyte A type of white blood cell that plays multiple roles in immune function.

monolayer A single, closely packed layer of atoms, molecules, or cells.

monomer (adjective, monomeric) An atom or a small molecule that may bind chemically to other monomers to form a polymer.

monomer-binding proteins Essential components of the actin polymerization.

monomeric G proteins GTP-binding proteins that regulate a variety of intracellular processes.

monosaccharide The most basic units of biologically important carbohydrates. Monosaccharides are the building blocks of disaccharides such as sucrose and polysaccharides.

motifs A nucleotide or amino-acid sequence pattern that has a biological significance. For proteins, a sequence motif is distinguished from a structural motif, a motif formed by the three-dimensional arrangement of amino acids, which may not be adjacent.

motor nerves Neurons located in the central nervous system (CNS) that project their axons outside the CNS and directly or indirectly control muscles.

motor proteins Molecular motors that are able to move along the surface of a suitable substrate powered by the hydrolysis of ATP.

MRN complex A heterotrimeric protein complex involved in DNA repair.

multipass region A protein with multiple regions that extend entirely through the membrane bilayer.

multivesicular body (MVB) Also known as late endosomes MVBs, are mainly spherical, lack tubules, and contain many close-packed lumenal vesicles.

muscle fibril A basic unit of a muscle composed of long proteins such as actin, myosin, and titin. Also called a myofibril.

muscle tissue A contractile tissue of animals derived from the mesodermal layer of embryonic germ cells; its function is to produce force and cause motion.

mutation Changes in a genomic sequence, often sudden and spontaneous.

myelin sheaths An insulating myelin layer around only the axon of a neuron.

myofibril See muscle fibril.

myosin ATP-dependent motor proteins that play a role in muscle contraction and other motility processes.

myosin light chain kinase A serine/threonine-specific protein kinase that phosphorylates the regulatory light chain of myosin II, important in muscle contraction.

myosin light chain phosphatase An enzyme that acts on phosphoric monoester bonds. The two substrates of this enzyme are myosin light chain phosphate and H_2O, whereas its two products are myosin light chain and phosphate.

N-oligosaccharyl transferase (OT) enzyme complex A complex of enzymes that transfers an oligosaccharide with high-mannose content from a lipid-linked oligosaccharide donor to an asparagine acceptor site on newly synthesized polypeptides to form glycoproteins.

nascent chain The polypeptide chain that is just beginning to grow.

necrosis The premature death of cells and living tissue. Necrosis is caused by factors external to the cell or tissue, such as infection, toxins, or trauma.

negative supercoiling Subtractive helical twisting.

nephron The basic structural and functional unit of the kidney whose chief function is to regulate the concentration of water and soluble substances by filtering the blood, reabsorbing what is needed and excreting the rest as urine.

nerve Cylindrical bundles of fibers that emanate from the brain and central spinal cord, and branch repeatedly to innervate every part of the body.

nervous tissue Tissue of nervous system, composed of neurons, which transmit impulses, and the glia cells, which assist propagation of the nerve impulse as well as provide nutrients to the neuron.

neurites A projection from the cell body of a neuron. This projection can be either an axon or a dendrite.

neuron An electrically excitable cell that processes and transmits information by electrical and chemical signaling.

neurotransmitter Endogneous chemicals that transmit signals from a neuron to a target cell across a synapse.

Nodes of Ranvier The gaps formed between the myelin sheaths.

noncovalent bonds A chemical bond, typically between macromolecules, that does not involve the sharing of pairs of electrons, but rather involves more dispersed variations of electromagnetic interactions.

nuclear envelope A double lipid bilayer that encloses the genetic material in eukaryotic cells. The nuclear envelope also serves as the physical barrier, separating the contents of the nucleus from the cytosol.

nuclear export sequences (NES) A short amino acid sequence of four hydrophobic residues in a protein that targets it for export from the cell nucleus to the cytoplasm through the nuclear pore complex.

nuclear localization sequences (NLS) An amino acid sequence that acts like a "tag" on the exposed surface of a protein. This sequence is used to target the protein to the cell nucleus through the nuclear pore complex and to direct a newly synthesized protein into the nucleus via its recognition by cytosolic nuclear transport receptors.

nuclear matrix The network of fibers found throughout the inside of a cell nucleus.

nuclear pore complex (NPC) Large protein complexes that extend across the nuclear envelope, forming a gateway that regulates the flow of macromolecules between the cell nucleus and the cytoplasm.

nuclear receptors A class of proteins found within cells that are responsible for sensing steroid and thyroid hormones and certain other molecules. Nuclear receptors have the ability to directly bind to DNA and regulate the expression of adjacent genes, hence these receptors are classified as transcription factors.

nucleating proteins Proteins that stabilize the formation of actin nuclei.

nucleic acid Biological molecules essential for life, including DNA (deoxyribonucleic acid) and RNA (ribonucleic acid). Together with proteins, nucleic acids make up the most important macromolecules; each is found in abundance in all living things, where they function in encoding, transmitting, and expressing genetic information.

nucleoid An irregularly shaped region within the cell of prokaryotes that has nuclear material without a nuclear membrane and where the genetic material is localized.

nucleolus (plural, nucleoli) A non-membrane-bound structure composed of proteins and nucleic acids found within the nucleus. ribosomal RNA (rRNA) is transcribed and assembled within the nucleolus.

nucleoplasm A highly viscous liquid that surrounds the chromosomes and nucleoli.

nucleoporins Proteins that are the constituent building blocks of the nuclear pore complex.

nucleosome The basic unit of DNA packaging in eukaryotes, consisting of a segment of DNA wound around a histone protein core.

nucleotide Molecules that, when joined together, make up the structural units of RNA and DNA.

nucleus A membrane-enclosed organelle found in eukaryotic cells that contains most of the cell's genetic material, organized as multiple long, linear DNA molecules.

obligate heterodimers Integrins are obligate heterodimers because they contain two distinct chains, called the α and β subunits.

occludins A protein that is an integral plasma-membrane protein located at the tight junctions.

octet rule A chemical rule of thumb that states that atoms tend to combine in such a way that they each have eight electrons in their valence shells, giving them the same electronic configuration as a noble gas.

Okazaki fragments Short molecules of single-stranded DNA that are formed on the lagging strand during DNA replication.

oligodendrocytes A type of glia whose main function is to insulate axons in the central nervous system.

oligomerization domain A protein domain involved in the formation of a polymer by combining monomers.

oligopeptide A peptide that consists of between 2 and 20 amino acids.

operons A functioning unit of genomic DNA containing a cluster of genes under the control of a single regulatory signal or promoter.

organ A collection of tissues joined in a structural unit to serve a common function.

organelle A specialized membrane-bound structure within a cell that has a specific function.

origin of replication A particular sequence in a genome at which replication is initiated.

origin recognition complex A multi-subunit DNA binding complex that binds in an ATP-dependent manner to origins of replication.

osteoblast Mononucleate cells that are responsible for bone formation

osteocyte A star-shaped cell, abundantly found in compact bone. When osteoblasts become trapped in the matrix they secrete, they become osteocytes.

outer dense plaque On the cytoplasmic side of the plasma membrane where the cytoplasmic domains of the cadherins attach to desmoplakin via plakoglobin and plakophillin.

oxidative respiration The metabolic processes in living cells by which molecular oxygen is taken in, organic substances are oxidized, free energy is released, and carbon dioxide, water, and other oxidized products are given off by the cell.

oxidized energy carriers Energy carrier that have released their electrons.

P site An RNA binding site on a ribosome that binds a peptidyl-tRNA.

pacemaker cells Cells that create rhythmical impulses.

pannexins A family of vertebrate proteins that predominantly exist as large transmembrane channels connecting the intracellular and extracellular space, allowing the passage of ions and small molecules between these compartments.

paracellular transport The transfer of substances between cells of an epithelium.

peptide bond A covalent chemical bond formed between two molecules when the carboxyl group of one molecule reacts with the amino group of the other molecule, causing the release of a molecule of water.

peptidyl transferase An enzyme that forms peptide links between adjacent amino acids using tRNAs during the translation process of protein synthesis.

pericentriolar material An amorphous mass of protein that makes up the part of the centrosome that surrounds the two centrioles.

peripheral membrane proteins Proteins that adhere only temporarily to the biological membrane with which they are associated.

peripheral nervous system The peripheral nervous system consists of the nerves and ganglia outside of the brain and spinal cord. The main function of the PNS is to connect the central nervous system to the limbs and organs.

peristalsis A radially symmetrical contraction and relaxation of muscles that propagates in a wave down the muscular tube, in an anterograde fashion.

peroxins A protein found in peroxisomes.

peroxisomal matrix The matrix of a peroxisome.

peroxisomal targeting signals (PTS) A region of the peroxisomal protein that receptors recognize and bind to.

peroxisome Organelles that are involved in the catabolism of very-long-chain fatty acids, branched chain fatty acids, D-amino acids, polyamines, and biosynthesis of plasmalogens, and etherphospholipids critical for the normal function of mammalian brains and lungs.

phagosome A vacuole formed around a particle absorbed by phagocytosis.

phosphatase and tensin homolog (PTEN) A protein acts as a tumor suppressor gene through the action of its phosphatase protein product. This phosphatase is involved in the regulation of the cell cycle, preventing cells from growing and dividing too rapidly.

phosphatases An enzyme that removes a phosphate group from its substrate by hydrolysing phosphoric acid monoesters into a phosphate ion and a molecule with a free hydroxyl group.

phosphate group A function group that contains phosphate and is important for energy transfer.

phosphodiester A group of strong covalent bonds between a phosphate group and two five-carbon-ring carbohydrates over two ester bonds.

phosphodiesterase An enzyme that breaks a phosphodiester bond.

phosphoester bond The covalent chemical bond that holds together the polynucleotide chains of RNA and DNA.

phosphoglycerides Glycerol-based phospholipids that are the main component of biological membranes.

phospholipases An enzyme that hydrolyzes phospholipids into fatty acids and other lipophilic substances.

phospholipid asymmetry The compositional differences of the inner and outer leaflets of a biological membrane.

phospholipid kinases An enzyme that regulates actin organization.

phospholipids A class of lipids that is a major component of all cell membranes as they can form lipid bilayers.

phosphoprotein phosphatases An enzyme that dephosphorylates certain phosphorylated proteins.

photosystem II The first protein complex in the light-dependent reactions that provides the electrons for all of photosynthesis to occur.

physical signals Signals such as pressure, temperature, and light.

physiology The study of the function of living systems.

pigment A material that changes the color of reflected or transmitted light as the result of wavelength-selective absorption.

pocket proteins Proteins that play crucial roles in the metazoan cell cycle through interaction with members of the E2F transcription factors family.

point mutation A type of mutation that causes the replacement of a single base nucleotide with another nucleotide of the genetic material, DNA or RNA.

pol I, pol II, and pol III Enzymes involved in the process of DNA replication.

polar microtubules Microtubules that interact with microtubules from the opposite centriole.

poly(A)-binding protein (PABP) An RNA-binding protein that binds to the poly(A) tail of mRNA.

poly(A) polymerase An enzyme that catalyzes the addition of adenosine to the 3′ end of mRNA forming the poly(A) tail.

polyadenylation complex A complex that adds a poly(A) tail to an RNA molecule.

polycistronic mRNA A polycistronic mRNA carries the information of several genes, which are translated into several proteins.

polypeptide Short polymers of amino acids linked by peptide bonds.

polypeptide exit tunnel The tunnel through which the elongating protein exits the ribosome.

polyubiquitin chain When a protein is tagged with a single ubiquitin molecule, it is a signal to other ligases to attach additional ubiquitin molecules. The result is a polyubiquitin chain that is bound by the proteasome, allowing it to degrade the tagged protein.

pore A small opening on the surface.

positive supercoiling Extra helical twists that are positive leading to positive supercoiling.

potential energy The energy stored in a body or in a system due to its position in a force field or due to its configuration.

prenyl groups A functional group that facilitates attachment to cell membranes, similar to lipid anchors like the GPI anchor.

prereplication complex A protein complex that forms at the origin of replication during the initiation step of DNA replication.

prereplicative complex A protein complex that forms at the origin of replication during the initiation step of DNA replication.

primary cilium Nonmotile cilia that serve as sensory organelles.

primary transcript An RNA molecule that has not yet undergone any modification after its synthesis.

primase An enzyme involved in the initiation of the synthesis of DNA.

programmed cell death The death of a cell in any form, mediated by an intracellular program.

proteasome Very large protein complexes that function to degrade unneeded or damaged proteins by proteolysis.

protein Biochemical compound consisting of one or more polypeptides typically folded into a globular or fibrous form, facilitating a biological function.

protein kinase A, or simply PKA An enzyme whose activity is dependent on cellular levels of cyclic AMP.

protein kinase G, or PKG A serine/threonine-specific protein kinase that is activated by cGMP.

protein targeting The mechanism by which a cell transports proteins to the appropriate positions in the cell or outside of it. This delivery process is carried out based on information contained in the protein itself.

proteinases protease An enzyme that conducts proteolysis, that is, begins protein catabolism by hydrolysis of the peptide bonds that link amino acids together in the polypeptide chain forming the protein.

proteoglycans Proteins that are heavily glycosylated. The basic proteoglycan unit consists of a "core protein" with one or more covalently attached glycosaminoglycan chains.

proteolytic enzymes Enzymes that catalyze the reaction of hydrolysis of various bonds with the participation of a water molecule.

protofilaments Tubulin dimers that have polymerized end to end.

proton pump proteins Integral membrane proteins capable of moving protons across a cell membrane, mitochondrion, or other organelles.

proximal tubule The portion of the duct system of the nephron of the kidney that leads from Bowman's capsule to the loop of Henle.

pump proteins Proteins involved in the active transport of a substance against its concentration gradient.

purines A heterocyclic aromatic organic compound, consisting of a pyrimidine ring fused to an imidazole ring.

pyrimidines A heterocyclic aromatic organic compound similar to benzene and pyridine, containing two nitrogen atoms at positions 1 and 3 of the six-member ring.

pyruvate An organic acid, a ketone. The key intersection in several metabolic pathways. It can be made from glucose through glycolysis, converted back to carbohydrates via gluconeogenesis, or to fatty acids through acetyl-CoA.

pyruvate dehydrogenase complex A complex of three enzymes that transform pyruvate into acetyl-CoA by a process called pyruvate decarboxylation.

quaternary complex The coiling of multiple folded proteins into a multi-subunit arrangement.

R, or R group A term denoting virtually any organic substituent.

Ran A small GTP-binding protein that is involved in transport into and out of the cell nucleus during interphase and is also involved in mitosis.

random coil A polymer conformation where the monomer subunits are oriented randomly while still being bonded to adjacent units.

Ras A monomeric member of the GTPase superfamily of switch proteins that is tethered to the plasma membrane by a lipid anchor and functions in intracellular signaling pathways.

reaction center A complex of several proteins, pigments, and other cofactors assembled together to execute the primary energy conversion reactions of photosynthesis.

receptor A receptor is a molecule found on the surface of a cell that receives specific chemical signals from neighboring cells or the extracellular environment.

reactive oxygen species (ROS) Chemically reactive molecules containing oxygen that are highly reactive due to the presence of unpaired valence shell electrons. ROS form as a natural byproduct of the normal metabolism of oxygen and have important roles in cell signaling and homeostasis.

receptor guanylyl cyclase The only known receptor for NO. It is most notably involved in vasodilation.

receptor-mediated endocytosis A process by which cells internalize molecules by the inward budding of plasma membrane vesicles containing proteins with receptor sites specific to the molecules being internalized.

receptor proteins Proteins found on the surface of a cell that receive specific chemical signals from neighboring cells or the extracellular environment.

receptor serine/threonine kinases A kinase enzyme that modifies other proteins by chemically adding phosphate groups to them.

receptor tyrosine kinases The high-affinity cell surface receptors for polypeptide growth factors, cytokines, and hormones that are key regulators of normal cellular processes but also have a critical role in the development and progression of many types of cancer.

recombination repair system Complexes that function to cleave DNA and rejoin the fragments in new combinations.

recycling endosomes A membrane-bound compartment that recycles materials recycles to the plasma membrane.

redox potential A measure of the tendency of a chemical species to acquire electrons and thereby be reduced. Reduction potential is measured in volts (V), or millivolts (mV).

reduced energy carriers Carriers that store potential energy (NADH and $FADH_2$).

refractory period A period of time during which a cell is incapable of repeating a particular action. More precisely, the amount of time it takes for an excitable membrane to be ready for a second stimulus once it returns to its resting state following an excitation.

regulated secretion A process in which proteins are packaged but only secreted in response to a specific signal, such as neural or hormonal stimulation.

regulatory domain A sequence that regulates protein function.

regulatory promoter A sequence upstream of the core promoter where transcriptional activators bind.

regulatory sequences A segment of DNA where regulatory proteins such as transcription factors bind preferentially.

release factors A protein that allows for the termination of translation by recognizing the stop codon in an mRNA sequence.

replication bubble A molecular structure that occurs during the replication of DNA when DNA helicase and DNA topoisomerase "unzip" the DNA double strand.

replication fork A Y-shaped structure generally found at the point where DNA is being synthesized.

replisome A complex molecular machine that carries out replication of DNA at the replication fork.

repressors A DNA-binding protein that regulates the expression of one or more genes by binding to the operator and blocking the attachment of RNA polymerase to the promoter, thus preventing transcription.

rescue The reversal of depolymerization of a microtubule.

response elements Short sequences of DNA within the gene promoter region that are able to bind a specific transcription factor and regulate transcription of genes.

resting potential The membrane potential of a cell that is not exhibiting the activity resulting from a stimulus.

restriction point / START point A G_1 phase checkpoint in the cell cycle of animal cells. Cells that progress past the restriction point are committed to enter S phase, where DNA synthesis and replication occurs.

reticular connective tissue A type of connective tissue that has a network of reticular fibers, made of type III collagen.

retrograde transport The movement of materials from the axon back to the soma.

reverse transcriptase A DNA polymerase that transcribes single-stranded RNA into double-stranded DNA.

rho protein Signaling G proteins that regulate many aspects of intracellular actin dynamics. Rho proteins have been described as "molecular switches" and play a role in cell proliferation, apoptosis, gene expression, and multiple other common cellular functions.

ribose A five-carbon monosaccharide found in RNA.

ribosomal binding site A sequence on mRNA that is bound by the ribosome when initiating protein translation.

ribosomal subunits RNA molecules that are the functional and structural components of ribosomes.

ribosome release factor A protein found in bacterial cells as well as eukaryotic organelles, specifically mitochondria and chloroplasts, that functions to recycle ribosomes after completion of protein synthesis.

rigor A condition when myosin remains attached to actin until an ATP molecule binds to it.

RING complex A protein complex that nucleates microtubules at the centrosome.

RNA primer A strand of nucleic acid that serves as a starting point for DNA synthesis.

RNA processing The removal of introns and joing of exons from pre-mRNA.

rolling stop Selectin-mediated stop of leukocytes How might a cell "roll" to a stop in a blood vessel? The cell must be able to form transient, reversible, adhesive interactions with the endothelial cells lining the vessel. As the leukocyte forms these associations, it drags (or rolls) along the vessel wall until it forms enough associations to come to a complete stop. This is known as discontinuous cell–cell adhesion.

RuBisCo An enzyme involved in the Calvin cycle that catalyzes the first major step of carbon fixation.

sarcomere The basic unit of a muscle composed of long, fibrous proteins that slide past each other when the muscles contract and relax.

sarcoplasmic reticulum A special type of smooth ER found in smooth and striated muscle that functions to store and pump calcium ions.

saturated fatty acid Fat that consists of triglycerides containing only saturated fatty acids (having no double bonds between the individual carbon atoms of the fatty acid chain).

scaffold A temporary structure used for mechanical and functional support.

Schwann cells The principal glia of the peripheral nervous system that function to support neurons.

scramblases Proteins responsible for the translocation of phospholipids between the two leaflets of a lipid bilayer of a cell membrane.

sebaceous glands Glands in the skin that secrete sebum to lubricate the skin and hair.

sebum An oily waxy substance secreted by sebaceous glands.

second messengers Molecules that relay signals from receptors on the cell surface to target molecules inside the cell, in the cytoplasm or nucleus.

secreted (or secretory) protein Any protein, whether it be endocrine or exocrine, that is secreted by a cell. Secretory proteins include many hormones, enzymes, toxins, and antimicrobial peptides.

secretion The process of releasing chemical substances from a cell or gland.

secretory vesicles Small membrane-bound compartments derived from the trans-Golgi network that transport molecules destined to be released from the cell.

securin A protein involved in control of the metaphase-anaphase transition and anaphase onset.

selectins Single-chain transmembrane glycoproteins that bind oligosaccharides on other cells to carry signals across the plasma membrane.

selectivity filter A constriction of the channel ringed by negatively charged carbonyl oxygens, which repel anions but attract cations.

separase A cysteine protease responsible for triggering anaphase by hydrolysing cohesin, which is the protein responsible for binding sister chromatids during metaphase.

sequon A sequence of three consecutive amino acids in a protein that can serve as the attachment site to a polysaccharide.

seven transmembrane spanning receptors A large protein family of transmembrane receptors that sense molecules outside the cell and activate inside signal transduction pathways.

severing proteins Proteins that fragment actin.

SH2 A structurally conserved protein domain that helps the protein "find its way" to another protein by recognizing phosphorylated tyrosine on the other protein.

Shine-Delgarno sequence A ribosomal binding sequence (AGGAGG) in prokaryotic mRNA, generally located eight basepairs upstream of the start codon AUG. The Shine-Dalgarno sequence exists only in prokaryotes. This sequence helps recruit the ribosome to the mRNA to initiate protein synthesis by aligning it with the start codon.

short, interfering RNA (siRNA Small interfering RNA (siRNA) Double-stranded RNA molecules that interfere with the expression of a specific gene.

sialyl Lewis(x) (sLex) A tetrasaccharide carbohydrate that is usually attached to glycans on the surface of cells and plays a vital role in cell-to-cell recognition processes.

sickle-cell disease An autosomal recessive genetic blood disorder characterized by red blood cells that assume an abnormal, rigid, sickle-cell shape

signal anchor sequence The sequence that allows a protein to insert and cross a membrane.

signal peptidase An intramembrane aspartyl protease with conserved active site motifs YD and GxGD in adjacent transmembrane domains.

signal recognition particle (SRP) A cytosolic ribonucleoprotein that recognizes and targets specific proteins to the endoplasmic reticulum.

signal sequences A short amino acid sequence within a protein that directs the protein to a specific location within the cell.

signal transduction The process by which a chemical or physical signal is converted into another form.

signal transduction pathway The process by which an extracellular signaling molecule activates a membrane receptor that in turn alters intracellular molecules eliciting a physiological response.

signaling Inducing a cellular response.

signaling proteins Extracellular or intracellular proteins involved in mediating the response of a cell to its external environment or to other cells.

signaling scaffold A scaffold that interacts with singaling pathways regulated signal transduction.

simple columnar epithelia A single-layered columnar epithelium that lines most organs of the digestive tract including the stomach, small intestine, and large intestine.

simple cuboidal epithelia Cube-like epithelial cells in a single layer found on the surface of ovaries, the lining of nephrons, the walls of the renal tubules, and parts of the eye and thyroid.

simple diffusion The spontaneous net movement of particles from an area of high concentration to an area of low concentration.

simple epithelia Epithelia that is one cell thick; that is, every cell is in direct contact with the underlying basement membrane.

simple squamous epithelia An epithelium characterized by one layer of squamous epithelial cells.

single-strand binding protein (SSBP) Proteins that bind to single stranded regions of DNA to prevent premature annealing.

singlepass region A single region that extends entirely through the membrane bilayer.

sister chromatids Two identical copies of a chromatid connected by a centromere.

size exclusion chromatography A chromatographic method in which molecules in solution are separated by their size, not by molecular weight.

skeletal muscle A form of striated muscle tissue existing under control of the somatic nervous system.

small ubiquitin-like modifiers (or SUMOs) Ubiquitin-like proteins that modify cellular targets in a pathway that is parallel to, but distinct from, that of ubiquitin.

SMC (structural maintenance of chromosome) proteins ATPases that participate in many aspects of higher-order chromosome organization and dynamics.

smooth muscle cell A cell of involuntary nonstriated muscle.

sodium-potassium pump (Na+/K+ ATPase) An enzyme located in the plasma membrane that facilitates active transport and pumps three sodium ions out of the cell for every two potassium ions pumped in.

soma The bulbous end of a neuron, containing the cell nucleus.

somatic cells Any plant or animal cell other than a germ cell.

specific heat The amount of energy (in joules or calories) needed to raise the temperature of 1 g of a pure substance by 1°C.

sphinganine A dihydroxy derivative of sphingosine, commonly occurring in sphingolipids.

sphingomyelin A type of sphingolipid found in animal cell membranes, especially in the membranous myelin sheath that surrounds some nerve cell axons.

spindle assembly checkpoint Monitors chromosome alignment and the structure of the mitotic spindle during mitosis.

spindle equator The portion of the spindle where all the chromosomes are aligned.

spliceosomes A large ribonucleoprotein complex that removes introns from a transcribed pre-mRNA segment.

splicing Modification of RNA after transcription in which introns are removed and exons are joined.

SRP receptor Signal recognition particle (SRP) receptor is a docking protein that identifies SRP units and helps ribosome-mRNA-polypeptide complexes settle on the membrane of the endoplasmic reticulum.

stains Artificial coloration of a substance to facilitate examination.

stem bodies Small bundles of microtubules.

steroid response element (SRE) DNA sequences that are bound by steroid receptors.

Stokes' radius (S) The radius of a hard sphere that diffuses at the same rate as the molecule.

stop codons A nucleotide triplet (UAA, UAG, UGA) within mRNA that signals a termination of translation.

stratified epithelia Epithelia that is multilayered.

stress fibers High-order structures in cells consisting of actin filaments, crosslinking proteins, and myosin II motors.

stretch-gated (or stretch-activated) channels Ion channels that open their pores in response to mechanical deformation of the plasma membrane.

striated Containing a series of ridges, furrows, or linear marks.

striated muscle Muscle tissue in which the contractile fibrils in the cells are aligned in parallel bundles, so that their different regions form stripes visible in a microscope.

structure—function relationship A relationship in which the structure imparts function and function imparts structure.

subunit A single protein molecule that assembles with other protein molecules to form a protein complex.

SWI/SNF A nucleosome remodeling complex that possesses DNA-stimulated ATPase activity and can destabilize histone-DNA interactions.

symport An integral membrane protein that is involved in movement of two or more different molecules or ions across the plasma membrane in the same direction.

synapse Specialized junction through which neurons signal to each other and to nonneuronal cells.

synaptic cleft The narrow space between the membranes of the pre- and postsynaptic cells.

TAP A member of the ATP-binding-cassette transporter family that delivers cytosolic peptides into the endoplasmic reticulum.

TATA box A DNA sequence found in the promoter region of genes in archaea. Considered to be the core promoter sequence, it is the binding site of either general transcription factors or histones and is involved in the process of transcription by RNA polymerase.

telomerase An enzyme that adds DNA sequence repeats to the 3′ end of DNA strands in the telomere regions found at the end of eukaryotic chromosomes.

telomere A region of repetitive DNA sequences at the end of a eukaryotic chromosome that protects the end of the chromosome from deterioration or fusion with neighboring chromosomes.

terminally differentiated Mature cells that become postmitotic but continue to perform their main functions for the duration of the organism's life.

terminator A genetic sequence that marks the end of the operon on genomic DNA for transcription.

TFIIB recognition element (BRE) A core promoter element that occurs downstream of the TATA box and is recognized by TFIIB that is involved in the selection of the transcription start site.

three-dimensional matrix adhesions Elongated integrin clusters that most likely resemble the type of integrin clustering found *in vivo*.

threshold potential The membrane potential to which a membrane must be depolarized to initiate an action potential.

thylakoid A membrane-bound compartment inside chloroplasts and cyanobacteria that is the site of photosynthesis.

thylakoid lumen The thylakoid lumen is the compartment bounded by the thylakoid membrane that plays a vital role for photophosphorylation during photosynthesis.

tight junction Closely associated areas of two cells whose membranes join together forming a virtually impermeable barrier to fluid. It is a type of junction present only in vertebrates.

tissue An ensemble of cells, not necessarily identical, but from the same origin, that together carry out a specific function.

topoisomerase Enzymes that unwind and wind DNA, in order for DNA to control the synthesis of proteins, and to facilitate DNA replication.

topoisomerase II An enzyme that cuts both strands of the DNA helix simultaneously in order to manage DNA tangles and supercoils.

torsional rotation The twisting of an object due to torque.

transconfiguration Orientation of the functional groups on opposite sides of a molecule.

trans face of the Golgi apparatus The side of the organelle where substances leave the endoplasmic reticulum.

trans fatty acid The common name for unsaturated fat with trans-isomer fatty acid configuration. Trans fats may be monounsaturated or polyunsaturated but never saturated.

trans-Golgi network, or TGN This area of the Golgi that is the point at which proteins are sorted and shipped.

transamidase complex An enzyme complex that catalyzes the posttranslational attachment of a precursor protein to a GPI-anchor.

transautophosphorylation Phosphorylation by the other kinase in the dimer.

transcription The process of creating a complementary RNA copy of a sequence of DNA.

transcription factor A protein that binds to specific DNA sequences, thereby controlling the transcription of genetic information from DNA to mRNA.

transcription startpoint The 5′ untranslated region that starts at the +1 position.

transesterification The process of exchanging the organic group R≤ of an ester with the organic group R′ of an alcohol.

transfer RNAs, or tRNAs An adaptor molecule composed of RNA that bridges the 3-letter genetic code in mRNA with the 20-letter code of amino acids in proteins.

transit peptides Proteins that have an N-terminal sequence that directs them to an organelle.

transitional epithelia A type of tissue consisting of multiple layers of epithelial cells that can contract and expand.

transitional ER A regoin of the ER that is partly smooth and partly rough. It contains ER exit sites from which transport vesicles carrying newly synthesized proteins and lipids bud off for transport to the Golgi apparatus.

translation The third stage of protein synthesis when messenger RNA produced by transcription is decoded by the ribosome to produce a specific amino acid chain, or polypeptide, that will later fold into an active protein.

transmembrane domain The portion of the protein that is stable in a membrane.

transmembrane proteins Proteins that span a membrane and often function as gateways to deny or permit the transport of specific substances.

treadmilling A phenomenon observed in many cellular cytoskeletal filaments, especially in actin filaments and microtubules, that occurs when one end of a filament grows in length while the other end shrinks.

triglyceride An ester derived from glycerol and three fatty acids.

tropocollagen A subunit of larger collagen fibrils.

tropoelastin A water-soluble molecule that covalently binds together to form the elastin.

tubulins Globular cytoskeletal proteins that polymerize to form the cylindrical walls of microtubules.

type I transmembrane proteins A single-pass molecule anchored to the lipid membrane with a stop-transfer anchor sequence and that have their N-terminal domains targeted to the ER lumen during synthesis.

type II transmembrane proteins A single-pass molecule targeted to the ER lumen with its C-terminal domain.

type III transmembrane proteins A single-pass molecule targeted to the N-terminal domains to the ER lumen.

type IV transmembrane proteins A multiple-pass molecule with its N-terminal domains targeted to the cytosol and lumen.

ubiquitin A small regulatory protein that binds to proteins and labels them for destruction by the proteasome.

ubiquitin ligases A protein that, in combination with an ubiquitin-conjugating enzyme, causes the attachment of ubiquitin to a lysine on a target protein via an isopeptide bond.

ubiquitin proteases A family of proteases that regulate ubiquitin-dependent metabolic pathways by cleaving ubiquitin-protein bonds.

ubiquitination An enzymatic, posttranslational modification to proteins in which the carboxylic acid of the terminal glycine from the di-glycine motif in the activated ubiquitin forms an amide bond to the epsilon amine of the lysine in the modified protein.

unfolded protein response A cellular stress response activated by the accumulation of unfolded or misfolded proteins in the lumen of the endoplasmic reticulum that aims to restore normal function of the cell by halting protein translation and activating the signaling pathways that leads to increasing the production of molecular chaperones involved in protein folding.

uniport The transport of a single ligand by a carrier protein in either direction down the concentration gradient.

unsaturated fatty acid A fatty acid that has one or more double bonds between carbon atoms.

van der Waals forces Weak noncovalent attractive forces between atoms, molecules, and surfaces that are due to the fluctuating polarizations of nearby particles.

vascular tissue A complex conducting tissue whose primary components are the xylem and phloem.

vesicle A small membrane-bound compartment that can store or transport material.

vesicle-mediated transport The movement of materials within the cell via vesicles.

vesicle shuttling The two-way (anterograde and retrograde) movement of vesicles.

vesicular tubular cluster A membrane-bound organelle in eukaryotic cells that mediates trafficking between the endoplasmic reticulum and Golgi complex, facilitating the sorting of cargo.

voltage The electric potential difference that is a measure of the energy of electricity; specifically, it is the energy per unit charge.

voltage-gated K+ channels Transmembrane channels specific for potassium and sensitive to voltage changes in the cell's membrane potential.

voltage-gated Na+ channels Transmembrane channels specific for sodium and sensitive to voltage changes in the cell's membrane potential.

volts The unit of measurement for electromotive force.

water shell Also called a hydration cell, this occurs when a chemical, regardless of its chemical makeup, must interact with the nearest water molecules, which imparts some degree of structural organization on the water molecules. For example, the hydrophobic portions of phospholipids aggregate because this reduces the number of water molecules in the water shell.

Z-scheme A diagram of the redox from P680 to P700 of the electron transport chain of photosynthesis that resembles the letter z.

zinc finger Small protein structural motifs that can coordinate one or more zinc ions to help stabilize their folds and typically function as interaction modules that bind DNA, RNA, proteins, or small molecules.

zipcode sequence Specific sequences on RNA that motor proteins bind in order to transport the RNA to a specific part of the cytoplasm.

zonula adherens Protein complexes that occur at cell–cell junctions in epithelial tissues whose cytoplasmic face is linked to the actin cytoskeleton.

zymogen granules Inactive precursor enzymes synthesized in the pancreas to avoid autodegradation.

Index

Page numbers followed by b, t or f indicate boxes, tables, or figures, respectively.

RNA polymerase complexes, 231
RNA polymerase I (pol I), 259
RNA polymerase II (pol II), 259, 261, 262f
    gene transcription by, 405–406, 405f, 406f
RNA polymerase III (pol III), 259
RNA primer, 237, 237f
RNA processing, 264–270, 265f, 266f
RNA splicing, 265–266, 265f
RNase, 96, 96f
"Rolling stop" function of selectins, 222, 222b, 222f
Rope analogy, 263b
ROS. *See* Reactive oxygen species
Rough endoplasmic reticulum (RER), 28–30, 30–31, 30f
Rubisco, 349
Ryk, 398–399, 399f

S

S-unit, 150b
Saccharomyces cerevisiae (yeast), 40
Sarcomeres, 36, 464
Sarcoplasmic reticulum, 464–465
SARs (scaffold attachment regions), 69
Saturated fatty acids, 121, 122f
Schwann cells, 460
Scientific names, 386b. *See also* Terminology
Scission, 301f, 303, 311–312, 312f
Scramblases, 137, 137f
Scruin, 175f
Sebaceous glands, 454
Sebum, 454
Second messengers, 371f, 383
Secreted (secretory) proteins, 30
Secretion, 299
    constitutive, 313
    regulated, 313
Secretory proteins, 30, 299
Secretory vesicles, 299
Securin, 250
Selectins, 204b, 211, 222
    cell adhesion regulation by, 222–223
    "rolling stop" function of, 222, 222b, 222f
Selectivity filter, 335, 336f
Self-defense, 440
Senescent cells, 419
Sensory neurons, 378
Separase, 250
Septate junctions, 212f, 213, 214f
Sequons, 288
SER. *See* Smooth endoplasmic reticulum
Serine, 22f
Serine/threonine kinase receptors, 374, 375f
Seven transmembrane spanning receptors, 373, 374f
Severing proteins, 174–175
SH2, 386
Sheet-like collagens, 190, 190f
Shine-Delgarno sequence, 272
Short interfering RNAs (siRNAs), 51
Sialyl Lewis(x) (sLex), 223
Sickle-cell disease, 54–55, 55f
Sickle-cell mutation, 114
Signal anchor sequences, 286–287, 288f
Signal peptidase, 286
Signal proteins, 370b
    membrane-impermeable, 371f
    membrane-permeable, 371f

Signal recognition particle (SRP), 285, 286b
Signal recognition particle (SRP) receptor, 285, 286b
Signal sequences, 278–279, 279b
Signal transduction, 26, 367–391, 386b
Signal transduction pathways, 370, 371f, 384–390
    feedback loops, 410, 411f
Signaling domains, 383–384, 385t
Signaling molecules, 372–376
Signaling networks, 369–371, 376–377
    components of, 370–371
    function of, 369–370
    memorization warning, 372b
    mutations in, 384
Signaling pathways, 384–390
    branching, 371f
    study tip for, 385b
Signaling proteins, 370, 370b, 371f, 376–384
    calcium-sensitive, 382, 382f
    common features of, 377
Signaling scaffolds, 376, 377f
Signals and signaling, 20, 26
    membrane-impermeable, 372
    membrane-permeable, 372
    physical, 372–373
Singer, S. Jonathan, 127
Single-strand binding protein (SSBP), 238
Singlepass regions, 129
Sinoatrial (SA) node, 467
Sir3, 74, 74f
Sir4, 74, 74f
Sister chromatids, 246
Size exclusion chromatography, 271
Skeletal muscle, 36, 463–465, 464f
    contraction of, 464–465, 466f, 467f
    structural organization of, 464, 465f
    structure of, 181–182, 181f
Skin, 453–454, 454f
Skou, Jens C., 364
Slow response, 458
Small molecules
    permeability of, 125
    transport across membranes, 335–341, 350–353
Small nuclear RNAs (snRNAs), 51
Small ubiquitin-like modifiers (SUMOs), 105
SMC (structural maintenance of chromosomes) family of proteins, 71–73
Smooth endoplasmic reticulum (SER), 28–30, 133–140
    functions of, 28–29
    membrane assembly in, 135–139
Smooth muscle, 463, 464f
    airway, 471
    vascular, 470–471, 471f
Smooth muscle cells, 36, 469–472
    contraction of, 469, 470f
    vascular, 470–471
SNAREs, 301f, 303
Soap bubble demonstration, 125b
Sodium (Na+)/glucose symporter, 341, 341b, 352–353
Sodium (Na+) channels, voltage-gated, 458–459
Sodium-potassium pump (Na+/K+ ATPase), 339, 341b, 351
Soluble proteins, 29–30

Soma, 459, 459f
Somatic cells, 49
Sonic Hedgehog, 90b
Specific heat, 11f, 12
Spectrin, 175f
Spelling, 60b
Sphinganine, 461–462, 461f
Sphingomyelin, 120
Spindle assembly checkpoint, 435
Spindle equator, 248
Spliceosomes, 231, 265–266
Splicing, 265–266, 265f
Squamous epithelia
    keratinized stratified, 453, 454f
    simple, 453, 453f
    stratified, 453, 453f
SREs. *See* Steroid response elements
SSBP. *See* Single-strand binding protein
Stains, 35
START point, 420, 420f
Ste20-like kinase (Slk), 435
Steel cable / highrise building analogy, 147b
Stem bodies, 251
Steroid hormones, 389–390
Steroid response elements (SREs), 377f, 390
Sterol carrier protein-2, 138, 138f
Stokes' radius (S), 271
Stomach, 454–455, 455f
Stop codons, 276
Storage components, 20
Storage proteins, 109
Stratified cuboidal epithelia, 453, 453f
Stratified epithelia, 453, 453f
Stratified squamous epithelia, 453, 453f, 454f
Strength, 177
Stress fibers, 178, 178f
Stretch-gated (or stretch-activated) channels, 338
Striated muscle, 36, 464
Striated muscle cells, 181–182, 181f
Striated muscle contraction, 181–182, 181f
Structural components, 20
Structural polarity, 166b
Structural proteins, 108
Structure-function relationship, 10
Study tips, 47b, 57b, 60b, 120b, 329b, 385b, 395b, 435b
Subcellular structures, 7–8, 9f
Subunit (term), 150b, 229b, 260b
Sucrose
    glycosidic bonds in, 19–20, 19f
    synthesis of, 349–350, 350f
Sugars, 18–20, 18f, 105
    disaccharides, 19–20
    monosaccharides, 18f, 19
Sulfide/sulfydryl, 16f
SUMOs. *See* Small ubiquitin-like modifiers
Sunlight, 342–350, 344f
Supercoiling
    negative, 263
    positive, 263, 263f
Surfing, 184
SWI/SNF proteins, 67–68, 68f
Symport, 338, 339b
Synapses, 459, 459f, 462, 463f
Synaptic clefts, 462, 463f